ANNUAL REVIEW OF
EARTH AND
PLANETARY SCIENCES

ANNUAL REVIEW OF EARTH AND PLANETARY SCIENCES

FRED A. DONATH, *Editor*
University of Illinois — Urbana

FRANCIS G. STEHLI, *Associate Editor*
Case Western Reserve University

GEORGE W. WETHERILL, *Associate Editor*
Carnegie Institution of Washington

VOLUME 5

1977

ANNUAL REVIEWS INC.   4139 EL CAMINO WAY   PALO ALTO, CALIFORNIA 94306

ANNUAL REVIEWS INC.
Palo Alto, California, USA

International Standard Book Number 0-8243-2005-0
Library of Congress Catalog Card Number 72-82137

REPRINTS

The conspicuous number aligned in the margin with the title of each article in
this volume is a key for use in ordering reprints. Available reprints are priced at
the uniform rate of $1 each postpaid. The minimum acceptable reprint order
is 10 reprints and/or $10, prepaid. A quantity discount is available.

FILMSET BY TYPESETTING SERVICES LTD, GLASGOW, SCOTLAND
PRINTED AND BOUND IN THE UNITED STATES OF AMERICA

CONTENTS

SOME RELATED ARTICLES APPEARING IN OTHER ANNUAL REVIEWS

From the *Annual Review of Astronomy and Astrophysics,* Volume 14 (1976)
 Chemical Evolution of Galaxies, Jean Audouze and Beatrice M. Tinsley
 Chemistry of Primitive Solar Material, Stephen S. Barshay and John S. Lewis
 Physical Processes in Comets, F. L. Whipple and W. F. Huebner
 Orbital Resonances in the Solar System, S. J. Peale

From the *Annual Review of Biochemistry,* Volume 45 (1976)
 Calcium-binding Proteins, Robert H. Kretsinger

From the *Annual Review of Ecology and Systematics,* Volume 7 (1976)
 Genetic Polymorphism in Heterogeneous Environments, Philip W. Hedrick, Michael E. Ginevan, and Evelyn P. Ewing
 Reproductive Strategies as Adaptations to Life in Temporally Heterogeneous Environments, James T. Giesel
 Population Responses to Patchy Environments, John A. Wiens
 The Size-Efficiency Hypothesis and the Size Structure of Zooplankton Communities, Donald J. Hall, Stephen T. Threlkeld, Carolyn W. Burns, and Philip H. Crowley
 Evolution and Systematic Significance of Polyploidy, R. C. Jackson
 Bacterial Substrates and Productivity in Marine Ecosystems, John McNeill Sieburth
 Population Dynamic Models in Heterogeneous Environments, Simon A. Levin
 Evolution and Ecological Value of Duplicate Genes, Ross J. MacIntyre
 The Evolution of Inbreeding in Plants, S. K. Jain

From the *Annual Review of Fluid Mechanics,* Volume 9 (1977)
 Compressible Turbulent Shear Layers, P. Bradshaw
 Incompressible Boundary-Layer Separation, James C. Williams, III
 Bubble Dynamics and Cavitation, Milton S. Plesset and Andrea Prosperetti
 Hydrodynamics of the Universe, Ya. B. Zel'dovich
 Particle Capture from Low-Speed Laminar Flows, Lloyd A. Spielman

From the *Annual Review of Materials Science,* Volume 6 (1976)
 High-Resolution Electron Microscopy of Inorganic Materials, John M. Cowley
 Mechanical Spectroscopy: An Introductory Review, Adi Eisenberg and Byung C. Eu

From the *Annual Review of Microbiology,* Volume 30 (1976)
 Denitrification, C. C. Delwiche and Barbara A. Bryan

From the *Annual Review of Nuclear Science,* Volume 26 (1976)
 Global Consequences of Nuclear Weaponry, J. Carson Mark
 Fossil Nuclear Reactors, Michael Maurette

From the *Annual Review of Physical Chemistry,* Volume 27 (1976)
 Liquid Theory and the Structure of Water, Mu Shik, Jhon and Henry Eyring
 Kinetics of Crystallization, M. Kahlweit
 Phase Transitions—Beyond the Simple Ising Model, J. F. Nagle and J. C. Bonner

ANNUAL REVIEWS INC. is a nonprofit corporation established to promote the advancement of the sciences. Beginning in 1932 with the *Annual Review of Biochemistry*, the Company has pursued as its principal function the publication of high quality, reasonably priced Annual Review volumes. The volumes are organized by Editors and Editorial Committees who invite qualified authors to contribute critical articles reviewing significant developments within each major discipline.

Annual Reviews Inc. is administered by a Board of Directors whose members serve without compensation.

Annual Reviews are published in the following sciences: Anthropology, Astronomy and Astrophysics, Biochemistry, Biophysics and Bioengineering, Earth and Planetary Sciences, Ecology and Systematics, Energy, Entomology, Fluid Mechanics, Genetics, Materials Science, Medicine, Microbiology, Nuclear Science, Pharmacology and Toxicology, Physical Chemistry, Physiology, Phytopathology, Plant Physiology, Psychology, and Sociology. The *Annual Review of Neuroscience* will begin publication in 1978. In addition, two special volumes have been published by Annual Reviews Inc.: *History of Entomology* (1973) and *The Excitement and Fascination of Science* (1965).

James Gilluly

Ann. Rev. Earth Planet. Sci. 1977. 5:1–12

AMERICAN GEOLOGY SINCE 1910 – A PERSONAL APPRAISAL

✳10064

James Gilluly
975 Estes Street, Lakewood, Colorado 80215

The editors have asked me to submit a personal evaluation of the development of the geological sciences during the fifty-five years I have been active in the field. They have encouraged autobiography, but I hope to limit this to such aspects of personal history as throw light on the "sociology" of the science. A most striking aspect of the earth sciences is the tremendous expansion of the field in the last sixty years.

In 1910, while a freshman at Franklin High School, Seattle, I enrolled in a class in "Physiography"—really elementary geology. The instructor was J. J. Runner, a most inspiring and considerate teacher, later to be known to generations of students at the University of Iowa as "Uncle Joe." At the end of the term I had become so fascinated by the subject that I asked Professor Runner (in those days any high school teacher was ex officio a professor) how one could earn a living in geology. He replied that about the only employment for a geologist was as a teacher in high school or college or as a geologist with a state survey or the Federal Survey. He estimated that there were only about a thousand people in the United States who considered themselves professional geologists. It was a tight profession with very few openings.

Runner was probably right at the time. Only a handful of geologists were employed in those days by either mining or petroleum companies. The membership of the Geological Society of America that year was only 305. If there were indeed a thousand geologists in the country there was one in 92,000 of the population. Today there is at least one geologist in 7,000 people, 13 times as many proportionately. In 1910 I felt the profession was entirely too crowded to be counted on as a career; I dismissed it from my plans.

Three years later I entered the University of Washington in engineering. I tried this and two or three other majors without developing much enthusiasm for any of them. By the beginning of my senior year I had plenty of credits for graduation but no major. My close friend William T. Nightingale, later to become a highly successful petroleum geologist, was graduating in geology and was already employed by

1

a major international oil company. He took me aside one day: "Since you have jumped around from one major to another, sampling them, why don't you jump once more and sample geology? You like the out-of-doors and you told me you liked the sample you had in high school. You have nothing to lose if you don't stay with it."

I decided it was worth the gamble; I dropped all the courses I had started and enrolled in elementary geology, mineralogy, and paleontology all at once. Under the flexible quarter system, I was able to graduate at the end of the year. My main instructor was that outstanding teacher, George E. Goodspeed, Jr.

Bill Nightingale was right; jobs were indeed "hanging on the bushes." Within two weeks of leaving the University I had employment with a small oil company mapping in the Judith Basin of Montana.

Some months before, at Goodspeed's suggestion, several of us had taken the two-day examination for Geologic Aid, the lowest rung on the U.S. Geological Survey's professional ladder. We were warned that Survey jobs were very few, but were urged to take the test as a measure of what we had learned.

Much to my surprise, less than two months after I had joined the oil company, I received an offer of employment from the Geological Survey. Although the salary offered was notably less than that of the company, a senior employee of the company advised me to take the Survey position: "The Survey training is much better than this company can give you." I followed his advice and accepted—a most fortunate choice.

I reported to Frank Reeves for a field season in central Montana, largely in the "Missouri Breaks" and adjoining country to the south. The work was hard, nearer seventy than forty hours a week. The "government sinecure," as my oil company boss called it with scorn as I resigned, was at least twice as demanding as the work with the oil company had been. But it was fascinating work and highly instructive.

Good fortune had it that a fellow neophyte, M. N. Bramlette, was also in the party. Bram is the keenest stratigrapher I have ever worked with, and though he had had little more training than I, working with him was a real education. He has been a close friend for fifty-five years, and during all that time I have never written a paper without thinking how Bram would appraise its logic and pertinence.

In 1921 the Geological Survey had about 60 field geologists, including such stars as Darton, Ransome, Keith, Paige, Spencer, Hewett, and Ferguson. There were eight or ten paleontologists, including Girty, White, Stanton, Woodring, Reeside, and Kirk. Of the four or five chemists, the leader was F. W. Clarke; the mineralogists were Schaller and Henderson; the petrologists Esper Larsen and Clarence Ross. Compare that small band with the present staff of nearly or quite a thousand geologists, more than a hundred chemists, and scores of physicists, seismologists, mineralogists, and geochronologists.

The scientific leaders of the "old Survey" included many outstanding men such as those just mentioned; no matter what the subject, there was nearly always a very knowledgeable scholar "just down the hall." Though there were a few exceptions, most of them were more than willing to discuss problems with the younger staff, loan specimens, give literature citations, and help in other ways. It was an outstanding school for a beginner; one wonders whether the present gigantic organization has retained this quality. (It is the prerogative of age to look back on the "good old days".)

During the spring semester of 1922, Bramlette, Rubey, and I used to commute Saturday mornings to Baltimore for the memorable morning-long course in stratigraphy taught by E. W. Berry at the Johns Hopkins. We had to catch the 6:00 AM interurban to make the 8:00 o'clock class but we all thought it worth the effort. Berry had a sardonic humor that kept one on his toes.

The Survey has always encouraged graduate study, so in the fall of 1922 Rubey and I took leave and entered the Yale graduate school. This was a most challenging experience. There is an old saying that one learns more from his fellow students than from even the most distinguished faculty. The faculty was indeed distinguished, Knopf, Schuchert, Gregory, Dunbar, Ford, and Longwell among the leaders. But my fellow students were of equal caliber and several became even more distinguished in later life. There were only about a dozen, but among them were T. B. Nolan, later to become Director of the Survey, William W. Rubey, George Gaylord Simpson, J. Frank Shairer, Carl H. Dane, and Ludlow J. Weeks. Of these, Nolan and Rubey were to become presidents of the Geological Society of America; Nolan, Rubey, Simpson, and Shairer all became members of the National Academy of Sciences; Rubey and Simpson became Penrose medalists; Shairer, a Roebling medalist and President of the Mineralogical Society; Dane, Chief of the Fuels Branch of the Geological Survey; and Weeks a group leader of the Geological Survey of Canada. Shall we say that the competition was keen? All these friends have continued to influence my work ever since; I owe much to them all, but especially to Nolan and Rubey, with whom it has been my good fortune to be closely associated throughout my professional life.

The most influential teachers in physical geology at Yale were Knopf and Gregory, with wholly contrasting styles. Knopf would generally give a single weekly assignment, perhaps only five or six pages long, to his graduate class in petrography. He would insist on our analyzing the article sentence by sentence until a clear consensus was arrived at: the author was either right, perhaps right but unproven, or definitely wrong. In the lab no mineral was satisfactorily identified with less than four diagnostic properties. (Simpson, who was to become the leading vertebrate paleontologist of his time, was far and away the best of us with the petrographic microscope.)

Gregory, on the other hand, with an innocent air, would assign each of us a handful of references on which to report in his seminar in physiography. When the class met we would find that the assignments were loaded. One of us would have read nothing but papers urging that glaciers protect the landscape from erosion; another only papers arguing that drastic glacial erosion occurs; one nothing but papers urging that erosional slopes all flatten with time; another that all slopes retreat parallel to the original slope. Obviously, when the reports were given, arguments were hot and heavy. Greg would say little, but he had a marvelous skill in prolonging the debates. One found oneself reading all the other assignments as well as one's own, simply in self-defense. When, after long debate—the "two-hour" seminar seldom lasted less than three or four hours—someone would try to get Greg to commit himself, his reply was always, "It doesn't matter what I think: I try to assign representative papers. What is important is, what do you fellows think?" Great teachers, both.

In the 1920s nearly every graduate student sat for his qualifying orals at the end of his second year of study, unless his undergraduate training was unusually weak. This was a far cry from the situation today, when many students delay for five or even eight years, thereby cutting off as much as twenty percent of their productive lives. One wonders whether many of the additional techniques—not theory—so acquired could not be picked up during, rather than before, a productive professional career. After all, few petrologists are expert in paleontology or vice versa. Why should every petrologist learn to operate high pressure furnaces? Of course he should learn to interpret the data gathered by the experimentalists, but he need not be an experimentalist himself.

Be that as it may, most of our group tried the oral hurdle after two years and nearly all passed creditably. The general expectation was to complete a thesis within the next few years, while earning a living, if one could find suitable employment. This was the road I took: although I had concentrated on petrology and mineralogy under Knopf and Ford, my thesis, presented two years later, was a stratigraphic and structural study of the San Rafaėl Swell, Utah, done under the auspices of the U.S. Geological Survey.

In the 1920s the environment of the Geological Survey was as stimulating to its younger employees as that of any university. For the beginners there was the Geologues Club, which met monthly and consisted of the twenty-five most junior employees of the Geological Survey, Geophysical Laboratory, and the earth science division of the National Museum. When a new man joined one of these institutions he was invited to join the club. If he accepted, the currently senior member was ejected, with due ceremony. Generally, at each meeting two members would present formal papers. Criticism of both content and presentation was always searching and commonly effective. The quality of papers presented generally improved notably during the four or five years a tyro was a member of the club.

For the entire geological group in Washington there was the venerable Geological Society of Washington, which met, as it does today, on the second and fourth Wednesdays of each month between October and May. In the 1920s the Society was rather formal, with the President, speakers, and Secretaries in formal dress. The formality gradually disappeared during that decade and the meetings have been quite informal ever since. Any speaker before this Society may expect a highly skeptical audience and if he is not a foreigner the skepticism finds ready expression.

For those interested in "hard-rock" geology, there was the Petrologists Club, which met monthly at the Geophysical Laboratory or the Geological Survey during Prohibition, but exclusively at the Geophysical Laboratory after beer became legal. (Of course alcohol is never allowed on government premises.) The club consisted of such men as Cross, Washington, Bowen, Fenner, Zies, Shairer, Larsen, Schaller, Wright, Nolan, Greig, Ross, Burbank, Lovering, Tunell, Spencer, and others. It was a very lively place.

A somewhat smaller group, the Paleontological Society, commonly met at the National Museum; no mere field geologist could invade this arcane body.

Finally there was the Pick and Hammer Club, principally composed of Survey employees, though open to any geologist in the area. It met on no fixed schedule, but only when the program committee could persuade some member or visiting geologist to talk about his research. As most foreign geologists visiting the United States eventually reach Washington, it was a rare year when half a dozen international figures failed to appear on one or another of the programs.

Once a year the Pick and Hammer Club presents a show, pointing out and exaggerating the foibles of the Survey brass, in the tradition of the Gridiron Club of Washington. During the 1920s this was an elaborate occasion; everyone dressed formally for the show and the dance that followed it. Through the years the formality grew less and less until for the last thirty years it has disappeared completely.

This growing informality has been paralleled by forms of salutation. In the 1920s all the paleontologists were addressed as "Doctor," the rest of us simply by the surname: "Good morning, Hewett." Today everyone, with the possible exception of the Director, is hailed by his given name: "Hi, George." A remarkable change within a generation.

The Survey had very small funds for training in new techniques. In 1928 permission and a small field grant were given me to work for a month with Robert Balk in the Adirondacks in order to pick up the then-new techniques of the Cloos school of structural studies. Robert was most hospitable and cooperative; we became and remained close friends until his untimely death twenty years later.

A few years later, during the winter of 1931–1932, at the suggestion of Dr. T. W. Stanton, the Chief Geologist, I was permitted to utilize leave accumulated from past years and leave in advance of 1932 to spend a winter in Europe. This included a month in Innsbruck, studying petrofabrics with that most generous and cooperative scholar, Bruno Sander. Sander gave me two hours of his time every afternoon for a solid month, going over his very complex textbook, line by line.

As it happened, I was to utilize Sander's techniques in only one small project, but for forty years we have been friends and I still consider myself deeply in his debt. His writings are difficult, even for his Austrian compatriots, but this is chiefly because of his painstaking care to avoid going beyond the evidence. His logic seems to me impeccable.

In 1938, I accepted a professorship in the University of California, Los Angeles, where I remained until 1950, except for three years during World War II. I was able to retain "when actually employed" status on the Survey and thus continue field work during the summers.

During the war I returned to the Survey on a full-time basis, first as a roving inspector of the many war mineral projects necessarily manned by inexperienced geologists quickly assembled from all over the country.

Later assignment was to the Southwest Pacific Command as a military geologist. Our headquarters at first were in Brisbane, then Hollandia, New Guinea, studying maps and air photos in search of landing beaches, potential air strips, water supplies, and road metal in the Philippines. We landed on Leyte on $D+1$. I may say that we were successful with our first two assignments, but much less so in the others; deep weathering made good road metal hard to find in the Philippines.

Aside from the wartime interval I was associated with UCLA for nine years. Graduate teaching I found very rewarding but I was never happy with my undergraduate classes. I felt that I was either over the heads of the students or insulting their intelligence by laboring the obvious. I could never seem to find the happy medium. Finally, when the Regents of the University insulted the faculty by demanding new oaths of loyalty (though all of us had previously made such oaths) I realized that I had always been happier doing geology on the Survey than talking about it in the University. I resigned from the University and returned to the Survey, absorbing a pay-cut exactly equal to the increase I had received in the reverse transfer twelve years before. Many other Survey geologists have gone through similar cycles. Woodring, Lovering, Sims, Sheldon, Cloud, James, and Reynolds come readily to mind. Similar motivation?

From 1950 until my retirement in 1966, I was involved in the fascinating geology of north central Nevada. Here, with many able assistants, I was able to map four fifteen-minute quadrangles of highly complex, and therefore very interesting, geology. These were the most enjoyable years of my professional life; my one regret is that the hills are still rising at a rate faster than I can climb them. Since 1966 field work in rough terrane has become impossible for me.

MINERALOGY SINCE 1920

In the 1920s the best equipped mineralogical laboratories in the United States had, perhaps, a chemical bench with a research assistant, a goniometer, a petrographic microscope, immersion oils, and blowpipe equipment. Funds might be available for purchase of a few specimens and perhaps for a collecting trip every few years. Of course the Braggs had already gone far with their X-ray studies, but such equipment in the United States was confined to a few of the more advanced physics laboratories. No mineralogical department had X-ray equipment until late in the decade, most not until the 1940s. Even a universal stage was a rarity in America, though standard equipment in Europe. No one had ever dreamed of the electron probe, electron microscopes, or Mossbauer. Today the only goniometer one finds in a mineralogical laboratory is an X-ray machine. The last blowpipe expert in America, Foster Hewett, died in 1970. Were Palache or Schaller to enter a modern laboratory they wouldn't know where to begin. My guess, though, is that it wouldn't take them long to find out!

PETROLOGY

Fifty years ago petrology was mainly a field and microscope science, as, of course, much of it still is and always will be. Chemical analyses were so expensive that they were carefully rationed. H. S. Washington made an international reputation by compiling all the "superior analyses" in the world literature into a single volume.

The major figures were Sederholm, V. M. Goldschmidt, Harker, Escola, Niggli, La Croix, and Becke in Europe; Daly, Lindgren, Ransome, Larson, Knopf, Grout, and Buddington in the United States; and Coleman and Adams in Canada.

The only experimental work of consequence was being done at the Geophysical Laboratory of the Carnegie Institution of Washington. Here Bowen, Fenner, Merwin, Sosman, Greig, Goranson, Schairer, and a few others were engaged in a well-conceived program of experiments on simple systems of components of the rock-forming minerals. They were making considerable progress, though they were restricted to pressures of only one atmosphere, and hence to systems free from water and carbon dioxide, known to be important constituents of magmas.

Despite the limitations of these experiments as guides to igneous processes, Bowen, in 1928, wrote what is still a very significant contribution to petrology: *The Evolution of the Igneous Rocks*. The main concept is that crystallization differentiation dominates the evolution of magmas from a primary basalt through more and more salic differentiates to granite and aplite. Although Daly and Grout soon pointed out that the volumes of the successive differentiates are not consistent with dominance of Bowen's suggested processes, they may still be significant.

When experiments at high pressures became available twenty years later it was shown that the incongruent melting of olivine, a major element in Bowen's scheme, disappears at the pressures to be expected in the lower crust. The theory cannot then be adequate as presented—though some trends of rock evolution may conform to it.

Although some work on evaluating the solubility of water in molten silicates at high pressures and temperatures had been carried out by Goranson in the early 1930s it was not until new alloys had been developed that routine work with apparatus of Tuttle's design could be done on such systems. From the late 1940s, experimental petrology could go on under more realistic pressure conditions. It became possible to reach pressures and temperatures corresponding to those at the highest metamorphic grade.

No one, I suppose, would claim that experimental petrology has solved all the important problems in igneous and metamorphic petrology, but everyone must agree that tremendous advances have been made in the last decade by the brilliant work of Tuttle, Schairer, Ringwood, Green, O'Hara, Boyd, Yoder, J. V. Smith, and their confreres.

Many constraints on petrologic theory have been imposed by isotopic studies, which of course had to wait until the development of mass spectrometers in the late 1930s and early 1940s. Clearly many of the developments that have revolutionized both mineralogy and petrology have depended more on new instrumentation than upon new theory. After all, Gibbs and Van't Hoff are nineteenth century names.

PHYSIOGRAPHY—GEOMORPHOLOGY

For some unknown reason, the subscience concerned with the evolution of land forms, which had been known for at least a century as physiography, has become better known as geomorphology. Perhaps this is a more impressive name, as it has an extra syllable; certainly the words are synonyms. Whatever the name, the methods and content of the field have drastically changed for the better during the last fifty years.

In the 1920s most physiographic papers were almost entirely deductive—based on hypotheses largely developed by William Morris Davis. In his hands the results seemed quite reasonable, even though often incorrect, but most practitioners were far less able than he and few of the papers were based on sound stratigraphy. Peneplains were multiplied so that a single landscape might contain three or four. This is still advocated in Africa and Arkansas but fortunately it is going out of style. The measurements of erosion rates have generally made the recognition of Jurassic or even Oligocene landscapes somewhat less popular than it was.

The quantitative studies of stream regimens by Richardson, Simon, Rubey, Leopold, Langbein, Wolman, Schumm, and many others has revolutionized the study of stream erosion just as Bagnold and Hack have clarified the work of the wind. Nye, Carol, Seligman, and Sharp have elucidated many aspects of glacial action.

Fifty years ago the problems of pediment formation were just being noted by Bryan and Passarge. The studies by Denny, Bull, Sharp, and others have gone far to solve them.

Fifty years ago the Pleistocene Atlantic shorelines were thought to be horizontal from Florida to New Jersey; Hack has shown that they vary notably in elevation, even along this relatively stable shore.

STRUCTURAL GEOLOGY

Hans Cloos began his investigations of the internal structure of plutons during the depression in Germany following the first World War, when travel was difficult. He had to work on granite within walking distance, therefore the refined observations! His techniques became standard practice only in the late 1920s and early 1930s. Sander's petrofabric studies began in 1911, but it was well into the 1930s before his methods became widely applied. Under favorable conditions these methods permit extrapolation of surface structures to considerable depths.

In the early 1920s the only geophysical tool in routine usage was the magnetic dip needle, used in prospecting for iron ore. With the refinement of equipment during the antisubmarine war, a new tool—aeromagnetics—became available for seeking out magnetic anomalies. During the last 30 years most of the United States and Canada have been covered by aeromagnetic traverses and many patterns of anomalies worked out. While the interpretation of these anomalies is commonly controversial, many of them have disclosed structural trends at depth that would not otherwise be discovered.

Other geophysical techniques have been developed, mostly in the last fifty years. The first tool used in prospecting for oil was the torsion balance, followed by the gravimeter. The gravimeter was used in searching for salt domes in the Gulf coastal plains and in northern Germany. Seismic prospecting was about a decade behind; it is now standard practice throughout the petroleum industry, since virtually all favorable structures determinable from surface studies have been thoroughly tested.

Other geophysical tools in wide use in prospecting for radioactive minerals are the Geiger counter and the scintillometer. They are less useful in structural studies than are seismic methods, but are largely responsible for most of the uranium reserves thus far identified.

Geophysical methods have proven useful in many other ways as well. Simmons was able by gravity measurements to outline the shape of the Adirondack anorthosite mass at depth and to locate the feeding conduit. Mott has used gravity and seismic equipment to show that the Cornish and Devon plutons coalesce at depth and the combined mass extends to a nearly uniform depth of 12 km. Seismic studies of the Yellowstone Park area have shown that a siliceous pluton underlies much of the park and extends to a depth of 20 km. This salic intrusive is underlain by much denser rocks and quite surely does not overlie a "hot spot" extending deep into the mantle as has been suggested by several workers.

Of course the greatest development in structural geology has been the evolution of the idea of plate tectonics. This, too, depends largely upon geophysical data. Fifty years ago, or even twenty, few earth scientists in the northern hemisphere accepted the idea of continental drift; today few deny it. Several steps were involved in this reversal of opinion. First came Hugo Benioff's discovery of the alignment of intermediate and deep focus earthquake foci into zones that now take his name. Then came Mason's discovery of linear magnetic anomalies on the floor of the northeastern Pacific and the later finding that these anomalies parallel the oceanic ridges. Cox and Doell found a recognizable sequence of magnetic reversals and it was soon noted that the spacing of the sea floor anomalies was such as to strongly suggest that they record these reversals. Harry Hess first suggested that the sea floor is spreading and that the mid-ocean ridges are fault blocks bordering rift valleys. In short, the most convincing evidence for continental drift is found not on the continents but on the sea floor.

True, many reconstructions of former arrangements of continental plates are highly subjective and quite unsupported by geologic evidence, especially those for pre-Jurassic time. The most quoted such reconstruction, that by Bullard, Everett, and Smith, 1965, a computer-guided matching of the continental slopes bordering the Atlantic, produces an excellent fit between South America and Africa—a fit strongly supported by the match of geologic provinces—but a far less convincing fit of North America with Europe and North Africa. This reconstruction allows no room for the Paleozoic rocks of Central America. Nor does it take account of the post-Jurassic crustal shortening of the Betic and Atlas ranges. Before the narrowing of the Strait of Gibraltar by the counter-clockwise rotation of the Iberian peninsula, the strait was at least 300 km wide, as shown by paleomagnetic measurements in Spain and Morocco and the Betic-Atlas shortening. The poor fit is readily understood when one recalls that the present continental slope of North America was formed long after Jurassic time and the location of a pre-Jurassic slope is pure guess work. It may have been several hundred miles east or west of the present slope which was the line matched with Europe and Africa. In fact few pre-drift reconstructions are as convincing as that of Africa and South America. Yet the mutual relations of ridges, magnetic stripes, and sea floor ages as found by JOIDES drilling completely confirm sea floor spreading and continental drift.

The existence of large plates mutually displaced horizontally does not, however, eliminate important intra-plate orogeny. The southern Rocky Mountains lie, for example, several hundred kilometers from any plate boundary; the horizontal movement must have been passive with respect to adjoining crustal segments. But the vertical excursions of the mass are measured in kilometers. The Cretaceous strata of the Rockies from New Mexico to northern Montana are in places as much as 5 km thick and overlie continental Jurassic. The post-Cretaceous uplift has been such that the Jurassic has been eroded from a surface commonly 5 km or more above the level of the sea, a vertical excursion of more than 10 km over large areas, and in places even more. Plate tectonics obviously does not account for all orogeny. even though it does well with the Andes and Coast Ranges.

SEISMOLOGY

The Seismological Society of America was founded in 1910, at a time when there were probably fewer than thirty people in America who considered themselves primarily seismologists. Recently the Society admitted more than 180 new members in a single year! Here again the great advances have come from the improvement of instrumentation. The military objective of monitoring nuclear tests led, during the 1950s, to the great expansion of the world-wide instrumental network and to a very considerable improvement in the sensitivity of the instruments. The brilliant insight of Beno Gutenberg led him to the suggestion of a Low Velocity Zone in the upper mantle, but its existence could not be demonstrated until much better instruments, more advantageously located, could become available.

I have already mentioned the great improvement in structural interpretations made possible by the advances in seismology. Marine geology is largely dependent on seismic work, which has disclosed the layering in the sea floor.

STRATIGRAPHY AND SEDIMENTATION

Stratigraphy is the backbone of the science. Without sound stratigraphy structural studies are impossible, and all historical and most of economic geology depend upon it. In the last fifty years advances have been great. Although exceptions are becoming more abundant, the distinction between miogeosynclinal and eugeosynclinal suites of strata is very useful, both in historical and economic geology. Not all flysch is demonstrably antecedent to orogeny, but it is surely highly suggestive. Molasse deposition has long been recognized as following mountain building. Pebbly mudstones are now recognized as slide deposits, as are olistoliths and olistostromes. Stromatoliths are becoming useful as crude guides to age in the Precambrian.

Paleotectonic maps have become extremely valuable in petroleum exploration, and their compilation unravels much geologic history otherwise unsuspected. It is unfortunate that the Geological Survey has abandoned its project to compile maps for each of the Systems. I think it quite safe to say that with the approaching scarcity of economic mineral resources of all kinds, this useful tool must be ever more intensively pursued. Correlation charts, too, need frequent revision if they are to keep up with progress.

GEOCHRONOLOGY

Fifty years ago, a dozen or so crude measurements of lead-uranium ages had been made. K/Ar and Rb/Sr were decades in the future, and ^{14}C and fission track studies still further. Although there was only a handful of workers in the field, optimism was great. It wasn't until the 1960s that we learned about concordia, cooling ages, and other complexities. Here again improvements depended upon better instrumentation, notably Nier's improvements of mass spectrometers. Decay constants are still somewhat uncertain. To an outsider, lead-uranium ages on zircons seem to be least subject to "resetting." It is unfortunate that the Canadian Survey has classified Precambrian time on a scale based on K/Ar dates—the method most readily subject to resetting. It seems safe to guess that a revision will soon be in order.

Of course ^{14}C dates have revolutionized the study of the later Pleistocene and Holocene, even though we have been disappointed to find that radiocarbon dates are not calendar dates.

MAPPING

All must be delighted at the development of so many new tools in so many aspects of geology. But the working out of geologic history, structure, and resource potential basically still rests on field mapping. The laboratory is only a valuable supplement to working out the mutual relations of the various rock masses, which can only be done by field mapping. Unfortunately with the development of these ancillary techniques there has resulted not merely a relative but an actual lessening of field mapping. Stratigraphic sections are published without map control and are thus completely unreliable.

I don't know what the situation is today, but when the National Science Foundation began operations it would award no grants for field mapping, only for laboratory equipment. To an onlooker, this still appears to be the case. More surprising is the diminution of the mapping program of the Geological Survey, whose much increased staff carries on much less mapping than the smaller staff of twenty years ago. I think it is demonstrable that field study is infinitely more likely to locate an ore body than any number of Landsat views from 500 miles away. As with the human eye, the discrimination of a photo lens diminishes with the square of the distance.

The Landsat picture of the Front Range west of Colorado Springs covers the area of major mining districts: Cripple Creek, Leadville, and Climax. None of these districts is identifiable on the Landsat picture, but a group of rusty-weathering pegmatites are conspicuous. It is only a confirmed optimist who could think the money spent on Landsat can be justified by its value to metalliferous prospecting. A small fraction of the cost of the satellite would fund detailed mapping on a scale of 1/24,000.

METALLIFEROUS GEOLOGY

Geology of metals remains largely as it was fifty years ago: primarily structural geology. One major development has been the discovery of sensitive field tests for metals, thereby lessening the dependence on the placer pan and leading to many discoveries, some of major importance. The use of the Geiger counter and the scintillometer in uranium prospecting has been mentioned.

Much progress has been made in studies of mineralizing solutions, fluid inclusions, and alteration geochemistry, all of theoretical interest, if not of much economic importance. Isotopic studies of O and H yield values equivalent to ground water, as would be expected because of long cooling times; they throw no light on the original metalliferous solutions.

PALEONTOLOGY

Being innocent of paleontological skills, I cannot pretend to evaluate progress on a critical level. It is obvious, nevertheless, to even the most ill-informed, that foraminiferal research was in its infancy in the 1920s, and is now one of the most highly developed branches of the science. A few people were working on conodonts then; now they are among the most useful fossils. Diatoms were then beautiful curiosities; today, under the microscope of Lohman, they are good guide fossils to the Cenozoic. Coccoliths have become, under the critical eye of Bramlette, excellent guide fossils and perhaps the best of all groups for intercontinental correlations. Fossil spore and pollen assemblages have become highly useful in testing former continental assemblies. Thick sequences of marine rocks have been broken down into distinct formal zones that have been assigned absolute age spans by isotopic studies. These studies not only permit accurate correlations of individual zones, but also give valuable data on rates of deposition. My paleontological friends insist, too, that ecological interpretations have become much more reliable than of yore.

SUMMARY

With the relative explosion of workers in the earth sciences, it is not surprising to observe the great forward movement of the sciences concerned. There are obviously many more able scholars at work all through the field than ever in history. Clearly the next fifty years should see more advance than ever.

Ann. Rev. Earth Planet. Sci. 1977. 5 : 13–33
Copyright © 1977 by Annual Reviews Inc. All rights reserved

BIOSTRATIGRAPHY OF THE CAMBRIAN SYSTEM— A PROGRESS REPORT

×10065

Allison R. Palmer
Department of Earth and Space Sciences, State University of New York, Stony Brook,
New York 11790

INTRODUCTION

The ultimate goal of biostratigraphy is the development of a sound basis for correlation of Phanerozoic rocks. Good biostratigraphy is needed to deal effectively with practical geological problems and with the more purely abstract problems of earth history. On a local scale, biostratrigraphic data can often be integrated with physical stratigraphic data to provide extremely precise correlation of particular historical events. On a global scale, biostratigraphic correlation may be less precise because it depends on the presence of rapidly evolving, widely distributed, and morphologically distinctive organisms—often an ideal rather than a practical reality. Nevertheless, precision of biostratigraphic correlation in the best circumstances for Early Paleozoic time is almost an order of magnitude better than by radiometric means. The potential for providing a precise independent check on biostratigraphically derived correlations comes from the rapidly developing field of magnetostratigraphy. However, until a larger body of data from this source is available for Cambrian time, biostratigraphic correlation is the most powerful available tool for unraveling Cambrian history. This article evaluates the current status of research on Cambrian biostratigraphy from a global viewpoint.

Except for the use of Archaeocyatha, Hyolitha, and acritarchs in zonal divisions of the Early Cambrian in the USSR, Cambrian biostratigraphy is almost entirely based on the spatial distribution of trilobites. This situation has some practical advantages. In younger systems, biostratigraphers have the constant problem of weighing the biostratigraphic value of a variety of coexisting phyla. Only the inarticulate brachiopoda, which are moderately abundant, but whose study is still in its infancy, and the Archaeocyatha, which are restricted to the early Cambrian, have some realistic potential in biostratigraphic competition with trilobites through significant parts of the Cambrian system.

Figure 1 shows the present distribution of occurrences of fossiliferous Cambrian rocks. There is an obvious potential for globally integrated biostratigraphic

13

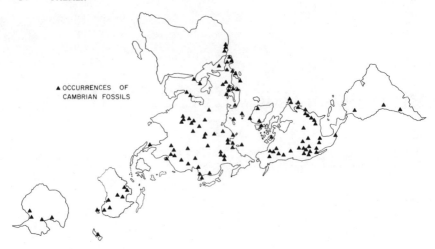

Figure 1   Map showing distribution of known occurrences of Cambrian fossils for the World.

schemes at some level. The present status, current active research, and future possibilities are discussed here for four levels in the hierarchy of biostratigraphic units: system, series, stage, and zone. It is hoped that this paper conveys something of the variety of research currently under way in this small but essential corner of geoscience.

SYSTEM BOUNDARIES AND SERIES SUBDIVISIONS

The Cambrian Subcommission of the International Stratigraphic Commission is reviewing the large-scale biostratigraphic problems of system boundaries and series subdivisions. Desire for greater precision in correlation increases the need to define system boundaries in some reproducible manner. The lower boundary of the Cambrian system has been a philosophical problem because of lack of biostratigraphic definition for the uppermost Precambrian. Although a variety of metazoans, without mineralized exoskeletons, has now been described from Precambrian rocks, there is no Precambrian biostratigraphy. Nevertheless, Ediacaran has been introduced as a name for the time interval immediately preceding the Cambrian—an interval to which others have ascribed the terms Eocambrian and Infracambrian. However, there is no consensus on criteria for an exact definition of the Ediacaran-Cambrian boundary.

The group working on the Cambrian-Precambrian Boundary under the leadership of John Cowie, University of Bristol, England, is an international group of paleontologists and stratigraphers charged with responsibility for establishing criteria that will be widely acceptable for this boundary, and for designating a stratotype locality where these criteria are well displayed. This group has examined localities in Siberia, North Africa, and Australia that might be potential stratotypes; American and other Soviet localities are still to be visited. Following the field studies, all evidences and philosophical arguments regarding the boundary will be reviewed and recommendations for criteria for definition of the boundary will be submitted to the International Stratigraphic Commission.

A separate international working group, under the leadership of W. T. Dean, Geological Survey of Canada, Ottawa, Canada, is beginning the task of stabilizing the Cambrian-Ordovician boundary. This boundary has different problems. Faunal data in some areas are plentiful. However, in Great Britain, which is the type region for both systems, critical faunal data for precise correlation are inadequate. The key to the problem is the Tremadocian Series, which has been considered by most British geologists to be Cambrian (Cowie, Rushton & Stubblefield 1972), but by most other geologists to be Ordovician in whole or in part. In northwestern Europe, the Tremadocian interval includes predominantly clastic rocks that bear faunas of low phyletic diversity, possibly representing cool water biofacies. In most of North America, Asia, and Australia, this interval includes predominantly carbonate rocks that bear faunas with moderate to high phyletic diversity, which may represent a warmer water biofacies. Resolution of this boundary involves both historical and paleoecological factors, and a program of field studies and philosophical discussions similar to that for the lower boundary of the system will be used to produce a consensus concerning the boundary stratotype and the criteria for its definition.

Within the Cambrian System, the traditional division has been tripartite—with Early, Middle, and Late Cambrian series—some of which have received local names. At present, international correlation of the boundaries of the Cambrian series is uncertain because of strong provinciality in Cambrian faunas and lack of reliable interprovincial correlation criteria.

Repina (1974) has summarized the problem faced by Soviet biostratigraphers concerned with the lower boundary of the Middle Cambrian, which has been placed at four different levels in Siberia. Identification of the correct one cannot be made until an international standard is identified. Öpik (1968) discussed this problem with relation to the Cambrian of Australia and proposed the Ordian stage for the earliest interval of the Middle Cambrian. He correlated this stage with the uppermost intervals of the European and Siberian Early Cambrian and presented a variety of reasons why these post-olenellid beds should be incorporated in the Middle Cambrian.

The upper boundary of the Middle Cambrian has seemed to be the least disputed of all of the intercontinentally recognized boundaries. It was established in the thin but richly fossiliferous Scandinavian sections where the widely distributed Agnostida are well developed. However, difficulty in establishing the biozones of some of the critical agnostid species has raised questions about some details of correlation that particularly affect the North American and Australian sequences (Daily & Jago 1975). Precise definition of an intercontinentally recognizable boundary may be difficult.

Robison & Rowell (1976) have suggested that there may be better chronohorizons for intercontinental correlation within the Cambrian system than the traditional series boundaries. At the International Geological Congress in August, 1976, the Cambrian subcommission established a single working group to examine the whole problem of internal correlations within the Cambrian system and to recommend to a future meeting of the subcommission the best intercontinentally correlatable chronohorizons that could serve as boundaries for internal subdivisions of the system. This working group is under the leadership of John Shergold of the Australian Bureau of Mineral Resources at Canberra.

All working groups will report on their activities at International Geological Congresses until the problems have been resolved. As a result of these international efforts there is an increasing number of biostratigraphers with experience in Cambrian terrains in many parts of the world. It is this real interchange of information, as well as the philosophical discussions, that holds some hope of stabilization of most critical international Cambrian boundaries within the next decade.

STAGES

Biostratigraphy at the stage level is still nationalistic. Although it is theoretically desirable to have a global system of stages, this may be a practical impossibility because of the strong provinciality of Cambrian faunas and the inadequate development of planktonic and pelagic elements that might provide effective intercontinental criteria. The best possibility for the development of intercontinental units on this scale will come through further knowledge of the biostratigraphy of the Agnostida, Archaeocyatha, and Acrotretida. Fossils of these groups are widespread in deposits accumulated on the unrestricted ocean-facing margins of the Cambrian continents. However, relatively little of this environment has survived Phanerozoic tectonic events unscathed.

As a result, separate stages (in Scandinavia, superzones) are recognized in some countries for some parts of the Cambrian. In some parts of the system, and in some countries, formal biostratigraphic divisions at the stage level are not defined. The current schemes of stages are shown on Tables 1-11. A correlation chart is not given because it would suggest more accuracy in correlation of stage boundaries than is warranted.

ZONES

Biostratigraphic activity and lively philosophical discussion now and in the near future will focus on the increasing sophistication of biostratigraphic units at the zonal level. This activity is centered on North America, Asia, and Australia, where good evidence of strong environmental control over trilobite distribution is creating pressure for major changes in the existing schemes. Elsewhere in the world, various national schemes seem to be stabilized or else basic information about the successions of trilobites is still being gathered. Because all of these schemes form the bases for establishing larger-scale biostratigraphic units, their current status and contribution to problems of general concern are discussed individually. But first, some general comments on the nature of biostratigraphic boundaries at the zonal level.

Evaluation of Zonal Boundaries

In any fossiliferous stratigraphic section, the boundaries between successive zones are of differing quality. In many discontinuously fossiliferous Cambrian sequences zonal nomenclature refers to distinctive fossiliferous intervals that are regionally separated by intervening barren intervals. In other instances, "boundaries" are in reality transition intervals between assemblages that differ in the dominance of particular taxa. Such "boundaries" may reflect either evolutionary changes within a single depositional regime or a gradual change of environment. In still other instances, zone or subzone boundaries may be marked by very abrupt faunal changes. Some abrupt changes can be attributed to stratigraphic juxtaposition of different depositional environments and associated biofacies, and the faunas representing each depositional environment may recur within a section with little sign of evolutionary change or appear in different stratigraphic orders in different parts of a region. In other cases, the abrupt change can be attributed to some subtle happening in the environment of continent-sized areas that is not reflected in the sediments, yet causes extinction of nearly all trilobites in almost all regions of the epicontinental seas. At present, there is no consistent procedure for distinguishing between boundaries of different quality that separate zones in fossiliferous Cambrian successions. Thus some boundaries are better than others.

North America

The prevailing biostratigraphic philosophy governing zonal division of the Cambrian of North America is that of assemblage zones. These zones are given names of characteristic species or genera and conventionally have been viewed as parts of a static succession applicable across the continent. However, increased understanding of biogeography and biofacies is forcing reconsideration. To understand the nature of the problem, one must look briefly at North American Cambrian paleogeography and some of the more recent developments of the biostratigraphical scheme.

Throughout Cambrian time, the present continent of North America, with the exception of the Avalonian belt along the eastern coast, was a distinct paleogeographic entity. Regional stratigraphic analysis (Holland 1971) shows a gradual marine inundation of the continental interior that began in late Precambrian time and continued into early Ordovician time. The shallow North American epeiric sea extended outward to the ocean-facing margins of a carbonate platform that may have fringed the entire continent (Palmer 1974). Beyond the platform deeper water conditions of the outer shelf or open ocean prevailed and can be documented in the Cordilleran region from Alaska to southern Nevada, in the Appalachian region from Newfoundland to Alabama (Palmer 1971a, b, North 1971), and in northern Greenland (Cowie 1971). Trilobite faunas reflect this paleogeography. Strikingly different co-existing trilobite complexes are now known from the epeiric seas, the platform margins, and the outer-shelf or deep-ocean regions (Palmer & Campbell 1976, Cook & Taylor 1975, Palmer & Taylor 1976) for most parts of the Cambrian system. More subtle biogeographic differentiation within the epeiric seas and over the open shelf regions is indicated for Late Cambrian faunas. The development of Cambrian biostratigraphy in North America reflects a growing appreciation of the major features of paleogeography.

In the first modern synthesis of North American Cambrian biostratigraphy (Howell et al 1944), much of the present biostratigraphic framework was established. This framework developed as a product of local biostratigraphic studies through the 1930s in the midcontinent and Cordilleran regions of the United States. Because the midcontinent studies were concerned only with Late Cambrian beds and the Cordilleran studies were concerned largely with Middle Cambrian beds, there was no appreciation of biofacies contrasts and each of the proposed zones was assumed to be continentwide except for the Atlantic Province along the east coast.

The fourteen-year interval between the publication of Howell et al's annotated correlation chart and the next important North American summary by Lochman-Balk & Wilson (1958) increased our understanding of the Middle Cambrian faunas of the Cordilleran region and the Late Cambrian faunas from limestones in Montana, Texas, and the northern Appalachian region. Much information about Late Cambrian faunas in the Cordilleran region remained unpublished, but the faunas were known to show important contrasts to those of the midcontinent region. The major contribution of the Lochman-Wilson summary was the recognition of contrasting biofacies realms that were designated as *cratonic, extracratonic (intermediate)*, and *extracratonic (euxinic)*. These were thought to characterize respectively the shallow-shelf, intermediate, and open-ocean environments that concentrically surrounded the North American continent. The principal changes from Howell et al (1944) were reductions in the number of regionally recognizable units in all series; some important substitutions in the Middle Cambrian, based principally on work by Rasetti (1951) in the southern Rocky Mountains of Canada; and addition of a new Late Cambrian zone for a faunal interval in the Cordilleran region that is represented by a hiatus in the midcontinent region. Despite the faunal contrasts between the cratonic and extracratonic intermediate regions, a single uniform zonation for the Cambrian of most of North America was retained.

The current view of North American Cambrian biostratigraphy (Table 1) is now being re-evaluated. Early Cambrian faunal sequences described by Fritz (1972) from open-shelf or platform-margin environments in northwestern Canada clearly contrast with contemporaneous faunal sequences from the restricted shelf in south-western United States. The medial and late Middle Cambrian zones described by Rasetti (1951) and Robison (1964) are now known to reflect the faunas of the platform margin and outer shelf, and the contrasting zonation for contemporaneous faunas from the epi-continental sea of Montana (Deiss 1939, Howell et al 1944) that was discarded by Lochman-Balk and Wilson may have validity for that region. Similar contrasts are apparent among Late Cambrian faunas now being described by several workers. Recommendations for the construction of separate biostratigraphies for each major biofacies region were discussed at the meeting of the Cambrian subcommission of the International Stratigraphic Commission in Australia in 1976 (Palmer & Taylor 1976, Robison & Rowell 1976). However, an

Table 1   Cambrian biostratigraphy of North America[a]

Late Cambrian:	*Dicanthopyge* Zone
Trempealeauian Stage	*Aphelaspis* Zone
	Crepicephalus Zone
Saukia Zone	*Cedaria* Zone
Corbinia apopsis Subzone	
Saukiella serotina Subzone	Middle Cambrian:
Saukiella junia Subzone	No stages named
Rasettia magna Subzone	
Franconian Stage	*Bolaspidella* Zone
	Bathyuriscus-Elrathina Zone
Saratogia Zone	*Glossopleura* Zone
Drumaspis Subzone	*Albertella* Zone
Idahoia lirae Subzone	*Plagiura-Poliella* Zone
Taenicephalus Zone	
Parabolinoides Subzone	Early Cambrian:
Elvinia Zone	No stages named
Dresbachian Stage	*Bonnia-Olenellus* Zone
	Nevadella Zone
Dunderbergia Zone	*Fallotaspis* Zone
Prehousia Zone	

[a] Lochman-Balk & Wilson (1958), Palmer (1965), Stitt (1971), Fritz (1972).

improved biostratigraphy must await further data, particularly from the Cordilleran region.

A further factor of importance in American Cambrian biostratigraphy is the existence of several striking extinction events among the trilobites. These events, marking the boundaries of Biomeres (Palmer 1965), occurred at infrequent intervals between 10 and 30 million years apart. Following each extinction event, the major elements of the newly appearing trilobite fauna are unrelated at the family level to any trilobites among the immediately preceding fauna. However, they are very similar to the trilobites that arrived after the previous extinction event.

Biomere boundaries are now well documented in the Late Cambrian between the *Aphelaspis* and *Crepicephalus* zones and the *Elvinia* and *Taenicephalus* zones, and at the Cambrian-Ordovician boundary (Palmer 1962, 1965, Longacre 1970, Stitt 1971). The events marking the boundaries seem to be most consistently recorded in the sedimentary record of the epeiric seas. There is suggestive evidence for perhaps two more events in the Middle Cambrian and one at the boundary between the Lower and Middle Cambrian in North American epeiric environments. As these events become better understood, they could serve as primary horizons for boundaries between major biostratigraphic subdivisions of the Cambrian deposits of the epeiric sea regions of North America.

Soviet Union

To the outsider, Soviet Cambrian biostratigraphy is often a bewildering complex of local sequences of "zones" and "horizons" coupled with an elaborate set of "stages" and "substages" that seems to be in continuous ferment. This is complicated by rules that require all state geological organizations to use an "official" nomenclature that was last revised in 1956, while various publications of the Academy of Sciences or Universities may propose and use revisions of this nomenclature.

According to the Soviet ideal, small-scale *biostratigraphic units* (s.s.) are the observable units in real sections and are identified as *zones* (which bear names of characteristic fossils) and *horizons* (which bear geographic names and which may include more than one zone). These units may have only local significance although within intervals representing the same depositional environments the units may be widely recognized.

Chronostratigraphic units, on the other hand, are abstract time intervals that are applicable to all areas independent of the specifics of depositional environments. These are series, stages, and substages. Most of the lively debates in Cambrian biostratigraphic literature from the USSR involve conflicts between local and regional biostratigraphic schemes and difficulties in precise correlation of the boundaries

Table 2   Cambrian biostratigraphy of Asiatic USSR

Late Cambrian			
Ivshin & Pokrovskaya (1968) (Stages)		Rosova (1964) (Superhorizons)	
Shidertinian		Khantayka Tukalanda Gorbiyachin	
Tuorian		Kulyumbe	
Middle Cambrian			
Chernysheva (1965) (Stages) Mayan Amginian			
Early Cambrian			
Cherynysheva (1965) (Stages) Lenian	Khomentovskiy & Repina (1965) (Stages) Lenian	Rozanov (1973) (Stages) Elankian	Repina et al (1964) (Horizons) Obrutchev Solontsov
———————	Botomian	Lenian	Sanashtykgol'
Aldanian	Aldanian	Atdabanian ——————— Tommotian	Kameshki Bazaikha Kundat
Precambrian (Judomian)			

of stages and substages. Savitskiy (1969a, b) has been a particularly vocal proponent of the importance of establishing boundary stratotypes for chronostratigraphic units in monofacial sequences. Thus, he is adding an important voice to international discussion about the increasing importance of trilobite paleoecology in small- and intermediate-scale biostratigraphy.

There are presently many schemes extant for various parts of the Cambrian of asiatic USSR (Table 2). A revision of the "official" scheme for Soviet Cambrian biostratigraphy (Chernysheva 1965) is currently in progress and should incorporate much new data and fresh thinking about biofacies. A few notes illustrating some of the biostratigraphic complexity are given below.

The Soviet Union has by far the most richly fossiliferous and fully described Early Cambrian sections and faunas in the world. This situation has resulted in a complex of biostratigraphic units in which archaeocyathids and other abundant nontrilobite components play a significant role (Rozanov 1973). In the earliest part of the Early Cambrian of the Siberian Platform, a richly fossiliferous interval below the oldest trilobites has been separately designated as the Tommotian stage (Rozanov et al 1969). The validity of a pretrilobite Early Cambrian stage outside of the Soviet Union is presently being debated. With regard to Early Cambrian trilobite biostratigraphy, the complications introduced by biofacies contrasts have been developed by Khomentovskiy & Repina (1965). They showed that the trilobite zones developed for the limestone facies of the southeastern part of the Siberian platform could not be applied to the largely dolomitic facies of the southwestern part of the platform.

Traditionally, Middle Cambrian deposits in the USSR have been assigned to two stages, the Amga (older) and the Maya. However, the exact base of the Middle Cambrian is in dispute (Repina 1974) and both boundaries of the Maya stage need to be clarified in the light of biofacies considerations (Savitskiy et al 1974). In and near the Siberian platform, the faunas from the type section of the Maya stage and most of its divisions are developed in rocks from unrestricted ocean-facing regions. However, the described faunas from the underlying Amga stage (Chernysheva 1961) and the overlying Late Cambrian beds (Rosova 1964, 1968) are developed in the restricted carbonate platform regions.

Biostratigraphic division of the Late Cambrian deposits of the USSR was first attempted in the 1960s. Rosova (1964) described the succession of faunas in a reference section of restricted platform carbonates in the northwestern part of the Siberian Platform and recognized four superhorizons, each composed of two horizons. In contrast, Ivshin & Pokrovskaya (1968) have proposed a two-stage division of the Late Cambrian based on analysis of the Late Cambrian faunas from unrestricted ocean-facing regions in Kazakhstan and the northeastern part of the Siberian Platform. These classifications are not yet reconciled.

Northern Europe and Avalonia

In contrast to the American and Soviet biostratigraphies that seem to be poised for major reorganization, the largely trilobite-based biostratigraphies of the "classical" Cambrian areas of Scandinavia, England, Newfoundland and paleogeographically related localities in eastern North America, Estonia, and Poland are stable and show little sign of impending major modification. In this region, stages and zones are chronostratigraphic units.

The tabular summaries given below (Tables 3-5) illustrate the uniformity in nomenclature over the European region. The contrasts are almost entirely in rocks

Table 3　Cambrian Biostratigraphy of Scandinavia (Martinsson 1974)

Late Cambrian:

　No stages named

　　Acerocare Zone (4 subzones)
　　Peltura scarabeoides Zone (4 subzones)
　　Peltura minor Zone (4 subzones)
　　Protopeltura praecursor Zone (5 subzones)
　　Leptoplastus Zone (6 subzones)
　　Parabolina spinulosa–Orusia lenticularis Zone (2 subzones)
　　Homagnostus obesus–Olenus Zone (6 subzones)
　　Agnostus pisiformis Zone

Middle Cambrian:

　Paradoxides forchammeri Stage

　　Lejopyge laevigata Zone
　　Jincella brachymetopa Zone

　Paradoxides paradoxissimus Stage

　　Triplagnostus lundgreni–Goniagnostus nathorsti Zone
　　Ptychagnostus punctuosus Zone
　　Tomagnostus fissus–Ptychagnostus atavus Zone
　　Triplagnostus gibbus Zone

Early Cambrian:

　No stages named

　　Strenuaeva linnarssoni Zone
　　Holmia kjerulfi Zone
　　Volborthella–Schmidtiellus Zone
　　Mobergella holsti Zone

Table 4　Cambrian biostratigraphy of Great Britain (Rushton 1974)

　　Late Cambrian: Same as Scandinavia
　　Middle Cambrian: Same as Scandinavia
　　Early Cambrian:
　　　Protolenid–Strenuellid Zone
　　　Olenellid Zone
　　　Nontrilobite Zone

Table 5   Cambrian biostratigraphy of the East European Platform

Orlowski 1964, 1968, 1974	Rozanov 1973
Late Cambrian: Same as Scandinavia Middle Cambrian: Same as Scandinavia Early Cambrian: *Protolenus* zone	Zansve Horizon
Holmia zone	Vergala horizon Lyukati (Talsy) horizon
sub-*Holmia* zone	Glebovo horizon Lontova horizon Rovno horizon

of Early Cambrian age where the faunas surrounding Avalonia, a linear paleo-geographic entity of uncertain dimensions lying between Europe, Africa, and North America and probably a fundamental part of a Gondwana plate, are notably different from those of mainland northern Europe in the kinds of olenellids that are present. These contrasts will not be fully appreciated or evaluated until new studies of the Cambrian of eastern Newfoundland (Fletcher, T. P., submitted for publication) appear. In Poland, a biostratigraphy for the Early Cambrian of the East European Platform, based largely on acritarchs and subsurface work, has recently been proposed (Rozanov 1973). The relations of this biostratigraphy to the trilobite based biostratigraphy are shown in Table 5.

The Middle Cambrian biostratigraphy derived from the thin platform cover of southern Sweden and Norway has been a standard against which many of the world's sections have been measured. This is largely because detailed studies here established the ranges of many of the agnostid species and genera (Westergard 1946) that characterize all the unrestricted ocean-facing sequences of the Cambrian.

The Late Cambrian biostratigraphy established in Scandinavia (Westergard 1922, Henningsmoen 1957) has been effectively applied from eastern Newfoundland to southern Poland and the Middle Cambrian biostratigraphy is applicable to all of northwestern Europe.

In eastern Newfoundland, representing Avalonia, a more elaborate biostratigraphy has been developed, but not yet published, for both the Early and Middle Cambrian sections. This area has the best development of fossiliferous Early Cambrian rocks in the Acado-Baltic region and additional detail for Middle Cambrian rocks will resolve many problems of correlation between the northern European and southern European subprovinces of the classic Acado-Baltic province (Sdzuy 1972). Thus, the published Early and Middle Cambrian biostratigraphies for Avalonia (Hutchinson 1952, 1962, North 1971) are outdated.

Southern Europe and North Africa

From the Hercynian massifs of central Europe southward to North Africa, scattered occurrences of fossiliferous Cambrian rocks are known. In southern East Germany, western and southern France, Spain, Sardinia and western Morocco, the Cambrian sections are characterized by a significant development of Early Cambrian limestones, often with archaeocyathids, and Middle Cambrian marine clastic sequences. Late Cambrian fossiliferous rocks have been identified only from one locality in northern Spain. Middle Cambrian or younger unfossiliferous marine clastics are generally overlain unconformably by Tremadocian rocks. The entire Cambrian section in Czechoslovakia is composed of marine clastics.

The southern European–North African region has been considered to be a southern subprovince of the Acado-Baltic Province and its biostratigraphy has been well summarized by Sdzuy (1972). Two subsequent publications by Courtessole (1973) and Rasetti (1972) have contributed some new information about the Middle Cambrian of the Montagne Noire of southern France, and the Early and Middle Cambrian of Sardinia, respectively.

Because the outcrop areas, except for Morocco, are small and widely scattered, there is no regional synthesis of the faunas, and no clear sense of paleogeographic relations between the areas. Thus lithofacies and biofacies remain to be analyzed and the biostratigraphies (Tables 6-10) reflect primarily local faunal successions. However, within each area the successions are composite and thus all units are chronostratigraphic.

Early Cambrian faunas are known throughout most of the region, but faunal sequences have been worked out only in Morocco, Spain, and Sardinia, and even these are incomplete. Hupé (1952) described many of the Early Cambrian trilobites of Morocco and proposed a series of eight zones. In a subsequent paper (Hupé 1960), he rejected the zones and substituted the succession of stages and substages shown in Table 6. Sdzuy (1971a, 1972) proposed a three-stage division for the Early Cambrian of Spain (Table 7) based primarily on data from the Iberian Chain. Neither

Table 6   Cambrian biostratigraphy of Morocco (Hupé 1960)

Late Cambrian: Absent

Middle Cambrian: Faunas present but largely undescribed

Early Cambrian:

    Isafenian Stage

      Aguilizian Substage
      Tasousekhtian Substage

    Soussian Stage

      Timghitian Substage
      Amouslekian Substage
      Assadasian Substage

succession has been tested for its regional utility. Rasetti (1972) has recorded a succession of Early Cambrian faunas from Sardinia that seems to represent the Ovetum Stage of Spain (Table 8). The faunas of North Africa, Spain, and Sardinia seem to represent a single faunal province different from 'that of northern Europe. The

Table 7 Cambrian biostratigraphy of Spain (Sdzuy 1971a, b; 1972)

Late Cambrian:
   Prochuangia–Chuangia fauna

Middle Cambrian:

   No stages named

   Solenopleuropsis Zone
   Pardailhania Zone
   Badulesia Zone
   Acadolenus Zone
   Conocoryphe ovata Zone
   Paradoxides mureroensis Zone

Early Cambrian:

   Biblis Stage

   Mariani Stage

   Ovetum Stage

Table 8 Cambrian biostratigraphy of Sardinia (Rasetti 1972)

Middle Cambrian: 3 unnamed faunules, cf. Montagne Noire, Spain
Early Cambrian:
   Dolerolenus fauna
   Hebediscus fauna

Table 9 Cambrian biostratigraphy of Czechoslovakia (Snadjr 1958)

Late Cambrian: Not known

Middle Cambrian:

   No stages named

   Hydrocephalus lyelli Zone
      Lingulella matthewi Subzone
      Ellipsocephalus hoffi Subzone
   Paradoxides gracilis Zone
   Stromatocystites pentagngularis Zone
   Eccapardoxides pusillus Zone

Early Cambrian: Known only from single aglaspid. No zonation.

Table 10   Cambrian biostratigraphy of the Montagne Noire, southern France (Courtessole 1973)

Late Cambrian: Absent
Middle Cambrian:
    Nine lettered zones with faunas corresponding to those of Spain.
Early Cambrian: Not subdivided. Faunas present but largely undescribed.

---

detailed documentation from well integrated fossiliferous sections that is necessary to establish a precise regional biostratigraphy is still lacking.

Middle Cambrian faunas are well described from Spain (Sdzuy 1958a, 1961, 1968 1969), Czechoslovakia (Snajr 1958), Sardinia (Rasetti 1972), and the Montagne Noire of southern France (Courtessole 1973), although only the lower part of the Middle Cambrian is present throughout the region. The faunas from Spain, Sardinia, and the Montagne Noire (Table 9) are very similar. The faunas from Morocco are largely undescribed but contain many elements in common with Spain (Sdzuy 1972). The Middle Cambrian zonation proposed by Sdzuy (1971b, 1972) for Spain can probably be applied throughout the Mediterranean region. The Czechoslovakian succession (Table 10), although distinct, contains some representatives of genera and families characteristic of the Mediterranean region. Sdzuy (1972) recognized this and included Czechoslovakia in his southern Acado-Baltic subprovince.

The single Late Cambrian fauna from the Sierra de la Demanda in northeastern Spain, reported but not yet described or illustrated, (Colchen 1967) contains trilobites tentatively assigned to the asiatic genera, *Prochuangia* and *Chuangia*. This further emphasizes the contrast between the northern and southern subprovinces of the Acado-Baltic region and strengthens the biogeographic relations of the western Mediterranean Cambrian faunas to those of the eastern Mediterranean and the Middle East. However, there is insufficient information to establish a Late Cambrian biostratigraphy for the southern-European–North-African region.

China and Southeastern Asia

This region and Australia, which is discussed separately, constitute a large area of varied biofacies and lithofacies that is also about ready to have its Cambrian biostratigraphy rephrased in more dynamic terms. Prior to an important recent summary paper by Lu et al (1974), most of the easily available information about the Cambrian of China, Korea, and southeast Asia came from Kobayashi (1967, 1971). His conclusions about both paleogeography and biostratigraphy are strongly criticized by Lu et al who present a revised biostratigraphy and paleogeography based on more recent work. Their interpretations of biofacies and lithofacies relationships of the trilobite faunas are remarkably similar to those derived from analysis of the North American data and their conclusions are followed here.

During Cambrian time, the present areas of China and Korea were parts of a paleogeographic unit that had its core area in Mongolia (Palmer 1974, Lu et al 1974). Lu et al point out the fundamental similarity of the Cambrian faunas of

China, Korea, southeast Asia and Australia and assign the faunas from these areas to two principal biofacies designated as North China and Southeast China types. The North China biofacies is characterized by benthonic or nektobenthonic trilobite assemblages in shallow-water shale or limestone lithofacies. The Southeast China biofacies is characterized by agnostid-rich pelagic faunas in dark limestones and black shales considered to represent a deeper water lithofacies. The boundary between these two biofacies is transitional, and stratigraphic sequences may contain intervals with the fauna of each biofacies. Lu et al argue that the faunal similarities between China, southeast Asia, and Australia demonstrate the coexistence of these areas on Pangaea during Cambrian time, and they reject all hypotheses suggesting that Pangaea was formed by the amalgamation of a number of separate early Paleozoic paleogeographic elements as proposed by Palmer (1974). However, they do not bring strong geological evidence to bear on their conclusions and do not discuss some critical areas. Thus, Cambrian paleogeographic relations of the area now represented by China and Korea to areas in southeast Asia, India, Australia, and the Middle Asia part of the USSR remain unresolved.

Lu et al have made an important contribution to Cambrian geology by clarifying in modern terms both the biostratigraphy and paleogeography of China. The biostratigraphy that they present is a chronostratigraphy based on the faunal sequences of the North China biofacies (Table 11), and faunal successions from regions with transitional and southeastern-type faunas are correlated to this succession. This biostratigraphy may also be applicable in many parts of southeast Asia and

Table 11   Cambrian biostratigraphy of China (Lu et al 1974)

Late Cambrian:	*Crepicephalina* Zone
Fengshan Stage	*Liaoyangaspis* Zone
	Hsuchuan Stage
Tellerina–Calvinella Zone	
Ellesmeroceras–Dictyella Zone	*Bailiella* Zone
Quadraticephalus Zone	*Poriagraulos* Zone
Ptychaspis–Tsinania Zone	*Sunaspis* Zone
	Kochaspis hsuchuangensis Zone
Changshan Stage	
	Early Cambrian:
Kaolishania Zone	
Changshania Zone	Maochuan Stage
Chuangia Zone	
	Shantungaspis Zone
Kushan Stage	
	Manto Stage
Drepanura Zone	
Blackwelderia Zone	*Redlichia chinensis* Zone
	Tsangpin Stage
Middle Cambrian:	
	Megapalaeolenus Zone
Changhia Stage	*Palaeolenus* Zone
	Drepanuroides Zone
Damesella Zone	*Malungia* Zone
Taitzuia Zone	two unnamed zones
Amphoton Zone	

Australia and in the Middle East, where most of the described faunas have a strong Chinese aspect. It is clearly distinct from the biostratigraphy of the Siberian platform and its adjacent orogens. However, the synthesis necessary to establish the geographic limits within which the Chinese Cambrian biostratigraphy can be applied must await

Table 12   Biostratigraphy of the Cambrian of Australia

References to system and series boundaries		Stages	Zones	References to zones
Jones, Shergold & Druce 1971	L. Ord.	L. Ord.		-----
	UЄ	PAYNTONIAN	Mictosaukia perplexa Pseudagnostus quasibilobus	Shergold 1975
		pre-PAYNTONIAN A	Sinosaukia impages	
		pre-PAYNTONIAN B	Pseudagnostus bifax Pseudagnostus clarki prolatus Pseudagnostus clarki patulus	
		NO NAME	Undesignated zones	
		IDAMEAN	Irvingella tropica Stigmatoa diloma Erixanium sentum Proceratopyge cryptica Glyptagnostus reticulatus	Öpik 1963 Henderson 1976
		MINDYALLAN	Glyptagnostus stolidotus Cyclagnostus quasivespa	Öpik 1967
Daily & Jago 1975	UЄ MЄ	NO NAMES	Erediaspis eretes Damesella toyosa- Ascionepea janitrix	
			Holteria arepo (L. laevigata III) Proampyx agra (L. laevigata II) Ptychagnostus cassis (L. laevigata I) Ptychagnostus nathorsti	Öpik 1961
			Ptychagnostus punctuosus- P. nathorsti Ptychagnostus punctuosus Euagnostus opimus Ptychagnostus atavus	Öpik 1970
		TEMPLETONIAN	Ptychagnostus gibbus Xystridura	Öpik & Pritchard 1960, Öpik 1968
Öpik 1968	MЄ	ORDIAN	Redlichia	
	LЄ	NO NAMES	Faunal assemblages 1–12	Daily 1956

more data from all of the areas mentioned. Some of the Australian work, discussed below, is establishing a Middle and Late Cambrian biostratigraphy based primarily on sequences from regions with the transitional and southeastern-type faunas of Lu et al. Thus, the final resolution of the Cambrian biostratigraphy of the present regions of southern Asia and Australia may be the development of parallel biostratigraphies for the two principal biofacies.

Australia

The present Cambrian biostratigraphy of Australia results from geological explorations largely undertaken within the past 20 years (Table 12). The primary data have been obtained from two principal areas: western Queensland and South Australia.

The Early Cambrian data are derived predominantly from the folded rocks of the Flinders Ranges in South Australia where the stratigraphic control is very good. However, there is no modern systematic treatment of the faunal assemblages listed by Daily (1957).

The Middle and Late Cambrian biostratigraphy is based on data from western Queensland, which is a region of low relief and predominantly subhorizontal Phanerozoic rocks so that much of the Cambrian section has been pieced together from scattered exposures of only a few meters' vertical extent. Only the Idamean and younger parts of the Cambrian are represented by measured stratigraphic sections with controlled fossiliferous horizons, and the faunas of the Templetonian stage and of the Middle Cambrian zones up to the base of the *Lejopyge laevigata* zone are largely undescribed in modern terms. Thus, the biostratigraphic scheme for Australia is in reality a compendium of information from limited areas and a variety of sources, and much descriptive documentation is lacking. Nevertheless, fossiliferous Cambrian occurrences within all of Australia are correlated to this general scheme.

Miscellaneous Areas

Cambrian sections with the potential to contribute to global Cambrian biostratigraphy exist in western Argentina (Borrello 1971) and parts of the Middle East (Kushan 1973, Dean 1975). However, much more biostratigraphic work is needed in these areas before the bits of information now known about their faunas become generally useful. Other areas such as India, New Zealand, and Antarctica will probably always be passive users of global biostratigraphies developed elsewhere.

CONCLUSIONS

Although Cambrian biostratigraphy for some parts of the world seems to have stabilized, the largest regions of Cambrian exposures such as North America, Asia, and Australia are about ready for major biostratigraphic revisions. These will reflect increasing sophistication of knowledge about trilobite paleoecology and biogeography, and about Cambrian paleogeography. However, there are some particular dangers involved in geographic syntheses that can contribute to potential misinterpretations.

In this era of evolution of ideas about plate tectonics and about the location and constitution of former continents, particularly in the Early Paleozoic, there is great danger in the assumption that regions that are biogeographically similar must of necessity have been geographically close together. The unrestricted ocean-facing margins of Cambrian continents share many nearly cosmopolitan taxa, whereas the restricted epicontinental seas have much more endemic populations. Global studies of Cambrian faunas that use cluster analysis techniques for biogeographic interpretation (Whittington & Hughes 1974, Jell 1974) have chosen operational units of continental or subcontinental size for comparison, but have not distinguished within these units between the more cosmopolitan and endemic biofacies. This practice has obscured some obvious qualitative differences between regions and has distorted some of the clusters. Furthermore, proper paleogeographic analysis requires that biogeographic data must be considered together with lithofacies and geophysical information. To date, this has not been done satisfactorily by those using quantitative approaches on a global scale.

Based on this analysis of the present status of Cambrian biogeography and of the principal problems facing Cambrian biostratigraphers, there are a number of areas of research that must be explored before a satisfactory global biostratigraphy can develop. The potential is there, as indicated by the wide distribution of fossiliferous Cambrian outcrops and the general richness and diversity of Cambrian faunas. However, new information is needed about the precise occurrence of fossiliferous horizons in many areas, about the quality of biostratigraphic boundaries in different regions and about biogeography, paleogeography, and trilobite paleoecology. In addition, as new biostratigraphies are proposed, the distinction should be made and clearly drawn between biostratigraphic units (s.s.) and chronostratigraphic units. There is still much to do!

ACKNOWLEDGMENTS
Special thanks are due to Dr. A. Yu. Rozanov, Geological Institute, Academy of Sciences USSR for review of this manuscript and for particular advice regarding the construction of Table 2. Work on this paper was completed during tenure of NSF grant DES7520173.

Literature Cited

Borrello, A. V. 1971. The Cambrian of South America. See Holland 1971, pp. 385–438

Chernysheva, N. E. 1961. Stratigrafiya kembriya aldanskoy anteklizy i paleontologicheskoe obosnovanie vydeleniya amginskogo yarusa. *Trudy VNIGI* 49:1–347

Chernysheva, N. E., ed. 1965. *Kembriyskaya Sistema.* Moscow: Nedra. 596 pp.

Colchen, M. 1967. Sur la présence du Cambrien supérieur a *"Prochuangia"* et a *"Chuangia"* dans la Sierra de la Demanda. *C. R. Acad. Sci. Ser. D* 264:1687–90

Cook, H. E., Taylor, M. E. 1975. Early Paleozoic continental margin sedimentation, trilobite biofacies, and the thermocline, western United States. *Geology* 3:559–62

Courtessole, R. 1973. *Le Cambrien Moyen de la Montagne Noire.* Toulouse: Imprimerie d'Oc. 248 pp

Cowie, J. W. 1971. The Cambrian of the North American Arctic Regions. See Holland 1971, pp. 325–84

Cowie, J. W., Rushton, A. W. A., Stubblefield, C. J. 1972. *A Correlation of Cambrian Rocks in the British Isles. Geol. Soc. London Spec. Rep.* 2:1–42

Daily, B. 1957. The Cambrian in South

Australia. In *The Cambrian Geology of Australia*, ed. A. A. Öpik, *Bur. Miner. Resour. Aust. Bull.* 49:91–148

Daily, B., Jago, J. B. 1975. The trilobite *Lejopyge* Hawle and Corda and the middle-upper Cambrian boundary. *Palaeontology.* 18:527–50

Dean, W. T. 1975. Cambrian and Ordovician correlation and trilobite distribution in Turkey. *Fossils and Strata.* 5:353–73

Deiss, C. F. 1939. Cambrian stratigraphy and trilobites of northwestern Montana. *Geol. Soc. Am. Spec. Pap.* 18:1–135

Fritz, W. H. 1972. Lower Cambrian trilobites from the Sekwi Formation type section, Mackenzie Mountains, northwestern Canada. *Geol. Surv. Can. Bull.* 212:1–58

Henderson, R. A. 1976. Upper Cambrian (Idamean) trilobites from western Queensland, Australia. *Palaeontology.* 19:325–64

Henningsmoen, G. 1957. The trilobite family Olenidae. *Skr. Nor. Vidensk. Akad. Oslo, 1.* 1:1–303

Holland, C. H., ed. 1971. *Cambrian of the New World* New York: Wiley. 456 pp.

Hupé, P. 1952. Contribution à l'étude du Cambrien Inférieur et du Précambrien III de l'Anti-Atlas Marocain. *Serv. Géol. Maroc. Notes Memo.* 103:1–402

Hupé, P. 1960. Sur le Cambrien Inférieur du Maroc. *21st Int. Geol. Cong. Proc.* 8:75–85

Howell, B. F., Bridge, J., Deiss, C. F., Edwards, I., Lochman, C., Raasch, G. O., Resser, C. E. 1944. Correlation of the Cambrian formations of North America. *Geol. Soc. Am. Bull.* 55:993–1003

Hutchinson, R. D. 1952. The stratigraphy and trilobite faunas of the Cambrian sedimentary rocks of Cape Breton Island, Nova Scotia. *Geol. Surv. Can. Mem.* 263:1–124

Hutchinson, R. D. 1962. Cambrian stratigraphy and trilobite faunas of southwestern Newfoundland. *Geol. Surv. Can. Bull.* 88:1–156

Ivshin, N. K., Pokrovskaya, N. V. 1968. Stage and Zonal Subdivision of the Upper Cambrian. *23rd Int. Geol. Çong. Proc.* 9:97–108

Jell, P. A. 1974. Faunal provinces and possible planetary reconstruction of the Middle Cambrian. *J. Geol.* 82:319–50

Jones, P. J., Shergold, J. H., Druce, E. C. 1971. Late Cambrian and Early Ordovician stages in western Queensland. *J. Geol. Soc. Aust.* 18:1–32

Khomentovskiy, V. V., Repina, L. N. 1965. *Nizhniy Kembriy stratotipicheskogo razreza Sibiri.* Moscow: Nauka. 199 pp.

Kobayashi, T. 1967. The Cambrian of Eastern Asia and other parts of the Continent. *J. Fac. Sci. Univ. Tokyo.* 16:381–534

Kobayashi, T. 1971. The Cambro-Ordovician Faunal Provinces and the interprovincial correlation. *J. Fac. Sci. Univ. Tokyo. Sect. 2.* 18:129–299

Kushan, B. 1973. Stratigraphie und trilobitenfauna in der Milaformation (Mittelkambrium-Tremadoc) im Alborz-Gebirge (N-Iran). *Paleontographica* 144:113–65

Lochman-Balk, C., Wilson, J. L. 1958. Cambrian biostratigraphy in North America. *J. Paleontol.* 32:313–50

Longacre, S. A. 1970. Trilobites of the Upper Cambrian Ptychaspid Biomere, Wilberns Formation, central Texas. *J. Paleontol.* 44:1–70

Lu, Y. H., Chu, C. L., Chien, Y. Y., Lin, H. L., Chow, T. Y., Yuan, K. S. 1974. Bio-environmental control hypothesis and its application to the Cambrian biostratigraphy and paleozoogeography. *Nanking Inst. Geol. Paleontol. Mem.* 5:27–110

Martinsson, A. 1974. The Cambrian of Norden. In *Cambrian of the British Isles, Norden and Spitzbergen*, ed. C. H. Holland, pp. 185–283. New York: Wiley. 300 pp.

North, F. K. 1971. The Cambrian of Canada and Alaska. See Holland 1971, pp. 219–324

Öpik, A. A. 1961. The geology and paleontology of the headwaters of the Burke River, Queensland. *Bur. Miner. Resour. Aust. Bull.* 53:1–249

Öpik, A. A. 1963. Early Upper Cambrian fossils from Queensland. *Bur. Miner. Resour. Aust. Bull.* 64:1–133

Öpik, A. A. 1967. The Mindyallan fauna of northwestern Queensland. *Bur. Miner. Resour. Aust. Bull.* 74:1–167

Öpik, A. A. 1968. The Ordian Stage of the Cambrian and its Australian Metadoxididae. *Bur. Miner. Resour. Aust. Bull.* 92:133–65

Öpik, A. A. 1970. Nepeid trilobites of the Middle Cambrian of northern Australia. *Bur. Miner. Resour. Aust. Bull.* 113:1–48

Öpik, A. A., Pritchard, P. W. 1960. Cambrian and Ordovician. *J. Geol. Soc. Aust.* 7:89–114

Orlowski, S. 1964. Middle Cambrian and its fauna in the eastern part of the Holy Cross Mts. *Stud. Geol. Pol.* 16:1–94

Orlowski, S. 1968. Upper Cambrian fauna of the Holy Cross Mts. *Acta Geol. Pol.* 18:257–91

Orlowski, S. 1974. Lower Cambrian biostratigraphy in the Holy Cross Mts, based on the trilobite family Olenellidae. *Acta Geol. Pol.* 24:1–16

Palmer, A. R. 1962. *Glyptagnostus* and associated trilobites in the United States. *US Geol. Surv. Prof. Pap.* 374-F, 1–49

Palmer, A. R. 1965. Trilobites of the Late Cambrian pterocephaliid biomers in the Great Basin, United States. *US Geol. Surv. Prof. Pap.* 493:1–105

Palmer, A. R. 1971a. Cambrian of the Great Basin and adjacent areas, western United States. See Holland, 1971, pp. 1–78

Palmer, A. R. 1971b. Cambrian of the Appalachian and eastern New England regions, eastern United States. See Holland 1971, pp. 169–218

Palmer, A. R. 1974. Search for the Cambrian world. *Am. Sci.* 62:216–24

Palmer, A. R., Campbell, D. P. 1976. Biostratigraphic implications of trilobite biofacies: Albertella Zone, Middle Cambrian, western United States. In *Paleontology and Depositional Environments: Cambrian of western North America,* ed. R. A. Robison, A. J. Rowell, *Brigham Young Univ. Res. Stud. Geol. Ser.* 23:39–50

Palmer, A. R., Taylor, M. E. 1976. Biostratigraphic implications of shelf-to-basin changes in Cambrian trilobite biofacies. *25th Int. Geol. Cong. Abstr.* 1:276–77

Rasetti, F. 1951. Middle Cambrian stratigraphy and faunas of the Canadian Rocky Mountains. *Smithson. Misc. Collect.* 116:1–277

Rasetti, F. 1972. Cambrian trilobite faunas of Sardinia. *Atti Del. Accad. Naz. dei Lincei Mem.* 9:1–100

Repina, L. N. 1974. K voprosy o granitse Nizhnego i Srednego Kembriya Sibirskoy Platformy i sopredel'nykh territory. In *Biostratigrafiya i Paleontologiya Nizhnego Kembriya Évropy i Severnoy Asii,* ed. I. T. Zhuravleva, A. Yu. Rozanov, pp. 76–103. Moscow: Nauka 311 pp.

Repina, L. N., Khomentovskiy, V. V., Zhuravleva, I. T., Rozanov, A. Yu. 1964. *Biostratigrafiya Nizhnego Kembriya Sayano-Altayskoy Skladchatoy Oblasti.* Moscow: Nauka 364 pp.

Robison, R. A. 1964. Late Middle Cambrian faunas from western Utah. *J. Paleontol.* 38:510–66

Robison, R. A., Rowell, A. J. 1976. Subdivision of the Cambrian system—are traditional boundaries appropriate? *25th Int. Geol. Cong. Abstr.* 1:279–80

Rosova, A. V. 1964. *Biostratigrafiya i opisanie trilobitov srednego i verkhnego kembriya severo-zapada Sibirskoy Platformy.* Moscow: Nauka 148 pp.

Rosova, A. V. 1968. Biostratigrafiya i trilobity verkhnego kembriya i niznego ordovika severo-zapada Sibirskoy Platformy. *Trudy Inst. Geol. Geof. Sib. Otd. Akad Nauk SSSR* 36:1–196

Rozanov, A. Yu. 1973. Zakonomernosti morfologicheskoy evolyutsii Arkheotsiat i voprosy yarusnogo raschleneniya nizhnego Kembriya. Moscow: Nauka. 164 pp.

Rozanov, A. Yu., Missarzhevsky, V. V., Volkova, N. A., Voronova, L. G., Krylov, I. N., Keller, B. M., Korolyuk, I. K., Lendzion, K., Michniak, R., Pychova, N. G., Sidorov, A. D. 1969. Tommotskiy yarus i problema nizhney granitsa Kembriya. *Trudy Geol. Inst. Akad. Nauk SSSR,* 206:1–380

Rushton, A. W. A. 1974. The Cambrian of Wales and England. See Martinsson 1974, pp. 43–122

Savitskiy, V. E. 1969a. O pravilakh o stratigraficheskoy klassifikatsii i terminologii i o prirode khronostratigraficheskikh podrazdeleniy. *Trudy SNIIGGIMS* 94:84–99

Savitskiy, V. E. 1969b. O yarusnom raschlenenii srednego kembriya Sibiri i nekotorykh obshchikh voprosakh razrabotki etalonnoy shkaly yarusnykh podrazdeleniy. *Trudy SNIIGGIMS* 94:140–49

Savitskiy, V. E., Yegorova, L. I., Shabanov, Yu. Ya. 1974. Probel v khronostratigraficheskoy shkale crednego kembriya Sibiri. *Trudy SNIIGGIMS* 173:22–29

Sdzuy, K. 1958a. Neue Trilobiten aus dem Mittelkambrium von Spanien. *Senckenbergiana Lethaea* 39:235–53

Sdzuy, K. 1961. Das Kambrium Spaniena. Teil II: Trilobiten. *Akad. Wiss. Lit. Mainz Abh. Math. Naturwiss. Kl.* 7, 8:217–408

Sdzuy, K. 1968. Trilobites del cámbrico medio de Asturias. *Trab. Geol.* 1:77–133

Sdzuy, K. 1969. Biostratigrafia de la griotte cámbrica de los Barrios de Luna (Léon) y de otras sucesiones comparables. *Trab. Geol.* 2:45–58

Sdzuy, K. 1971a. Acerca de la correlación del Cámbrico inferior en la Peninsula Ibérica. *Publ. I, Congr. hispano-luso-americ. Geol. econ.* 2:763–68

Sdzuy, K. 1971b. La subdivisión biostratigráfica y la correlación del Cambrico medio. *Publ. I, Cong. hispano-luso-americ. Geol. econom.* 2:769–82

Sdzuy, K. 1972. Das Kambrium der acado-baltischen Faunenprovinz. *Zentralbl. Geol. Paläontol. Teil II* 1/2:1–91

Shergold, J. H. 1975. Late Cambrian and Early Ordovician Trilobites from the

Burke River Structural Belt, Western Queensland, Australia: *Bur. Miner. Resour. Aust.* 153:1–251

Snajr, M. 1958. Trilobiti Ceského stredniho kambria. *Rozpr. Ústred. Ústavu Geol.* 24:1–280

Stitt, J. H. 1971. Late Cambrian and earliest Ordovician trilobites, Timbered Hills and Lower Arbuckle Groups, western Arbuckle Mountains, Murray County, Oklahoma. *Okla. Geol. Surv. Bull.* 110:83

Westergard, A. H. 1922. Sveriges Olenidskiffer: *Sver. Geol. Unders.* 18:1–205

Westergard, A. H. 1946. Agnostidea of the Middle Cambrian of Sweden. *Sver. Geol. Unders.* 477:1–140

Whittington, H. B., Hughes, C. P. 1974. Geography and faunal provinces in the Tremadoc Epoch. In *Paleogeographic Provinces and Provinciality,* ed. C. A. Ross, *Soc. Econ. Paleontol. Mineral. Spec. Pub.* 21:203–18

Ann. Rev. Earth Planet. Sci. 1977. 5 : 35–64

UNDERSTANDING INVERSE THEORY �ష10066

Robert L. Parker

Institute of Geophysics and Planetary Physics, Scripps Institution of Oceanography, University of California, San Diego, La Jolla, California 92093

INTRODUCTION

Much of our knowledge of the Earth's interior is perforce based on the interpretation of measurements made at the surface, rather than direct sampling of the material in the interior. In the past few years there have been great advances in the mathematical aspects of this problem, and the topic has come to be called geophysical inverse theory. To apply these ideas, there must be a valid mathematical model of the physics of the system under study, so that one would be able to calculate the values of observations made on an exactly known structure: the calculation of the behavior of a specified system is the solution of the "forward" or "direct" problem. Frequently it is the forward problem that presents a difficult challenge to the theoretical geophysicist. Illustrations include the mechanism of earthquake rupture or the generation of the Earth's magnetic field; in problems like these, inverse theory is normally quite inappropriate. When the forward problem has been completely solved, there are of course unknown parameters in the mathematical model representing physical properties of the Earth such as Lamé parameters, density or electrical conductivity. The goal of inverse theory is to determine the parameters from the observations or, in the face of the inevitable limitations of actual measurement, to find out as much as possible about them. The quality that distinguishes inverse theory from the parameter estimation problem of statistics (Bard 1974, Rao 1973) is that the unknowns are *functions,* not merely a handful of real numbers. This means that the solution contains in principle an infinite number of variables, and therefore with real data the problem is as under-determined as it can be. Naturally, there are geophysical problems containing a relatively small number of free parameters: for example, in describing the relative instantaneous motion of N lithospheric plates, we find that the assumption of internal rigidity reduces the number of unknowns to $3N-3$ for the $N-1$ relative angular velocity vectors (McKenzie & Parker 1974). Sometimes, however, unknown structures are conceived in terms of small numbers of homogeneous layers for reasons of computational simplicity rather than on any convincing geophysical or

35

geological grounds. Such simplification may lead to false confidence in the solution because the true amount of freedom has not been allowed in the parameters.

The theory falls into two distinct parts. One deals with the ideal case in which the data are supposed to be known exactly and as densely as desired, and is of course mainly the province of the applied mathematician. The other treats the practical problems that are created by incomplete and imprecise data. It might be thought that an exact solution to an inverse problem with perfect data would prove extremely useful for the practical case. Actually, this often turns out to be untrue because geophysical inverse problems are almost always unstable in a sense to be defined more precisely later; when this is so, the solution obtained by the analytic technique is very sensitive to the way in which the data set is completed and to the errors in it. In my view analytical studies are more valuable for their results concerning uniqueness and conditions for existence and stability.

The advances mentioned earlier arise primarily from a recognition that practical inverse problems never possess unique solutions and that an honest attempt to interpret the data must appraise the variety of compatible solutions. In the class of linear problems there is now a very satisfactory body of theory for achieving this end, but work on the larger and more prevalent class of nonlinear problems is still in a relatively primitive state and perhaps always will be.

The plan of this article is as follows. In the next section I outline and discuss some mathematical questions in the analysis of idealized data. Although many of the points raised by the geophysical problems are fascinating to the mathematician, I have tried to keep in mind that the aim of our science is to learn about the Earth and therefore I attempt to restrict discussion to matters of practical consequence. After this, I deal directly with the problem of using actual measurements. First, there is a fairly extended account of the linear theory in which I permit myself occasionally to give some of the mathematical details. Here the principles of the celebrated Backus-Gilbert method are described and a particular formulation is given of the spectral expansion method which has made possible the solution of so many practical problems. Finally, there is a section on the analysis of nonlinear systems, consisting of a discussion of the linearization approximation and a brief sketch of some of the commendable efforts to do without approximation.

ANALYSIS OF PERFECT DATA

The task of retrieving model parameters from a complete and precise set of data is clearly a mathematical undertaking. There is no single simple method of attacking the various aspects of inverse theory, but instead methods are drawn from almost every branch of applied mathematics. Therefore, I make no attempt to give any derivations in this section, but rather to highlight the fundamental concepts and the way that they impinge upon the more empirical side of geophysics. I consider the questions of existence, uniqueness, construction, and stability, which are some of the concerns that must be dealt with in a complete solution of an inverse problem. Sabatier (1971) has given a more extended account of these topics in a wide-ranging comparative review.

Mathematicians take the view, logically enough, that before attempting to calculate parameters, one ought to define the class of possible data that are associated with the model.

This is the question of *existence* and, although it receives little attention in the geophysical literature, it is of great importance in testing the assumptions behind any mathematical model. Every model contains simplifications and approximations, some of which may be hard to justify initially; for example, in many inverse problems the idealization of horizontal (or radial) stratification is introduced at the outset. To give a concrete instance, consider the problem of electrical conductivity sounding with the magneto-telluric method (Cagniard 1953). Here recordings are made in one place of orthogonal components of the horizontal electric and magnetic field (E_x and B_y say, with z vertical). The ratio of the Fourier transforms of the two signals is related to the electrical conductivity profile of a horizontally layered medium beneath the observing station. If the medium is truly layered, the orientation of the x and y axes will not affect the results, i.e. the magneto-telluric ratio is isotropic. Mathematically this is a statement of a *necessary condition* for the existence of a solution to the inverse problem of finding $\sigma(z)$ from the ratio. When the data are not isotropic we know that there is no solution to the problem as posed and that one of our model assumptions is false. Under these circumstances it would be foolish to continue with an interpretation based on the original model. Suppose now that an isotropic ratio was indeed observed; does this in itself guarantee a solution of the kind proposed? This is equivalent to asking whether isotropy is a *sufficient condition* for existence. In fact, isotropy is by no means enough to insure a corresponding layered solution: a particularly thorough exploration of the various restraints imposed upon the data by the layering assumption has been given by Weidelt (1972). All of these are further necessary conditions; a sufficient condition for their problem has never been derived to my knowledge. Another geophysical inverse problem which has been considered from this viewpoint is that of body-wave travel times; Gerver & Markusevitch (1966) give a detailed account.

A different mathematical matter of great geophysical importance is the question of *uniqueness*: if it is granted that there is a solution for a given set of data, is there only one such solution? Profound consequences follow if the answer is no, for it is then established that even perfect data (complete and exact) do not contain enough information to recover the Earth's structure. Several courses are then open: one can explore what further assumptions can be plausibly made about the Earth to narrow the class of solutions (perhaps down to uniqueness); alternatively, additional kinds of measurements might be introduced; or finally, one may be willing to tolerate the ambiguity if the class of admissible solutions still contains decisive information about the Earth. The last course is indeed the one that must be followed when actual observations are to be analyzed.

Uniqueness is often rather difficult to prove and, in the absence of a proof, practising geophysicists naturally proceed without one. This has sometimes led to surprises. Consider the problem of modelling the geomagnetic field. For technical reasons the intensity of a magnetic field is far easier to measure accurately than its three components; these reasons become even more compelling when the instruments are in orbit about the Earth. Therefore our knowledge of the global field depends heavily upon intensity values. The question arises whether one can uniquely[1] recover a harmonic vector field when only its intensity is known. To simplify the problem assume $|\mathbf{B}|$ is known everywhere on the surface of a sphere and that the field has an internal source. Here the forward problem is trivial: calculation of $|\mathbf{B}|$ from $\mathbf{B}$. Backus (1970) was able to devise a doubly infinite family of pairs of vector fields, in which the members of each pair generate identical intensities on a sphere: this counter-example demonstrates the impossibility of a general uniqueness proof. Backus (1968) also proved a series of uniqueness results, however, one being that the vector acceleration of gravity, $\mathbf{g}$, can be determined uniquely from $|\mathbf{g}|$ alone. Models of the geomagnetic field largely based on the intensity observations had of course been constructed, but a problematic inaccuracy in predicting the components was soon discovered (Cain 1971), and the poor performance of models based upon intensity became a major worry (Stern & Bredekamp 1975). It is now widely agreed that lack of uniqueness in the analytic problem is a major factor (called the Backus Ambiguity). In this case the only way to improve the solution appears to be measurement of the component values themselves.

An example of nonuniqueness, known for three centuries, is one that arises in the interpretation of gravitational fields: whereas a knowledge of the density structure of a system completely specifies its external gravitational field, a complete knowledge of the external field does not specify the density structure uniquely. In this case further information can be supplied by our knowledge of geology: for example, in exploration work and crustal studies it may often be assumed that the buried systems consist of relatively homogeneous units, perhaps with known densities; then the problem is to determine the shapes of these units. With several more assumptions about the buried body (e.g. that it is finite in extent and that a vertical line never intersects the body more than once) Smith (1961) was able to prove a uniqueness theorem stating that there is only one uniform body responsible for a given gravity anomaly, when the density is specified.

Uniqueness has been established for a number of geophysical inverse problems, e.g. the body-wave travel-time problem (Gerver & Markusevitch 1966) and the magneto-telluric problem mentioned earlier (Bailey 1970). Perhaps the most important exception is in the normal-mode inverse problem: the frequencies of free oscillation of the Earth have in the past five years yielded a precise, detailed picture of the mechanical structure of the deep interior (e.g. Gilbert & Dziewonski 1975), but there is no proof that the totality of such frequencies actually defines only one Earth model. Progress has been made on some analogous systems, however (Barcilon 1976).

We come next to the matter of *construction,* which, as Sabatier (1971) has remarked, receives an undue degree of attention. The uniqueness and existence of a solution are granted; what is now required is a procedure that will, in a finite number of steps, produce the solution with any specified finite precision. If iterative

---

[1] There is of course a trivial ambiguity of sign since both $+\mathbf{B}$ and $-\mathbf{B}$ give rise to the same $|\mathbf{B}|$. A single additional measurement of $\mathbf{B}$ serves to resolve this ambiguity.

methods are involved, a global convergence proof is required, and on this point most purely numerical schemes founder.

The classical solution of a geophysical inverse problem is the renowned Herglotz-Wiechert formula for obtaining the velocity-depth function from travel-time distance measurements in seismology (Bullen 1965). The idealized problem is as follows. A surface source emits signals travelling as seismic body waves through a spherically symmetric Earth, in which the continuous velocity v increases monotonically with depth. The time, T, taken for the earliest impulse to travel from the source to an observer at a range Δ (measured as an angle at the center of the Earth) is assumed known everywhere in some finite range, including $\Delta = 0$. If the rate of increase of v with depth is great enough but not too great, T is a monotonically increasing, differentiable function of Δ, and from the ray geometry we have

$$\frac{r_\Delta}{v_\Delta} = \frac{\mathrm{d}T}{\mathrm{d}\Delta},$$

where r_Δ is the radius of greatest penetration of a ray arriving at range Δ and v_Δ is the velocity at that radius. Then r_Δ can be found from the integral

$$\ln(a/r_\Delta) = \frac{1}{\pi} \int_0^\Delta \cosh^{-1}\left[\frac{\mathrm{d}T}{\mathrm{d}\Delta'} \bigg/ \frac{\mathrm{d}T}{\mathrm{d}\Delta}\right] \mathrm{d}\Delta',$$

where a is the Earth's radius. This formula combined with the first gives a means for constructing v at a radius of r. A similar formula can be developed for a flat Earth model. If the velocity fails to increase quickly enough with depth or there is a decreasing velocity, the derivative $\mathrm{d}T/\mathrm{d}\Delta$ ceases to exist and the method fails. Indeed there is no longer only one solution in the latter case, and then arrival-time functions of buried sources are needed to find the velocity (Gerver & Markusevitch 1966). This classical method is one that has been used with actual measurements, and it is the basis of a novel technique devised to account for data inadequacy, as we shall see later.

The geophysical inverse problem that seems to have collected the largest number of distinct methods of construction is the magneto-telluric problem, or its close relative, the problem of electromagnetic induction where only magnetic fields are recorded. The method of Siebert (1964) relies on the behavior of the data as the frequency tends to infinity and constructs a power series for the solution. Bailey's approach (Bailey 1970) uses the response data over the whole frequency range to develop a nonlinear integro-differential equation based upon the principle of causality (i.e. that the currents in the Earth must flow after the forcing field has been applied, never before). Weidelt (1972) employs a modification of the Gel'fand-Levitan method (1955). This method has the advantage of a degree of generality, for inverse problems where unknown parameters appear as coefficients in a Sturm-Liouville differential equation can usually be cast into the required form. A Fredholm integral equation is derived from the spectrum of the differential operator. Weston (1972) uses the Gel'fand-Levitan approach, but on data recorded in the time domain, rather than using the frequency-domain Fourier transform.

Having applied some of these methods to field observations, Bailey (1973) concluded that they were not particularly successful. The reason for this, in simple terms, is that the relevant equations describe magnetic and electric fields diffusing into the Earth, and that the information about deep structures is returned to the surface by strongly attenuated fields. Thus, unless the measurements are of astronomical precision, that information is lost. The analytic solutions seem to rely heavily upon the infinite density and precision of the idealized data and are therefore unsuitable for application to actual measurements.

There is a way to predict whether, in a particular problem, the solution depends upon the data in this rather unsatisfactory way: it is by deciding the question of *stability* of the problem. Mathematically, a problem is said to be stable if the solution depends continuously on the data and unstable if it does not. In simple terms this means that for all data sets lying close[2] to a particular set the solutions fall close to each other. This concept was introduced by Hadamard (1902) in connection with the study of boundary-value problems and he designated unstable problems "ill-posed." Very many inverse problems in geophysics are unstable in this sense. The best known example is that of downward continuation of harmonic fields (Bullard & Cooper 1948). Here the data are measurements of a harmonic function (typically a gravity or magnetic anomaly) taken on a particular level: the field values are then required at a deeper level that is nearer the sources. It is easy to show that two fields may differ by an arbitrarily small amount at the upper level, yet be quite different at the lower one, if the difference between the two original signals is confined to short enough wavelengths. This indicates the instability of downward continuation, which has long been recognized. One consequence of this instability is the corresponding lack of stability in any procedure to construct the source structure from potential field observations because downward continuation is implicit in all such processes.

Almost no work has been done on stability in the Western literature on analytical geophysical inverse problems, but the Russian school of applied mathematicians has made some important contributions here (e.g. see Lavrentiev 1967). Numerical methods for mitigating the undesirable effects of instability are discussed when we consider linear inverse problems in the next section.

ANALYSIS OF EXPERIMENTAL DATA: LINEAR PROBLEMS

Introduction

The class of linear inverse problems is particularly simple and it is the class about which the most is known; indeed, nonlinear problems are often treated by making approximations that reduce them locally to linear ones. A linear inverse problem is defined as one in which the data are linear functionals[3] of the model. Fortunately

---

[2] The concept of closeness requires the introduction of a metric onto the space of functions defining the model and the observations. Usually a measure based on a norm is meant, like the two-norm: then the distance between f_1 and f_2 is $\| f_1 - f_2 \|_2$.

[3] Mathematically, a functional is a mapping that maps a set of functions into the real numbers.

we can be more specific, since in geophysics it is almost always true that the model is related to the data via an integral transform:

$$e(x) = \int_I G(x, y)m(y)\,dy, \tag{1}$$

where m is the unknown function we are seeking, e a function representing the observations, and G a kernel derived from theory (for definiteness the independent variables may be thought of as one-dimensional and I as a real interval). Viewed as an equation for m, (1) is a Fredholm integral equation of the first kind. The two analytical questions of greatest importance for practical situations are those of uniqueness and stability.

The problem of uniqueness of solutions to equation (1) can be boiled down to the question: are there any nontrivial solutions $a(y)$ to the equation

$$0 = \int_I G(x, y)a(y)\,dy?$$

If the answer is no, then $m(y)$ is unique. If it is yes, then the class,[4] A, of all such solutions (so that $a \in A$) is called the annihilator of $G(x, y)$. Our knowledge of $e(x)$ can tell us nothing whatsoever about those parts of m that belong to A, and therefore these parts must be deduced from information other than that contained in e. For the complex kernels G of physical processes it is relatively rare that uniqueness can be established; in those cases that can be handled, a common procedure is to transform (1) into one of the well-studied integral transforms, for example:

Laplace: $G(x, y) = e^{-xy}$, $0 \leqq y < \infty$,

Fourier: $G(x, y) = e^{2\pi ixy}$, $-\infty < y < \infty$,

Hankel: $G(x, y) = J_\nu(xy)(xy)^{1/2}$, $0 \leqq y < \infty$.

All of these have unique solutions for a sufficiently well-behaved class of model functions. Furthermore, many Volterra equations [which contain the Heaviside function $H(x-y)$] can be shown to possess unique solutions, e.g. the Abel equation with

$$G(x, y) = H(x-y)(x-y)^{-\nu}, \qquad 0 < \nu < 1.$$

Suppose an annihilator exists for a particular $G(x, y)$; its presence will not necessarily be revealed by numerical solutions of (1) because the (necessarily finite-dimensional) representation of $G(x, y)$ may not be singular, even though the true kernel is.

The matter of stability is also of great importance from a practical viewpoint. As we have seen in the previous section, the construction of solutions to an unstable problem is difficult: the smallest error in e may result in a wild excursion of m that bears no relation to the true solution. The numerical inversion of the Fourier and Laplace transforms is a good illustration of the influence of stability. Both

---

[4] A is actually a linear vector space; it is also sometimes called the null space of $G(x, y)$.

transforms possess unique inverses and analytic methods for their construction. However, the inversion of the Fourier transform is stable (in the two-norm) while the inversion of the Laplace transform is not (in any conventional norm); this fact accounts for the relative ease with which numerical inversion can be performed on the Fourier transform, while the Laplace transform is notoriously difficult to invert.

For linear problems the study of stability is not difficult. Roughly speaking, if the kernel G tends to "smooth" m, the inversion is unstable. A fairly widely applicable result is that, if I is finite and G is continuous, inversion for m is not stable.

In addition to downward continuation, some other unstable linear problems of geophysics are inversion of surface strain for fault displacement (Weertman 1965) and calculation of density contrasts arising from isostatic compensation (Dorman & Lewis 1972). Actually, differentiation of a data series is an unstable process also, and since differentiation enters some nonlinear construction methods (e.g. the Gel'fand-Levitan method) they may be unstable also.

Limitations of Experimental Data

The first limitation of actual observations that comes to mind is their imprecision: everyone knows that experimentally determined numbers are inexact. From a mathematical viewpoint, an equally important property is availability of only a finite number of measurements; indeed we shall maintain that it is *always* realistic to replace (1) with

$$e(x_i) = \int_I G(x_i, y)m(y)\,dy, \qquad i = 1, 2 \ldots N,$$

or

$$e_i = \int_I G_i(y)m(y)\,dy, \qquad i = 1, 2 \ldots N. \tag{2}$$

This manifestation of observational inadequacy is immediately obvious in some measurements, such as digital samples in the time domain or the determination of free periods of vibration of the Earth. When continuous records are considered, such as seismograms, it is still possible to reduce the continuous curves to a finite list of numbers. One way of achieving this is to construct a Fourier series[5] for the record; we may discard coefficients corresponding to frequencies higher than some finite limit because the sensing and recording instruments cannot respond to arbitrarily high frequency inputs.

The inadequacy of real measurements as expressed by the finiteness of N in (2) is in principle independent of their inaccuracy, and it is useful initially to study the idealized situation where the numbers e_i are perfectly accurate. This enables us to deal at once with the question of uniqueness and existence in regard to $m(y)$ determined from (2). It is easy to show that there are infinitely many different

---

[5] All actual records are finite in duration and of bounded variation—these conditions guarantee the existence of the Fourier series.

solutions $a(y)$ to

$$\int_I G_i(y)a(y)\,\mathrm{d}y = 0, \qquad i = 1,2\ldots N;$$

in fact the annihilator here is infinite-dimensional for any finite collection of $G_i(y)$ whatever. Therefore models constructed from actual measurements can never be unique. Furthermore, if the kernels G_i are linearly independent (as they usually are in practical problems), solutions to (2) always exist.

Experimental error contributes additional indeterminacy to that caused by incompleteness of the data. In the following discussions of various methods, we assume that the statistical errors in e_i can be adequately described by a Gaussian distribution with known parameters. Although this may frequently be a poor approximation, there are two reasons for retaining it: first, the analysis can be carried out exactly under this assumption; second, knowledge of the true statistical distribution in a set of measurements is usually very poor so that an elaborate treatment based on a perfectly general error law seems hardly justified. Here we shall consider only statistically independent data, but there is no great difficulty in generalizing to the case of a joint-normal distribution over the N-data set (Gilbert 1971).

Since the actual solution $m(y)$ cannot be recovered from our measurements, there are in fact two courses open to us: (a) we can derive properties of $m(y)$ that all solutions share, which then must be properties of the true solution; (b) we can introduce assumptions about m to restrict the class of admissible solutions. In our development we begin by following the first course and this leads us naturally to suggestions of what useful yet plausible assumptions might be made.

Backus-Gilbert Formulation

Backus & Gilbert (1968) suggested that a very simple type of model-property be calculated, namely linear functions of m. By taking linear combinations of the data we can compute all functionals of the form

$$l = \int_I F(y)m(y)\,\mathrm{d}y, \tag{3}$$

where l is calculated from the data (assumed error-free for the moment). Thus we have

$$l = \sum_{i=1}^{N} \alpha_i e_i, \tag{4}$$

and $F(y)$ is given by

$$F(y) = \sum_{i=1}^{N} \alpha_i G_i(y). \tag{5}$$

The weights α_i are of course entirely arbitrary. If these weights could be chosen so that $F(y)$ approached the Dirac function $\delta(y-y_0)$, the value of l would become $m(y_0)$. This is usually not feasible but Backus & Gilbert showed how to construct functions $F(y)$ that are concentrated as much as possible on a chosen y_0 and are small elsewhere. One way of doing this is to define a delta-function quality measure such as

$$S[\tilde{\delta}] = 12 \int_I [\tilde{\delta}(y, y_0)(y - y_0)]^2 \, dy, \tag{6}$$

with $\int_I \tilde{\delta}(y, y_0) \, dy = 1$.

Here the approximation $\tilde{\delta}(y, y_0)$ is intended to focus upon the position y_0. The measure S has dimensions of length and is indeed a rough estimator of the width of peak in $\tilde{\delta}$ around y_0; the smaller S can be made, the more closely $\tilde{\delta}(y, y_0)$ resembles the improper function $\delta(y - y_0)$. We now identify $\tilde{\delta}$ with F in (5) and choose α_i so as to minimize S for a particular point y_0. Then the model property l in (4) gives us a smoothed estimate of the true $m(y_0)$, the degree of smoothing depending on the shape of $\tilde{\delta}(y, y_0)$. Further, we can loosely identify the length S with the averaging width associated with our estimate l of $m(y_0)$; when S is large we obtain only a blurred picture of m at y_0, whereas small S indicates good resolution. Recall that these averages are universal properties of any solution to (2) and that such averages of any member of the annihilator would always vanish.

It should also be understood that (6) is only one of a variety of ways to define deviation of $\tilde{\delta}$ from a delta function. One important reason for choosing this particular definition is that it leads to *linear* equations for the coefficients α_i when the smallest S is sought. It shares this almost essential property with several other definitions of "deltaness," for example,

$$W = 12 \int_I \left[\int^y \tilde{\delta}(y', y_0) \, dy' - H(y - y_0) \right]^2 dy,$$

where $H(y)$ is the Heaviside function. A general, unifying theory of the criteria for "deltaness" appears in a rather difficult paper by Backus (1970b). He considered different operations for mapping the linear space of functionals containing distributions into a "smoother" Hilbert space and thereby generated a very wide (but not exhaustive) class of quality measures. Sometimes an appropriate choice of criterion can result in substantial reduction in numerical work; for example, by using the W-measure above, Oldenburg (1976) was able to arrive at equations for α_i that were particularly simple to solve.

Before discussion of the role of experimental error in the Backus-Gilbert formulation, it is worthwhile asking how l is related to the true value of $m(y_0)$. In fact, $m(y_0)$ can be arbitrarily far from l. The only circumstance in which the two numbers will be close is that when the model $m(y)$ is itself smooth in some sense. If we are willing to assume a certain degree of smoothness for all admissible models, we can calculate the greatest deviation between our estimator and the true value. These ideas are made precise by Backus (1970a,b) who gives a very general treatment of the problem. Actually this approach has enjoyed almost no practical application, perhaps because of the difficulty in estimating the numbers that quantify smoothness.

When statistical uncertainty is included, we find that the number l in equation (4) has an error that is easily computed when Gaussian statistics describe the randomness in e_i. This error is given by

$$\sigma^2 = \text{var}\,[l] = \sum \alpha_i^2 \,\text{var}\,[e_i], \tag{7}$$

when the e_i are uncorrelated. We should like to make σ^2 in (7) and S in (6) simultaneously as small as possible because they both quantify deficiencies in l as an estimator of $m(y_0)$. Backus & Gilbert (1970) showed how to obtain the best possible values of α_i, but a unique set is not possible: there is a single degree of freedom in the choice, and this leads us to the concept of a trade-off diagram.

Consider the plane defined by the variables σ^2 and S (Figure 1). It is possible to show that, for an arbitrary set of α_i (subject only to the condition $\int \tilde{\delta}\,dy = 1$), there is a region $\mathscr{R}$ inside of which all associated pairs (S, σ^2) must lie. Clearly, to do the best job we should like to be on the lower-left edge of the region, because then, for a given value of σ^2, there would be no way to obtain a smaller (i.e. better) value of S. The curve on the lower left side of the region $\mathscr{R}$ is called the *trade-off curve*. Backus & Gilbert proved that decreasing S always increased σ^2 on this part of the boundary of $\mathscr{R}$, so that improved resolution can only be obtained at the price of degraded statistical reliability. There is no "best" point on the curve. In some applications a poor resolution may be acceptable if good statistical precision is thereby guaranteed: for example, when the solution is very flat, one may be content

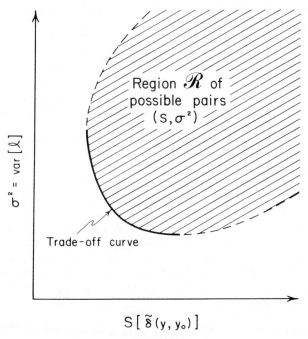

Figure 1   Region of feasible pairs (S, σ^2) shown shaded. The trade-off curve on which the best σ^2 is found for a given S is on the extreme left-lower edge of the region.

to average over a fairly broad interval to obtain precise estimates of the model value, while with rapidly varying solutions one might wish to obtain high resolution, even with relatively poor precision.

The Backus-Gilbert approach is not really a method for finding solutions $m(y)$ to (1); indeed, considered as a function of y_0, l in (4) does not usually satisfy the original data, even when they are error-free. Rather, it is a technique for assessing the significance of a solution. One may judge whether a particular feature (like a jump or spike) is really resolved by the data (Dorman & Lewis 1972; Hobbs 1973); one can compute how much detail is available in principle from a given distribution of observations. To calculate solutions $m(y)$ [especially when (1) is a linearization of a nonlinear problem] a different method has been found that offers distinct advantages, mainly with regard to numerical stability: this is the spectral expansion method.

A Spectral Expansion

The procedure to be described is a particular formulation of the numerical solution to practical linear inverse problems advocated by many authors (Gilbert 1971; Jackson 1972; Wiggins 1972; Jupp & Vozoff 1975), all of whom based the solution ultimately upon the philosophy of Lanczos (1961). Most workers have dealt with finite-dimensional models because this allows a direct description in terms of matrices and makes possible a limited form of the Backus-Gilbert interpretation of the results. The approach developed here, following Gilbert (1971), is in greater harmony with the cardinal factor distinguishing inverse theory from conventional parameter estimation, namely, that the space of unknowns is infinite-dimensional.

Analytic instability appears as extremely poor conditioning in the matrices encountered in a numerical solution, even if nonuniqueness is evaded by reducing the number of unknowns sought. The spectral expansion explicitly isolates those parts of the solution that are well determined by the data and those that are not; furthermore, the statistically reliable component of m (as derived from the known error estimates in the data) are also those with the highest numerical reliability in actual, finite-accuracy computations. These features of the method make it the most suitable when very large data sets are handled, even though it is up to ten times slower computationally than simple matrix inversion.

We treat noisy data from the beginning: consider (2) weighted by the inverse of the standard error of each measurement (recall that statistically independent e_i are assumed with individual Gaussian error distributions of zero mean and standard error σ_i)

$$\frac{e_i}{\sigma_i} = \int_I \frac{G_i(y)}{\sigma_i} m(y)\, \mathrm{d}y, \qquad i = 1, 2 \dots N,$$

which we shall write as

$$e'_i = \int_I G'_i(y)m(y)\, \mathrm{d}y, \qquad i = 1, 2 \dots N. \tag{6}$$

Thus e'_i is dimensionless with unit variance. Define the matrix Γ with elements Γ_{ij}

thus:

$$\Gamma_{ij} = \int_I G_i'(y)G_j'(y)\,dy.$$

Γ is easily shown to be positive-definite and symmetric, so that it may be diagonalized with an orthogonal matrix $\mathbf{O}$ thus:

$$\mathbf{O}^T\Gamma\mathbf{O} = \Lambda,$$

where

$$\Lambda = \mathrm{diag}\,\{\lambda_1, \lambda_2 \ldots \lambda_N\},$$

and

$$\lambda_1 \geq \lambda_2 \geq \lambda_3 \geq \cdots \lambda_N > 0.$$

Some of the eigenvalues of Γ may be zero if the $G_i(y)$ are linearly dependent, but we shall ignore this case for the moment. The set of eigenvalues will be called the spectrum of the problem defined by (6). This spectrum is not the same as that of the analytic problem (1), since the values λ_i depend critically upon the estimated errors in the data and the distribution of the observations. Now consider the functions $\psi_i(y)$ defined by

$$\psi_i(y) = \lambda_i^{-1/2} \sum_j O_{ji} G_j'(y).$$

It can easily be verified that $\psi_i(y)$ are an orthonormal set, i.e.

$$\int_I \psi_i \psi_j \, dy = \delta_{ij}.$$

Therefore we may consider an expansion of m in terms of these orthogonal functions

$$m = \sum_{i=1}^N a_i \psi_i + \psi_*, \tag{7}$$

where $\psi_* \in A$ (the annihilator)[6] and $\int \psi_* \psi_i \, dy = 0$. The coefficients of this expansion are obviously

$$a_i = \int_I \psi_i m \, dy$$

$$= \lambda_i^{-1/2} \sum_j O_{ji} e_j' \tag{8}$$

and are common properties of all solutions satisfying the data. It requires a little algebra to show that the standard error of each coefficient a_i is $\lambda_i^{-1/2}$ and that the a_i are *statistically independent*. Thus the expansion of m is in terms of functions whose coefficients increase in uncertainty; after ψ_N we reach ψ_*, whose uncertainty is total.

[6] If any zero eigenvalues occur their (associated) functions should be assimilated into ψ_*, not normalized by the factor $\lambda^{-1/2}$.

When analytically unstable problems are treated in this way, it is almost invariably the case that the functions $\psi_i(y)$ become more oscillatory as i increases; i.e. the largest eigenvalues are associated with the smoothest functions. Then the smoothest parts of m are most accurately determined because their coefficients are most precise. However, such behavior is not universal, because examples can be constructed with the opposite pattern,[7] but these exceptions are so rare in practice that we shall proceed on the assumption that the expansion (7) is in order of decreasing smoothness. What can we say about ψ_*, the part of m quite undetermined by the data? It is possible to verify that this too is an oscillatory function by asking for the smoothest possible member of A: we might define this function to be the one with the least value of $\int (d^2f/dy^2)^2\, dy$ and calculate its shape with the usual calculus of variations. No one has ever done this to my knowledge. A geophysical example with relatively smooth members in the annihilator is given by Parker & Huestis (1974), where the annihilator contains all the components that do not average to zero.

Another common property of the spectra obtained in practical problems is the condensation of the eigenvalues towards zero; this means that the spacing between eigenvalues gets smaller as their magnitude decreases. When many of the eigenvalues are small (i.e. a small fraction of λ_1) this indicates a numerically (and physically) very poorly posed problem, since now many of the coefficients exhibit large uncertainties.

Let us now return to (7) and regard it as an expansion in the natural "modes" of the data. Perhaps we should like to "filter out" the noisy components of the series and thus gain some statistical stability in our estimate of m. Clearly we should then discard those functions ψ_i where coefficients exhibit unacceptably large variance; if at the same time this means throwing out high frequency components, our solution will also be a smoothed version of the true m (provided ψ_* is not a smooth function). This is of course exactly analogous to the Backus-Gilbert formulation. The more reliable, truncated version of (7) can be accurately represented as that part of m expansible with the smoother functions ψ_i. We may further say either (a) that the high frequency parts are less interesting, or perhaps (b) that they can be assumed to be of small amplitude anyway.

The precise number of components to be accepted in the expansion is still largely a matter of judgement. One factor that should be taken into account is that the truncated series (7) no longer fits the original data precisely. Because of the random component in e_i, we should not expect or demand exact agreement. If we define χ^2 to be the squared two-norm misfit to the data thus:

$$\chi^2 = \sum_{i=1}^{N} (\tilde{e}_i - e_i')^2,$$

where $\tilde{e}_i$ is the value obtained by substituting the truncated series into the right side of (6), we find after a little algebra that

---

[7] For example, consider the set of kernels $G_n'(y) = n \sin \pi n y$, $n = 1, 2 \ldots N$ with $0 \leq y \leq 1$; here we have $\lambda_n = n^2/2$ and $\psi_n(y) = 2^{-1/2} \sin \pi n y$.

$$\chi^2 = \sum_{j=L+1}^{N} \lambda_j a_j^2 \, ;$$

here L terms have been accepted in (7). This equation shows that those coefficients a_i associated with small eigenvalues contribute relatively little to the misfit compared with their contributions to the solution or its uncertainty.

Now χ^2 is the standard statistic, since e_i' has unit variance. It is possible to pick L so as to make χ^2 equal to its expected value; the difficulty here is in selecting the number of degrees of freedom, because the underlying model m has in principle an infinite number of parameters. If one regards the truncated series as the "model," one has $v = N - L$, and L can be picked to make $P(\chi^2) = 0.5$. In fact v so defined is regarded by many (see Richards 1975) to be the number of independent data in the original set: often v is only 10% of N.

Other methods of stabilizing (7) have been suggested; for example one can choose m with the smallest norm, $\| m \|_2$ (here $\sqrt{\sum a_i^2}$), subject to its not exceeding a certain misfit level (Jackson 1973). This has the effect of replacing λ_i by $\lambda_i + C$ in (8), the equation for a_i, where C is a positive constant determined by a nonlinear equation. Evidently this has the same effect of reducing the influence of the functions associated with small eigenvalues. The method of Twomey (1963), which seeks the smoothest model in another sense, has the same sort of properties, as does the method of Franklin (1970).

An Illustration

The following example is based upon a solution to a geophysical inverse problem given by Dorman (1975). The objective is to determine the density structure within the Earth from gravity observations but, as noted earlier, further restrictions of some kind are always required to reduce the large degree of ambiguity in the general problem. Here very strong geometric constraints are imposed: the two-dimensional system (Figure 2) consists of two quarter-spaces, each horizontally layered; gravity anomaly gradients $\partial \Delta g / \partial x$ are measured on the surface $y = 0$. This is a model for a vertical fault with vertical displacement or the figuration that may be encountered across an oceanic fracture zone where two lithospheric plates of different ages meet. Gravity gradients are no longer commonly measured, but this type of data is assumed here for mathematical convenience. The solution to the

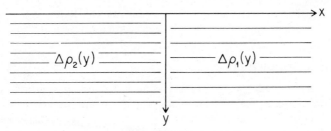

Figure 2 Geometry of gravitational edge effect problem. Each quarter-space is horizontally stratified and they meet along the plane $x = 0$.

forward problem is found to depend only upon the difference in density $\Delta\rho$ between the layers on each side at a given level:

$$\Delta g'(x) = \int_0^\infty \frac{2\mathscr{G}y}{y^2+x^2}\,\Delta\rho(y)\,\mathrm{d}y, \tag{9}$$

where $\mathscr{G}$ is Newton's gravitation constant and $\Delta g'$ is written for $\partial\Delta g/\partial x$. Our aim is to recover $\Delta\rho$ from $\Delta g'$.

In order to justify some later steps, it is useful to restrict $\Delta\rho$ to a particular class: $\Delta\rho \in L^1(0, \infty)$. This means that the integral of the magnitude of $\Delta\rho$ is assumed to be bounded. We could reasonably assume also that $\Delta\rho$ itself is everywhere bounded, then $\Delta\rho \in L^2(0, \infty)$ also, but this additional restriction is not particularly valuable in the analysis.

Let us first observe some of the conditions for existence of a solution. From (9) it is obvious that $\Delta g'(x) = \Delta g'(-x)$, i.e. Δg is an even function and this is a necessary condition for existence. Indeed, the approximate symmetry of the measured anomalies over a transform fault is a reason for proposing a model of this kind for that system. Consider now complex values of x; we must of course keep y real, but (9) allows us to *calculate* $\Delta g'$ for any x real or complex. It is easy to show that $\Delta g'$ is an analytic function of a complex variable whose only singularities are on the imaginary x axis. This implies that $\Delta g'$ is infinitely differentiable on the real axis (which is accessible to observation!) except possibly at the origin, $x = 0$. The analyticity conditions on $\Delta g'$ are clearly necessary from (9). A little strengthening of these restrictions concerning the decay of $\Delta g'$ at large $|x|$ results in a sufficient condition but the details will not detain us.

Dorman's development proceeds as follows. The Fourier transform of $\Delta g'$ is taken:

$$\begin{aligned}\overline{\Delta g'}(k) &= \int_{-\infty}^\infty \Delta g'(x)\,e^{ikx}\,\mathrm{d}x \\ &= \int_{-\infty}^\infty \mathrm{d}x\,e^{ikx} \int_0^\infty \frac{2\mathscr{G}y\Delta\rho(y)}{x^2+y^2}\,\mathrm{d}y \\ &= \int_0^\infty 2\pi\mathscr{G}\Delta\rho(y)\,e^{-|k|y}\,\mathrm{d}y\end{aligned} \tag{10}$$

The validity of interchanging the integrations and the existence of the transform follow from Fubini's theorem and the fact that $\Delta\rho \in L^1$. Restricting ourselves to positive k, we note that (10) is simply a Laplace transform so that it may formally be inverted:

$$\Delta\rho(y) = \mathscr{L}^{-1}[\overline{\Delta g'}(k)]/2\pi\mathscr{G}.$$

This is Dorman's solution. Uniqueness of $\Delta\rho$ is a direct consequence of Lerch's lemma which states that two functions with the same Laplace transform are identical up to a set with measure zero. The analytical machinery for finding $\Delta\rho$ from $\Delta g'$ is an integral in the complex k plane on the Bromwich contour. This procedure is of little account practically because of its inherent instability.

This brings us to the question of stability. Rather than appealing to the well-known instability of the inverse Laplace transform, let us deal directly with (9). We construct an example of noncontinuous dependence of $\Delta\rho$ on $\Delta g'$. First we need a distance measure for the space of data and density functions. In the latter case, a natural choice is based upon the one-norm:

$$\|\Delta\rho\|_1 = \int_0^\infty |\Delta\rho(y)|\,dy;$$

this is because $\Delta\rho \in L^1(0,\infty)$ and therefore all $\Delta\rho$'s of interest possess bounded one-norms. We shall use a one-norm on the space of analytic functions containing $\Delta g'$ just for symmetry. Now a density function $\delta\rho \in L^1$ must be found for which the corresponding $\|\Delta g'\|_1$ can be made as small as desired, while $\|\delta\rho\|_1$ remains bounded away from zero. The smoothing action of the kernel in (9) suggests that an oscillating function might be suitable, and after some experimentation I discovered the function

$$\delta\rho(y) = \frac{\rho_0 a^2 \sin\mu y}{(a^2 + y^2)}$$

with $\rho_0, a, \mu > 0$. Since $|\sin\mu y| \geq \sin^2\mu y$, it is easily shown that

$$\|\delta\rho\|_1 \geq \pi\rho_0 a(1 - e^{-2\mu a})/4.$$

The form of $\delta\rho$ makes it easy to find $\Delta g'$ from (9) with the calculus of residues: the expression is not important, only the fact that $0 \leq g'(x)$ for all x. Obviously then we have

$$\|\Delta g'\|_1 = \int_0^\infty \Delta g'(x)\,dx$$

$$= \pi\mathscr{G} \int_0^\infty \delta\rho(y)\,dy;$$

this last result comes from (10) with $k = 0$. Thus we have

$$\|\Delta g'\|_1 = \pi\mathscr{G}\rho_0 a^2 \int_0^\infty \frac{\sin\mu y}{a^2 + y^2}\,dy.$$

By the Riemann-Lebesgue theorem, $\|\Delta g'\|_1$ can be made as small as we please by choosing a large enough μ, but from the earlier inequality, $\|\delta\rho\|_1$ does not decrease below some fixed value. Because of the linearity of the problem, this $\delta\rho$ perturbation can be added to any other $\Delta\rho$, and it will be seen that arbitrarily close functions $\Delta g'$ are associated with functions $\Delta\rho$ that never come closer together than some fixed separation. The same test function also serves to demonstrate instability with the two-norm or the uniform norm.

One final analytic point I cannot resist including is the equivalence of (9) to the equation of Weertman (1965) which relates fault displacement to surface strain in a simple system. Weertman found a different way of constructing solutions. His results show that, if $\Delta g'$ is regarded as the real-axis realization of a complex function of complex x, the corresponding density is simply

$$\Delta\rho(y) = -\operatorname{Im} \Delta g'(iy)/\pi\mathscr{G},$$

that is, $\Delta\rho$ can be obtained by analytic continuation of $\Delta g'$ onto the imaginary x axis. Analytic continuation in the complex plane can be cast into a form that requires the solution of Laplace's equation with over-determined boundary conditions, the unstable problem that Hadamard (1902) first used to illustrate the concept.

We come now to a demonstration of the spectral expansion method. Gravity gradients were computed for a density difference model of 100 kg m^{-3} between 5 and 15 km, zero elsewhere; then zero-mean, normally distributed random errors were added to the 20 exact values. The values are plotted in Figure 3. Note that gradients have been provided on only one side of the density interface; this avoids certain difficulties of exact linear dependence in the Backus-Gilbert analysis, and it would be natural with real observations to average the two half-profiles together to get the best estimate of the symmetric function. The range of these "observations" is from 0–15 μm s^{-2} km^{-1}, and the standard error of the noise is 0.1 μm s^{-2} km^{-1} (there appears to be no definitive statement about gravity units in the SI system; therefore I have picked these. Note 1 μm s^{-2} km$^{-1} \equiv 0.1$ mgal km^{-1}). The scaled data and their kernels are calculated next in order to find the orthogonal kernels. ψ_i. The matrix Γ is easily shown to be

$$\Gamma_{ij} = 2\pi\mathscr{G}^2/(x_i+x_j)\sigma^2,$$

where x_i is the coordinate of the ith measurement and σ^2 the variance of each datum. The eigenvalues of Γ cover an enormous range, as can be seen from Table 1; the number of very small values indicates the very poor conditioning of the matrix. Some of the corresponding orthogonal functions are shown in Figure 4 where the standard pattern, increasingly oscillatory functions with decreasing eigenvalue, is

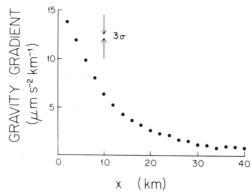

Figure 3 Artificial data values used in numerical examples. There are twenty gravity gradients computed at 2 km intervals from $x = 0$; the value at $x = 0$ is not included because the associated data kernel is singular.

Table 1 Selected values of parameters derived in the numerical solution of the gravitational edge-effect problem: λ_i stands for eigenvalues of matrix $\mathbf{\Gamma}$; a_i, spectral expansion coefficient in (7); $\lambda_i^{-1/2}$, standard error of a_i; $\chi^2(i)$, normalized squared misfit of expansion when all terms up to and including the ith have been summed in (7).

i	λ_i (m^6 kg^{-2} km^{-1})	a_i (kg m^{-3} km$^{1/2}$)	$\lambda_i^{-1/2}$ (kg m^{-3} km$^{1/2}$)	$\chi^2(i)$
1	2.090	166.7	0.692	4441.0
2	0.3721	89.3	1.64	1474.0
3	4.683×10^{-2}	-175.1	4.62	37.8
4	4.855×10^{-3}	-63.1	14.3	18.5
5	4.253×10^{-4}	47.1	48.5	17.6
6	3.179×10^{-5}	302.0	177.4	14.6
7	2.037×10^{-6}	491.0	700.7	14.2
8	1.122×10^{-7}	-4186.0	2986.0	12.2
9	5.312×10^{-9}	-1022.0	13720.0	12.2
10	2.158×10^{-10}	36972.0	68075.0	11.9
20	1.930×10^{-29}	2.976×10^{14}	2.276×10^{14}	0

evident. In Table 1 the column headed $\chi^2(i)$ gives the value of the misfit when all terms in (7) up to i have been included. Since there are 20 data, a value of about 20 is to be expected, a smaller value indicating over-fitting. Thus an expansion with only four functions would seem appropriate; we are led to the same conclusion by examining $\lambda_i^{-1/2}$, the errors in the coefficients, since after $i = 4$ there is a rapid deterioration in the accuracy of a_i. Figure 5 shows the result of taking four terms in (7).

If we had ignored the problems of error and instability and tried to fit the data exactly (recall there are infinitely many such solutions), we would find that the rather small errors in the data are magnified grotesquely in the solution. For example the $\Delta\rho$ with the smallest two-norm is found by summing (7) all the way up to N: this yields a density contrast function with oscillations exceeding $\pm 5 \times 10^{14}$ kg m^{-3} and because that solution minimizes $\|\Delta\rho\|_2$ it represents one of the smallest possible functions that fits the data precisely.

The Backus-Gilbert analysis tells the same story as the spectral expansion. Ignoring the noise in the data, I constructed two delta-function approximations, one for a depth of 15 km, the other for 50 km (Figure 6). There is a little trouble with (6) over the infinite interval, because the integral is divergent for general linear contributions of G_i. This can be remedied in several ways, some more elegant (Backus 1970b) than others; I chose the crudest device of restricting the depth to 100 km, equivalent to the assumption that there are no density variations below that level. As can be seen from the figures, the distribution of observations is capable in principle of resolving rather fine details around 15 km, and the resolution is not too bad even at 50 km. However, when the standard error of the linear estimate is calculated we find that it exceeds 10^{12} kg m^{-3} for both cases. Clearly it is desirable to sacrifice resolving ability for improved statistical properties. Figure 7 is the trade-off diagram for the two depths. This shows that to obtain a standard error of ± 20 kg m^{-3} (recall from Figure 5 typical values are around 50 kg m^{-3} at 15 km) we need a resolving width of about 11 km, which is perhaps geophysically interesting. The curve for $y_0 = 50$ km leads to the discouraging conclusion that little useful information is contained in the data about densities so deep in the Earth.

ANALYSIS OF EXPERIMENTAL DATA: NONLINEAR PROBLEMS

Linearization and the Fréchet Derivative

Many of the most important geophysical problems are nonlinear, which means that the solution to the forward problem cannot be expressed with linear functionals

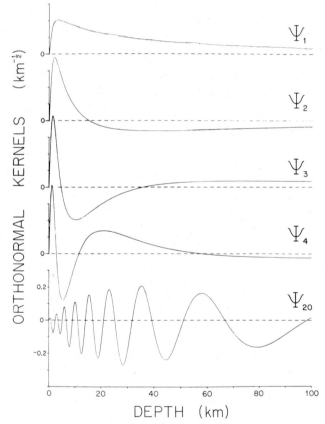

Figure 4 Orthonormal kernels found by diagonalization of Γ. The function ψ_i exhibits i zero-crossings (counting the one at $y = 0$). Thus ψ_{20} has three more zero-crossings than are shown in the figure, all deeper than 100 km.

over the model space. However, the powerful tools of linear theory are so attractive that geophysicists are often willing to make approximations allowing their use. One difficulty is that of establishing the validity of the approximation; cautionary comments (e.g. Anderssen 1975) and examples of the failure of linearization (Wiggins

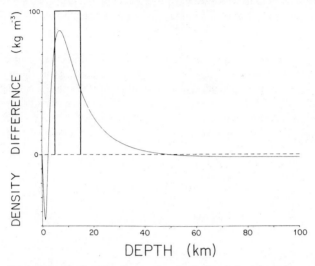

Figure 5 Original density function (box-car) and solution derived from four orthonormal kernels (smooth curve).

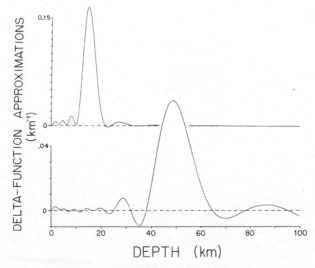

Figure 6 Best delta-function approximations according to the *S*-criterion for target depths of 15 km and 50 km.

et al 1973, Sabatier 1974) have appeared in the literature warning of the misleading results that can sometimes occur. In some cases, unfortunately, there appears to be no satisfactory alternative for the construction of a solution from real observations or assessment of its significance.

The approximation to which we have been alluding is *linearization*. Suppose the ith observation e_i is related to the model via a (nonlinear) functional F_i, formally:

$$e_i = F_i[m] \qquad i = 1, 2 \ldots N. \tag{11}$$

This deceptively simple equation may hide such complex calculations as the solution of many coupled, ordinary differential equations, as in the case of the inverse problem for the mechanical structure of the Earth from free oscillation data (Backus & Gilbert 1967); it represents the fact that the forward problem is completely solved and that when a particular m is given we can, somehow, find all the appropriate data associated with it. Next, consider a second model $m + \delta m$

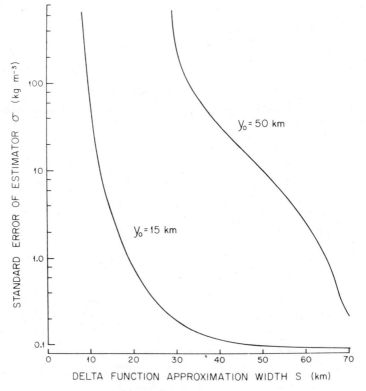

Figure 7   Part of the trade-off curves of standard error σ, and resolution width S according to the S-criterion for target depths of 15 km and 50 km. The complete diagram extends to 10^{14} kg m^{-3} in σ and 200 km in S but such extreme values are meaningless.

close[8] to m and with associated data $e_i + \delta e_i$. To find the new data we might solve a perturbation problem, since m is in a certain sense small; it will be assumed that the solution of this perturbation problem can be written

$$\delta e_i = \int_I D_i(m; y)\, \delta m(y)\, dy + O \int_I [\delta m(y)]^2\, dy \qquad (12)$$

(once more, a simple single-parameter, one-dimensional problem has been chosen for simplicity). Then F_i is said to be *Fréchet differentiable* at m and the function D_i is the *Fréchet derivative* of F_i at m. Another name for D_i, sometimes used in the applied mathematical literature, is the "sensitivity function." Equation (12) can be compared with the more familiar expression obtained when there are a finite number of parameters, μ_j, governing e_i:

$$\delta e_i = \sum_j \frac{\partial F_i}{\partial \mu_j} \delta\mu_j + O\left(\sum_j \delta\mu_j^2\right). \qquad (13)$$

This reveals a correspondence between the partial derivatives $\partial F_i/\partial \mu_j$ and the Fréchet derivative $D_i(m; y)$ if the parameter μ_j is the value of the model m in the jth layer.

Quite surprisingly, perhaps, the solution to perturbations of almost all forward problems in geophysics can be written as (12), and therefore these problems are Fréchet differentiable. One way of arriving at a perturbed solution in the required form is of course through standard perturbation analysis (e.g. Parker 1970 or Johnson & Gilbert 1972); another powerful technique particularly useful in certain seismic problems utilizes the variational formulation of the equations (e.g. Backus & Gilbert 1967 or Gilbert 1976), but there is insufficient space here to deal with these ideas as they deserve.

Having obtained (12) we can see that, if the term $O \int \delta m^2\, dy$ is dropped, it is identical with (2), where $D_i(m; y)$ plays the role of $G_i(y)$. Thus we treat the system as if it were locally linear. Using (12) and the spectral expansion method we could attempt to discover the perturbation δm that would bring an initial guess model into satisfactory agreement with the real observations. If the guess solution and a satisfactory one are not sufficiently close, the neglected term in (12) may not be truly negligible and the new solution will not be satisfactory when substituted into the full nonlinear equation (11). Then the process should be repeated, linearizing about the most recent estimate of m, until improvement ceases or a satisfactory solution is obtained. There is of course no guarantee that this procedure will work at all: it may be that no solution exists matching the observations to an adequate degree, and then it is obvious that the iterative process must fail (recall, however, that in a *linear* problem there are always an infinite number of "good" models, each fitting the observations arbitrarily well, provided the $G_i(y)$ are linearly independent); the other alternative is that, even though satisfactory solutions exist, the initial guess is so far from any of them that divergence ensues. The reader will perhaps recognize this iterative procedure as analogous to Newton's solution of nonlinear equations in several unknowns (Ortega & Rheinboldt 1970).

---

[8] See footnote 2.

Consider now the situation in which a satisfactory solution has been arrived at; to assess its significance one can simply apply the Backus-Gilbert method to the problem linearized about the best-fitting solution, in the hope that the neglected nonlinear terms do not make an important contribution. One serious problem is that there might be several satisfactory solutions widely separated in the model space, but the linearization affords approximate description of the neighborhood of only one of them. Nothing can be done about this except to attempt a more exhaustive search of the model space for other possible solutions.

There is one final remark I wish to make concerning the use of the Fréchet derivative (12) rather than the corresponding approximate representation (13). Some authors prefer to parameterize their models through layers or cells so that (13) seems a very natural way to express the solution to the perturbation problem. It is apparently not well appreciated that the Fréchet derivative D_i is often far easier to find both algebraically and computationally than the corresponding partial differential coefficients. Even if a finite representation is chosen, there are still great advantages to finding $\partial F_i/\partial \mu_j$ from D_i. A case in point is the following: consider the inverse problem of discovering the shape of crystalline basement that is everywhere buried under sediments. The data are the gravity anomalies caused by the (known) density contrast between the two materials, $\Delta\rho$. If we let $h(\mathbf{x})$ be the depth at a horizontal position $\mathbf{x} = (x, y)$, then the solution to the forward problem is

$$\Delta g(\mathbf{x}_i) = \int_S \frac{\mathscr{G}\Delta\rho \, dS}{[|\mathbf{x}_i - \mathbf{x}|^2 + h(\mathbf{x})^2]^{1/2}}, \qquad i = 1, 2 \ldots N,$$

where $\Delta g(\mathbf{x}_i)$ is the anomaly observer at $\mathbf{x}_i$ on a horizontal level and $\mathscr{G}$ is Newton's gravitational constant. Clearly this is a nonlinear problem for h. It is easy to show that

$$D_i(h; \mathbf{x}) = - \frac{\mathscr{G}\Delta\rho h(\mathbf{x})}{[|\mathbf{x}_i - \mathbf{x}|^2 + h(\mathbf{x})^2]^{3/2}}, \qquad i = 1, 2 \ldots N$$

and that these Fréchet derivatives exist for all h with $h(\mathbf{x}) > 0$ for all $\mathbf{x} \in S$. In contrast, the formula for $\partial\Delta g(\mathbf{x}_i)/\partial h_j$, when the plane is divided into square cells and h is then forced to be constant within each cell, apparently extends over many lines (Burkhard & Jackson 1976). It would be very simple to derive an approximation for $\partial\Delta g(\mathbf{x}_i)/\partial h_j$ by rough numerical quadrature of D_i.

Fully Nonlinear Treatment

Linearization is very successful at discovering an acceptable solution to a nonlinear inverse problem, but a finite number of data do not specify a unique solution (except in certain pathological cases). It is not enough to possess one adequate solution, because, in order to draw valid conclusions about the Earth, we need in principle to consider every possible solution, not just one of them. Linearization provides only a partial answer to this question by giving an approximate description of the neighborhood near the preferred model. The difficulty of describing the complete class of viable solutions makes it convenient to consider a simpler problem whose solution is geophysically just as important: the derivation of *properties* that all solutions have in common. The Backus-Gilbert local averages are an example of a shared property in the linear case. The properties that seem to be most amenable to mathematical analysis in nonlinear systems are upper and lower bounds of various parameters, and these are extremely important geophysical quantities. In the next few paragraphs we discuss three different approaches to the practical calculation of the limits placed by the data on geophysical models.

The method of widest applicability is Monte Carlo modelling (Keilis-Borok & Yanovskaja 1967; Anderssen et al 1972). The principle is simple; one parameterizes the model space with a large but finite number of unknowns and then generates a sequence of structures at random, testing each one against the observations. While most structures will fail the test, some will pass and these form a family of solutions whose common properties are chosen to represent those of the complete class. Normally the structures investigated are profiles of a single parameter, e.g. seismic velocity or density as a function of depth; the property invariably investigated is the "corridor" of upper and lower limits varying with depth within which all solutions are hypothesized to lie. There is unfortunately an objection to the use of a corridor that is seldom mentioned. Consider for example a layer 1 mm thick with density 10^6 kg m^{-3}; should such a layer exist within the Earth it would be undetectable by seismic or gravimetric means because it is so thin. Therefore the addition of such a layer to an acceptable solution would not upset agreement with the observations. Since the layer could be placed at any depth, the upper bound in density cannot be lower than 10^6 kg m^{-3} anywhere, a geophysically uninformative value. There are of course ways of eliminating improbable extreme cases like this. One way is suggested by the Backus-Gilbert trade-off diagram: the corridor is defined to concern the average value of density in a succession of layers with specified thicknesses. Alternatively, gradients larger than a certain amount can be rejected a priori as "geophysically unreasonable." Whatever method is chosen, the fact remains that the corridor is sensitive to additional information not present in the original observations. Practical applications of the Monte Carlo approach usually impose other restrictions as well, such as an a priori corridor of reasonable maximum and minimum parameter values; these assumptions are made to reduce the Herculean effort of conducting a complete search, since they decrease the size of the parameter space. The possibly strong dependence on seemingly unimportant computational details must be kept in mind when one wishes to interpret the results of these studies.

The method has been applied notably to seismic problems, for example the determination of the interior mechanical structure using the dispersion of surface waves (e.g. Jackson 1973, who uses a refinement based upon local linearization to ease the computational labor) or from the frequencies of free oscillation of the Earth (e.g. Press 1968), but the great number and precision of recent observations precludes any further direct application in the latter problem.

The next nonlinear treatment is in some respects the antithesis of the Monte Carlo method: it is an elegant treatment of a specific problem and is based on an analytic construction algorithm. The problem is that of inverting travel-time distance data for seismic velocity, and the underlying analysis is of course the

Herglotz-Wiechert solution discussed earlier. There are different variations (McMechan & Wiggins 1972; Bessonova et al 1974) depending upon what type of data are assumed (array studies can give direct estimates of $dT/d\Delta$, for example), but the objective is again to discover a corridor within which all permissible velocities must fall. The principle of the method is to use the Herglotz-Wiechert formula (in spherical or flat-Earth form) to give the greatest or least depths at which a specified velocity can occur. Considerable complications ensue when the possibility of low-velocity zones is permitted, but only then is the method useful practically, for such layers are commonly encountered in crustal and upper mantle studies. When the velocity decreases (or fails to increase rapidly enough with depth in the sphere), a unique solution no longer is possible even with perfect data. To cope with this difficulty a limit is placed a priori on the permissible magnitude of any velocity reversals, which has the gratifying effect of automatically eliminating troublesome thin layers with high velocities. The size of the bound can be fairly well estimated from a knowledge of the materials likely to be encountered. Unlike Monte Carlo searching the method is cheap computationally, so that different assumed values can be tried if necessary. Clear comparative reviews of this extremal technique and the corresponding linearization have been given by Wiggins et al (1973) and Kennett (1976).

The last approach is one I have personally been working on for several years, and I believe it is the one with the greatest potential. The principle is to choose a scalar property of the model and then to maximize or minimize its value subject to the constraints that the solution fit the data and that any other necessary restrictions be taken into account. If we return briefly to a linear inverse problem and choose a *linear* functional of the model, we discover that there is no upper or lower limit to this value of the functional unless it happens to be made up of a linear combination of data kernals (leading to Backus-Gilbert theory). But when a nonlinear functional of m is used, bounds can often be discovered. This remark suggests that property extremization is a process that must contain some nonlinear element to succeed. The most satisfying application of the method so far has been to the inverse problem of determining the shape of a buried body of known density from the gravity anomalies it produces, a nonlinear problem mentioned several times already. My solution (Parker 1975) depends upon discovering the body with the smallest possible maximum density fitting the observations, which is a nonlinear functional of the model in a linear problem. An additional constraint is frequently needed, namely, the density contrast should be positive. It turns out that the body of least density is uniform and its shape is unique, i.e. there is only one body with that density fitting the data.[9] These two properties enable us to find restrictions on the shape of a uniform body if we now assume the density is known but the shape is not. For example, one can give a level, above which any buried uniform body with the specified density must protrude or another level which it must penetrate from above. Similarly, constraints on lateral extent can be obtained. Very

---

[9] Here is an example of an inverse problem where a finite number of data are compatible with only one solution, but of course this happens only when the buried body has a very special shape and the density is a very special value.

recently, Sabatier (1976) has shown that my approach, which seemed very specific to this one problem, is in fact an example of a problem that can be dealt with by means of convex analysis and linear programming theory.

The foregoing theory centered upon finding the minimum of a nonlinear functional in a linear forward problem. In principle there is no reason why nonlinear forward problems cannot be treated directly, as I have shown (Parker 1972). Now the Fréchet derivatives D_i enter in place of the linear kernels G_i; when this happened during linearization an approximation was involved, but here none is made. Unfortunately, as with nearly all nonlinear analysis, the resultant equations are difficult to solve even numerically and it is often almost impossible to show rigorously that a global extremum has been found, not merely a local one. Both of these troubles diminish if only a very small number of observations are available.

The place where parameter extremization may make the greatest impact in future is in the testing of hypotheses. Often a physical model of a geological system is arrived at by extrapolation or the synthesis of circumstantial evidence; sometimes (but not as often as one would wish perhaps) the model can be tested against geophysical data. Naturally the test cannot prove the correctness of the model, but it may be able to reject it or choose between two competing hypotheses. One powerful way of making a test is to use the physical model to predict, say, a lower limit on a property such as the mean seismic velocity in a certain depth range or the maximum value of S-wave attenuation. Then the solution to the appropriate inverse problem is found to maximize the chosen parameter; if the greatest value is smaller than that predicted, the model must be rejected; if it is within the predicted range there is no conflict with the geophysical data.

A recent unusual example of this approach has been given by Gubbins (1975), who, using the spherical harmonics of the geomagnetic field, derives a lower limit on the energy requirements of the geomagnetic dynamo to test a mechanism invoking earthquakes to maintain the field. He then compares this with the available energy. Unfortunately the test was inconclusive because of the uncertainty in that figure, but a stronger test involving gravity anomalies (but not using parameter extremization) was more decisive in ruling against the model.

Another ingenious application is that of Jordan (1975): he uses the observed travel-time differences of shear waves between continent and ocean and the Love wave dispersion data over corresponding pure oceanic and continental paths to argue for continental influences as deep as 400 km into the mantle. In this case the parameter minimized is the depth to which velocity differences are permitted between the two regimes. Jordan finds that with limits on only two data (the mean travel time difference in the heterogeneous region and the difference in Love wave dispersion at about 180 s) the smallest allowed layer thickness is 310 km. Additional data force even larger values.

CONCLUSIONS

Indirect measurements will continue to be the major source of information about the Earth's deep interior and about much shallower regions also. The problem of providing some model consistent with the data has for practical purposes been solved by the use of the spectral expansion (or its finite-dimensional equivalents) and linearization if the equations are nonlinear. However, most geophysicists are now conscious of their obligation to assess the range covered by all satisfactory solutions. In the linear case the elegant concept of resolution and its inverse relation with statistical error has made the Backus-Gilbert analysis a most appealing treatment. With nonlinear equations the two general techniques, linearization and the Monte Carlo search, suffer from drawbacks that, to me at least, make them somewhat unattractive.

One of the most promising alternative strategies is to use the data directly to test a hypothesis. This requires the construction of only one solution, the one violating the hypothesis the least, yet still satisfying the demands of observation. As yet few geophysicists have tried the idea, but it offers a means of bypassing the difficult problem of completely characterizing an infinite set of models.

Literature Cited

Anderssen, R. S., Worthington, M. H., Cleary, J. R. 1972. Density modelling by Monte Carlo inversion—I Methodology. *Geophys. J. R. Astron. Soc.* 29 : 433–44

Anderssen, R. S. 1975. On the inversion of global electromagnetic induction data. *Phys. Earth Planet. Inter.* 10 : 292–98

Backus, G. E. 1968. Application of a non-linear boundary-value problem for Laplace's equation to gravity and geomagnetic intensity surveys. *Q. J. Mech. Appl. Math.* 21 : 195–221

Backus, G. E. 1970. Non-uniqueness of the external geomagnetic field determined by surface intensity measurements. *J. Geophys. Res.* 75 : 6339–341

Backus, G. E. 1970a. Interference from inadequate and inaccurate data, I. *Proc. Nat. Acad. Sci. USA* 65 : 1–7

Backus, G. E. 1970b. Interference from inadequate and inaccurate data, II. *Proc. Nat. Acad. Sci. USA* 65 : 281–87

Backus, G. E., Gilbert, F. 1967. Numerical application of a formalism for geophysical inverse problems. *Geophys. J. R. Astron. Soc.* 13 : 247–76

Backus, G. E., Gilbert, F. 1968. The resolving power of gross Earth data. *Geophys. J. R. Astron. Soc.* 16 : 169–205

Backus, G. E., Gilbert, F. 1970. Uniqueness in the inversion of inaccurate gross Earth data. *Phil. Trans. R. Soc. London Ser. A* 266 : 123–92

Bailey, R. C. 1970. Inversion of the geomagnetic induction problem. *Proc. R. Soc. London* 315 : 185–94

Bailey, R. C. 1973. Global geomagnetic sounding—methods and results. *Phys.* *Earth Planet. Inter.* 7 : 234–44

Barcilon, V. 1976. A discrete model of the inverse Love wave problem. *Geophys. J. R. Astron. Soc.* 44 : 61–76

Bard, Y. 1974. *Nonlinear Parameter Estimation*. New York : Academic. 341 pp.

Bessonova, E. N., Fishman, V. M., Ryaboyi, V. Z., Sitnikova, G. A. 1974. The tau method for the inversion of travel times—I Deep seismic sounding data. *Geophys. J. R. Astron. Soc.* 36 : 377–98

Bullard, E. C., Cooper, R. I. B. 1948. The determination of the masses necessary to produce a given gravitational field. *Proc. R. Soc. London A* 194 : 332–47

Bullen, K. E. 1965. *An Introduction to the Theory of Seismology*. Cambridge Univ. Press. 381 pp.

Burkhard, N., Jackson, D. D. 1976. Application of stabilized linear inverse theory to gravity data. *J. Geophys. Res.* 81 : 1513–32

Cagniard, L. 1953. Basic theory of the magneto-telluric method of geophysical prospecting. *Geophysics* 18 : 605–35

Cain, J. C. 1971. Geomagnetic models from satellite surveys. *Rev. Geophys. Space Phys.* 9 : 259–73

Dorman, L. M. 1975. The gravitational edge effect. *J. Geophys. Res.* 80 : 2949–50

Dorman, L. M., Lewis, B. T. R. 1972. Experimental isostasy (3). *J. Geophys. Res.* 77 : 3068–77

Franklin, J. N. 1970. Well-posed stochastic extension of ill-posed problems. *J. Math. Anal. Appl.* 31 : 682–716

Gel'fand, I. M., Levitan, R. M. 1955. On the determination of a differential equation by its spectral function. *Am. Math. Soc. Transl. Ser. 2.* 1:253–304

Gerver, M., Markusevitch, V. 1966. Determination of seismic wave velocity from the travel-time curve. *Geophys. J. R. Astron. Soc.* 11:165–73

Gilbert, F. 1971. Ranking and winnowing gross Earth data for inversion and resolution. *Geophys. J. R. Astron. Soc.* 23: 125–28

Gilbert, F. 1976. Differential kernels for group velocity. *Geophys. J. R. Astron. Soc.* 44:649–60

Gilbert, F., Dziewonski, A. M. 1975. An application of normal mode theory to the retrieval of structural parameters and source mechanisms from seismic spectra. *Phil. Trans. R. Soc. London A* 278:187–269

Gubbins, D. 1975. Can the Earth's magnetic field be sustained by core oscillations? *Geophys. Res. Lett.* 2:409–12

Hadamard, J. 1902. Sur les problèmes aux derivées partielles et leur significations physiques. *Bull. Univ. Princeton* 13:1–20

Hobbs, B. A. 1973. The inverse problem of the moon's electrical conductivity. *Earth Planet. Sci. Lett.* 17:380–84

Jackson, D. D. 1972. Interpretation of inaccurate, insufficient and inconsistent data. *Geophys. J. R. Astron. Soc.* 28:97–110

Jackson, D. D. 1973. Marginal solutions of quasi-linear inverse problems in geophysics: the edgehog method. *Geophys. J. R. Astron. Soc.* 35:121–36

Johnson, L. E., Gilbert, F. 1972. Inversion and inference for teleseismic ray data. *Methods of Computational Physics,* 12: 231–66. New York: Academic

Jordan, T. H. 1975. The continental tectosphere. *Rev. Geophys. Space Phys.* 13:1–12

Jupp, D., Vozoff, K. 1975. Stable iterative methods for the inversion of geophysical data. *Geophys. J. R. Astron. Soc.* 42:957–76

Keilis-Borok, V. I., Yanovskaja, T. B. 1967. Inverse problems of seismology (structural review). *Geophys. J. R. Astron. Soc.* 13: 223–34

Kennett, B. L. N. 1976. A comparison of travel-time inversions. *Geophys. J. R. Astron. Soc.* 44:517–36

Lanczos, C. 1961. *Linear Differential Operators.* London: van Nostrand. 564 pp.

Lavrentiev, M. M. 1967. *Some Improperly Posed Problems of Mathematical Physics.* Heidelberg: Springer. 72 pp.

McKenzie, D., Parker, R. L. 1974. Plate tectonics in ω-space. *Earth Planet. Sci. Lett.* 22:285–93

McMechan, G. A., Wiggins, R. A. 1972. Depth limits in body wave inversions. *Geophys. J. R. Astron. Soc.* 36:239–43

Oldenburg, D. W. 1976. Calculations of Fourier transforms by the Backus-Gilbert method. *Geophys. J. R. Astron. Soc.* 44: 413–31

Ortega, J. M., Rheinboldt, W. C. 1970. *Iterative Solution of Nonlinear Equations in Several Variables.* New York: Academic. 572 pp.

Parker, R. L. 1970. The inverse problem of electrical conductivity in the mantle *Geophys. J. R. Astron. Soc.* 22:121–38

Parker, R. L. 1972. Inverse theory with grossly inadequate data. *Geophys. J. R. Astron. Soc.* 29:123–38

Parker, R. L. 1975. The theory of ideal bodies for gravity interpretation. *Geophys. J. R. Astron. Soc.* 42:315–34

Parker, R. L., Huestis, S. P. 1974. The inversion of magnetic anomalies in the presence of topography. *J. Geophys. Res.* 79:1587–93

Press, F. 1968. Earth models obtained by Monte Carlo inversion. *J. Geophys. Res.* 73:5223–34

Rao, C. R. 1973. *Linear Statistical Inference and its Applications.* New York: Wiley. 625 pp.

Richards, P. G. 1975. Theoretical seismology. *Rev. Geophys. Space Phys.* 13:295–98

Sabatier, P. C. 1971. Comparative evolution of inverse problems. *NASA Tech. Memo., X-62, 150*

Sabatier, P. C. 1974. Remarks on approximate methods in geophysical inverse problems. *Proc. R. Soc. London A* 337: 49–71

Sabatier, P. C. 1976. Positivity constraints in linear inverse problems. *Geophys. J. R. Astron. Soc.* In press

Siebert, M. 1964. Ein Verfahren zur un Mittelbaren Bestimmung der Vertikalen Leitfähigkeitsverteilung im Rahmen der Erdmagnetischen Tiefensonderung. *Nachr. Akad. Wiss. Göttingen Math. Phys. Kl. 2*

Smith, R. A. 1961. A uniqueness theorem concerning gravity fields. *Proc. Cambridge Phil. Soc.* 57:865–70

Stern, D. P., Bredekamp, J. H. 1975. Error enhancement in geomagnetic models derived from scalar data. *J. Geophys. Res.* 80:1776–82

Twomey, S. 1963. On the numerical solution of Fredholm integral equations of the first kind by the inversion of the linear system produced by quadrature. *J. Assoc. Comput. Mach.* 10:97–101

Weidelt, P. 1972. The inverse problem of geomagnetic induction. *Z. Geophys.* 38: 257–89

Weertman, J. 1965. Relationship between displacements on a free surface and the stress on a fault. *Bull. Seismol. Soc. Am.* 55:945–53

Weston, V. H. 1972. On the inverse problem for a hyperbolic dispersive partial differential equation. *J. Math. Phys.* 12: 1952–56

Wiggins, R. A. 1972. The general linear inverse problem: implication of surface waves and free oscillations for Earth structure. *Rev. Geophys. Space Phys.* 10: 251–85

Wiggins, R. A., McMechan, G. A., Toksöz, M. N. 1973. Range of Earth-structure non-uniqueness implied by body wave observations. *Rev. Geophys. Space Phys.* 11: 87–113

Ann. Rev. Earth Planet. Sci. 1977. 5 : 65–110

A REVIEW OF HYDROGEN, CARBON, NITROGEN, OXYGEN, SULPHUR, AND CHLORINE STABLE ISOTOPE FRACTIONATION AMONG GASEOUS MOLECULES[1]

✲10067

P. Richet, Y. Bottinga, and M. Javoy
Laboratoire de Géochimie et Cosmochimie,[2] Universités de Paris 6 et 7,
4 Place Jussieu, 75230 Paris Cedex 05, France

1 INTRODUCTION

Fairly soon after the discovery of the existence of stable isotopes, it was noticed that significant stable isotope fractionation occurred in nature for the isotopes of the light elements (Maillard 1936, Nier & Gulbransen 1939, Murphey & Nier 1941). But the real development of stable isotope geochemistry started when Urey (1947) published his paper on the thermodynamic properties of isotopic substances, and McKinney et al (1950) described a mass spectrometer with which one could measure on a routine basis, with high precision, the small differences in isotopic composition of naturally occurring substances. Soon after these developments a series of basic papers on stable isotope geochemistry was published by Silverman (1951) for oxygen, Friedman (1953) for deuterium, Craig (1953) for carbon, Tudge & Thode (1950) for sulphur and Epstein & Mayeda (1953) for water. In these early papers the isotopic composition of naturally occurring substances was surveyed and discussed, and with the theoretically calculated fractionation factors published by Urey (1947) and Tudge & Thode (1950), it was possible to study quantitatively the observed isotopic variations in certain cases.

In geochemistry the light stable isotopes are useful as tracers as well as palaeo-thermometers. Their palaeothermometric characteristics are a result of the fact that the isotopic composition of two cogenetic phases depends on the temperature at which these phases were or are in equilibrium. To use effectively the tracer and thermometric properties of stable isotope distributions one needs to know the isotopic fractionation factors and their temperature dependence. As is virtually

---

[1] Contribution I. P. G. NS 213
[2] Laboratory associated with CNRS No. 196

always the case the measurement of these fractionation factors is very tricky. Fortunately isotopically substituted molecules form perfect solutions; it is this aspect that makes it possible to calculate without too much difficulty stable isotope fractionation factors. The light stable isotopes are very useful as tracers and palaeo-temperature indicators, because they are ubiquitous in nature, very often in coexisting phases, which frequently are simple enough that one may calculate isotopic fractiona-tion factors for them.

The theory of stable isotope fractionation probably starts with the work by Lindemann (1919) and Lindemann & Aston (1919), who applied statistical thermo-dynamics to calculate the difference in vapor pressure among lead isotopes. The method used by Lindemann has remained essentially the same since then. Following Lindemann, Urey & Rittenberg (1933) calculated equilibrium constants for exchange reactions involving hydrogen, deuterium and hydrogen and deuterated halogen compounds. Farkas & Farkas (1934) did the same for hydrogen–deuterium exchange with water. This work was followed by the well-known contribution of Urey & Greiff (1935) who did calculations for hydrogen, lithium, carbon, nitrogen, oxygen, and halogen isotope exchanges between various diatomic and polyatomic molecules. In these early days such calculations were very tedious because of the lack of modern computing facilities. Waldmann (1943) developed probably the first con-venient equations for isotopic exchange calculations. Bigeleisen & Mayer (1947) derived in part (independently) the same equations as Waldmann. Urey (1947) in his classical survey of the stable isotope geochemistry developed a somewhat different formulation. Since then many authors have used the methods of Urey (1947) or Waldmann (1943) and Bigeleisen & Mayer (1947). These methods have been refined and applied in different disciplines.

It is correct to conclude that the theory of stable isotope fractionation is "old hat." What is then the reason to review it once more? It is our experience that geochemists, even those who specialize in stable isotope geochemistry, are often badly informed on the theory of this subject. Another reason is that in geochemistry, more than in nearly any other discipline where stable isotopes are used, one insists on the availability of equilibrium constants that are as accurate as possible. This is of course due to the fact that the naturally observed fractionations are frequently very small. For these reasons we give only a condensed review of the theory but discuss in detail possible sources of errors in the calculations. We also discuss briefly the spectroscopic data, which are the input for the calculations.

We limit ourselves to a discussion of isotopic fractionation between gaseous substances. Various reviews have been published that deal with the isotopic fractionation among solids (see, for example, Taylor 1968, Bottinga & Javoy 1975). Fractionation factors for the exchange of hydrogen, carbon, nitrogen, oxygen, sulphur and chlorine isotopes among gases of geochemical interest have been calculated for this contribution and are given in the form of tables at convenient temperature intervals. One may wonder why it is necessary to calculate these fractionation factors in certain cases, because some of them have already been published previously, notably by Urey (1947), Tudge & Thode (1950, Saxena et al (1962), and Bottinga (1968a, 1969). Our reasons for doing this have been several:

1. We wanted to obtain a uniform set of fractionation factors, all calculated in the same way, in order to prevent errors that may result from combining calculated data from different authors.

2. Since the heretofore mentioned calculations, theoretical developments have taken place. Their importance has been evaluated for several cases.

3. New spectroscopic data, as well as more sophisticated analyses of old spectroscopic data, have been published, making more precise calculation possible.

4. Certain simplifying but also error-introducing approximations in previously published calculations are avoided.

5. In several of the older compilations of calculated fractionation factors the temperature intervals for which results are given are too small for modern geochemical needs.

The calculated results are compared with the available laboratory-measured fractionation factors. Lack of space prevents us from a discussion of naturally occurring isotopic fractionation in the light of our data. However, we hope that a uniform and readily available compilation of fractionation factors will stimulate geochemists to measure isotopic ratios in volcanic, hydrothermal, and bacterial gases, and most interestingly perhaps in the gases occurring in inclusions in many minerals.

2   CALCULATIONS—DIATOMIC MOLECULES

In this section we give an outline of the theory, in order that the general reader may obtain an idea of how these calculations are done and be aware of the approximations that are applied. For a full account of the theory we refer the reader to the papers by Mayer (1958) or Vojta (1960).

The simplest possible exchange reaction is between diatomic molecules and atoms. Such systems serve very well to illustrate the physical principles involved in isotopic fractionation. For reaction (1),

$$ZY + Y' = ZY' + Y, \tag{1}$$

where the element Y has two isotopes Y and Y', the equilibrium constant K_1 is given by

$$K_1 = \frac{[ZY']}{[ZY]} \bigg/ \frac{[Y']}{[Y]}, \tag{2}$$

where $[ZY]$ is the concentration of ZY. A well-known result is (see, e.g., Denbigh 1971, Chap. 12)

$$K_1 = \left(\frac{Q'}{Q}\right)_{ZY} \bigg/ \left(\frac{Q'}{Q}\right)_Y, \tag{3}$$

where $(Q'/Q)_{ZY}$ stands for the molecular partition-function ratio of molecules ZY' and ZY. The molecular partition function Q is defined as

$$Q = \sum_i g_i \exp(-E_i/kT), \tag{4}$$

where the summation is over all quantum states i accessible to the system. E_i is the energy of state i and g_i is the degeneracy of state i. To evaluate Equation 4 one has to know E_i and g_i. This information is contained in the Schrödinger equation

$$H\psi = E\psi$$

for the constituents of reaction (1). H is the Hamiltonian operator and ψ is the wave function.

The energy of a molecule may be split up into two components,

$$E = E_{tr} + E_{int}, \tag{5}$$

where E_{tr} is the energy associated with the movements of the center of mass of the molecule and E_{int} is the internal energy. We do not take into consideration the potential energy the molecule may have due to the earth's gravity field; because of the weakness of the interaction this energy can be safely ignored. Hence the E_{int} will be independent of the location of the molecule and thus we have that the wave function $\psi = \psi_{tr}\psi_{int}$. Consequently the translations of the molecule are governed by the translational Schrödinger equation

$$-\frac{\hbar^2}{2M}\nabla^2\psi_{tr} = E_{tr}\psi_{tr}, \tag{6}$$

where M is the mass of the molecule and $\hbar$ is the Planck constant divided by 2π. Solution of Equation 6 gives

$$E(n_x, n_y, n_z) = \frac{h^2}{8Ma^2}(n_x^2 + n_y^2 + n_z^2), \tag{7}$$

where n_x, n_y, n_z are zero or positive integers, the quantum numbers for translations in the X, Y, or Z directions, and a is the dimension of the cubic box in which the molecule is placed. To calculate E_{int} one has to solve the Schrödinger equation

$$\sum_n \frac{\hbar^2}{2M_n}\nabla_n^2\psi_{int} + \sum_r \frac{\hbar^2}{2m}\nabla_r^2\psi_{int} + (E_{int} - V_{kk} - V_{ee} - V_{ke})\psi_{int} = 0, \tag{8}$$

where M_n is the mass of nucleus n and m is the mass of an electron. V_{kk} is the potential for the interaction between nuclei, V_{ee} for electrons, and V_{ke} for electron-nuclei interaction. For nearly all molecules of interest in geochemistry Equation 8 cannot be solved, and approximations have to be applied to obtain values for E_{int}.

The general practice is to apply the Born-Oppenheimer approximation (Born & Oppenheimer 1927). In this approximation one considers the nuclei to be fixed in space. The justification of this method is that the movements of the heavy nuclei are very small with respect to the rapid movements of the light electrons. As a result of this we can write

$$\psi_{int} = \psi_e\psi_k,$$

where ψ_e is the wave function for the electron system and ψ_k is the wave function for nuclei. Equation 8 can now be split into Equations 9 and 10

$$\sum_r \frac{\hbar^2}{2m} \nabla_r^2 \psi_e + (E_e - V_{ee} - V_{ke})\psi_e = 0 \qquad (9)$$

and

$$\sum_n \frac{\hbar^2}{2M_n} \nabla_n^2 \psi_k + (E_{int} - E_e - V_{kk})\psi_k = 0, \qquad (10)$$

where because of the Born-Oppenheimer approximation we have neglected derivatives like $\partial\psi_e/\partial x_n$. In Equations 9 and 10, we have E_e = energy eigenvalue of the electron system. Ideally one would like to solve Equation 9 for E_e and subsequently solve Equation 10 for E_{int}. However, in general the molecule with which one is concerned contains too many electrons for Equation 9 to be solved. E_e will depend on the distance between the nuclei only.

2.1 Vibrational and Rotational Molecular Energy

In order to solve Equation 10 without having a solution for E_e available, one assumes that the term $E_e + V_{kk}$ in Equation 10 may be represented by a potential like, for instance, the Morse potential (Morse 1929),

$$V(r - r_e) = D_e[1 - \exp - a(r - r_e)]^2, \qquad (11)$$

where r is the internuclear distance, r_e, the equilibrium internuclear distance, D_e, the dissociation energy of the molecule measured with respect to the minimum of the potential energy curve (i.e. when $r = r_e$), and a, a constant. Solving Equation 10 for E_{int}, while making use of the Morse potential, one obtains as a solution for the energy of the molecule in excess of the minimum of the electronic energy (T_e)

$$E_{v,J}/hc = \omega_e(v + 1/2) - \omega_e x_e(v + 1/2)^2 + B_e J(J+1) - DJ^2(J+1)^2 \\ - \alpha_e(v + 1/2)J(J+1) \qquad (12)$$

for the energy level $E_{v,J}$ characterized by the vibrational quantum number v and the rotational quantum number J. The constants in Equation 12, ω_e, x_e, B_e, D, and α_e, are functions of the parameters D_e, r_e, and a in the Morse potential and the masses of the two nuclei in the diatomic molecule. We further define ω_e as $\omega_e = v/c$, where v is the vibrational frequency and c is the speed of light; v can be expressed as $v = a/2\pi\sqrt{2D_e/\mu}$, where $\mu = M_1 M_2/(M_1 + M_2)$, the reduced mass of the molecule with nuclei 1 and 2. We also have $B_e = h/(8\pi^2 I_e c)$, where I_e is the moment of inertia; $D = 4B_e^3/\omega_e^2$; and

$$\alpha_e = 3\hbar^2\omega_e\left[\frac{1}{ar_e} - \left(\frac{1}{ar_e}\right)^2\right] \bigg/ 4\mu r_e^2 D_e.$$

The first and third terms on the right-hand side (RHS) of Equation 12 would have been obtained if a harmonic potential $V(r) = \frac{1}{2}K(r - r_e)^2$ with $r - r_e \ll r_e$ was used instead of the Morse potential; these terms correspond to a rigid rotator executing simple harmonic oscillations. The second RHS term is an anharmonicity correction to the first RHS term. The fourth RHS term is due to the nonrigidity of the rotator and represents rotational stretching of the molecule. The last RHS term describes the interaction between vibration and rotation. By fitting Equation 12 to the observed vibrational and rotational spectrum of a diatomic molecule, the constants occurring in Equation 12 and hence the Morse potential parameters for the molecule can be obtained. These aspects have been extensively described by Herzberg (1950).

As is shown later, in the calculation of isotope effects, the zero-point energy, i.e. the energy difference between the minimum of the potential energy curve and the level $E_{0,0}$, plays a predominant role. For Equation 12 $E_{0,0} = \frac{1}{2}\omega_e - \frac{1}{4}\omega_e x_e$. This energy is not observable in spectroscopy, where one can measure only energy differences between excited states or between an excited state and the ground state. Hence the calculation of the zero-point energy amounts to an extrapolation of the observed spectrum. To arrive at the correct value of $E_{0,0}$ it is of great importance that Equation 12, and thus also the Morse potential, are valid expressions. It has been found that to describe adequately the vibrational part of diatomic spectra, the first two RHS terms of Equation 12 are not always sufficient. To obtain a good fit to the observed vibrational energies expression 13 has also been used:

$$E_v/hc = \omega_e(v+1/2) - \omega_e x_e(v+1/2)^2 + \omega_e y_e(v+1/2)^3 + \omega_e z_e(v+1/2)^4 \ldots \tag{13}$$

In order to have an idea of the importance of the various terms in Equations 12 and 13, we give the molecular constants for H_2 and $^{16}O_2$ in Table 1.

As suggested by the diminishing values for the constants ω_e, $\omega_e x_e$, and $\omega_e y_e$ in Table 1, the term $\omega_e z_e$ is not important. The data of Table 1 show also that vibrational energy is significantly greater than rotational energy. The term D (Table 1) is much smaller than the other constants in Equation 12, meaning that the energy associated with the rotational-stretching effect is at best only of minor importance.

In general (see e.g. Figure 1), the Morse potential reproduces the observations quite well; however, there remain a few discrepancies. In order to resolve these, Dunham (1932) has attempted to solve Equation 10 with a more general potential. For a general potential expanded about the equilibrium internuclear distance r_e one can write

$$V(r) = V(r_e) + \frac{dV}{dr}\bigg|_{r=r_e} (r-r_e) + \frac{1}{2}\frac{d^2V}{dr^2}\bigg|_{r=r_e} (r-r_e)^2$$

$$+ \frac{1}{6}\frac{d^3V}{dr^3}\bigg|_{r=r_e} (r-r_e)^3 + \frac{1}{24}\frac{d^4V}{dr^4}\bigg|_{r=r_e} (r-r_e)^4 + \cdots \tag{14}$$

By definition we have $V(r_e) = 0$ and $(dV/dr)_{r=r_e} = 0$. Hence Equation 14 may be written as

$$V(r) = K_2(r-r_e)^2 + K_3(r-r_e)^3 + K_4(r-r_e)^4. \tag{15}$$

Using the WKB method, Dunham (1932) solved Equation 10 with the potential (15) obtaining Equation (16),

$$E_{v,J} = hc \sum_{j=0} \sum_{k=0} Y_{jk}(v+1/2)^j [J(J+1)]^k. \tag{16}$$

Table 1 Molecular constants (cm^{-1})

	H_2	$^{16}O_2$
ω_e	4401.118	1580.360
$\omega_e x_e$	121.284	12.073
$\omega_e y_e$	0.804	0.047
B_e	60.8380	1.4456
α_e	3.0258	0.1593
D	0.0463	5×10^{-6}

The coefficients Y_{jk} were calculated to second order. The resulting zero-point energy is

$$E_{0,0} = hc\left(G_0 + \frac{\omega_e}{2} - \frac{\omega_e x_e}{4} \right) \tag{17}$$

with

$$G_0 = \frac{B_e}{4} + \frac{\alpha_e \omega_e}{12 B_e} + (\alpha_e \omega_e/12 B_e)^2/B_e - \frac{\omega_e x_e}{4}. \tag{18}$$

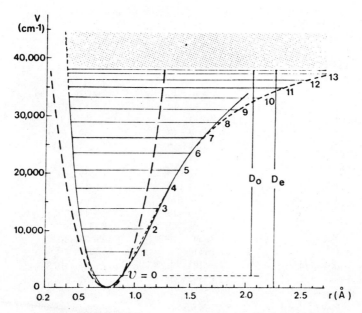

Figure 1 Potential energy curves and vibrational levels for the electronic ground state of H_2 (Herzberg 1950, modified): ——, real potential; ----, Morse potential; — — —, harmonic potential.

Equations 17 and 18 can be obtained from Dunham (1932), and they are also given in Herzberg (1950). Wolfsberg (1969b) has given numerical values for G_0 for several diatomic hydrides.

More recently Huffaker (1976a, b) has solved Equation 10, using a perturbed Morse oscillator,

$$V(r) = V_e\left(F^2 + \sum_{n=4} b_n F^n\right),$$ (19)

as potential. In Equation 19 we have $F = 1 - \exp a(r - r_e)$. The important aspect for us is that Huffaker (1976a) derived the same expression (16) as Dunham (1932) and that he continued the evaluation of the coefficients Y_{jk} to a higher perturbation order (Huffaker 1976b) than Dunham. Huffaker expressed the coefficients Y_{jk} as

$$Y_{jk} = \sum_{r=0} Y_{jk}^{2r}.$$ (20)

The importance of the various terms for the zero-point energy is illustrated by the data listed in Table 2 for HF and CO. The zero-point energy in terms of Dunham coefficients (Y_{jk}) is given by

$$E_{0,0} = hc \sum_{r=0} \sum_{j=0} Y_{j0}^{2r}/2^j.$$ (21)

Table 2 Molecular constants and Dunham coefficients (Y_{jk}^{2r}) in cm^{-1} for CO and HF[a]

	CO	HF
Y_{00}^0	0.	0.
Y_{00}^2	1.9004×10^{-1}	3.90777
Y_{00}^4	8.5781×10^{-7}	4.5212×10^{-2}
$Y_{10}^0 = \omega_e$	2169.8181	4139.357
Y_{10}^2	-4.4372×10^{-4}	-3.2532×10^{-1}
Y_{10}^4	7.5284×10^{-6}	-6.4393×10^{-3}
$Y_{20}^0 = -\omega_e x_e$	-13.2906	-90.5660
Y_{20}^2	-3.9577×10^{-5}	1.3529×10^{-1}
Y_{20}^4	-3.0584×10^{-6}	-8.5241×10^{-3}
$Y_{30}^0 = \omega_e y_e$	1.0964×10^{-2}	1.2001
Y_{30}^2	1.3416×10^{-5}	-2.2417×10^{-2}
$Y_{40}^0 = \omega_e z_e$	2.29622×10^{-5}	-7.32256×10^{-2}
Y_{40}^2	-2.6547×10^{-8}	9.5086×10^{-4}
B_e	1.931265	20.94948
D (Eq. 12)	6.1198×10^{-6}	2.1461×10^{-3}
α_e	1.7505×10^{-2}	0.7961
$E_{0,0}/hc$ (Eq. 12)	1081.59	2047.04
$E_{0,0}/hc$ (Eq. 17)	1081.78	2050.94
$E_{0,0}/hc$ (Eq. 21)	1081.78	2051.00

[a] After Huffaker (1976b), with molecular constants slightly different from those we use (cf Table 5).

We have gone into more detail than is customary to discuss the zero-point energy. The reasons for this are that the zero-point energy plays an important role in the calculation of isotopic exchange equilibrium constants and that the evaluation and importance of G_0 for isotopic calculations is somewhat controversial (Richet & Bottinga 1976, Bottinga 1968b).

The data of Table 2 show clearly that for our purposes Equation 12 plus the term G_0 (Equation 18) are adequate. The higher perturbation contributions to the zero-point energy are insignificant. This will become clearer somewhat later on when we discuss the importance of the uncertainty in the spectroscopic data and evaluate the consequences of a 1 cm^{-1} error in zero-point energy.

2.2   Born-Oppenheimer Approximation, Electronic Energy

The Born-Oppenheimer (B-O) approximation has been recently discussed in many papers. An excellent review of the current literature is given by Kolos (1970). The

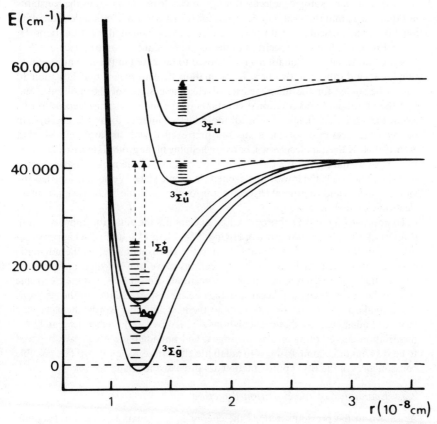

Figure 2   Potential energy curves and vibrational levels for various electronic states of O_2, plotted versus internuclear distance (Herzberg 1950).

Table 3 Electronic states and molecular constants for $^{16}O_2$, in cm^{-1} (Herzberg 1950)

Electronic state	T_e^a	ω_e	B_e	$N(T_e)/N(0)^b$
$b^1\Sigma_g^+$	13195	1433	1.40	7×10^{-6}
$a^1\Delta_g$	7918	1509	1.43	8×10^{-4}
$X^3\Sigma_g^-$	0	1580	1.45	—

[a] Minimum electronic energies, measured with respect to the minimum of the electronic ground state.
[b] Fraction of the molecules $^{16}O_2$, which is in the indicated state at 1600°K.

importance of the B-O approximation is that it is related to the assumption that the potential energy curve for the electronic ground state, as it is given for instance in Figure 1, is the same for isotopically substituted molecules. This assumption is of great importance because spectroscopic data are usually available only for the most abundant isotopic molecule (for this molecule one knows thus the constants in Equation 12 and the values of $E_{v,J}$), but for the rare isotopic molecule these data must be usually calculated with the parameters of the potential (i.e. Equations 11, 15, or 19) known from the spectrum of the most abundant isotopic molecule.

As already discussed, Equation 9 is supposed to be solved in the B-O approximation for E_e and subsequently one solves Equation 10. This means that Equation 10 has to be solved for the different excited electronic states of the molecule, and that the vibrational and rotational constants of the molecule must depend on its electronic state (see Figure 2). For all molecules of interest to us the energy gap between the electronic ground state and first excited state is large enough that even at 1600°K these molecules will be overwhelming in the ground state (see Table 3). In Table 3 we have given the data for oxygen because, of the molecules we discuss in this review, it has the lowest first excited electronic state. We have also listed the vibrational (ω_e) and rotational (B_e) constants for the molecule $^{16}O_2$ when it is in an electronic excited state.

To improve on the B-O approximation, one applies the adiabatic approximation (Van Vleck 1936, Born 1951, Born & Huang 1954, Chap. IV). In the B-O approximation the nuclei are in a fixed position while the electrons move; in the adiabatic approximation the nuclei move and the electrons move too, but their movement is as if the nuclei were fixed in instantaneous positions. This means that the electronic state of the molecule changes because of the nuclear displacements, but the electrons do not make a transition from one state to the next; in other words, the electronic motion is adiabatic. Adiabatic calculations (see Kolos 1970) predict accurately the dissociation energy of molecular hydrogen, and nonadiabatic effects are believed to be very minor in comparison with adiabatic effects (see Kolos 1970). The possible consequences of the B-O approximation are discussed in a later section.

2.3   Evaluation of the Partition Function

As shown in the previous sections the energy E_i in Equation 4 for the partition function is made up of a series of terms

$$E = E_{tr} + E_0 + E_{vib} + E_{anh} + E_{rot} + E_{rot-str} + E_{rot-vib} + T_e, \qquad (22)$$

where E_{tr} stands for translational energy (see Equation 7); E_0, zero-point energy (for instance, Equation 17); E_{vib}, vibrational energy, from which is substracted the zero-point energy, in the harmonic approximation; E_{anh}, anharmonic vibrational energy, without zero-point energy; E_{rot}, rotational energy (third term, RHS, Equation 12); $E_{rot-vib}$, energy associated with the rotation-vibrational interaction (fifth term, RHS, Equation 12); and T_e, electronic energy.

Combining Equations 22 and 4, and neglecting the term T_e, because all of the molecules we are dealing with are overwhelmingly in their ground state, we can write

$$Q = Q_{tr} Q_0 Q_{vib} Q_{anh} Q_{rot} Q_{rot-vib}. \qquad (23)$$

In Equation 23 we have neglected the contribution to the partition function of the rotational stretching term. Later on when we evaluate the isotopic exchange equilibrium constant K_1 for Equation 1 we justify this omission. From Equations 4 and 7 we obtain the translational partition function

$$Q_{tr} = (2\pi M k T/h^2)^{3/2} V, \qquad (24)$$

where V is the volume in which the molecule is enclosed. Q_0 is given by

$$Q_0 = \exp\left[-(G_0 + \omega_e/2 - \omega_e x_e/4)hc/kT\right] \qquad (25)$$

and

$$Q_{vib} = 1/(1 - e^{-u}), \qquad (26)$$

where $u = hc\omega_e/kT$. A reader unfamiliar with Equations 24 and 26 is advised to consult Denbigh (1971, Chap. 12). We also have

$$Q_{anh} = 1 - 2x_e u\, e^{-u}/(1 - e^{-u})^2 \qquad (27)$$

as derived by Vojta (1961). The rotational partition function is given by

$$Q_{rot} = \sum_{J=0} (2J+1) \exp\left[-B_e J(J+1)hc/kT\right]. \qquad (28)$$

The factor $2J+1$ in Equation 28 is the degeneracy for the rotational state J. When B_e is small or when the temperature is high the rotational levels will be closely spaced, and in that case the summation in Equation 28 may be replaced by an integration, and the rotational partition function will have its classical value (Equation 29):

$$(Q_{rot})_{class} = 1/s\sigma, \qquad (29)$$

where s is the symmetry number (see Mayer & Mayer 1940) and $\sigma = B_e hc/kT$. In order to include the rotation-vibration interaction for the vibrational ground state one can replace B_e in Equation 28 by B_0, where $B_0 = B_e - \alpha_e/2$. The rotation-vibration interaction when the molecule is not in the vibrational ground state, i.e. $v \neq 0$ in Equation 12, gives rise to the factor

$$Q_{rot-vib} = 1 + \delta/(e^u - 1), \qquad (30)$$

where $\delta = \alpha_e/B_0$. When B_e is too large or T too small in Equation 28, the rotational partition function will deviate significantly from its classical value (Equation 29). In that case one can either sum Equation 28 directly or use Equation 31, derived by Mulholland (1928) and Viney (1933):

$$Q_{rot} = \frac{1}{s\sigma}(1 + \sigma/3 + \sigma^2/15 + 4\sigma^3/315 + \cdots). \tag{31}$$

The term between brackets at the RHS of Equation 31 is a quantum mechanical correction to Equation 29.

2.4 Partition Function Ratios

In order to calculate K_1 (Equation 2) we evaluate the ratio $(Q'/Q)_{ZY}$ of Equation 23:

$$\left(\frac{Q'}{Q}\right)_{ZY} = \left(\frac{Q'}{Q}\right)_{tr}\left(\frac{Q'}{Q}\right)_0\left(\frac{Q'}{Q}\right)_{vib}\left(\frac{Q'}{Q}\right)_{anh}\left(\frac{Q'}{Q}\right)_{rot}\left(\frac{Q'}{Q}\right)_{rot\text{-}vib}. \tag{32}$$

The ratio $(Q'/Q)_Y$ of Equation 3 is simply given by the ratio of translational partition functions, because the atoms have only translational energy. Hence, using Equation 24 as an analogy, we find that

$$\left(\frac{Q'}{Q}\right)_Y = (m'/m)^{3/2}, \tag{33}$$

where m' and m are the atomic masses of Y' and Y respectively. Similarly, the translational term in Equation 32 is given by

$$\left(\frac{Q'}{Q}\right)_{tr} = (M'/M)^{3/2}, \tag{34}$$

where now the M' and M are the molecular masses of ZY' and ZY respectively. From Equation 25 we find that

$$\left(\frac{Q'}{Q}\right)_0 = [\exp - (G_0' - G_0)hc/kT][\exp - (u' - u)/2][\exp(u'x_e' - ux_e)/4], \tag{35}$$

and from Equation 26 we obtain

$$\left(\frac{Q'}{Q}\right)_{vib} = (1 - e^{-u})/(1 - e^{-u'}). \tag{36}$$

The contribution to the partition function ratio of the anharmonicity of vibrationally excited molecules is given by

$$\left(\frac{Q'}{Q}\right)_{anh} = \frac{1 - 2x_e'u' \, e^{-u'}/(1 - e^{-u'})^2}{1 - 2x_e u \, e^{-u}/(1 - e^{-u})^2}. \tag{37}$$

To evaluate the rotational partition function ratio for the diatomic molecules, except for CS and S_2 we summed Equation 28 directly for the normal and isotopically substituted molecule and then calculated the ratio. The summation was continued until the fourth figure after the decimal point did not change with increasing J. For CS and S_2 we used Equation 38:

$$\left(\frac{Q'}{Q}\right)_{\text{rot}} = \left(\frac{s\sigma}{s'\sigma'}\right)\left(\frac{1+\sigma'/3+\sigma'^2/15+\cdots}{1+\sigma/3+\sigma^2/15+\cdots}\right). \tag{38}$$

In both cases we replaced B_e by B_0 and multiplied the obtained rotational partition function ratio by the factor

$$\left(\frac{Q'}{Q}\right)_{\text{rot-vib}} = [1+\delta'/(e^{u'}-1)]/[1+\delta/(e^{u}-1)] \tag{39}$$

to take into account the rotation-vibration interaction.

Of course we are well aware that for many calculations simpler expressions for the partition function ratios would be adequate. However, with modern computing equipment available it is no longer difficult to evaluate Equations 32–39. Moreover, for certain of the molecules treated, less complete equations would not suffice. The importance of the various terms in Equation 32 is illustrated by the calculated results given in Table 4, where we list the various contributions to the ratios $Q(HD)/Q(H_2)$, $Q(DF)/Q(HF)$, and $Q(C^{18}O)/Q(C^{16}O)$ at 0°C and 1000°C. It is evident from the values listed in Table 4 that the zero-point energy contribution (line 5 of Table 4, Equation 35) is very important in the calculation of the partition function ratio (line 12, Table 4). It was for that reason that in Section 2.2 so much attention was payed to the correct evaluation of the zero-point energy. Of the terms making up the zero-point energy, the harmonic vibration term (cf line 3, Table 4) is considerably greater than the other terms. However, its importance diminishes with increasing temperature. Under all conditions the G_0 term in the zero-point energy is of the least significance (see Table 4, line 2). A geochemist would like to know stable-isotope equilibrium constants with the same relative accuracy as he can measure isotopic ratios, which is often better than $0.5°/_{\circ\circ}$. A 1 cm^{-1} error in the difference $|\omega'_e - \omega_e|$ will already cause an uncertainty of $2.6°/_{\circ\circ}$ in the value of the partition function ratio at 273°K and $0.6°/_{\circ\circ}$ at 1273°K. Such uncertainties in the vibrational frequency differences are not excluded. Inspection of line 2, Table 4, shows that the G_0 effect is often comparable to the uncertainty introduced by the limits on the accuracy of the difference $|\omega'_e - \omega_e|$. This feature is also evident from the G_0 values computed by Wolfsberg (1969b). Calculations of G_0 for diatomic molecules not containing H show that the G_0 term can be safely neglected in such cases (see, for instance, the values on line 4, columns 3 and 7 of Table 4). In the calculation we report in this paper the G_0 effect has been ignored, because it appears only to be of interest in H exchange, but in that case the fractionations are so large that the G_0 effect becomes relatively unimportant in geochemistry. Moreover, G_0 cannot in general be calculated for polyatomic molecules, and thus for reasons of homogeneity we have also ignored the G_0 contribution in diatomic calculations.

The values of line 6, Table 4, indicate that the vibrational partition function ratio becomes significantly different from unity only at higher temperatures when the vibrational excited states become occupied. As shown by line 7, the importance of the anharmonic correction increases with temperature. Lines 8, 9 and 10 (Table 4) give calculated results for $(Q'/Q)_{\text{rot}}$. For hydrogen-bearing diatomic molecules it is recommended that Equation 28 be used rather than the classical ratio or the quantum

Table 4 Contributions to the partition function ratio

	Eq.	0°C				1000°C			
		1 HD/H_2	2 DF/HF	3 $C^{18}O/C^{16}O$	4[b] $2/1$	5 HD/H_2	6 DF/HF	7 $C^{18}O/C^{16}O$	8[b] $6/5$
1 $(M'/M)^{3/2}$	34	1.83561	1.07638	1.10929	0.58639	1.83561	1.07638	1.10929	0.58639
2 $\exp-(G'_0-G_0)hc/kT$	35	1.00910	1.01192	1.00005	1.00280	1.00194	1.00255	1.00001	1.00061
3 $\exp-1/2(u'-u)$	35	4.71546	20.1428	1.14804	4.27165	1.39477	1.90456	1.03006	1.36550
4 $\exp 1/4(x'_e u' - x_e u)$	35	0.96079	0.94356	0.99917	0.98207	0.99146	0.98761	0.99982	0.99612
5 $(Q/Q)_0$	35	4.57179	19.2325	1.14714	4.20677	1.38554	1.88576	1.02988	1.36103
6 $(Q/Q)_{vib}$	36	1.00000	1.00000	1.00000	1.00000	1.00663	1.02531	1.00578	1.01856
7 $(Q/Q)_{anh}$	37	1.00000	1.00000	1.00000	1.00000	0.99908	0.99818	0.99993	0.99910
8 $(s\sigma/s'\sigma')$	38	2.65666	1.89381	1.05001	2.85142	2.65666	1.89381	1.05001	0.71285
9 $(Q/Q)_{rot}$	38	2.58778	1.86165	1.04985	2.87758	2.64196	1.88691	1.04997	0.71421
10 $(Q/Q)_{rot}$	28	2.59246	1.86165	1.04985	2.87234	2.64196	1.88680	1.04997	0.71417
11 $(Q/Q)_{rot-vib}$	39	1.00000	1.00000	1.00000	1.00000	1.00025	1.00058	1.00003	1.00033
12 $(Q/Q)^a$	32	21.7560	38.5389	1.33595		6.75935	3.92187	1.20640	
13 $K = \dfrac{[DF]/[HD]}{[HF]/[H_2]}$					1.77141				0.58021

[a] The product of lines 1, 5, 6, 7, 10 and 11; the values given are not the same as in Table 9 because, as explained in the text, in the tables the G_0 effect (line 2) was not included.

[b] In columns 4 and 8 we give the quotient of values listed in columns 2 and 1, and columns 6 and 5, respectively.

mechanically corrected classical ratio (Equation 38). However, for the non-hydrogen-bearing diatomic molecules Equation 38 and even the classical ratio are usually adequate. The rotation-vibration interaction is always of only marginal interest. As is to be anticipated, this term increases in significance with temperature. We have investigated numerically the rotational stretching effect which increases with temperature. Though it may be not completely negligible when the absolute value of the partition function ratio is calculated, its contribution to the partition function ratio is too small to be of interest.

The importance of the B-O approximation for isotopic calculations for diatomic molecules has been discussed in a series of papers by Kleinman & Wolfsberg (1973, 1974a, 1974b). These authors followed the procedure developed by Van Vleck (1936), Born (1951), and Born & Huang (1954, Chap. IV) and did an approximate calculation within the framework of the adiabatic approximation. They evaluated the energy difference (C) for the minimum of the potential-energy function, such as is shown in Figure 1, between the adiabatic approximation and the B-O approximation for various diatomic molecules. For reaction 1 it is thus important to know the quantity $\Delta(\text{B-O}) = C(ZY') - C(ZY)$. The quantity $\Delta(\text{B-O})$, which is zero in the B-O approximation but may be nonzero in a more exact treatment, enters the equilibrium constant calculation just as the term $G'_0 - G_0$ enters Equation 35. Unfortunately, to calculate the terms $C(ZY')$ and $C(ZY)$ one needs the wave functions for these molecules, and they can only be approximated for most of the molecules that are of interest to us. This limits severely the accuracy to which one can calculate the $C(ZY)$. The $\Delta(\text{B-O})$ for an exchange reaction is always very small in comparison with the C terms for reactants and reaction products for such a reaction.

Van Vleck (1936) has shown that for the reaction $H_2 + D_2 = 2HD$, $\Delta(\text{B-O}) = 0$; Kleinman & Wolfsberg (1974a) showed that this is always the case for internal equilibria involving diatomic molecules. The mass dependence of the C terms (see Bardo & Wolfsberg 1975) indicates also that a deviation from the B-O approximation has the most serious consequences for the case of deuterium-hydrogen exchange. A comparison of the rigorously derived C terms by Kolos & Wolniewicz (1964) for molecular hydrogen and its isotopically substituted varieties and those calculated by Kleinman & Wolfsberg (1973, 1974a) suggests that the uncertainty in the Kleinman & Wolfsberg results for the individual C terms is comparable to the $\Delta(\text{B-O})$ for a given exchange reaction. Hence in conclusion the effect may be of importance for H-D exchange, but otherwise it will be probably minor.

3   CALCULATIONS—POLYATOMIC MOLECULES

A possible exchange reaction involving polyatomic molecules is

$$ZY_n + nY' = ZY'_n + nY. \tag{40}$$

We have now

$$K_{40} = \left(\frac{Q'}{Q}\right)_{ZY_n} \bigg/ \left(\frac{Q'}{Q}\right)_Y^n, \tag{41}$$

which is analogous to Equation 3. The partition function Q is given by Equation 4. As in the diatomic case (Equation 22) the energy in Equation 4 is given by

$$E = E_{tr} + E_0 + E_{vib} + E_{anh} + E_{rot} + E_{rot\text{-}vib}. \tag{42}$$

We right away neglect the term T_e (Equation 22) because all polyatomic molecules we are discussing in this review will be in the ground state for the temperature interval 273 to 1600°K. $E_{rot\text{-}str}$ is not appearing in Equation 42, because we have already found that in the diatomic case this energy term did not contribute to the isotopic fractionation. Moreover in most cases the polyatomic spectra are not well enough known to determine the needed constants.

The equation for the translational energy is the same as for the diatomic case and hence the translational partition function ratio is given by Equation 24.

A diatomic molecule has only one mode of vibration but a nonlinear polyatomic molecule has $3n - 6$ modes of vibration, where n is the number of atoms in the molecule. Linear polyatomic molecules have $3n - 5$ modes of vibration. The vibrational energy is the sum total of the vibrational energies associated with the different modes of vibration. We do not dwell on these aspects because they do not introduce uncertainties in the calculation of K_{40} (Equation 41), and excellent discussions on these topics are available (see for instance Mayer & Mayer 1940). For the zero-point energy one has

$$E_0 = hc\left(G_0 + 1/2 \sum_i d_i \omega_i + 1/4 \sum_{i \leq j} \sum d_i d_j x_{ij}\right). \tag{43}$$

As before, the zero-point energy is not observable in molecular spectroscopy and to evaluate it correctly one should solve the polyatomic equivalents of Equations 10 and 15. This subject has been treated by Nielsen (1951), but while he mentions the existence of a term like G_0 (Equation 43), he does not give an expression for it. For the time being we will neglect G_0. The summations i and j are over the normal modes of vibration of the molecule. The coefficients d_i and d_j are the degeneracies of the modes of vibration i and j. The coefficient x_{ij} is due to anharmonicity and should be compared with the coefficient $-x_e \omega_e$ in Equation 17. The anharmonicity of normal mode i is expressed by x_{ii}, while x_{ij} is due to coupling between the modes i and j, as a result of the anharmonicity of modes i and j.

The vibrational energy associated with the excited vibrational states in the harmonic approximation is given by

$$E_{vib} = hc \sum_i d_i v_i \omega_i, \tag{44}$$

where the ω_i are the zero-order frequencies and are comparable to ω_e for the diatomic molecule. The anharmonicity correction to (44) is

$$E_{anh} = hc \sum_{i \leq j} \sum d_i d_j x_{ij} [v_i v_j + 1/2(v_i + v_j)]. \tag{45}$$

Since molecular rotations are in most cases classical at room temperature and above, and the quantum mechanical expression for E_{rot} is rather complicated (see for instance Nielsen 1951), we only give the classical expression for the asymmetric top

$$E_{rot} = \frac{I_A \Omega_A^2}{2} + \frac{I_B \Omega_B^2}{2} + \frac{I_C \Omega_C^2}{2},$$ (46)

where I_A, I_B, and I_C are the three principal moments of inertia and Ω_A, Ω_B, and Ω_C are the x, y, and z components of the angular velocity.

By means of Equations 43, 44, 45, and 46 we can evaluate the partition function 4 and calculate the following partition function ratios. For polyatomic molecules, $(Q'/Q)_{tr}$ is given by Equation 32. The ratio $(Q'/Q)_0$ resembles very much Equation 35.

$$\left(\frac{Q'}{Q}\right)_0 = \left[\exp - (G_0' - G_0)\frac{hc}{kT}\right]\left[\exp - \frac{1}{2}\sum_i (u_i' - u_i)d_i\right]$$

$$\cdot \left[\exp - \frac{1}{4}\frac{hc}{kT}\sum_{i \leq j}\sum d_i d_j (x_{ij}' - x_{ij})\right]$$ (47)[3]

Likewise, the harmonic vibrational partition function ratio for polyatomic molecules,

$$\left(\frac{Q'}{Q}\right)_{vib} = \prod_i \frac{(1 - e^{-u_i})^{d_i}}{(1 - e^{-u_i'})^{d_i}},$$ (48)[3]

resembles Equation 36. The expression for the contribution due to the anharmonicity of the excited vibrational states is quite complicated; it is expressed by Vojta (1961) as

$$\left(\frac{Q'}{Q}\right)_{anh} = 1 + \frac{hc}{kT}\sum_i d_i(d_i + 1)\{x_{ii} e^{u_i}/(e^{u_i} - 1)^2 - x_{ii}' e^{u_i}/(e^{u_i} - 1)^2\}$$

$$+ \frac{hc}{2kT}\sum_{i < j}\sum d_i d_j \{x_{ij}(e^{u_i} + e^{u_j})/[(e^{u_i} - 1)(e^{u_j} - 1)]$$

$$- x_{ij}'(e^{u_i} + e^{u_j})/[(e^{u_i} - 1)(e^{u_j} - 1)]\}$$ (49)[3]

The rotational partition function ratio is virtually equal to the classical ratio. It is given by

$$\left(\frac{Q'}{Q}\right)_{rot} = \frac{s}{s'}\left[\frac{I_A' I_B' I_C'}{I_A I_B I_C}\right]^{1/2},$$ (50)

where as before s and s' are the symmetry numbers for the normal and isotopically substituted molecules. Common practice is to apply the Teller-Redlich product rule, which gives a relation between the normal-mode frequencies and the moments of inertia:

$$\left[\frac{I_A' I_B' I_C'}{I_A I_B I_C}\right]^{1/2}\left[\frac{M'}{M}\right]^{3/2}\left[\frac{m}{m'}\right]^{3n/2}\prod_i \frac{u_i}{u_i'} = 1,$$ (51)

where n is the number of exchangeable atoms.

---

[3] As is generally done, we have in writing Equations 47–49 assumed that the normal modes of the normal and isotopically substituted molecules have the same degeneracies.

For the calculations reported in this paper we have not made use of Equation 51; for all nonlinear molecules not involved in hydrogen–deuterium exchange, Equation 52 was used. For polyatomic molecules the quantum-mechanical correction to the classical rotational partition function given by Stripp & Kirkwood (1951) was used:

$$Q_{rot} = (Q_{rot})_{class}Q_{corr}, \tag{52}$$

where

$$Q_{corr} = 1 + \left[2\sigma_A + 2\sigma_B + 2\sigma_C - \frac{\sigma_A\sigma_B}{\sigma_C} - \frac{\sigma_B\sigma_C}{\sigma_A} - \frac{\sigma_A\sigma_C}{\sigma_B} \right] \bigg/ 12 + \cdots \tag{53}$$

and

$$\sigma_A = \frac{\hbar^2}{2I_A kT}.$$

For linear polyatomic molecules Equation 38 was used.

The evaluation of the rotation–vibration interaction is analogous to the diatomic case (Equation 30). In Equations 50 and 53, the moments of inertia (I_A, I_B, I_C) or the coefficients σ_A, σ_B, σ_C, derived for the rotating molecule in the vibrational ground state, are used, and Equation 52 is multiplied by the polyatomic equivalent of Equation 30 (see Vojta 1960). We then have

$$Q_{rot\text{-}vib} = \prod_i \left\{ 1 + \frac{1}{2} \frac{\delta_i}{\exp(u_i) - 1} \right\}^{d_i}. \tag{54}$$

The multiplication over i is over the normal modes of vibration and the δ_i are the rotation–vibration interaction constants. For all molecules $Q_{rot\text{-}vib}$ was calculated. In general one can safely ignore it; only in the case of hydrogen isotope exchange did this term become limited in interest.

The energy effects discussed in this section on polyatomic molecules are the only ones of importance, sometimes only marginally, in the calculation of the needed partition function ratio. Splitting of energy levels of polyatomic molecules such as may be caused by Fermi resonance in CO_2 or Darling-Dennison resonance in H_2O (see Herzberg 1945), does not affect our calculations in a measurable way. Similarly, the quantification of the rotation, which results from the composite motion of the degenerate modes of vibration, such as in a linear triatomic molecule, has no observable consequences for our calculations. Also in the temperature interval we are interested in, a phenomenon like ortho- and para-hydrogen has no measurable influence on our calculated results. The G_0 term occurring in Equation 43 is discussed in Section 6; in our calculations it has been neglected. The B-O approximation for polyatomic molecules has been treated by Bardo & Wolfsberg (1975); it does not seem to cause observable consequences in the exchanges discussed in this review. As in diatomic molecules the B-O approximation does not affect internal isotopic equilibria (Bardo & Wolfsberg 1975).

4   MOLECULAR CONSTANTS

In the discussion of the Born-Oppenheimer approximation mention was made that frequently not all molecular constants are available for all the isotopic molecules. If

Table 5 Molecular constants of diatomic molecules (in cm⁻¹)

	ω_e	$\omega_e x_e$	B_0	α_e	Ref.
H_2	4401.118	121.28	59.3251	3.0258	a
HD	3812.293	90.908	44.6613	2.0034	b
D_2	3113.259	60.687	29.9069	1.0711	c
HF	4138.32	89.88	20.5567	0.798	d
DF	2998.192	45.761	10.8546	0.2907	e
$H^{35}Cl$	2990.946	52.819	10.44020[g]	0.30718	f
$D^{35}Cl$	2145.115	27.169	5.392271[g]	0.11332	c
$H^{37}Cl$	2988.681	52.738	10.42451[g]	0.30668	c
$D^{37}Cl$	2141.957	27.089	5.376489[g]	0.11282	c
$^{35}Cl_2$	559.72	2.67	0.24325		h
$^{37}Cl_2$	544.37	2.53	0.23001		c
$^{14}N_2$	2358.027	14.135	1.98914		i
$^{14}N^{15}N$	2318.512	13.665	1.92325		c
$^{15}N_2$	2278.313	13.196	1.85721		c

	ω_e	$\omega_e x_e$	B_0	Ref.
$^{14}N^{16}O$	1904.087	14.044	1.696284	j
$^{15}N^{16}O$	1869.997	13.535	1.636369	j
$^{14}N^{18}O$	1853.941	13.314	1.608332	c
$^{12}C^{16}O$	2169.817	13.290	1.922529	k
$^{13}C^{16}O$	2121.430	12.698	1.837973	l
$^{12}C^{18}O$	2117.400	12.655	1.830967	c
$^{16}O_2$	1580.361	12.073	1.438225[n]	m
$^{16}O^{18}O$	1535.737	11.401	1.358365	c
$^{18}O_2$	1489.777	10.729	1.278475	c
$^{12}C^{32}S$	1285.08	6.46	0.81708	o
$^{13}C^{32}S$	1248.51	6.10	0.77133	o
$^{12}C^{34}S$	1274.74	6.36	0.80401	c
$^{32}S_2$	725.680	2.852	0.2948	p
$^{34}S_2$	704.040	2.684	0.2775	c

a Buijs & Gush (1971).
b Durie & Herzberg (1960).
c Calculated with Equations 55–58.
d Webb & Rao (1968).
e Spanbauer & Rao (1965).
f Rank et al (1965).
g De Lucia et al (1971).
h Clyne & Coxon (1970).

i Benesh et al (1965)
j Olman et al (1964).
k Roh & Rao (1974).
l Johns et al (1974).
m Babcock & Herzberg (1948).
n Amano & Hirota (1974).
o Barrow et al (1960).
p Herzberg (1950).

the B-O approximation is valid and the isotopic molecules have the same potential-energy function then it is simple to calculate the constants needed for the isotopically substituted molecule from those known for the normal molecule. For instance, Dunham (1932) has shown that for diatomic molecules

$$\omega'_e = \omega_e \rho, \tag{55}$$

where $\rho = \sqrt{\mu/\mu'}$, and μ is the reduced mass. Furthermore, we have

$$\omega'_e x'_e = \omega_e x_e \rho^2, \tag{56}$$

$$\omega'_e y'_e = \omega_e y_e \rho^3,$$

$$B'_e = B_e \rho^2, \tag{57}$$

and

$$\alpha'_e = \alpha_e \rho^3. \tag{58}$$

These equations were employed to complete the data set needed for the calculations. The molecular constants for diatomic molecules are listed in Table 5. Measurements for certain diatomic molecules such as H_2 have been repeated many times, but sometimes the agreement among these measurements is poor. In our selection of the data given in Tables 5–8 we have favored the most complete and most recent observations over incomplete and older ones, and we preferred data that obeyed Equations 55 to 58.

For polyatomic molecules the molecular constants are less well known than for the diatomic ones. In particular, measurements for isotopically substituted molecules are relatively rare when the number of atoms in a molecule becomes larger than three. To complete the data, force fields derived from the normal molecules were used to calculate zero-order frequencies for the isotopically substituted molecules. The semi-empirical Darling-Dennison rule (see Burneau 1974),

$$x'_{ij} = x_{ij} \omega'_i \omega'_j / \omega_i \omega_j, \tag{59}$$

was employed to calculate anharmonicity constants for the isotopically substituted molecules. For isotopic molecules with the same symmetry Equation 59 works well. Because of lack of anything better we have also used Equation 59 when the symmetry of the isotopic molecules is not the same. Pragmatically, we observed that the agreement between measured and calculated isotopic equilibrium constants does not seem to be affected by this possible misuse of Equation 59. The empirical relation

$$\delta'_i = \delta_i \omega'_i / \omega_i \tag{60}$$

introduced by Haar et al (1955) was used. However, the reduced significance of rotation–vibration interaction in isotopic fractionation calculations has a consequence that this will not be an important error source.

In the following paragraphs we furnish, briefly, detailed information on how the calculations indicated in Tables 5–8 were made.

Hydrogen   There are no recent observations of the spectrum of D_2. We have used the data of Buijs & Gush (1971) to calculate these constants with Equations 55–58.

The resulting constants give for the internal equilibrium $H_2 + D_2 = 2HD$ and $G_0 = 2G_0(HD) - G_0(H_2) - G_0(D_2) = 0.86$ cm^{-1}. This suggests a certain lack of internal consistence in the constants. This is also observable when one compares the data for H_2 (Buijs & Gush 1971) and those for HD (Durie & Herzberg 1960) with the predictions of Equations 55–58.

Other diatomic molecules   Equations 55–58 are well obeyed.

Carbon dioxide   The constants for $^{12}C^{18}O_2$ were calculated from the data of Courtoy (1957, 1959) for $^{12}C^{16}O_2$ by means of the Urey-Bradley force field equations given in Herzberg (1945). These equations were also used to calculate the vibrational frequencies for $^{12}C^{34}S_2$ from the data for $^{12}C^{32}S_2$ of Giguère et al (1973).

Bent triatomic molecules XY_2   Zero-order frequencies for the isotopically substituted molecules were calculated by means of the equations given by Shaffer & Schuman (1944).

Ammonia   The equations of Rosenthal (1935) and the force constants deduced by Benedict & Plyler (1957) for NH_3 were used to calculate the frequencies for the isotopic molecules of ammonia. Internuclear distances for NH_3 from Benedict & Plyler (1957) were used to calculate the rotational constants for the isotopic molecules of ammonia. The anharmonicity constants x_{11} and x_{22} of NH_3 have not been observed, but values of -30 and -51.8 cm^{-1} were assumed, respectively, because they are in agreement with the observed fundamental frequencies. Equation 59 was used to calculate the remaining unknown anharmonicity constants.

Sulphur-trioxide   The data of Dorney et al (1973) for $^{32}S^{16}O_3$ were used with the equations of Anderson et al (1936) to calculate the vibrational constants for $^{34}S^{16}O_3$ and $^{34}S^{18}O_3$.

Methane   Zero-order frequencies for the isotopic methane molecules were calculated as in Bottinga (1969). Anharmonicity constants were calculated by means of Equation 59, and experimental rotational constants for CH_4, CH_3D, CH_2D_2, and CD_4 were employed.

5   CALCULATED RESULTS

Normal practice in geochemistry is to report the isotopic composition of a substance VY_n in the delta notation,

$$\delta(VY_n) = \left| \frac{R_Y(VY_n)}{R_Y(ST)} - 1 \right| \times 1000,$$

where R_Y stands for the atomic ratio Y'/Y and ST refers to a standard. In case there are two substances VY_n and WY_n that are in isotopic exchange equilibrium one can define a fractionation factor

$$\alpha(VY_n, WY_m) = R(VY_n)/R(WY_m). \tag{61}$$

Table 6 Molecular constants of linear triatomic molecules (in cm^{-1})

	H^{12}C^{14}N	D^{12}C^{14}N	H^{13}C^{14}N	H^{12}C^{15}N	Ref.	^{14}N$_2$^{16}O	^{15}N$_2$^{16}O	^{14}N$_2$^{18}O	Ref.
ω_1	2128.67	1952.12	2094.16	2095.12		1298.66	1280.50	1255.30	
ω_2	727.10	579.85	720.71	726.01	a	596.36	579.09	591.65	b
ω_3	3441.16	2703.34	3422.04	3439.85		2282.00	2209.00	2275.60	
x_{11}	−10.45	−6.84	−10.27	−10.13		−4.30^{d}	−4.18	−4.02	
x_{12}	−3.61	3.01	−4.13	−3.46		0.20	0.19	0.19	
x_{13}	−14.61	−32.40	−12.90	−14.08	a	−25.56	−24.39	−24.63	c
x_{22}	−2.44	−2.08	−2.31	−2.46		−0.27	−0.25	−0.26	
x_{23}	−18.98	−15.80	−18.38	−18.91		−14.63	−13.75	−14.47	
x_{33}	−51.71	−20.50	−51.87	−51.88		−15.19	−14.24	−15.10	
B_o	1.47822	1.20775	1.43999	1.43527	a	0.419011	0.404859	0.395577	e

	^{16}O^{12}C^{32}S	^{16}O^{13}C^{32}S	^{18}O^{12}C^{32}S	^{16}O^{12}C^{34}S	Ref.	^{12}C^{32}S$_2$	^{13}C^{32}S$_2$	^{12}C^{34}S$_2$	Ref.
ω_1	875.70	871.15	854.13	863.89		671.89	671.89	651.39^{f}	
ω_2	523.62	508.03	518.09	522.83	f	398.38	385.22	396.52	h
ω_3	2092.46	2037.45	2055.86	2091.77		1558.73	1507.25	1551.48	
x_{11}	−4.32^{f}	−4.28	−4.11	−4.20		−0.83	−0.83	−0.78^{c}	
x_{12}	−7.16	−6.91	−6.91	−7.05		−1.75	−1.68	−1.64	
x_{13}	−2.97	−2.88	−2.85	−2.93	c	−7.59	−7.36	−7.32	h
x_{22}	1.78	1.68	1.74	1.77		0.82	0.77	0.81	
x_{23}	−6.53	−6.17	−6.35	−6.52		−6.49	−6.04	−6.43	
x_{33}	−10.83	−10.27	−10.45	−10.82		−6.54	−6.10	−6.48	
B_o	0.202856	0.202202	0.190192	0.197897	g	0.109092	0.109092	0.102686	i

Table 6 (continued)

	$^{12}C^{16}O_2$	$^{13}C^{16}O_2$	Ref.	$^{12}C^{16}O_2$	$^{12}C^{18}O_2$	Ref.
ω_1	1353.637	1353.680		1354.91[1]	1277.25	m
ω_2	672.625	653.771	j	673.00	662.70	
ω_3	2396.269	2328.037		2396.49	2359.81	
x_{11}	−2.885	−2.885		−3.75[1]	−3.33	
x_{12}	−5.416	−5.247		3.62	3.39	
x_{13}	−19.205	−18.731	j	−19.37	−17.98	c
x_{22}	1.642	1.546		−0.63	−0.61	
x_{23}	−12.515	−11.728		−12.53	−12.15	
x_{33}	−12.564	−11.797		−12.63	−12.25	
B_0	0.390210	0.390250	k	0.390210	0.346820	k

[a] Nakagawa & Morino (1969).
[b] Jones (1970).
[c] Calculated with Equation 59.
[d] Pliva (1968).
[e] Griggs et al (1968).
[f] Foord et al (1975).
[g] Morino & Nakagawa (1968).
[h] Giguère et al (1973).
[i] Foss Smith & Overend (1971).
[j] Jobard & Chedin (1975).
[k] Courtoy (1959).
[l] Courtoy (1957).
[m] Calculated (see text).

Table 7 Molecular constants of bent triatomic molecules (in cm^{-1})

	$H_2{}^{16}O$	$HD^{16}O$	$D_2{}^{16}O$	$H_2{}^{18}O$	Ref.	$H_2{}^{32}S$	$HD^{32}S$	$D_2{}^{32}S$	$H_2{}^{34}S$	Ref.
ω_1	3835.37	2823.19	2762.84	3827.59	a	2722.28	1958.34	1952.27	2720.07	d
ω_2	1647.59	1444.53	1206.72	1640.62		1214.38	1057.29	872.10	1213.24	
ω_3	3938.74	3888.63	2885.99	3922.69		2733.75	2727.91	1964.21	2731.20	
x_{11}	−44.18	−41.51	−21.94	−44.01	a	−25.09[e]	−12.98	−12.91	−25.05	f
x_{12}	−16.20	−16.93	−8.77	−16.11		−19.69	−12.33	−10.14	−19.65	
x_{13}	−164.49	−12.91	−85.76	−163.49		−94.68	−67.96	−48.80	−94.51	
x_{22}	−17.22	−11.90	−9.46	−17.07		−5.72	−4.34	−2.95	−5.71	
x_{23}	−20.11	−20.08	−10.17	−19.94		−21.09	−18.32	−10.88	−21.05	
x_{33}	−45.12	−82.34	−24.99	−44.74		−24.00	−23.90	−12.39	−23.95	
A_o	27.88039	23.41107	15.41988	27.53087	b	10.34656	9.75179	5.48950	10.44493[h]	g
B_o	14.51317	9.09802	7.27059	14.51234		9.03572	4.93214	4.51580	9.13083	
C_o	9.28551	6.41288	4.48756	9.24608		4.72680	3.22570	2.44316	4.87189	
δ_1	0.065	0.050	0.045		c	0.028[e]	0.020	0.020		i
δ_2	−0.102	−0.082	−0.077			−0.033	−0.029	−0.024		
δ_3	0.066	0.062	0.055			0.094	0.093	0.067		

	$^{14}N^{16}O_2$	$^{15}N^{16}O_2$	Ref.	$^{14}N^{16}O_2$	$^{14}N^{18}O_2$	Ref.	$^{32}S^{16}O_2$	$^{32}S^{18}O_2$	Ref.	$^{32}S^{16}O_2$	$^{34}S^{16}O_2$	Ref.
ω_1	1355.90	1338.50		1357.78	1306.71		1167.60	1113.98		1166.92	1160.37	
ω_2	756.80	747.10	j	756.90	728.73	l	526.27	505.30	n	523.39	518.55	p
ω_3	1663.50	1628.30		1665.82	1633.96		1380.91	1335.18		1380.96	1363.52	
x_{11}	−8.1	−7.4		−8.1[j]	−7.5	f	−3.99	−3.15	n	−3.99[n]	−3.94	f
x_{12}	−9.7	−9.5		−9.7	−9.0		−2.05	−1.20		−2.05	−2.02	
x_{13}	−28.7	−27.7	j	−28.7	−27.1		−13.71	−12.85		−13.71	−13.46	
x_{22}	−0.5	−0.4		−0.5	−0.5		−3.00	−3.20		−3.00	−2.94	
x_{23}	−2.7	−2.6		−2.7	−2.5		−3.90	−3.10		−3.90	−3.81	
x_{33}	−15.6	−15.3		−15.6	−15.0		−5.17	−4.65		−5.17	−5.04	
A_0	8.00251	7.63062	k			m	2.03735	1.91414	o	2.02735	1.96773	o
B_0	0.43367	0.43372					0.34417	0.30589		0.34417	0.34418	
C_0	0.41049	0.40949					0.29353	0.26317		0.29353	0.29225	

[a] Khachkuruzov (1959).

[b] Cook et al (1974).

[c] Benedict et al (1956).

[d] Calculated (see text) with force constants derived by Gamo (1967).

[e] Allen & Plyler (1956).

[f] Calculated with Equation 59.

[g] H_2S: Cook et al (1973); HDS: Helminger et al (1971); D_2S: Helminger et al (1972).

[h] Burrus & Gordy (1953); their constants for H_2 ^{32}S ($A_0 = 10.55676$, $B_0 = 9.22344$, $C_0 = 4.92127$) were used for sulphur exchange

[i] Calculated with Equation 60.

[j] Blanck & Hause (1970).

[k] Olman & Hause (1968).

[l] Calculated (see text) with force constants derived by Papousek & Pliva (1964).

[m] $\beta(NO_2)$ for nitrogen exchange was computed using the Teller–Redlich product rule.

[n] $S^{16}O_2$: Shelton et al (1953); $S^{18}O_2$: Barbe & Jouve (1971).

[o] Steenbeckeliers (1968) for $^{32}S^{16}O_2$; De Hemptinne et al (1963) for $^{32}S^{18}O_2$; Van Riet (1964) for $^{34}S^{16}O_2$.

[p] Calculated (see text) with force constants derived by Barbe & Jouve (1971).

When the delta values are small we have

$$1000 \ln \alpha(VY_n, WY_m) \simeq \delta(VY_n) - \delta(WY_m) = \Delta(VY_n, WY_m).$$

In the case of simple molecules, when n and m are unity, $\alpha(VY_n, WY_m)$ is equal to the equilibrium constant for the exchange reaction,

$$VY + WY' = VY' + WY,$$

but when the element V has one or more isotopes and n and/or m are not unity, the fractionation factor is not any longer equal to an equilibrium constant. Examples of this are carbon dioxide and methane discussed by Urey (1947) and Bottinga (1969), respectively.

In general we have

$$\alpha(VY_n, WY_m) = K(VY_n, WY_m)^{1/mn} X(VY_n)/X(WY_m), \tag{62}$$

where

$$K(VY_n, WY_m) = \left[\frac{(VY'_n)}{(VY_n)}\right]^m \bigg/ \left[\frac{(WY'_m)}{(WY_m)}\right]^n$$

is the equilibrium constant for the reaction

$$mVY_n + nWY'_m = mVY'_n + nWY_m.$$

Table 8   Molecular constants of tetra- and pentatomic molecules (in cm^{-1})[a]

	$^{32}S^{16}O_3$	$^{32}S^{18}O_3$	$^{34}S^{16}O_3$	Ref.	$^{14}NH_3$	$^{14}ND_3$	$^{15}NH_3$	Ref.
ω_1	1064.89	1003.85	1064.89		3505.7	2498.9	3503.9	
ω_2	505.97	494.58	496.97		1022.0	778.4	1016.4	
ω_3	1410.00	1368.08	1390.33	b	3573.1	2630.0	3563.4	b
ω_4	535.62	508.69	533.53		1689.7	1226.5	1686.4	
x_{11}	-1.4^d	-1.2	-1.4		-30.0^f	-15.2	-30.0	
x_{12}	5.1	4.7	5.0		20.6	11.2	20.5	
x_{13}	-6.9	-6.3	-6.8		-92.0	-48.3	-91.7	
x_{14}	10.3	9.2	10.2		-6.7	-3.5	-6.7	
x_{22}	-3.2	-3.1	-3.1		-51.8	-30.0	-51.2	
x_{23}	-2.4	-2.2	-2.3	c	32.4	18.2	32.1	c
x_{24}	-2.2	-2.0	-2.1		-10.7	-5.9	-10.6	
x_{33}	-4.6	-4.4	-4.5		-18.5	-10.0	-18.4	
x_{34}	-2.6	-2.4	-2.6		-17.3	-9.2	-17.2	
x_{44}	-2.6	-2.4	-2.6		-8.0	-4.6	-8.8	
A_o	0.1746	0.1551	0.1746		9.9444	5.1426	9.9209	
B_o	0.1746	0.1551	0.1746	e	9.9444	5.1426	9.9209	b
C_o	0.3491	0.3103	0.3491		6.1960	3.1170	6.1960	
δ_1					0.040	0.028		
δ_2					0.019	0.014		
δ_3					0.034	0.025		f
δ_4					-0.036	-0.026		

Table 8 (continued)

	$^{14}NH_2D$	$^{14}NHD_2$	Ref.	$^{12}CH_4$	$^{12}CD_4$	$^{13}CH_4$	$^{12}CH_3D$	Ref.
ω_1	2587.9	2544.1		3143.0	2223.3	3143.0	2306.1	
ω_2	943.5	857.2		1573.0	1112.7	1573.0	1512.1	
ω_3	3528.0	3550.2	b	3154.0	2333.0	3143.0	3146.0	b
ω_4	3572.8	2630.1		1357.0	1026.8	1348.5	3153.6	
ω_5	1657.2	1279.2					1352.2	
ω_6	1439.1	1513.3					1197.3	
x_{11}	-16.3	-15.8		-64.6	-32.3	-64.6	-34.8	
x_{12}	14.0	12.5		0.0	0.0	0.0	0.0	
x_{13}	-67.1	-66.3		-65.0	-34.0	-64.8	-47.6	
x_{14}	-67.9	-49.1		0.0	0.0	0.0	-47.7	
x_{15}	-4.9	-3.7					0.0	
x_{16}	-4.2	-4.4					0.0	
x_{22}	-44.1	-36.4		0.0	0.0	0.0	0.0	
x_{23}	29.5	27.0		-15.0	-7.8	-14.9	-14.4	
x_{24}	29.9	20.0		-11.2	-6.0	-11.1	-14.4	
x_{25}	-9.7	-6.8					-10.7	
x_{26}	-8.4	-8.0					-9.5	
x_{33}	-18.0	-18.3	c	-17.5	-9.6	-17.4	-17.4	b
x_{34}	-18.3	-13.5		-12.0	-6.7	-11.9	-17.5	
x_{35}	-16.8	-13.0					-11.9	
x_{36}	-14.5	-15.4					-10.6	
x_{44}	-18.5	-10.0		-6.0	-3.4	-5.9	-17.5	
x_{45}	-17.0	-9.6					-12.0	
x_{46}	-14.7	-11.4					-10.6	
x_{55}	-8.5	-5.0					-6.0	
x_{56}	-7.4	-6.0					-5.3	
x_{66}	-6.4	-7.1					-4.7	
A_o	9.6805	7.4460		5.2406	2.6330	5.2406	5.2507	
B_o	6.3936	5.3231	b	5.2406	2.6330	5.2406	3.8801	g
C_o	4.6620	3.7179		5.2406	2.6330	5.2406	3.8801	
δ_1	0.029	0.029		0.025^h	0.017^h		0.018	
δ_2	0.017	0.016		-0.050	-0.035		-0.048	
δ_3	0.033	0.034	i	0.015	0.011		0.015	i
δ_4	0.034	0.025		0.043	0.032		0.015	
δ_5	-0.035	-0.027					0.043	
δ_6	-0.030	-0.032					0.038	

^a ω_3 and ω_4 are doubly degenerate vibrations for XY_3 molecules; ω_2 is doubly, ω_3 and ω_4 are triply degenerate for XY_4 molecules; ω_2, ω_4, and ω_6 are doubly degenerate for XY_3Y' molecules.
^b Calculated (see text).
^c Calculated with Equation 59.
^d Dorney et al (1973).
^e Kaldor et al (1973).
^f Benedict & Plyler (1957).
^g CH_4: Thomas & Welsh (1960); CD_4: Olafson et al (1961); CH_3D: Olson (1972).
^h Calculated with the data of Ref. g and those of Kutchisu & Bartell (1962).
ⁱ Calculated with Equation 60.

We call the factors X in Equation 62 excess factors. They are functions of the internal equilibrium constants involving the molecule VY_n for $X(VY_n)$ and WY_m for $X(WY_m)$ and the natural abundance ratio Y/Y', and if V has isotopes the natural abundances of these isotopes enter also into the expression for $X(VY_n)$. The dependence of the excess factors on the natural abundance ratios is weak. To tabulate our results we have made use of the beta factor defined as

$$\beta(VY_n, Y) = K(VY_n, Y)X(VY_n),$$

where K is the equilibrium constant for the reaction

$$\frac{1}{n}VY_n + Y' = \frac{1}{n}VY_n' + Y.$$

Physically, the β factor is the fractionation factor between VY_n and Y, i.e.

$$\beta = \left[\frac{Y'}{Y}\right]_{VY_n} \Big/ \left[\frac{Y'}{Y}\right].$$

Table 9 β factors for hydrogen exchange

$T°C$	NH_3	H_2O	$CH_4{}^a$	$CH_4{}^b$	HF	HCN	H_2S	HCl	H_2
0	16.1336	15.1307	14.0578	12.8201	13.4815	11.0964	6.7076	6.3512	3.8161
10	14.3395	13.5308	12.5241	11.4585	12.1573	9.9831	6.1713	5.8795	3.6197
20	12.8490	12.1930	11.2473	10.3214	11.0407	9.0497	5.7107	5.4716	3.4460
30	11.5980	11.0635	10.1737	9.3626	10.0907	8.2597	5.3120	5.1163	3.2915
40	10.5385	10.1014	9.2629	8.5470	9.2755	7.5852	4.9644	4.8046	3.1533
50	9.6336	9.2753	8.4838	7.8475	8.5707	7.0045	4.6594	4.5295	3.0290
75	7.8753	7.6572	6.9666	6.4801	7.1754	5.8625	4.0416	3.9671	2.7674
100	6.6177	6.4873	5.8783	5.4943	6.1520	5.0319	3.5749	3.5370	2.5594
125	5.6866	5.6129	5.0708	4.7597	5.3775	4.4076	3.2128	3.1994	2.3905
150	4.9775	4.9414	4.4546	4.1970	4.7759	3.9255	2.9254	2.9288	2.2508
175	4.4242	4.4135	3.9733	3.7559	4.2981	3.5446	2.6930	2.7080	2.1336
200	3.9835	3.9903	3.5895	3.4033	3.9115	3.2377	2.5021	2.5249	2.0340
250	3.3326	3.3598	3.0221	2.8800	3.3285	2.7773	2.2088	2.2404	1.8741
300	2.8810	2.9177	2.6282	2.5152	2.9136	2.4519	1.9961	2.0311	1.7516
350	2.5534	2.5942	2.3424	2.2497	2.6059	2.2120	1.8363	1.8718	1.6550
400	2.3072	2.3492	2.1278	2.0497	2.3702	2.0293	1.7128	1.7474	1.5770
450	2.1169	2.1585	1.9622	1.8950	2.1848	1.8863	1.6152	1.6481	1.5128
500	1.9665	2.0069	1.8313	1.7725	2.0359	1.7721	1.5366	1.5674	1.4591
600	1.7458	1.7826	1.6398	1.5928	1.8131	1.6023	1.4189	1.4454	1.3748
700	1.5938	1.6269	1.5083	1.4690	1.6560	1.4837	1.3362	1.3586	1.3117
800	1.4842	1.5138	1.4138	1.3798	1.5405	1.3972	1.2757	1.2946	1.2632
900	1.4024	1.4289	1.3436	1.3133	1.4529	1.3320	1.2301	1.2461	1.2249
1100	1.2902	1.3120	1.2477	1.2221	1.3308	1.2419	1.1670	1.1785	1.1689
1300	1.2185	1.2370	1.1869	1.1638	1.2515	1.1837	1.1265	1.1348	1.1308

[a] The excess factor needed to calculate β was computed in the harmonic approximation, as explained in the text (see Section 7).

[b] The excess factor was calculated anharmonically.

In terms of β factors expression 62 becomes

$$\alpha(VY_n, WY_m) = \beta(VY_n, Y)/\beta(WY_m, Y). \tag{63}$$

In terms of partition functions β may be expressed as

$$\beta(VY_n, Y) = \left[\frac{Q(VY_n')}{Q(VY_n)}\right]^{1/n} \bigg/ \left[\frac{m'}{m}\right]^{3/2} X(VY_n)$$

$$= f^{1/n}(VY_n', VY_n)X(VY_n),$$

where f is the reduced partition function ratio (Urey 1947, Bigeleisen & Mayer 1947). The need for this terminology arises from the fact that stable isotope geochemists measure atomic isotopic ratios instead of isotopic molecular abundances. It should be appreciated that often the excess factors are very nearly unity.

In Tables 9–14, we list our calculated β factors for hydrogen, carbon, nitrogen, oxygen sulphur and chlorine isotope exchange. Geochemical applications of these data are the subject of a forthcoming publication. The extrapolation of these results to low or high temperatures is beset with problems. Frequently the β factors and the fractionation factors that one may calculate from them do not linearly decrease

Table 10 β factors for carbon exchange

$T°C$	CO_2	OCS	HCN	CS_2	CH_4	CO	CS
0	1.2171	1.1695	1.1431	1.1302	1.1262	1.1095	1.0692
10	1.2060	1.1608	1.1363	1.1232	1.1204	1.1045	1.0657
20	1.1958	1.1527	1.1301	1.1168	1.1150	1.0999	1.0624
30	1.1864	1.1453	1.1244	1.1109	1.1100	1.0957	1.0593
40	1.1777	1.1384	1.1190	1.1055	1.1053	1.0917	1.0565
50	1.1697	1.1321	1.1140	1.1005	1.1009	1.0880	1.0538
75	1.1519	1.1180	1.1030	1.0893	1.0912	1.0796	1.0480
100	1.1370	1.1062	1.0936	1.0800	1.0829	1.0725	1.0430
125	1.1243	1.0962	1.0854	1.0720	1.0758	1.0663	1.0387
150	1.1133	1.0875	1.0784	1.0651	1.0696	1.0609	1.0350
175	1.1038	1.0800	1.0722	1.0592	1.0641	1.0561	1.0318
200	1.0954	1.0734	1.0667	1.0540	1.0593	1.0519	1.0290
250	1.0815	1.0625	1.0576	1.0455	1.0512	1.0447	1.0244
300	1.0705	1.0538	1.0502	1.0388	1.0446	1.0389	1.0207
350	1.0615	1.0468	1.0441	1.0334	1.0393	1.0341	1.0178
400	1.0541	1.0410	1.0391	1.0290	1.0348	1.0302	1.0155
450	1.0480	1.0363	1.0349	1.0254	1.0311	1.0268	1.0135
500	1.0428	1.0323	1.0313	1.0225	1.0279	1.0240	1.0119
600	1.0346	1.0260	1.0255	1.0178	1.0228	1.0195	1.0094
700	1.0285	1.0213	1.0212	1.0145	1.0190	1.0161	1.0076
800	1.0238	1.0178	1.0179	1.0119	1.0160	1.0135	1.0063
900	1.0202	1.0151	1.0153	1.0100	1.0137	1.0114	1.0053
1100	1.0150	1.0110	1.0115	1.0072	1.0103	1.0085	1.0038
1300	1.0115	1.0086	1.0089	1.0054	1.0079	1.0066	1.0029

with T^{-2}, where T is the absolute temperature (see Stern et al 1968, Spindel et al 1970; Bottinga 1968a and 1969).

When using the β factors in Tables 9–14, the difference between equilibrium constants and β factors should be kept in mind. For example, for the simple reaction

$$HD + HF = H_2 + DF, \tag{64}$$

we have given on line 13 in Table 4 the values of the equilibrium constant K_{64} at 0°C. A geochemist who has in a container a gaseous mixture of hydrogen and hydrogen fluoride measures the ratios $(D/H)_{\text{hydrogen fluoride}}$ and $(D/H)_{\text{hydrogen}}$. Assuming that the container was at 0°C and the gases were in isotopic exchange equilibrium, the fractionation factor $\alpha = (D/H)_{\text{hydrogen fluoride}}/(D/H)_{\text{hydrogen}}$ (Equation 61) should be equal to the ratio of β factors for hydrogen fluoride over that of hydrogen (Equation 63), which at 0°C is equal to 3.533, while the equilibrium constant in Table 4 is 1.771 or roughly half of the fractionation factor. This is because we have

$$\alpha = \frac{[DF]}{[HF]} \bigg/ \frac{[HD] + 2[D_2]}{2[H_2] + [HD]} = \frac{[DF]}{[HF]} \bigg/ \left\{ \frac{[HD]}{[H_2]} \frac{1 + 2[D_2]/[HD]}{2 + [HD]/[H_2]} \right\},$$

and since $D/H \leqq 1$, we also have $\alpha \simeq 2K_{64}$.

Table 11 β factors for nitrogen exchange

$T°C$	NO_2	N_2O	N_2	HCN	NH_3	NO
0	1.1224	1.1213	1.0902	1.0801	1.0778	1.0743
10	1.1162	1.1153	1.0862	1.0765	1.0744	1.0709
20	1.1105	1.1097	1.0825	1.0731	1.0713	1.0677
30	1.1052	1.1046	1.0791	1.0700	1.0683	1.0648
40	1.1003	1.0998	1.0759	1.0671	1.0656	1.0620
50	1.0957	1.0954	1.0729	1.0643	1.0631	1.0594
75	1.0856	1.0856	1.0662	1.0582	1.0574	1.0537
100	1.0770	1.0772	1.0604	1.0530	1.0525	1.0488
125	1.0696	1.0701	1.0554	1.0485	1.0482	1.0445
150	1.0632	1.0640	1.0510	1.0445	1.0446	1.0407
175	1.0577	1.0586	1.0471	1.0410	1.0413	1.0374
200	1.0528	1.0539	1.0437	1.0379	1.0384	1.0345
250	1.0448	1.0460	1.0378	1.0327	1.0336	1.0296
300	1.0384	1.0397	1.0330	1.0284	1.0297	1.0256
350	1.0333	1.0346	1.0291	1.0249	1.0264	1.0224
400	1.0291	1.0304	1.0258	1.0221	1.0237	1.0197
450	1.0257	1.0269	1.0230	1.0196	1.0215	1.0174
500	1.0228	1.0240	1.0206	1.0176	1.0195	1.0155
600	1.0183	1.0193	1.0168	1.0143	1.0165	1.0125
700	1.0150	1.0159	1.0140	1.0119	1.0141	1.0103
800	1.0125	1.0132	1.0117	1.0100	1.0123	1.0086
900	1.0106	1.0112	1.0100	1.0085	1.0108	1.0073
1100	1.0079	1.0083	1.0075	1.0064	1.0088	1.0054
1300	1.0061	1.0063	1.0058	1.0050	1.0074	1.0041

Table 12 β factors for oxygen exchange

$T°C$	CO_2	CO	OCS	SO_3	NO	N_2O	SO_2	NO_2	O_2	H_2O
0	1.1333	1.1191	1.1296	1.1169	1.1118	1.1092	1.1053	1.1027	1.0874	1.0705
10	1.1265	1.1137	1.1225	1.1104	1.1066	1.1035	1.0997	1.0973	1.0831	1.0675
20	1.1203	1.1087	1.1168	1.1045	1.1017	1.0982	1.0945	1.0923	1.0791	1.0648
30	1.1145	1.1040	1.1111	1.0990	1.0972	1.0933	1.0896	1.0878	1.0754	1.0622
40	1.1091	1.0997	1.1059	1.0940	1.0930	1.0888	1.0852	1.0835	1.0719	1.0598
50	1.1041	1.0956	1.1010	1.0893	1.0891	1.0846	1.0810	1.0796	1.0687	1.0576
75	1.0931	1.0865	1.0902	1.0790	1.0803	1.0753	1.0719	1.0709	1.0614	1.0526
100	1.0837	1.0788	1.0811	1.0704	1.0728	1.0674	1.0642	1.0635	1.0553	1.0483
125	1.0757	1.0720	1.0734	1.0631	1.0663	1.0608	1.0577	1.0572	1.0500	1.0445
150	1.0688	1.0661	1.0667	1.0569	1.0607	1.0550	1.0521	1.0518	1.0453	1.0412
175	1.0628	1.0609	1.0609	1.0515	1.0557	1.0500	1.0472	1.0471	1.0413	1.0383
200	1.0575	1.0563	1.0558	1.0469	1.0513	1.0456	1.0430	1.0430	1.0378	1.0357
250	1.0488	1.0485	1.0474	1.0393	1.0439	1.0385	1.0361	1.0362	1.0318	1.0313
300	1.0419	1.0422	1.0408	1.0334	1.0380	1.0328	1.0307	1.0309	1.0271	1.0277
350	1.0363	1.0370	1.0354	1.0287	1.0331	1.0283	1.0264	1.0266	1.0234	1.0247
400	1.0318	1.0327	1.0310	1.0250	1.0291	1.0246	1.0230	1.0232	1.0203	1.0222
450	1.0280	1.0291	1.0273	1.0219	1.0258	1.0216	1.0202	1.0203	1.0177	1.0201
500	1.0249	1.0260	1.0243	1.0193	1.0230	1.0191	1.0178	1.0180	1.0156	1.0182
600	1.0200	1.0211	1.0195	1.0155	1.0185	1.0153	1.0143	1.0143	1.0123	1.0152
700	1.0163	1.0174	1.0160	1.0126	1.0152	1.0124	1.0117	1.0116	1.0098	1.0129
800	1.0136	1.0146	1.0133	1.0106	1.0127	1.0103	1.0097	1.0096	1.0080	1.0110
900	1.0115	1.0124	1.0112	1.0090	1.0107	1.0087	1.0083	1.0081	1.0066	1.0095
1100	1.0085	1.0092	1.0083	1.0067	1.0079	1.0064	1.0062	1.0059	1.0047	1.0073
1300	1.0065	1.0071	1.0063	1.0053	1.0061	1.0049	1.0049	1.0045	1.0034	1.0058

In Table 15, we have tabulated certain similarities and differences in the isotopic fractionation behavior of various molecules. Oxygen β values for CO_2, COS, and CO are very much alike, as are sulphur β values for CS_2, COS, and CS. Such similarities are due to the fact that by means of a valence force field one can adequately explain the observed vibrational frequencies for these molecules and that the valence force-field force constant for a certain bond is virtually the same for each molecule that possesses this bond. For instance, the force constant for the vibrational frequency of CO is nearly the same as that for the stretching vibrational mode for CO_2. Oxygen in CO, COS, and CO_2 and sulphur in CS, COS, and CS_2 are in equivalent positions with the result that the isotopic fractionation for sulphur or oxygen among these molecules is very small. In contrast with this one notices a strong fractionation for carbon isotopes between CO and CO_2 and for nitrogen isotopes between NO and NO_2. This can be understood as being due to the fact that the position of N in NO and NO_2 is different, as is the position of C in CO and CO_2.

Table 15 also shows clearly that oxygen β factors for O_2, H_2O, CO_2, COS, and CO are larger than sulphur β factors for the corresponding series of molecules $S_2, H_2S, CS_2, COS, and CS. This illustrates very well that stable isotope fractionation diminishes with a diminishing relative mass difference between the stable isotopes in equivalent positions.

When the results are compared with previously made calculations, then one may notice a certain amount of agreement as well as disagreement. The differences are due mainly to the following causes: 1. More recent and presumably more accurate spectroscopic data were used. 2. The spectroscopic data available at the moment are more complete than before. 3. Only approximations were applied, which were shown to be harmless. 4. The use of the Teller-Redlich product rule was avoided. 5. Excess factors were calculated. 6. No use was made of the hydrogen isotope data by Haar et al (1961) as was done for example by Bottinga (1969). 7. Anharmonicity of methane was taken care of, but in a different way from Bottinga (1969).

In Table 16 a survey of comparisons with previously calculated results is given.

6 COMPARISON WITH EXPERIMENTAL DETERMINATIONS

In the internal equilibrium $H_2 + D_2 = 2HD$ the G_0 energy contribution should cancel. As remarked before, the constants available for hydrogen and its deuterated variations do not result in a zero G_0 term in the zero-point energy. The calculated

Table 13 β factors for sulphur exchange

$T°C$	SO_3	SO_2	OCS	CS_2	CS	S_2	H_2S
0	1.0928	1.0507	1.0211	1.0222	1.0192	1.0147	1.0125
10	1.0879	1.0480	1.0199	1.0209	1.0182	1.0138	1.0119
20	1.0834	1.0456	1.0188	1.0198	1.0173	1.0130	1.0114
30	1.0793	1.0434	1.0178	1.0188	1.0165	1.0122	1.0110
40	1.0754	1.0413	1.0169	1.0178	1.0157	1.0116	1.0105
50	1.0718	1.0393	1.0160	1.0169	1.0150	1.0109	1.0101
75	1.0640	1.0350	1.0141	1.0149	1.0134	1.0096	1.0092
100	1.0573	1.0314	1.0125	1.0133	1.0121	1.0085	1.0084
125	1.0516	1.0283	1.0112	1.0119	1.0109	1.0075	1.0077
150	1.0467	1.0256	1.0100	1.0107	1.0099	1.0067	1.0071
175	1.0425	1.0233	1.0091	1.0097	1.0090	1.0060	1.0066
200	1.0388	1.0212	1.0082	1.0089	1.0082	1.0054	1.0061
250	1.0327	1.0179	1.0069	1.0074	1.0069	1.0045	1.0053
300	1.0279	1.0152	1.0058	1.0063	1.0059	1.0038	1.0047
350	1.0241	1.0131	1.0050	1.0055	1.0051	1.0032	1.0041
400	1.0210	1.0114	1.0043	1.0048	1.0044	1.0028	1.0037
450	1.0185	1.0100	1.0037	1.0042	1.0038	1.0024	1.0033
500	1.0164	1.0088	1.0033	1.0037	1.0034	1.0021	1.0030
600	1.0132	1.0070	1.0026	1.0030	1.0027	1.0017	1.0024
700	1.0108	1.0057	1.0021	1.0024	1.0022	1.0013	1.0020
800	1.0091	1.0047	1.0018	1.0020	1.0018	1.0011	1.0017
900	1.0077	1.0039	1.0015	1.0017	1.0015	1.0009	1.0015
1100	1.0059	1.0029	1.0011	1.0013	1.0011	1.0007	1.0011
1300	1.0047	1.0022	1.0008	1.0010	1.0008	1.0005	1.0009

Table 14 β factors for chlorine exchange

$T°C$	0	25	50	100	200	400	700	1000	1300
Cl_2	1.0086	1.0073	1.0063	1.0048	1.0031	1.0016	1.0008	1.0005	1.0003
HCl	1.0055	1.0050	1.0045	1.0038	1.0028	1.0017	1.0010	1.0006	1.0004

Table 15 Similarities in β factors for similar molecules at 0°C

β(Hydrogen)		β(Carbon)		β(Nitrogen)		β(Oxygen)		β(Sulphur)	
HF	13.5								
HCl	6.4					O_2	1.09	S_2	1.01
H_2O	15.1					H_2O	1.07	H_2S	1.01
H_2S	6.7							H_2S	1.01
		CO_2	1.22	NO_2	1.12	CO_2	1.13		
		COS	1.17			COS	1.12	COS	1.02
		CS_2	1.13					CS_2	1.02
		CO	1.11	NO	1.07	CO	1.12	CS	1.02

Table 16 Comparison of calculated β factors at 25°C.

Authors

Hydrogen exchange	H_2O	HCl	H_2	
Urey (1947)	12.518	5.306	3.735	
This work	11.605	5.288	3.367	

Carbon exchange	CO_2	HCN	CO	
Urey (1947)	1.191	1.121	1.097	
Bottinga (1968a)	1.192	—	—	
This work	1.191	1.127	1.098	

Nitrogen exchange	N_2	NH_3	NO	HCN
Urey (1947)	1.081	1.069	1.066	1.066
This work	1.082	1.070	1.066	1.072

Oxygen exchange	CO_2	CO	SO_2	O_2	H_2O
Urey (1947)	1.117	1.105	1.089	1.082	1.067
Bottinga (1968a)	1.117	—	—	—	1.064
Saxena et al (1962)	1.117	—	1.088	—	1.069
This work	1.117	1.106	1.092	1.077	1.064

Sulphur exchange	SO_3	SO_2	OCS	CS_2	S_2	H_2S
Tudge & Thode (1950)	1.084	1.045	1.019	1.019	1.013	1.013
Sakai (1957)	—	1.053	—	—	—	1.013
This work	1.081	1.044	1.019	1.019	1.013	1.011

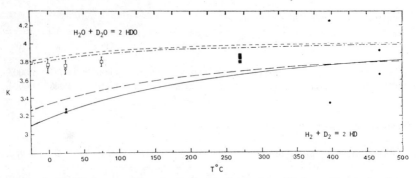

Figure 3 Equilibrium constant for hydrogen internal equilibrium. Observed values: ●, Rittenberg et al (1934); ■, Gould et al (1934); ▲, Niki et al (1965). Calculated curve: ——, this review. Equilibrium constant for water vapor internal equilibrium is also shown. Observed values: ○, Pyper et al (1967a); □, Friedman & Shiner (1966). Calculated curves (this review): · — · — ·, harmonic; — —, anharmonic without G_0; – – – –, anharmonic with G_0.

results are plotted on Figure 3 together with the experimental observations, and the agreement is good. Undoubtedly H_2 is spectroscopically one of the best known and most often looked at molecules. The molecular constants for H_2 derived in the past have shown a certain amount of divergence, which was well in excess of 1 cm^{-1}. At the moment the two most recent determinations are in almost perfect agreement (Foltz et al 1966, Buijs & Gush 1971); nevertheless, the difference in G_0 for these two sets amounts to 0.3 cm^{-1} while the difference in zero-point energy without G_0 contribution amounts to only 0.03 cm^{-1}. This illustrates very well how sensitive the computation of G_0 is to the input data. This sensitivity stems from the fact that G_0 is a small difference of large numbers (see Equation 17). In Table 17 we give certain values for G_0 and $\omega_e x_e/4$.

Table 17   Values for G_0 and $\omega_e x_e/4$ in cm^{-1}

Molecule	$\omega_e x_e/4$	G_0
H_2	30.32	8.60
HD	22.73	6.88
D_2	15.17	4.30
HF	22.47	4.13
DF	11.44	1.88
$H^{35}Cl$	13.20	1.60
$D^{35}Cl$	6.79	0.82
$^{14}N_2$	3.53	0.23
$^{15}N_2$	3.30	0.21
$^{16}O_2$	3.02	0.21
$^{18}O_2$	2.68	0.19
$C^{16}O$	3.32	0.19
$C^{18}O$	3.16	0.18

In Figure 3 we have also plotted our calculated results for the internal equilibrium

$$H_2O + D_2O = 2HDO. \tag{65}$$

It is evident that the anharmonic calculation, without taking into consideration G_0, does not agree with the experimental observation. For this reason Hulston (1969) and Wolfsberg (1969a) proposed that in the calculation of isotope effects the term G_0, neglected until then, should be included. For the bent triatomic molecule Wolfsberg (1969a) gave the equation

$$G_0 = G_{0av} + G_{0rv} \tag{66}$$

where

$$G_{0av} = \frac{1}{16} \left\{ \sum_i (6K_{iiii} - 7K_{iii}^2/\omega_i) - 4\omega_1\omega_2\omega_3 K_{123}^2 \Big/ \left[(\omega_1 + \omega_2 + \omega_3) \right. \right.$$
$$\left. \left. \times (\omega_1 - \omega_2 - \omega_3)(\omega_1 - \omega_2 + \omega_3)(\omega_1 + \omega_2 - \omega_3) \right] \sum_{i \neq j} \sum 3K_{iij}^2 \omega_j(4\omega_i^2 - \omega_j^2) \right\} \tag{67}$$

and

$$G_{0rv} = -(A_e + B_e + 3C_e)/4. \tag{68}$$

G_{0av} depends on the zero-order frequencies and the force constants while G_{0rv} depends on the rotational constants. Evaluation of Equation 68 does not pose any problem since the rotational constants are usually well known. Unfortunately it is the G_{0av} contribution that is important and difficult to evaluate in most cases because of lack of input information. As shown in Figure 3 the calculation in which G_0 has been used is in much better agreement than the calculation in which it has been ignored.

For the similar internal equilibrium $H_2S + D_2S = 2HDS$ we calculate an equilibrium constant of 3.88 (25°C). This coincides with the observed value by Pyper & Newbury (1970). Wolfsberg et al (1970), who included G_0 in their computation, have calculated a value of 3.93, which is outside the limits of the experimental accuracy. Because of the lack of detailed information in the paper by Wolfsberg et al (1970), it is difficult to discuss their calculated value, which would be of interest since they calculated an equilibrium constant for the hydrogen-sulfide internal equilibrium, excluding the G_0 effect, of 3.69; this differs considerably from our analogous value of 3.88.

The uncertainty we noted already in the calculated G_0 values for H_2 is appreciably larger for the triatomic molecules we have listed in Table 18. The maximum difference in ΔG_0 for water (Table 18) amounts to about 4 cm^{-1}, which means an uncertainty of 2% in the calculated equilibrium constant for reaction 65 at 25°C. Table 18 suggests that the uncertainty in the G_0 value of the individual molecules is important. ΔG_0 for water is appreciable because of the "anomalous" value for $G_0(HDO)$.

Table 18 G_0 values for triatomic molecules (cm^{-1})

		H_2O	HDO	D_2O	$\Delta G_0^{\,h}$	H_2S	D_2S	$^{14}NO_2$	$^{15}NO_2$	$^{32}S^{16}O_2$
G_{0rv}		-17.64	-12.87	-9.33	$+1.23$	-8.39	-4.33	-2.42	-2.32	-0.81
G_{0av}	a	27.33	7.08	14.48	-27.65	16.93		7.66		2.37
	b	21.46	3.15	10.99	-26.15	15.22		7.18		1.46
	c	22.02		7.05		14.85				1.69
	d	19.75		11.89						1.22
	e	22.39		11.39		11.92	6.05	5.05	5.08	1.29
	f	21.43	1.44	11.78	-30.33					
	g									1.16

a,b,c Papousek & Pliva (1964), with force fields a, SUBFF, b, MUBFF, c, EXP.
d,e Kutchisu & Morino (1965), with force fields d, OBS and e, CALC.
f Foss Smith & Overend (1972).
g Saito (1969).
h $\Delta G_0 = 2G_0(\text{HDO}) - G_0(\text{H}_2\text{O}) - G_0(\text{D}_2\text{O})$

Table 19 Ammonia internal equilibria

Equilibrium constant	Temperature °C	Experimental value	Calculated values		
		Pyper et al (1967[b])	Bron & Wolfsberg (1972)		This work
			with G_0	without G_0	without G_0
$K(69)$	0	2.90 ± 0.06	2.89	3.48	2.70
$K(69)$	25	2.92 ± 0.08	2.91	3.45	2.73
$K(70)$	0	2.86 ± 0.09	2.90	3.01	2.85
$K(70)$	25	2.89 ± 0.09	2.92	3.02	2.87

The internal equilibria

$$NH_3 + NHD_2 = 2NH_2D \tag{69}$$

and

$$ND_3 + NH_2D = 2NHD_2 \tag{70}$$

have been measured by Pyper et al (1967[b]).

Observed and calculated results for these equilibria are given in Table 19. For K_{69} the results by Bron & Wolfsberg (1972) with G_0 are in good agreement with the observations; the agreement with our calculations is less satisfactory. For K_{70} the agreement with the experimentally determined values of the Bron & Wolfsberg results with G_0 and our results without G_0 is to within the experimental error given by Pyper et al (1967[b]). There is a large disagreement between Bron & Wolfsberg's calculations without G_0 and ours. This is due to the fact that these calculations are done quite differently.

In Figure 4 we have plotted our calculated results for the equilibrium $HD + NH_3 = H_2 + NH_2D$ and the results computed by Bron & Wolfsberg (1972) together with various experimental measurements. The figure shows well the spread in values

Table 20 Deuterium fractionation among molecular hydrogen and hydrogen sulfide

Temperature °C	K(experimental)	K(calculated)
350	0.855	0.864
400	0.889	0.894
450	0.919	0.919
500	0.943	0.939
550	0.963	0.955
600	0.973	0.968

[a] Grafe et al (1939). $K = \dfrac{[D_2]}{[H_2]} \bigg/ \dfrac{[D_2S]}{[H_2S]}$

resulting from the various force fields used by Bron & Wolfsberg. Our calculations without G_0 agree with the observations, as is the case for Bron & Wolfsberg (1972) with force field II.

The experimental measurements of the equilibrium $H_2O + HD = HDO + H_2$ are plotted in Figure 5 together with calculated results. Calculations with and without G_0 are in equally good agreement with the observations. The perfect agreement

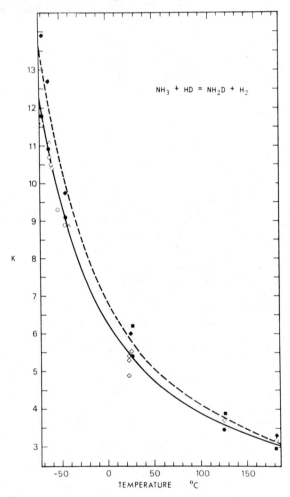

Figure 4 Equilibrium constant for $NH_3 + HD = NH_2D + H_2$. Observed values: $\Diamond$, Herrick & Sabi (1943); $\triangle$, Perlman et al (1953); $\bigcirc$, Ravoire et al (1963). Calculated values, Bron & Wolfsberg (1972): $\blacksquare$, anharmonic without G_0; $\bullet$, anharmonic with G_0 (force field II); $\blacklozenge$, anharmonic with G_0 (force field III). This review: $---$, harmonic; ———, anharmonic without G_0.

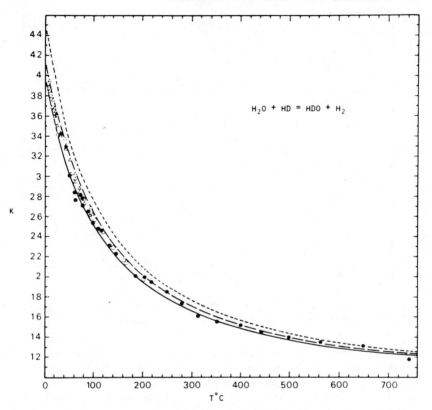

Figure 5   Equilibrium constant for $H_2O + HD = HDO + H_2$. Observed values: ○, Suess (1949); ●, Cerrai et al (1954); △, Rolston et al (1975). Calculated values (this review): – – – –, harmonic; ——, anharmonic without G_0; – –, anharmonic with G_0.

obtained with the calculations of Bottinga (1968b) is probably accidental; he used the now superseded free enthalpy tables of Haar et al (1961) for H_2 and HD in his calculations.

The equilibrium $D_2S + H_2 = H_2S + D_2$ has been measured indirectly by Grafe et al (1939). The measured values are compared with calculated values in Table 20, and the agreement is satisfactory. In Figure 6, we show experimental data for the equilibrium $H_2O + HDS = HDO + H_2S$ obtained by Marx (1960) and our calculations. Again the difference between observations and calculations is negligible.

The equilibrium

$$^{32}SO_2 + H_2^{34}S = {}^{34}SO_2 + H_2^{32}S \tag{71}$$

has been studied experimentally (Grinenko & Thode 1970, Thode et al 1971) and theoretically (Sakai 1968, Thode et al 1971). Previously obtained results and our

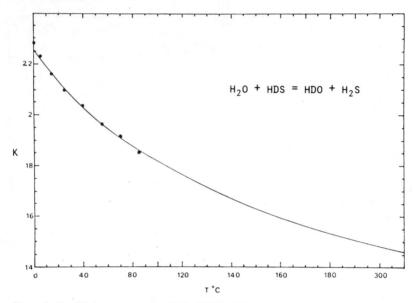

Figure 6   Equilibrium constant for $H_2O + HDS = HDO + H_2S$. Observed values: ●, Marx (1960). Calculated values:——, anharmonic without G_0, this review.

results are plotted in Figure 7. The experimental results show too much dispersion to allow a meaningful comparison with the calculations. Calculations by Thode et al and us are in good agreement. The small difference between these two calculations is due to the slightly different force fields used to compute the frequencies of the isotopically substituted molecules.

To compare our calculated results with equilibria involving liquid water instead of water vapor, we have made use of the most recent measurement of the fractionation factor α(steam-liquid water) = 1.00937 (Majoube 1971) at 25°C. In Table 21 we list the experimental values for the oxygen isotope exchange between CO_2 and liquid water and between NO and liquid water at 25°C. As is evident the calculations reproduce the observed results quite well. The difference between our result and the calculated $\alpha(CO_2 - H_2O) = 1.0411$ reported by Bottinga & Craig (1969) is mainly due to the use of the α(steam-liquid water) at 25°C of Majoube (1971), which is 0.2°/$_{oo}$ larger than the corresponding α used by Bottinga & Craig.

7   DISCUSSION AND CONCLUSIONS

At present the main error source in the calculation of fractionation factors is the uncertainty in the vibrational molecular constants. Errors as small as 1 cm^{-1} in the frequency difference between a vibrational mode of a normal and of an isotopically substituted molecule already have consequences in the evaluation of the β factor that are considerably larger than the accuracy limits with which one can

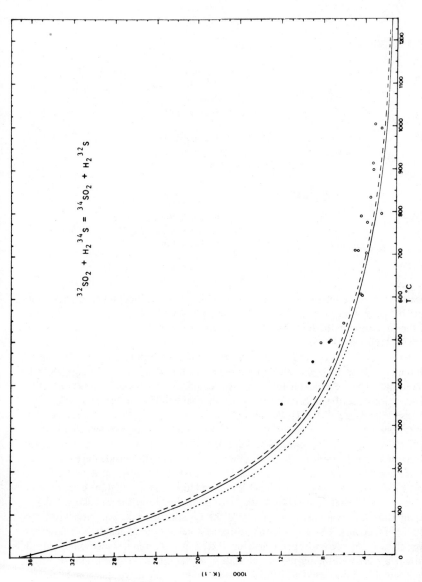

Figure 7 Equilibrium constant for $^{32}SO_2 + H_2{}^{34}S = {}^{34}SO_2 + H_2{}^{32}S$. Observed values: ●, Grinenko & Thode (1970); ○, Thode et al (1971). Calculated values:·····, Sakai (1968), — — —, Thode et al (1971), ———, this review.

Table 21 Oxygen isotope exchange between carbon dioxide, nitric oxide, and liquid water at 25°C

$\alpha(CO_2—H_2O)$

Observed	Calculated	Reference
1.0407		Compston & Epstein (1958)
1.0417		Majoube (1966)
1.0407		O'Neil & Epstein (1966)
1.0409		Bottinga & Craig (1969)
1.0414		Horibe et al (1973)
1.0417		Blattner (1973)
1.0412		O'Neil et al (1975)
	1.0408	This work

$\alpha(NO—H_2O)$

Observed	Calculated	Reference
1.025 ± 0.003		Jordan & Bonner (1971)
1.024 ± 0.003		Jordan & Bonner (1971)
1.027 ± 0.004		Jordan & Bonner (1971)
	1.024	This work

measure isotopic compositions. This is a very important restriction because spectral data of many molecules are only incompletely known, because of the complexity of the molecule or the fact that the spectrum is difficult to observe: such is the case for HDO and HDS.

The existence of a G_0 term in the zero-point energy is well established. However, its importance in isotopic calculations is less certain (Richet & Bottinga 1976a). The only exchange reaction for which we could not get agreement between the observations and the calculations was the internal equilibrium for water. Our imprecise knowledge of the HDO vibrational constants is here of importance and one also would like to see confirmed the experimental observations, which are quite difficult to perform. One notices (Figure 3) that even when G_0 is included in the calculation, the theoretical internal equilibrium constant falls outside the error limits of the experimentally determined one. In the case of the internal equilibrium of water it was observed by Wolfsberg (1969a) and Hulston (1969) that the contribution of the anharmonicity correction to the zero-point energy in the calculation of the equilibrium constant was comparable to the G_0 contribution. The data given by Wolfsberg et al (1970) suggest that this may be a general feature for internal equilibria involving hydrogen-deuterium exchange; see for instance on Figure 3 how relatively well the harmonic calculation of the internal equilibrium for water agrees with the measured equilibrium constant. Our calculated harmonic calculations for the internal equilibria of ammonia (Equations 69 and 70) are in agreement with the observations to within the experimental error limits. But for normal exchanges we find that our anharmonic calculation without G_0 is closer to the observed values than the harmonically calculated value. Similarly, we find that for hydrogen-deuterium fractionation between molecular hydrogen and water (Figure 4), the anharmonic calculations are superior to the harmonic ones. Therefore it seems advisable that when one cannot calculate G_0 for an internal equilibrium, one does the calculation in the harmonic approximation, i.e. one ignores in that case the terms on lines 2, 4, and 7 of Table 4 and does the same for polyatomic molecules. In the evaluation of excess factors, internal equilibrium constants are needed; for methane they were calculated in the anharmonic as well as in the harmonic approximation for this reason. But for normal isotopic exchange reactions anharmonicity should be taken into account.

The computation of G_0 demands a very precise spectral knowledge, because it is a small difference of large numbers. Often the required information is not available; hence for reasons of homogeneity we have ignored the G_0 effect. Comparison of our results with observations does not suggest that this neglect of G_0 has observable consequences.

For this review we have calculated β factors with the most recent data available and as accurately as is feasible at present. Comparison with previous calculations shows good agreement as well as divergencies. One of the reasons for this lack of agreement is that previous authors have sometimes applied approximations that are clearly not justified. It should be evident that the calculation amounts to the evaluation of a very small difference of two large energy terms. This is the free enthalpy difference between a molecule and its isotopically substituted equivalent. Such differences are far too small to be directly measurable with present day techniques of calorimetry. Because this energy difference is so very small one should be very careful when one applies approximations.

In the calculation of β factors there are no free parameters that one can change in order to obtain agreement with observations. However, there is still a choice of possible input data, i.e. the spectroscopic information. It was for this reason that we used the water data by Khachkuruzov (1959), rather than the different sets published by other authors [however, in this case, we have obtained with data by Benedict et al (1956) β factors for hydrogen exchange which are very similar to those listed in Table 9; see Richet & Bottinga (1976b)]. But once we made a choice, which most often was only based on the criteria given in Section V, we used the complete set of data for a given molecule, as given by the author. Notwithstanding these restrictions we have found that our calculated results are in good agreement with the observations, except for the internal equilibrium of water (Equation 65). But for other hydrogen–deuterium exchange reactions involving water no such problems were encountered.

To interpolate the values given in Tables 9–14, it is probably best to assume that the β factors vary linearly with T^{-2}. However, this linearity should not be relied upon to extrapolate the tabulated values. In many cases we noted that our results do not follow a T^{-2} law.

In general the value of the excess factors is very close to unity, except for the exchange of hydrogen–deuterium; in particular when molecular hydrogen, water, hydrogen sulphide, ammonia or methane are involved. This is caused by significant deviations from the classical value of the concerned internal equilibrium constants. As far as we are aware, our results form the most extensive and homogeneous set of stable isotope fraction factors that is available for gaseous substances. These data can be of use in many different fields. In volcanic gas research they may be applied in order to understand the genesis as well as the isotopic composition of the source region of these gases. In geothermal exploration the gas phase plays an important role; an understanding of its isotopic composition, without knowing the necessary fractionation factors, is not possible. In biochemistry the isotopic composition of gases produced by living organisms, such as CO_2 and CH_4 by bacteria, may be of interest. The listed fractionation factors serve to determine under which physical conditions these gases were formed. In cosmochemistry, one has measured the isotopic composition of certain interstellar gas clouds, stars, planets, and meteorites. Some of the observed variations may be explicable in terms of isotopic exchange reactions; in such case our listed fractionation factors should be of use.

Literature Cited

Allen, H. C., Plyler, E. K. 1956. *J. Chem. Phys.* 25:1132–36

Amano, T., Hirota, E. 1974. *J. Mol. Spectrosc.* 53:346–63

Anderson, T., Lassettre, E. N., Yost, D. M. 1936. *J. Chem. Phys.* 4:703–06

Babcock, H. D., Herzberg, L. 1948. *Ap. J.* 108:167–90

Barbe, A., Jouve, P. 1971. *J. Mol. Spectrosc.* 38:273–80

Bardo, R. D., Wolfsberg, M. 1975. *J. Chem. Phys.* 62:4555–58

Barrow, R. F., Dixon, R. N., Lagerquist, A., Wright, C. V. 1960. *Ark. Fys.* 18:543–62

Benedict, W. S., Gailar, N., Plyler, E. K. 1956. *J. Chem. Phys.* 24:1139–65

Benedict, W. S., Plyler, E. K. 1957. *Can. J. Phys.* 35:1235–41

Benesh, W. B., Van der Slice, J. T., Tilford, S. G., Wilkinson, P. G. 1965. *Ap. J.* 142:1227–40

Bigeleisen, J., Mayer, M. G. 1947. *J. Chem. Phys.* 15:261–67

Blanck, C. E., Hause, C. D. 1970. *J. Mol. Spectrosc.* 34:478–86

Blattner, P. 1973. *Geochim. Cosmochim. Acta* 37:2691–93

Born, M. 1951. *Nachr. Akad. Wiss. Göttingen Math. Phys. Kl.* p. 1

Born, M., Huang, K. 1954. *Dynamical Theory of Crystal Lattices.* London: Oxford Univ. Press. 430 pp.

Born, M., Oppenheimer, R. 1927. *Ann. Phys.* 84:457–84

Bottinga, Y. 1968a. *J. Phys. Chem.* 72:800–08

Bottinga, Y. 1968b. *J. Phys. Chem.* 72:4338–40

Bottinga, Y. 1969. *Geochim. Cosmochim. Acta* 33:49–64

Bottinga, Y., Craig, H. 1969. *Earth Planet. Sci. Lett.* 5:285–95

Bottinga, Y., Javoy, M. 1975. *Rev. Geophys. Space Phys.* 13:401–18

Bron, J., Wolfsberg, M. 1972. *J. Chem. Phys.* 57:2862–69

Buijs, H. L., Gush, H. D. 1971. *Can. J. Phys.* 49:2366–75

Burneau, A. 1974. *Spectrochim. Acta* 30A:1861–71

Burrus, C. A., Gordy, W. 1953. *Phys. Rev.* 92:274–77

Cerrai, E. Marchetti, C., Renzoni, R., Rosco, L., Silvestri, M., Villari, S. 1954. *Chem. Eng. Prog. Symp. Ser.* 50:271–80

Clyne, M. A. A., Coxon, J. A. 1970. *J. Mol. Spectrosc.* 33:381–406

Compston, W., Epstein, S. 1958. *Trans. Am. Geophys. Union* 39:511

Cook, R. L., De Lucia, F. C., Helminger, P. 1973. *J. Chem. Phys.* 56:4581–84

Cook, R. L., De Lucia, F. C., Helminger, P. 1974. *J. Mol. Spectrosc.* 53:62–76

Courtoy, C. P. 1957. *Can. J. Phys.* 35:608–48

Courtoy, C. P. 1959. *Ann. Soc. Sci. Bruxelles* 73:5–230

Craig, H. 1953. *Geochim. Cosmochim. Acta* 3:53–92

De Hemptinne, M., Van Riet, R., Defossez, A., Bruyninckx, F., Dachelet, P. 1963. *Ann. Soc. Sci. Bruxelles* 77:163–71

De Lucia, F. C., Helminger, P., Gordy, W. 1971. *Phys. Rev. A* 3:1849–57

Denbigh, K. 1971. *The Principles of Chemical Equilibrium.* London: Cambridge Univ. Press. 494 pp. 3rd ed.

Dorney, A. J., Hoy, A. R., Mills, I. M. 1973. *J. Mol. Spectrosc.* 45:253–60

Dunham, J. L. 1932. *Phys. Rev.* 41:721–31

Durie, R. A., Herzberg, G. 1960. *Can. J. Phys.* 38:806–18

Epstein, S., Mayeda, T. 1953. *Geochim. Cosmochim. Acta* 31:181–214

Farkas, A., Farkas, L. 1934. *Proc. R. Soc. London Ser. A* 144:467–80

Foltz, J. V., Rank, D. H., Wiggins, T. A. 1966. *J. Mol. Spectrosc.* 21:203–16

Foord, A., Smith, J. G., Whiffen, D. H. 1975. *Mol. Phys.* 29:1685–704

Foss Smith, D. Jr., Overend, J. 1971. *J. Chem. Phys.* 54:3632–39

Foss Smith, D. Jr., Overend, J. 1972. *Spectrochim. Acta Part A.* 28:471–83

Friedman, I. 1953. *Geochim. Cosmochim. Acta* 4:89–103

Friedman, L., Shiner, V. J. Jr. 1966. *J. Chem. Phys.* 44:4339–40

Gamo, I. 1967. *J. Mol. Spectrosc.* 23:472–75

Giguère, J., Wang, V. K., Overend, J., Cabana, A. 1973. *Spectrochim. Acta Part A.* 29:1197–1206

Gould, A. J., Bleakney, W., Taylor, H. S. 1934. *J. Chem. Phys.* 2:362–73

Grafe, D., Clusius, K., Kruis, A. 1939. *Z. Phys. Chem. Abt. B.* 43:1–19

Griggs, J. L. Jr., Rao, K. N., Jones, L. H., Potter, R. M. 1968. *J. Mol. Spectrosc.* 25:34–61

Grinenko, V. A., Thode, H. G. 1970. *Can. J. Earth Sci.* 7:1402–09

Haar, L., Bradley, J. C., Friedman, A. S. 1955. *J. Res. Nat. Bur. Stand.* 55:285–90

Haar, L., Friedman, A. S., Beckett, C. W. 1961. *Nat. Bur. Stand. U.S. Monogr.* 20

Helminger, P., Cook, R. L., De Lucia, F. C. 1971. *J. Mol. Spectrosc.* 40:125–36

Helminger, P., Cook, R. L., De Lucia, F. C. 1972. *J. Mol. Spectrosc.* 41:123–36

Herrick, C. E., Sabi, N. 1943. *SAM Rep. A 765*

Herzberg, G. 1945. *Molecular Spectra and Molecular Structure.* Vol. II, *Infrared and Raman Spectra.* New York: Van Nostrand. 632 pp.

Herzberg, G. 1950. *Molecular Spectra and Molecular Structure.* Vol. I, *Spectra of Diatomic Molecules.* New York: Van Nostrand. 658 pp. 2nd ed.

Horibe, Y., Shigehara, K., Takakuwa, Y. 1973. *J. Geophys. Res.* 78:2625–29

Huffaker, J. N. 1976a. *J. Chem. Phys.* 64:3175–81

Huffaker, J. N. 1976b. *J. Chem. Phys.* 64:4564–70

Hulston, J. R. 1969. *J. Chem. Phys.* 50:1483–84

Jobard, I., Chedin, A. 1975. *J. Mol. Spectrosc.* 57:464–79

Johns, J. W. C., McKellar, A. R. W., Weitz, D. 1974. *J. Mol. Spectrosc.* 51:539–45

Jones, L. H. 1970. *J. Mol. Spectrosc.* 34:108–12

Jordan, S., Bonner, F. T. 1971. *J. Chem. Phys.* 54:4963–64

Kaldor, A., Maki, A. G., Dorney, A. J., Mills, I. M. 1973. *J. Mol. Spectrosc.* 45:247–52

Khachkuruzov, G. A. 1959. *Gos. Inst. Prikl. Khim.* 42:51–131

Kleinman, L. I., Wolfsberg, M. 1973. *J. Chem. Phys.* 59:2043–53

Kleinman, L. I., Wolfsberg, M. 1974a. *J. Chem. Phys.* 60:4740–48

Kleinman, L. I., Wolfsberg, M. 1974b. *J. Chem. Phys.* 60:4749–54

Kolos, W. 1970. *Adv. Quantum Chem.* 5:98–133

Kolos, W., Wolniewicz, L. 1964. *J. Chem. Phys.* 41:3663–73

Kutchisu, K., Bartell, L. S. 1962. *J. Chem. Phys.* 36:2478–81

Kutchisu, K., Morino, Y. 1965. *Bull. Chem. Soc. Jpn.* 38:814–24

Lindemann, F. A. 1919. *Philos. Mag.* 38:173–81

Lindemann, F. A., Aston, F. W. 1919. *Philos. Mag.* 37:523–34

McKinney, C. R., McCrea, J. M., Epstein, S., Allen, H. A., Urey, H. C. 1950. *Rev. Sci. Instrum.* 21:724–30

Maillard, A. 1936. *C.R. Acad. Sci.* 203:804–06

Majoube, M. 1966. *J. Chim. Phys.* 63:563–68

Majoube, M. 1971. *J. Chim. Phys.* 68:1423–36

Marx, D. 1960. *C.E.A. Rep.* No. 1382. 58 pp.

Mayer, J. E. 1958. *Handb. Phys.* 12:73–204

Mayer, J. E., Mayer, M. G. 1940. *Statistical Mechanics.* New York: Wiley. 495 pp.

Morino, Y., Nagakawa, T. 1968. *J. Mol. Spectrosc.* 26:496–523

Morse, P. M. 1929. *Phys. Rev.* 34:57–64

Mulholland, H. P. 1928. *Proc. Camb. Philos. Soc.* 24:280–89

Murphey, B. F., Nier, A. O. 1941. *Phys. Rev.* 59:771–72

Nagakawa, T., Morino, Y. 1969. *Bull. Chem. Soc. Jpn.* 42:2212–19

Nielsen, H. H. 1951. *Rev. Mod. Phys.* 23:90–136

Nier, A. O., Gulbransen, E. A. 1939. *J. Am. Chem. Soc.* 61:697–98

Niki, H., Rousseau, Y., Mains, G. J. 1965. *J. Phys. Chem.* 69:45–52

Olafson, R. A., Thomas, M. A., Welsh, H. L. 1961. *Can. J. Phys.* 39:419–36

Olman, M. D., Hause, C. D. 1968. *J. Mol. Spectrosc.* 26:241–53

Olman, M. D., McNelis, M. D., Hause, C. D.

1964. J. Mol. Spectrosc. 14:62–78
Olson, W. B. 1972. J. Mol. Spectrosc. 43:
190–98
O'Neil, J. R., Epstein, S. 1966. J. Geophys.
Res. 71:4955–61
O'Neil, J. R., Adami, L. H., Epstein, S. 1975.
J. Res. US Geol. Survey 3:623–24
Papousek, D., Pliva, J. 1964. Collect. Czech.
Chem. Commun. 29:1973–97
Perlman, M., Bigeleisen, J., Elliott, N. 1953.
J. Chem. Phys. 21:70–72
Pliva, J. 1968. J. Mol. Spectrosc. 27:461–88
Pyper, J. W., Newbury, R. S. 1970. J. Chem.
Phys. 52:1966–71
Pyper, J. W., Newbury, R. S., Barton, G. W.
Jr. 1967a. J. Chem. Phys. 46:2253–57
Pyper, J. W., Newbury, R. S., Barton, G. W.
Jr. 1967b. J. Chem. Phys. 47:1179–82
Rank, D. H., Rao, B. S., Wiggins, T. A.
1965. J. Mol. Spectrosc. 17:122–30
Ravoire, J., Grandcollot, P., Dirian, G. 1963.
C.E.A. Rep. No. 2244. 32 pp.
Richet, P., Bottinga, Y. 1976a. C.R. Acad.
Sci. Ser. B 282:425–28
Richet, P., Bottinga, Y. 1976b. C.R. Acad.
Sci. Ser. D283:299–302
Rittenberg, D., Bleakney, W., Urey, H. C.
1934. J. Chem. Phys. 2:48–49
Roh, W. B., Rao, K. N. 1974. J. Mol.
Spectrosc. 49:317–21
Rolston, J. H., Den Hartog, J., Butter, J. B.
1975. A.E.C.L. Rep. 5025. 25 pp.
Rosenthal, J. E. 1935. Phys. Rev. 47:235–37
Saito, S. 1969. J. Mol. Spectrosc. 30:1–16
Sakai, H. 1957. Geochim. Cosmochim. Acta
12:150–69
Sakai, H. 1968. Geochim. Cosmochim. Acta
2:29–49
Saxena, S. C., Bhatnagar, D. N., Ramaswamy,
S. 1962. J. Chem. Eng. Data 7:240–42
Shaffer, W. H., Schuman, R. P. 1944. J.
Chem. Phys. 12:504–13
Shelton, R. D., Nielsen, H. H., Fletcher,
W. H. 1953. J. Chem. Phys. 21:2178–83
and 1954. J. Chem. Phys. 22:1791 (correc-
tion)

Silverman, S. R. 1951. Geochim. Cosmochim
Acta 2:26–42
Spanbauer, R. N., Rao, K. N. 1965. J. Mol.
Spectrosc. 16:100–02
Spindel, W., Stern, M. J., Monse, E. U.
1970. J. Chem. Phys. 52:2022–35
Steenbeckeliers, G. 1968. Ann. Soc. Sci.
Bruxelles 82:331–404
Stern, M. J., Spindel, W., Monse, E. U. 1968.
J. Chem. Phys. 48:2908–19
Stripp, K. F., Kirkwood, J. G. 1951. J.
Chem. Phys. 19:1131–33
Suess, H. 1949. Z. Naturforsch. Teil A 4:
328–32
Taylor, H. P. Jr. 1968. Contrib. Mineral.
Petrol. 19:1–71
Thode, H. G., Cragg, C. B., Hulston, J. R.,
Rees, C. E. 1971. Geochim. Cosmochim.
Acta 35:35–45
Thomas, M. A., Welsh, H. L. 1960. Can. J.
Phys. 38:1291–1303
Tudge, A. P., Thode, H. G. 1950. Can. J.
Res. Sect. B 28:567–78
Urey, H. C. 1947. J. Chem. Soc. pp. 562–81
Urey, H. C., Greiff, L. J. 1935. J. Am. Chem.
Soc. 57:321–27 ·
Urey, H. C., Rittenberg, D. 1933. J. Chem.
Phys. 1:137–43
Van Riet, R. 1964. Ann. Soc. Sci. Bruxelles
78:97–104
Van Vleck, J. H. 1936. J. Chem. Phys. 4:
327–38
Viney, I. E. 1933. Proc. Camb. Philos. Soc.
29:142–48
Vojta, G. 1960. Kernenergie 3:917–27
Vojta, G. 1961. Ann. Phys. 7:397–402
Waldmann, L. 1943. Naturwissenschaften 31:
205–06
Webb, D. U., Rao, K. N. 1968. J. Mol.
Spectrosc. 28:121–24
Wolfsberg, M. 1969a. J. Chem. Phys. 50:
1484–86
Wolfsberg, M. 1969b. Adv. Chem. 89:185–
91
Wolfsberg, M., Massa, A. A., Pyper, J. W.
1970. J. Chem. Phys. 53:3138–46

Ann. Rev. Earth Planet. Sci. 1977. 5:111–22
Copyright © 1977 by Annual Reviews Inc. All rights reserved

DISCRIMINATION BETWEEN   ✻10068
EARTHQUAKES AND
UNDERGROUND EXPLOSIONS

Robert Blandford
Teledyne-Geotech, Alexandria, Virginia 22314

INTRODUCTION

In 1963 a Limited Test Ban Treaty (LTBT) prohibiting tests in the oceans, atmosphere, and space was signed in Moscow by the United States, Great Britain, and the Soviet Union. Underground nuclear tests were not banned in part because after five years of geophysical research, beginning in 1958 with the Geneva Conference of Experts and with the founding of the Vela Uniform program [administered by the Advanced Research Projects Agency (ARPA) of the United States Department of Defense], the United States and USSR governments could not agree on teleseismic means to discriminate between earthquakes and explosions.

The Vela Uniform program of research has continued up to the present (July 1976) with the result that, in the opinion of many seismologists, very few earthquakes per year of body-wave magnitude greater than $m_b = 4.5$ will be indistinguishable from explosions if reliable network measurements can be made of Rayleigh waves (LR waves) near a period of 20 sec and of the body compressional waves (P waves) near a period of 1 sec. The P waves are initially used to determine depth by triangulation or by the time interval between arrival of the direct wave and the reflection from the earth's surface, and if the event cannot be said to lie below 10 km with 95% certainty, then the event magnitude determined from LR (M_s) is compared to the magnitude determined from P (m_b). The discriminant is referred to as the $M_s:m_b$ discriminant. M_s is larger compared to m_b for earthquakes than for explosions. This discriminant has proven to be generally reliable in practice and has the great virtue that a simple and convincing theoretical explanation is available to explain its success. That is: the long-period shear-wave amplitude emitted by an earthquake is approximately six times the compressional wave amplitude. An explosion, in theory, emits no shear waves. Let us suppose that an earthquake and explosion emit equal-amplitude P waves. The earthquake will also emit large shear waves, and these together with the compressional waves are converted to Rayleigh waves at the earth's surface. Thus the Rayleigh wave (M_s) from an earthquake is larger than from an explosion with m_b equal to that of the earthquake. There are occasional earthquakes (and explosions) for which the $M_s:m_b$ discriminant fails. Many of them have been intensively studied.

It may be expected from this argument that the amplitudes of short- and long-period shear waves, as compared to compressional waves, would be a powerful discriminant, and this is in fact the case. However, the detection threshold for shear waves is rather high due to high absorption of shear waves in many regions of the earth.

Recording of a single reliable rarefactional first motion has been taken to be a reliable indicator of an earthquake. Unfortunately, practice with both earthquakes and explosions has shown that apparently reliable short-period first motions are sometimes incorrect. Perhaps this may be traced to the arrivals of weak signals along high-velocity multipaths. Long-period first motions integrate over the arrivals and are more reliable, but the threshold for detection of long-period P from earthquakes is one-to-two magnitude units above that of short-period P. Only megaton-range explosions give detectable compressional first motions on the World-Wide Standard Seismograph Network (WWSSN) long-period system; the signal characteristically has a very short period compared to those from earthquakes. This weak high-frequency signal presumably reflects cancellation at low frequencies due to the opposite sign pP phase reflected from the surface.

Explosion short-period P waves are more easily detected than are Rayleigh waves. Thus a discriminant based on P-wave data would be very valuable. Several discriminants have been proposed; however, a simple and convincing theoretical explanation based on characteristics of the seismic sources is not available for them. In fact, the observed discriminatory power may readily be explained by a combination of differences in upper-mantle absorption near the sources, together with effects of the nulls due to pP. Since both of these test-site characteristics could be altered by a suitable choice of test site and event depth, discriminants of this sort seem unreliable even when their statistical success may be impressive for a particular data set.

There are several possible descriptions of the earthquake source that lead to the hypothesis that the P-wave signal from earthquakes should be much less impulsive, i.e. more complex than the signal from explosions. However, experience shows that explosions, for example those from Novaya Zemlya, also have complex signals.

Application of all discriminants requires that the event be detected, and that the relevant phase amplitudes be measurable. For many normally detectable events each year the Rayleigh waves are not measurable because they are hidden in the long-period signal from another event. This situation could be taken advantage of by a party to a comprehensive test-ban treaty, in that he may purposely detonate an explosion after a large earthquake. By using greater care it may be possible for him to obscure, or completely hide, even the short-period compressional waves. Countermeasures to this evasion technique include advanced array and three-component processing to filter out the interfering event and use of signals of stations where the earthquake signal is especially small (e.g. at stations in the earthquake P-core shadow zone) or where the signal from the explosion is especially large (e.g. at stations near the explosion PKP caustic, 142°).

Another evasion technique involves enhancing the amplitude of the 20-sec Rayleigh waves by setting off several large explosions within 10 sec or so of each other. The Rayleigh waves will add up approximately in phase, while the P-wave signal may be made complex and of small amplitude. The signal may be made to emerge slowly from the noise by the use of small precursor shots. This can obscure short-period first motion. But notice that the long-period shear-wave to Rayleigh-wave ratio will not be altered, nor will the short-period shear-wave to body-wave ratio. These ratios, then, remain as possible discriminants.

Evasion may also be carried out by decoupling the explosion by exploding it in a large hole.

In the remainder of this paper I elaborate on the points introduced above, making references to the literature as appropriate.

EVASION

It is good to begin the study of discrimination with a discussion of the possibility of evasion. In a realistic test-ban situation, a monitoring network must logically assume that attempts will be made to evade detection of an unallowed explosion. Otherwise the treaty would simply be renounced and open testing resumed. Discussion of evasion techniques helps to focus attention on those discriminants that will be most valuable in the event of a test ban.

Decoupling

Latter et al (1961) and Herbst et al (1961) raised the theoretical possibility of reducing the teleseismic amplitudes of explosions by setting them off in large underground holes. Studies by Springer et al (1968) and Healy et al (1971), using the nuclear explosions Salmon and Sterling, verified earlier results by others using the Cowboy series of chemical explosion and showed that the decoupling theory correctly predicted the observed low-frequency amplitude reductions: a factor on the order of 75. It was, however, also pointed out by these workers that high-frequency decoupling (10–30 Hz) was substantially less efficient; factors on the order of 10 were obtained. By cube-root scaling arguments one may deduce that for larger events this poor high-frequency coupling would appear at lower frequencies, perhaps 3–10 Hz for a 10 kton explosion.

Epicenters of Russian peaceful nuclear explosions listed by Nordyke (1973) fall in regions of the USSR north of the Caspian and Black Seas where geological maps of the USSR show there to be many large salt domes. One might speculate that preexisting solution-mined storage cavities could be enlarged to a size suitable for decoupling large explosions. Evernden (1976a) has discussed the question of the stability of the large (170-m diameter for 50 kton) hole needed for full decoupling. Although he reaches no definite conclusion he argues that it is unlikely that such a hole could be constructed and remain sufficiently stable.

Let us suppose for the moment, however, that these problems could be overcome. If the high-frequency decoupling observed for Sterling can be cube-root scaled, then a decoupling factor of only 10 would be available. However, if tests could be conducted at depths substantially greater than that for sterling (831 m), then the frequency at which decoupling weakens might be increased. It seems implausible that decoupling would be attempted without extensive experimentation at large yields; and on balance it seems that the outcome of these experiments would lead to rejection of decoupling as an evasion scheme at 50 kton. At 5 kton the problems seem more difficult.

Shot Arrays

The concept here is that a series of explosions would be set off in such a way as to yield an emergent, complex signal such that first motion would be ambiguous. The larger events in the series would be set off within 10 sec of each other so that the Rayleigh waves would add up. If small explosions were set off initially the body-wave magnitude would be small because it is conventionally taken as the maximum motion in the first 3–4 sec of the signal. Work along these lines by several scientists in the Vela Uniform program has not yet been declassified; however, Kolar & Pruvost (1975) have published a study of this kind suggesting that such evasion could be successful. Marshall & Hurley (1976) immediately put forward a counter-discriminant. They noted that on broadband recordings the characteristic period of an explosion P wave is 1–2 sec, whereas for earth-quakes it is typically 5–10 sec. (The same can be seen on the WWSSN long-period system.) Upon summation of signals as appropriate for Kolar and Pruvost's shot array the dominant period remained the same, showing that the discriminant would still work on the shot array. Application of this discriminant to 10–20 kton would require arrays of broad-band instruments with perhaps multichannel filter processing.

I believe that the reason for the success of this discriminant is that the pP reflec-tion has cancelled the low-frequency energy from the explosions. The same sug-gestion has been published by Molnar (1971) and by Helmberger & Harkrider (1972). If this is the reason for the success of the discriminant, then chances would seem slim to evade it. Any reduction of pP by means of venting might lead to detection by other means.

In studies using US explosions, von Seggern (1972) and Blandford & Clark (1974) have pointed out that the long-period shear to Rayleigh ratio and the short-period shear to compressional ratio are good discriminants. These discriminants should also work for shot arrays. The long-period shear will add up in exactly the same way as does the LR, so the ratio should be unchanged. The short-period shear signals should be mixed in the same way as the short-period compressional signals, again leaving the ratio unchanged.

Landers (1973) has pointed out that earthquakes of the complexity typical of the shot array signal typically have exceptionally large M_s values with respect to their m_b values. Thus, a shot array, while falling into the earthquake population, will not be as high up in that population as would be expected from its complexity. Thus a plot of $M_s - m_b$ versus complexity reveals suspicious earthquakes.

A related technique has been investigated by Blandford & Clark (1975). These authors examined events in the Kamchatka-Kurils region, some of which had $M_s - m_b$ values close to the explosion population. Some others of the events were complex; and, in accordance with Lander's (1973) observations, had high $M_s - m_b$ values. If the m_b were defined using the maximum of the signal in the first 10 sec these events moved toward, but not into, the explosion population. Applied to shot arrays, this definition of m_b would make it much more difficult to move the shot array into the earthquake population, while introducing very few or no earthquakes into the explosion population.

Blandford & Clark (1975) also showed that from three highly seismic 1° diameter regions of Kamchatka investigated, only the one at the intersection of Kamchatka with the Aleutian arc had events of the type of emergent complexity hypothesized by Kolar & Pruvost (1975). An event of this type from either of the other two regions would be highly suspicious. This shows that careful regionalization of seismogram character using years of data would be a useful countermeasure to shot array evasion.

Although there has been some success reported in detecting pP from single explosions, deconvolution of the received signal into a series of pulses characteristic of a shot array seems unpromising. The source pulses will be unalike because 1. the source spectra are yield-dependent (von Seggern & Blandford 1972), 2. the surface reflection pP may well have a different spectrum from the direct pulse, and 3. different amount of spall may be present. If the direct and reflected wave from each of a series of explosions are thought to comprise a series of single pulses to be match filtered with a signal from a single explosion, then the analyst must contend with the probability that the reflection times are significantly different from event to event within the series. Thus while several counter shot array evasion techniques appear very promising (but are not proven on large data sets), deconvolution is not among them.

Hide-in-Earthquake (HIE)

In this evasion technique the evader waits for a moderate earthquake near the test site, or a large earthquake farther away, and then fires, hoping that the signal from the test will be hidden in the coda of the earthquake. Detailed work evaluating evasion possibilities along these lines by several scientists in the Vela Uniform program has not yet been declassified; however, Evernden (1976a–c) has recently published a study of this kind.

Jeppsson (1975) has also considered the problem from a network point of view. Jeppsson concludes that not many opportunities exist for evasion at explosions $m_b = 5.0$. However, his conclusions rest on an assumed amplitude decay of short-period P coda after arrival time much more rapid than that found as an average by Sweetser et al (1973) or by Filson (1973).

The principle approach for countering this evasion technique must be to develop techniques for separating the mixed signals. Simple beamforming of an array is one technique; but there is a large literature of advanced maximum-likelihood techniques that have been evaluated by many workers and found to work better than beamforming. One method that is similar to beamforming and easy to implement in the seismic case has been developed by Blandford et al (1976) and evaluated by Cohen (1974). In this procedure one first beams on the signal to be rejected, subtracts it appropriately from the individual channels, and then beams on the target signal. This is then subtracted from the original traces and the process is iterated. One iteration is often enough.

Nonlinear processing of the three components of motion at a single station and combinations of pendulum and strain instruments have also been shown to be effective in rejecting mixed long-period LR signals, although there seem to be difficult technical problems in practice (Simons 1968, Woolson 1972).

Another technique for countering HIE evasion has been discussed by Blandford et al (1975). It takes advantage of the fact that there are large variations in the P-wave amplitude as a function of distance. The core shadow zone ($103°-112°$) will exhibit low earthquake coda levels. Thus a station this distance from the earthquake will easily detect an explosion not at this distance. Also, at the PKP caustic distance from the explosion ($142°-155°$), the explosion amplitude is very large. Thus it is relatively easy to detect in the coda of earthquakes at other distances. The counter-strategy then is to predict the locations of large differences in amplitude given the known earthquake epicenter and some hypothetical explosion epicenters, and then to critically examine data from stations near those locations. Surprisingly enough, because of the sharp changes of amplitude near the core shadow zone and the PKP caustic zone, the method is worth applying even when the explosion is within two to five degrees of the earthquake.

It is very difficult to carry out a clear and convincing theoretical discussion of the number of opportunities per year for evasion using the HIE technique. Many difficult operational, statistical, and almost psychological questions arise. The results of Evernden (1976a–c) are probably correct in that there are real possibilities for evasion and that stations close to possible test sites are probably required if one is to be confident that evasion is not being practiced. For most seismic areas in the USSR such locations are available near the borders, in particular for Kamchatka if ocean-bottom seismometers may be used. With the advent of worldwide digital Seismic Research Observatory (SRO) data, superposition of seismograms from real explosions and earthquakes is now possible for a comprehensive worldwide network, and many of the difficult points of judgment with respect to probability of detection and false alarm rates may be bypassed by use of real data. We may look forward to studies of this kind.

SHORT-PERIOD DISCRIMINATION

Depth

Depth determination by the time interval between the arrival of short-period P and pP is a particularly powerful discrimination tool. When the signals at a large number of stations extended in distance are examined simultaneously one may see the change in the time interval with distance characteristic of the event depth. This is as close to an infallible discriminant as we have at our disposal.

The P waves themselves may, of course, be used in a conventional iterative linearized nonlinear least-squares location program, using standard travel-time tables to determine depth. These depths are always less confidence-inspiring than those due to pP because the worldwide average travel-time tables are not sufficiently accurate for the purpose in most individual regions of interest. For example, despite the most careful and complete routine analysis the depth of the Longshot nuclear test event in Amchitka was determined to be 60 km (Lambert et al 1969). Veith (1975) has developed a systematic technique for building tabular travel-time corrections for Kamchatka as a function of latitude and longitude into the location program. The corrections are determined at each point as the residuals of earthquakes with a good P – pP depth constrained to that depth. New events, located with these residuals, are found to be very close to their pP depths when available. The procedure is similar in concept to one discussed and extensively applied Evernden (1969) to Kamchatka earthquakes.

Evernden (1969) himself found that this general approach can fail in critical applications. He applied correction factors for an Aleutian earthquake to the nearby Longshot explosion, and noted that the standard deviation of the travel-time residuals was not reduced. The problem almost certainly is that a shallow event will not have the same residuals as a deep event beneath it unless the earth between is horizontally homogeneous and similar to that implied by the travel-time table used. Location of earthquakes at the same latitude, longitude, and depth as the calibration events will, of course, be relatively accurate. We have the situation of a technique that works perfectly except when it is most needed.

Blandford (1975a) has applied the technique to NTS, using Pahute Mesa residuals to locate Yucca Flats events. In this application the calibration events are set at the known depth. Using well-distributed networks of five or nine stations with distances greater than 16° from the sources he found mean depths of zero (averaged over an ensemble of events), as appropriate, but found standard deviations of the depth of 30 km for five stations and 20 km for nine stations. This shows that, even with excellent data, depth is very uncertain for shallow events. This is due to the fact that, for near-surface events, as the depth changes there is very little differential change in travel time with respect to distance to the station if it is at teleseismic distances. If close-in stations are available then depth control is better, and S-P times may be used to constrain depth. The S-P technique has not, however, been sufficiently studied in the range 5°–20°

Spectral Ratios

Studies at Lincoln Labs by several workers culminated in a report by Lacoss (1969) showing good separation between Eurasian explosions and earthquakes using the ratio of energy in the P phase from 0.4 to 0.8 Hz to that from 1.4 to 1.8 Hz. Other workers have since computed moments of the spectra, which yield related numbers, and have obtained similarly encouraging results. The most serious difficulty with short-period discriminants is that there is no source theory that explains the success of this discriminant in a convincing manner. Blandford (1975b) has suggested that earthquakes may have almost any shape spectrum so long as it is monotonically decreasing, and eventually decays as ω^{-3} at some arbitrarily high frequency. Extra high-frequency energy is hypothesized to come from many smaller shocks triggered along with the main shock.

On the other hand, the apparent success of short-period spectral discriminants may be easily explained by considerations that expose the discriminants as useless in a test-ban situation. Noponen (1975) has calculated how the third moment of the power spectrum (TMF) varies as a function of absorption (t^*) for a rather narrow range of t^* as compared to that observed. As a result of the variations in t^* earthquakes TMF values for two regions completely span the TMF values for explosions from a third region. Discrimination for the southwest United States, where earthquakes and explosions are both in a region above an upper mantle with high absorption, is not good. A similar conclusion was reached by Bakun & Johnson (1970) with respect to explosion aftershocks, and by H. Israelson as cited by Bolt (1976).

Another possible reason for the success of short-period discriminants is the cancellation of the low-frequency energy by pP for explosions. With respect to broad-band and long-period data we have already discussed this point above in the section on shot arrays. For short-period data the cancellation cannot be so complete since the first spectral null is at zero frequency. In fact, elementary calculations show that a delay of approximately 0.6 sec between P and pP will completely destroy the Lacoss spectral ratio as a discriminant if we assume both earthquakes and explosions have the same source spectrum. This delay, related to depth and medium velocity, is large as compared to the typical Eurasian explosion, but it is rather typical for the Nevada Test Site (NTS). Thus, everything works against short-period spectral discriminants at NTS. In Eurasia, if we assume that earthquakes are in regions of greater absorption than are explosions, then everything works for short-period spectral discriminants. Yet, in a test-ban situation one might overbury an explosion in a high-absorption seismic region and completely foil the discriminant. Again we may have a very dangerous discriminant, one that works perfectly until it is most needed.

Shear Waves

As discussed in the section on shot arrays, short-period shear (S) to short-period P is an excellent discriminant. It follows from elementary source theory and, if regionalized, will work even in high absorption regions, provided S is detectable.

LONG-PERIOD DISCRIMINATION

$M_s - m_b$

Several recent workers, e.g. Evernden et al (1971), Marshall & Basham (1972), and Peppin & McEvilly (1974), have shown that earthquakes and explosions separate down to low magnitudes ($m_b \approx 4$) if the $M_s - m_b$ discriminant is used. In the Introduction was discussed a simple physical explanation for this separation, and Douglas et al (1973) have used a full mathematical apparatus to deduce a separation of about one-half magnitude unit, on the average, between point-source earthquakes and explosions. Blandford (1975b) has discussed other source-related reasons why large-fault-plane earthquakes may have larger values of $M_s - m_b$ than earthquakes with smaller fault planes. Douglas et al (1973) showed that point-source 45° dip-slip earthquakes will fall into the explosion population. A point-source earthquake (small fault area) may have a high magnitude if the stress is high. Thus, we may expect to find high-magnitude "anomalous" events in regions where the stress is exceptionally high.

Such events apparently do exist. The best-documented events are a 1968–1969 swarm of events near 30°N, 95°E in Tibet. A few additional events have occurred at this location since 1969. These have been studied intensively by Landers (1972), Der (1973), Clark et al (1975), and Tatham et al (1976). The events can be identified as earthquakes by virtue of short-period dilatational first motions, and by the short-period S/P ratio. Yet short-period first motions are notoriously unreliable, as discussed previously, and the short-period S/P ratio has not been sufficiently studied.

Tatham et al (1976) give an extensive discussion of the tectonics of the source region for the anomalous events and conclude that it is plausible that high stresses could be developed there. Blandford & Gurski (1975) find from examination of satellite pictures that the events fall precisely at the intersection of a major east-west fault with a major north-south trending valley that may be the surface expression of a fault. This certainly suggests a high-stress region. The coincidence of the epicenters with the fault traces also suggests that either the faults are steeply dipping or that the epicenters are shallow. Der (1973), Clark et al (1975), and Tatham et al (1976) also concluded that these events could not be deep.

Nuttli & Kim (1975) also found several 1971 Asian earthquakes that were explosion-like with respect to $M_s - m_b$. Tatham et al (1976) have asserted that these events, and several others, discussed by Der (1973) but not at 30°N 95°E, may be identified as earthquakes by using an M_s determined from the higher modes if it is larger than the fundamental mode M_s. This seems a promising approach to reducing the number of anomalous events, but I feel that it must still be regarded as experimental until we see how application of the new definition of M_s shifts the total populations of Southwest United States explosions and earthquakes.

Shear Waves

Many assertions in the literature to the contrary notwithstanding, the Love to Rayleigh ratio (LQ/LR) does not appear to be promising as a discriminant. Von Seggern (1972) has shown that the population of earthquake ratios completely overlaps the population of observed explosion ratios from NTS and Amchitka. While it is true that Love-wave generation is small from explosions detonated in the USSR to date, we must remember that these explosions are in aseismic regions. If tectonic-strain release is the cause of the Love waves, we must expect them in a realistic test-ban situation where an evader would try to mask the explosion character by firing in a seismically active region.

On the other hand the long-period body-wave shear to Rayleigh ratio (S/LR), long-period P-wave period, and long-period P to short-period P amplitude ratio (von Seggern 1972) appear to be good and reliable discriminants and should work against shot arrays, as discussed above in the section on shot arrays.

Depth from Rayleigh Waves

Tsai & Aki (1970) developed a method for determining focal depth given the focal mechanism by varying the focal depth until predicted Rayleigh-wave amplitude spectra matched the observed ones. Since in most cases of interest for discrimination the focal mechanism will not be known, later work, culminating in the study by Patton (1977), is more relevant. In this study both the phase and amplitude spectra are matched to theoretical predictions using a search grid of focal mechanisms and depths. For discrimination purposes an explosion mechanism would be one of the "focal mechanisms" considered, and if the fit to this mechanism were superior to all others we would have a discrimination. The technique is technically difficult and fraught with practical problems, however. Distortion of the spectrum by multipathing and by absorption are two problems to be overcome in the data base itself; these lead to the consideration that close-in data is much to be preferred over teleseismic data. The effects of strain release would certainly hinder the technique.

REMARKS

I have tried to emphasize in this review that application of discrimination in a test-ban situation requires that one assume that attempts are being made at evasion. This has led to greater emphasis on discriminants using long-period body waves and short-period shear waves. It calls into question the applicability of short-period spectral discriminants, no matter how well they may work for explosions in aseismic regions before a test ban. It should be emphasized that the newer discriminants have not been tested so extensively as $M_s - m_b$, and that their threshold of application is high with present systems.

We have seen that an important countermeasure to shot arrays involves a new definition of m_b to better measure the total short-period energy emitted by the source in the case of emergent events, in order to retain the $M_s - m_b$ discriminant as a serious constraint on evasion.

In a test-ban situation it will be vital to "pick apart" the codas of many larger events each year in order to look for possibly hidden events. Inspection of seismographs in the earthquake shadow zone and PKP caustic zone of selected test sites will be routine on the occasion of large earthquakes.

A data base of typical event waveforms needs to be built up over many years and made easily available to the on-line analyst so that he can decide if a waveform from a particular area is unusual. All of this requires a worldwide digital data collection and storage system, and a graphics-oriented network processor. Fortunately work along these lines is being carried forward under the Vela Uniform program.

Literature Cited

Bakun, W. H., Johnson, L. R. 1970. Short-period spectral discriminants for explosions. *Geophys. J. R. Astron. Soc.* 22: 139–52

Blandford, R. R. 1975a. Use of source-region-station time corrections at NTS for depth estimation. SDAC-TR-75-4, Teledyne Geotech, Alexandria, Va. ADA 025 349[1]

Blandford, R. R. 1975b. A source theory for complex earthquakes. *Bull. Seismol. Soc. Am.* 65: 1385–1405

Blandford, R. R., Clark, D. M. 1974. Detection of long-period S from earthquakes and explosions at LASA and LRSM stations with application to positive and negative discrimination of earthquakes and underground explosions.

[1] AD reports may be obtained from the National Technical Information Service, 5285 Port Royal Road, Springfield, Va. 22151.

SDAC-TR-74-15, Teledyne Geotech, Alexandria, Va. ADA 013 672[1]

Blandford, R. R., Clark, D. M. 1975. Variability of seismic waveforms at LASA from small subregions of Kamchatka. SDAC-TR-75-12, Teledyne Geotech, Alexandria, Va.

Blandford, R. R., Cohen, T. J., Woods, J. 1976. An iterative approximation to the mixed signal processor. *Geophys. J. R. Astron. Soc.* 45:677–88

Blandford, R. R., Gurski, J. 1975. Use of earth resources technology satellites (ERTS) to determine tectonic characteristics near low $M_s - m_b$ earthquakes in Tibet. SDAC-TR-75-13, Teledyne Geotech, Alexandria, Va. ADA 025 177[1]

Blandford, R. R., Sweetser, E. I., Cohen, T. J. 1975. Hide-in-earthquake countermeasures using earthquake P shadow zone and explosion PKP caustic zone. SDAC-TR-75-15, Teledyne Geotech, Alexandria, Va.

Bolt, B. A. 1976. *Nuclear Explosions and Earthquakes: The Parted Veil.* San Francisco: Freeman. 309 pp.

Clark, D. M., Sweetser, E. I., Der, Z. A. 1975. Additional investigation of earthquakes with low M_s to m_b ratios in the Tibet-Himalaya region. SDAC-TR-75-2, Teledyne Geotech, Alexandria, Va.

Cohen, T. J. 1974. Coda suppression capabilities of the beam and mixed signal processor. *Bull. Seismol. Soc. Am.* 64: 415–26

Der, Z. A. 1973. $M_s - m_b$ characteristics of earthquakes in the eastern Himalayan regions. SDL Rep. 296, Teledyne Geotech, Alexandria, Va. AD 759 835[1]

Douglas, A., Hudson, J. A., Blamey, C. 1973. A quantitative evaluation of seismic signals at teleseismic distances—III computed P and Rayleigh wave seismograms. *Geophys. J. R. Astron. Soc.* 28: 385–410

Evernden, J. F. 1969. Identification of earthquakes and explosions by use of teleseismic data. *J. Geophys. Res.* 75: 3828–56

Evernden, J. F. 1976a. Study of seismological evasion. Part I. General discussion of various evasion schemes. *Bull. Seismol. Soc. Am.* 66: 245–80

Evernden, J. F. 1976b. Study of seismological evasion. Part II. Evaluation of evasion possibilities using normal microseismic noise. *Bull. Seismol. Soc. Am.* 66: 281–324

Evernden, J. F. 1976c. Study of seismological evasion. Part III. Evaluation of evasion possibilities using codas of large earthquakes. *Bull. Seismol. Soc. Am.* 66: 549–94

Evernden, J. F., Best, W. J., Pomeroy, P. W., McEvilly, T. V., Savino, J. M., Sykes, L. R. 1971. Discrimination between small-magnitude earthquakes and explosions. *J. Geophys. Res.* 76: 8042–55

Filson, J. R. 1973. On estimating the effect of Asian earthquake codas on the explosion detection capability of LASA. Tech. Rep. 1973–29, Lincoln Lab., Mass. Inst. Technol., Lexington, Mass.

Healy, J. H., King, C., O'Neill, M. E. 1971. Source parameters of the Salmon and Sterling nuclear explosions from seismic measurements. *J. Geophys. Res.* 76: 3344–55

Helmberger, D. V., Harkrider, D. G. 1972. Seismic source descriptions of underground explosions and a depth estimate. *Geophys. J. R. Astron. Soc.* 31: 45–66

Herbst, R. F., Werth, G. C., Springer, D. L. 1961. Use of large cavities to reduce seismic waves from underground explosions. *J. Geophys. Res.* 66: 959–78

Jeppsson, I. 1975. Evasion by hiding in earthquake. FOA Rapp. C 20043–T1, Forsvarets Forskningsanstalk, Stockholm, Sweden

Kolar, O. C., Pruvost, N. L. 1975. Earthquake simulation by nuclear explosions. *Nature* 253: 242–45

Lacoss, R. T. 1969. LASA decision probabilities for $M_s - m_b$ and modified spectral ratio. Techn. Note 1969–40, Lincoln Lab., Mass. Inst. Technol., Lexington, Mass.

Lambert, D. G., von Seggern, D. H., Alexander, S. S., Galat, G. A. 1969. The LONGSHOT experiment. Volume I: Basic observations and measurements: Volume II: Comprehensive analysis. SDL Rep. 234, Teledyne Geotech, Alexandria, Va. AD 698 319[1]

Landers, T. E. 1972. Some interesting central Asian events on the $M_s : m_b$ diagram. *Geophys. J. R. Astron. Soc.* 31: 329–39

Landers, T. E. 1973. Multiple explosions that produce earthquake-like seismograms. ESD-TR-73-175 Semiannual Tech. Summary, 30 June 1973, MIT Lincoln Lab., Mass. Inst. Technol., Lexington, Mass.

Latter, A. L., LeLevier, R. E., Martinelli, E. A., McMillan, W. G. 1961. Method of concealing underground nuclear explosions. *J. Geophys. Res.* 68: 943–46

Marshall, P. D., Basham, P. W. 1972. Discrimination between earthquakes and underground explosions employing an improved M_s scale. *Geophys. J. R. Astron. Soc.* 28: 431–58

Marshall, P. D., Hurley, R. W. 1976.

Recognizing simulated earthquakes. *Nature* 259:378–80
Molnar, P. 1971. P-wave spectra from underground nuclear explosions. *Geophys. J. R. Astron. Soc.* 23:273–87
Noponen, I. 1975. Compressional wave power spectrum from seismic sources. Final Sci. Rep., Feb. 21, 1975. Inst. Seismol., Univ. Helsinki, Finland
Nordyke, M. C. 1973. A review of Soviet data on the peaceful uses of nuclear explosions. UCRL-51414, Lawrence Livermore Lab., Calif.
Nuttli, O. W., Kim, S. G. 1975. Surface-wave magnitudes of Eurasian earthquakes and explosions. *Bull. Seismol. Soc. Am.* 65:693–710
Patton, H. J. 1977. Source and path effects for Rayleigh waves from central Asian earthquakes. In press
Peppin, W. A., McEvilly, T. V. 1974. Discrimination among small magnitude events on the Nevada Test Site. *Geophys. J. Roy. Astron. Soc.* 37:227–43
Simons, R. 1968. PHILTRE—A surface wave particle motion discrimination process. *Bull. Seismol. Soc. Am.* 58:629–37
Springer, D. L., Denny, M., Healy, J., Mickey, W. 1968. The Sterling experiment: decoupling of seismic waves by a shot-generated cavity. *J. Geophys. Res.* 73:5995–6012
Sweetser, E. I., Cohen, T. J., Tillman, M. F. 1973. Average P and PKP codas for earthquakes. SDL Rep. 305, Teledyne Geotech, Alexandria, Va. ADA 004 960[1]
Tatham, R. H., Forsyth, D. W., Sykes, L. R. 1976. The occurrence of anomalous seismic events in eastern Tibet. *Geophys. J. R. Astron. Soc.* 45:451–82
Tsai, Y. B., Aki, K. 1970. Precise focal depth determination from amplitude spectra of surface waves. *J. Geophys. Res.* 75:5729–43
Veith, K. F. 1975. Refined hypocenters and accurate reliability estimates. *Bull. Seismol. Soc. Am.* 75:1199–1222
von Seggern, D. H. 1972. Seismic shear waves as a discriminant between earthquakes and underground nuclear explosions. SDL Rep. 295, Teledyne Geotech, Alexandria, Va. AD 743 072[1]
von Seggern, D. H., Blandford, R. R. 1972. Source time functions and spectra for underground nuclear explosions. *Geophys. J. R. Astron. Soc.* 31:83–97
Woolson, J. R. 1972. Combinations of strain and pendulum seismographs to enhance long-period P and Rayleigh waves at Queen Creek, Ari. SDL Rep. No. 290, Teledyne Geotech, Alexandria, Va. AD 743 007[1]

Ann. Rev. Earth Planet. Sci. 1977. 5 : 123–58

QUATERNARY VEGETATION ×10069 HISTORY—SOME COMPARISONS BETWEEN EUROPE AND AMERICA

H. E. Wright, Jr.
Limnological Research Center, University of Minnesota, Minneapolis, Minnesota 55455

INTRODUCTION

Great advances in paleoclimatology in the last few years have resulted from careful paleoecological and oxygen-isotope studies of Foraminifera from deep-sea cores. They provide the basis for a comprehensive global look at the entire Quaternary. Continuous stratigraphic sequences, dated by the paleomagnetic time scale, indicate that about 9 climatic cycles occurred during the last 700,000 years, with many more dating back to the beginning of Quaternary. They also indicate that the dominant climate of the Pleistocene was cold, and that only a very small portion of each cycle was as warm as it is today.

This picture prompts a fresh look at the terrestrial scene for a record of Quaternary climates. For glaciated areas the record is clearly incomplete, for only three or four major episodes of glaciation can usually be identified. For periglacial areas the record has more to offer, because both cold-time and warm-time intervals are represented by fossil-bearing sediments. Pollen grains are particularly useful, for they not only yield paleoclimatic and paleoecologic data but they also furnish an independent means of correlating undated with dated sites. Recent work in England and particularly The Netherlands shows that several climatic cycles occurred before the first continental glaciation, but a one-to-one correlation with the ocean-core cycles is certainly not yet possible, except for the very late Quaternary.

Regardless of the correlations and the problems of global climatic cycles for the entire Quaternary, the pollen profiles provide important insights into the dynamics of vegetation formations that are stressed by repeated climatic change as well as by progressive soil development and various biologic factors. These insights in turn furnish an historical perspective that contributes to an understanding of modern vegetation and the stability of ecosystems.

The review that follows concentrates on a few aspects of Quaternary vegetation history—aspects that involve new developments with potential significance beyond a local area. Among the subjects considered are the floristic changes at the beginning of the Quaternary in Europe, the nature and duration of a typical European vegetational cycle, and the details of late-Pleistocene and Holocene vegetational change in response to climatic and nonclimatic influences in western Europe and eastern North America, where most information is available.

PLIOCENE-PLEISTOCENE BOUNDARY

Recent paleobotanical studies of Pliocene sediments in The Netherlands and adjacent parts of Germany indicate several climatic fluctuations extending over millions of years (Zagwijn 1960, Menke & Behre 1973). The inception of the Pleistocene in this region, however, is marked by the first signs of open vegetation of a boreal character (Figure 1). Although Zagwijn (1974) suggests that for the first time the northern tree line shifted south of The Netherlands, a more conservative view is that the high pollen percentages of herbs merely indicate the widespread occurrence of wetlands in a boreal forest (Menke & Behre 1973). Of greater paleoclimatic and stratigraphic consequence is the extinction of several important trees throughout Europe—from England to Poland and Italy. About half the species in the rich Pliocene flora, as represented by seeds and fruits as well as pollen grains, became extinct at that time or were displaced to distant refuges such as the Black Sea area. Many of these forms are the genera that are conspicuous

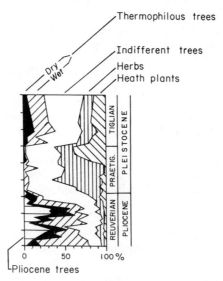

Figure 1 Pollen sequence across the Pliocene/Pleistocene boundary at Meinweg, The Netherlands. The Pliocene trees include *Liquidambar, Sequoia, Sciadopitys, Taxodium,* and *Nyssa.* Redrawn from Zagwijn (1975).

Table 1 Latin names for plants mentioned in the text and figures

Alder	*Alnus*
Black ash	*Fraxinus nigra*
Aspen	*Populus tremula* (European)
	P. tremuloides (American)
Beech	*Fagus*
Birch	*Betula*
Cedar	*Chamaecyparis*
Chenopods	Chenopodiaceae-Amaranthaceae
Chestnut	*Castanea*
Composites	*Compositae*
Cypress	*Taxodium*
Elm	*Ulmus*
Fir	*Abies*
Grasses	Gramineae
Sweet gum	*Liquidambar*
Hazel	*Corylus*
Heath plants	Ericales
Hemlock	*Tsuga*
Hickory	*Carya*
Holly	*Ilex*
Hornbeam	*Carpinus*
Horse-chestnut	*Aesculus*
Ironwood	*Ostrya, Carpinus*
Ivy	*Hedera*
Juniper	*Juniperus*
Larch	*Larix*
Lime, linden	*Tilia*
Red maple	*Acer rubrum*
Sugar maple	*A. saccharum*
Mistletoe	*Viscum*
Oak	*Quercus*
Jack pine	*Pinus banksiana*
Pitch pine	*P. rigida*
Red pine	*P. resinosa*
White pine	*P. strobus*
Plane tree	*Platanus*
Balsam poplar	*Populus balsamifera*
Ragweed	*Ambrosia*
Sedges	Cyperaceae
Sorrel	*Rumex*
Sour gum	*Nyssa*
Black spruce	*Picea mariana*
Red spruce	*P. rubens*
White spruce	*P. glauca*
Tupelo	*Nyssa*
Willow	*Salix*

today in the rich temperate forests of southeastern North America or the Far East, e.g. cypress, sweet gum, tupelo, horse-chestnut, and *Sequoia* (Table 1). A few others such as hickory, hemlock, and *Pterocarya* survived this first great climatic event, but they passed on during later cold intervals (Zagwijn 1974, 1975). At the same level the Foraminifera and marine mollusks change to cool-water forms.

The first cold period of the Pleistocene, called the Pretiglian in The Netherlands, was thus a major catastrophe for the warm-temperate forests that had prevailed in western Europe for millions of years. A similar sequence can be found in northern Italy; here the change takes place near the base of the Calabrian, concurrently with the first appearance of cold-water mollusks. This stratigraphic level marks the Pliocene/Pleistocene boundary according to a recommendation of the 1948 International Geological Congress. Paleomagnetic measurements in The Netherlands place this event near the end of the Gauss paleomagnetic epoch about 2.5 million years ago (Zagwijn 1974).

Although the paleobotanical evidence for climatic cooling at the Pliocene Pleistocene boundary as thus defined is substantial, the chronological relation to the first continental glaciation in northern Europe is uncertain. Several vegetational cycles can be identified before the first glacial deposit. Furthermore, all the early Quaternary preglacial cycles combined with all the glacial cycles do not add up to the total number of cycles evident in the oxygen-isotope profiles of deep-sea sediments—presumably because the terrestrial record contains some serious gaps. The oldest glacial deposits in northern Europe may not be much more than a million years old, and a case has been made by Menke & Behre (1973) for resubdividing the Quaternary into a preglacial epoch (the Cenocene), an epoch with glacial and interglacial stages (Pleistocene), and a postglacial epoch (Holocene).

QUATERNARY VEGETATIONAL CYCLES IN EUROPE

The abrupt disappearance of several important tree types at the beginning of the Quaternary, along with the disappearance or permanent emigration of others during succeeding cold intervals, were unique events of the early Quaternary. In general, however, the climatic cycles in northwestern Europe were all characterized by a grossly similar vegetation sequence. An exception is the first warm interval in The Netherlands (the Tiglian), in which there occurs pollen of *Quercus ilex*, an evergreen oak now confined to the Mediterranean region of summer drought (Zagwijn 1975). Otherwise, each cold phase, dominated by herb pollen, is succeeded by birch and pine and then by pollen of a series of temperate trees. The warm phase terminates with the reappearance of conifers. The general pattern is so faithful that one can infer similar climates for the several interglacials. It is even likely that the durations of the interglacials were somewhat the same. Pollen profiles for annually laminated sediments in England and Germany for the last three interglacials indicate that the temperate phase lasted 10,000 to 30,000 years (Wright 1972).

The usual sequence for the interglacials is closely similar to that for the Holocene, which has not yet run its course. The symmetrical development of the pollen curves for the Holocene prompted von Post (1930), the founder of stratigraphic pollen analysis, to emphasize a tripartite subdivision for the inferred climatic cycle, centering around a climatic optimum—a simpler climatic sequence than the then-popular Blytt-Sernander scheme, which involves two dry and two humid intervals during the Holocene. This scheme was founded on Scandinavian bog stratigraphy and is still used by some as a climatic framework.

Although a tripartite climatic classification for the interglacial cycles is valid in general, it is clear from recent work in Denmark and England particularly that the pollen sequence reflects regional factors other than climate. Iversen (1964, 1973), for example, shows the effects of progressive soil changes on the vegetation, and Andersen (1966) suggests that because of this four phases instead of three should be recognized in the interglacial vegetation cycle (Figure 2): the long climatic optimum in the middle part of von Post's cycle should be subdivided into an earlier portion, in which temperate trees were fully expanded on soils that still contained a substantial quantity of nutrients, and a later portion, in which some soils became excessively leached and favored the immigration of coniferous trees and heath shrubs. He points out that the conifer phase must be part of the climatic optimum because ivy, a plant restricted to temperate regions with oceanic climates, was still common during this phase. When the climate cooled at the end of the warm interval, the forest opened up further and herbs succeeded.

For England, West (1961) adopts a similar fourfold subdivision for the inter-glacial cycle. He emphasizes that the sequence of pollen maxima for the various

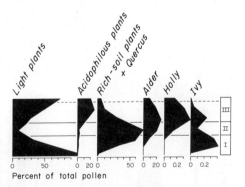

Percent of total pollen

Figure 2 Generalized pollen diagram for Herning, an Eemian (last-interglacial) site in Denmark, showing the generalized subdivision of interglacials into an initial phase dominated by light-demanding pioneer trees such as birch and pine (phase I), a phase dominated by shade-tolerant and shade-producing rich-soil forest trees (phase II), and a phase characterized by acidophilus plants that tolerate leached or degenerated soils (phase III). Alder lasted longer than other rich-soil trees because it favors moist areas, where the effects of soil-leaching are less pronounced. Ivy and holly together are the best indicators of the inter-glacial temperature profile, but ivy has its maximum early because it reaches the light by climbing to the tree tops in rich soils, and holly is a forest-floor shrub that needs a thinner canopy but can tolerate acid soils. Redrawn from Andersen (1966).

temperate trees is different for each interglacial. For example, in England in the present interglacial hazel was the first to expand in the temperate phase when pine and birch could no longer retain their prominence (Figure 3). Then other major trees were progressively added: elm, oak, lime, alder, and finally beech. Beech, incidentally, had not previously appeared in significant quantities in England or northwestern Europe since the Pliocene. During the last interglacial (Ipswichian), the sequence was elm–oak–alder–hazel–maple–hornbeam. During the penultimate interglacial (Holsteinian) the sequence was oak–elm–lime–alder–hazel–spruce–hornbeam–fir, and during the antepenultimate (Cromerian) it was elm–oak–alder–spruce–hornbeam.

Although all of these tree types have slightly different distributions today with respect to climate, soils, and landforms, the differences in order of immigration or

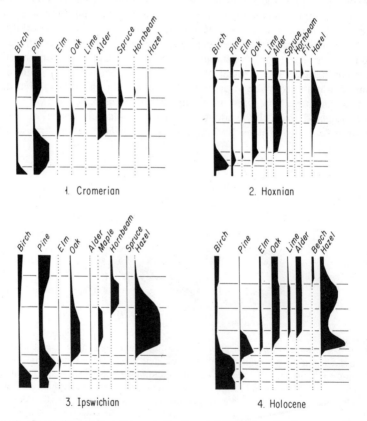

Figure 3 Generalized pollen diagrams for four interglacial cycles in eastern England, including the present (Holocene). Note that each begins and ends with birch and pine, and that the order of appearance of the deciduous-tree types—or the abundance or even the presence—differs from one cycle to the next. Redrawn from West (1961).

expansion cannot be explained solely on the basis of climate. Biologic factors such as seed dispersal, competition, succession, and predation also may play a role. Even evolution may have been important. It was suggested by Firbas (1958), for example, that the rapid expansion of beech in the mid-Holocene may have been the result of the evolution of a new ecotype, which for the first time since the Pliocene could flourish in the oceanic climate of western Europe.

In addition to these biologic factors, in some cases geologic factors may have been critical in the expansion of certain trees. For example, the development of heathlands in the late Holocene of northwestern Europe is considered by some to reflect the increased humidity related to the rise in sea level—a local climatic trend reflecting melting of the ice sheets rather than contemporaneous global climatic change. As another example, the absence of lime from the Ipswichian in England and its presence on the continent at this time might reflect an early rise in sea level to produce a migrational barrier (West 1961).

It is as if nature performed a grand experiment a number of times, with more or less the same primary boundary conditions, i.e. climatic change, and as if each time that the cards were shuffled the results were somewhat different, because of the complex interaction of the minor factors. We know so little about the magnitude of these minor biologic and geologic factors that it is difficult to understand all the implications of the differences. Many of the interactions can be sorted out by developing further knowledge of the autecology of key pollen-producing plants, and by investigating closely related paleoenvironmental features. For example, stratigraphic studies of other microfossils and chemical components of lake sediments can yield independent measures of the history of climate, soils, lake productivity, and other environmental factors (Andersen 1966).

The present interglacial (the Holocene) is the easiest to study, because in many cases the entire forest formations and their various subdivisions are here today to be studied in toto. Actually, this is not strictly true, because human disturbance in most areas has been so massive that many vegetation types have been drastically altered. This is true particularly in Europe, where man has had a role in changing the face of the earth for thousands of years (Mitchell 1965). Even if such areas were now left to themselves without disturbance for hundreds of years, it is doubtful that they would ever return to their predisturbance condition, partly because the climate has changed since that time, and partly because the loss of seed sources and the modification of the soil constitutes another great experiment—this time by man—and the plants that came to dominate in the area may reflect another accident. Nonetheless, the present is at least a generalized key to the past, and an understanding of the plant ecology and phytogeography of modern communities provides a starting point for paleoclimatic reconstructions.

PRINCIPLES AND PROBLEMS IN POLLEN ANALYSIS

The pollen produced by most temperate trees and shrubs and many herbs is dispersed by the wind and becomes incorporated in great abundance in the sediments of lakes and wetlands, where it remains preserved as long as it is not affected by circulating oxygen-bearing waters. The stratigraphic sequence of fossil pollen in a sediment core provides a picture of the vegetational history of the area. How accurate is this picture?

Because the pollen of some plant types is more abundantly produced or more widely dispersed than that of others, it is difficult to determine with great accuracy from an assemblage of fossil pollen just what kind of vegetation it represents. Even if it were possible to measure absolute pollen production from individual trees or from stands of trees, the source area of the pollen falling in a lake remains poorly defined. In general, dispersal from a tree or stand of trees follows an exponential curve, but the rate of fallout from this source varies greatly, depending not only on such factors as the buoyancy of the grain but also the wind patterns within the forest and other weather factors at the time of dispersal. On the basis of such exponential curves, Janssen (1967, 1973) draws a useful distinction between local pollen rain, derived from plants immediately adjacent to the site, and regional pollen rain, derived from a great distance and thus uniform over a broad area. Attention to such relations may yield insights necessary for the proper interpretation of pollen diagrams, so that broad climatic trends can be distinguished from local nonclimatic developments related to processes in lakes and wetlands.

The best way to approach the problem of representativity—the quantitative relation between vegetation and pollen rain—is to analyze the pollen of surface samples (moss polsters, core tops in lake sediments) in areas of known vegetation. Such a strategy was never used extensively in European pollen studies, primarily because the modern vegetation is so disturbed that it was considered a poor model for the past. American pollen analysts, however, who developed their own approaches to the problems of vegetation reconstruction in the 1960s, have made increasing use of surface-sample surveys, and today more than 1000 surface-pollen counts are available in eastern North America alone. Several reviews and comprehensive analyses also have been made (Wright 1967, 1968, Davis 1967a, Ritchie & Lichti-Federovich 1967, Lichti-Federovich & Ritchie 1968, Davis et al 1973, Webb 1974, Davis & Webb 1975, Birks et al 1975). Although the vegetation of North America has been disturbed by logging and farming, the disturbance level in a short core of lake or wetland sediment can be detected by the abrupt increase in pollen of *Ambrosia* (ragweed), which is a plant that thrives on ground from which the soil has been removed or diluted with excessive mineral matter. Comparison of modern and predisturbance levels in a pollen diagram indicates only modest differences in other pollen percentages across the boundary.

When the first surface-sample surveys were compiled, there was every expectation that modern analogues for fossil pollen assemblages could be found, and the lack of success was attributed to gaps in sampling all the modern forest types. It is now becoming increasingly apparent, however, that no extensive analogues exist for some common pollen assemblages. This realization opens the field for a much closer look at the developmental history of modern forest types, especially the problems of differential rates of migration of dominant tree types into well-established forests without the contemporaneous influence of climatic change, so that the composition of the major vegetational formations today cannot be considered stable in the long run.

In Europe the problem of differential migration in the Holocene has long been recognized, especially through comparisons with previous interglacial cycles, as discussed above. But European Holocene history is confused by the intrusion of man, who disrupted existing forests not only by cutting trees, farming, and disturbing habitats, but also possibly by importing seeds from southern Europe, thus speeding up a migration or introducing a plant that might not otherwise have reached a particular area.

In America, on the other hand, Indian influence on forest development had no demonstrable effect. The only documented example of prehistoric agriculture is at a locality near Toronto, where a pollen record of maize, grasses, and a few other plants can be closely correlated with Indian habitation 500 years ago (McAndrews 1976). The problem of differential migration of pine into the Great Lakes region from distant Pleistocene refuges in the Appalachian Mountains is perhaps the most striking example (Wright 1968). The pollen sequence in New England, previously attributed to climatic change, now is believed to record differential migration of various dominant tree types. Now the subject has been taken up more directly, and preliminary maps have been compiled for the northeastern United States to show the rates of migration of key trees such as white pine, beech, hemlock, hickory, and chestnut (Davis 1976, Bernabo & Webb 1977). The main task now is to cover the region with well-executed and well-dated pollen diagrams, so that the maps can be checked and refined.

Another way to approach the problem of representativity is through the determination of the absolute pollen influx (pollen grains cm^{-2} yr^{-1}). The usual pollen diagram is scaled to show the percentage of the different types at each sample level. An absolute increase in one pollen type (e.g. pine) must be matched by decreases in other types (e.g. spruce) even though the spruce may also increase slightly in absolute amounts. In this case the percentage diagram implies that spruce trees decreased in the area, even though they actually increased.

This problem can be examined by calculating the absolute influx as well as the percentages. The technique requires that the pollen concentration and the rate of sedimentation be measured. Concentration is usually determined by the addition of a known number of grains of an exotic pollen type to a known volume of sediment (thus grains cc^{-1}). Sedimentation rate (or its reciprocal, deposition time, yr cm^{-1}) is determined by interpolation between radiocarbon dates or other time markers in the sediment core. Influx can then be calculated and compared to the influx today in different vegetation types. Results in North America show, for example, that the pollen influx in tundra regions is very low, primarily because the pollen production from tundra herbs and low shrubs is very low. Consequently tree pollen blown from forest regions to the south has high pollen percentages, even though the absolute influx may be low. In the forest-tundra area, on the other hand, influx from tundra openings may be higher because of lusher growth than in the tundra proper, but tree-pollen influx is also higher. In such a situation the percentage values for herbs and trees may be similar to those in the tundra, but the total influx may be much greater, and a distinction can be made on this basis. Influx values in the open coniferous forest, for example, average about 6000 grains cm^{-2} yr^{-1}, compared to 2000 in the tundra (Lichti-Federovich & Ritchie 1968). In the closed coniferous forest, values rise to 20,000 grains cm^{-2} yr^{-1}, and in the Great Lakes forest, with such big pollen producers as white pine and red pine, to 40,000.

THE LAST MAJOR QUATERNARY CYCLE

General

As a starting point the vegetational development of the last cold period and the Holocene that follows is best related to the major vegetation formations that can be observed today. It cannot be overemphasized that these formations are not necessarily identical to those that prevailed in the past, not only because of the biological, edaphic, and disturbance factors mentioned above, but because during much of the time the great ice sheets may have exerted unique climatic influences that cannot be found today. Even with these potential problems in finding perfect modern analogues for past vegetation, comparison with modern vegetation is better than a reconstruction based purely on speculation, in part because it permits an evaluation of some errors inherent in pollen analysis.

Europe

WEICHSELIAN   The last major cold interval in Europe, the Weichselian, was marked generally by tundra vegetation. The time of its beginning is not known directly, because no materials are suitable for isotope dating, but a figure of about 100,000 years can be borrowed from the correlative paleoclimatic level of ocean cores. The Weichselian is interrupted during its early part by at least two interstadials, the Amersfoort and Brörup. These are dated as 65,000–60,000 BP by the method of enriched C-14, in which a small amount of contamination can reduce the accuracy of the date. During the Brörup interval, northwestern Europe (including at least southern England) was apparently covered with forests of pine, birch, spruce, and alder (Andersen 1961, Simpson & West 1958). The spruce involved was partly *Picea omoricoides*, which today as Serbian spruce (*P. omorika*) is confined to a few mountains in Yugoslavia. The contemporaneous herbaceous and aquatic flora implies a climate warmer than that indicated by the trees; perhaps the more temperate trees had not had time to immigrate (Andersen 1961). The Brörup has been identified at many other localities in Europe, including southern Poland, where a spruce–pine–fir–alder assemblage does not resemble any interglacial pollen zone and has a radiocarbon age too old for later Weichselian interstadials (Mamakowa et al 1975).

Following these early Weichselian interstadial intervals, the climate was apparently frigid in northwestern Europe for at least the next 45,000 years. Brief interstadial intervals are recognized in The Netherlands about 38,000 BP (Hengelo) and 30,000 BP (Denekamp), when the normal herb tundra changed temporarily to shrub tundra, according to the meagre record of fossil-bearing sediments (van der

Hammen et al 1967). The Stillfried B soil at Paudorf in the loess sections in Austria may be the record of at least the latter of these two interstadials; pollen content of the loess implies the presence of temperate hardwoods as well as spruce, pine, and birch, although open treeless vegetation dominated (Frenzel 1964). A correlative site in southern Poland with a date of about 26,000 BP has a record of open pine–larch–birch forest with grasses and steppe plants in the openings (Mamakowa 1968).

The severest cold then returned to most of western Europe, as the Weichselian ice reached its maximum extent about 20,000 years ago. In northwestern Europe in front of the ice the terrain is described as a polar desert, with strong winds and few plants. To the south in France the tundra was somewhat lusher, and it supported the great numbers of reindeer, horse, bison, mammoth, and other animals so beautifully depicted by late Paleolithic man in the caves of the Dordogne.

The final warming that. caused the fluctuating retreat of the Scandinavian ice sheet brought an end to the barren lands in Europe. The first sign, about 13,000 years ago, was an expansion of *Artemisia*, chenopods, and other steppe plants, which joined the cold-hardy arctic plants that had prevailed during the glacial maximum—plants like *Dryas octapetala, Salix herbacea, Saxifraga oppositifolia,* and *Armeria maritima.* Both the arctic and the steppe plants require abundant sunlight, and most of them need bare mineral soil low in nutrients. In fact some of them are specialized nitrogen fixers, which build up the nutrient content of the soil. Even though much of Europe had not been directly affected by the Weichselian ice sheet, the soils had been disturbed by frost action, and a blanket of sand or loess had been deposited in many areas. Many of the plants of the late-glacial tundra are found today primarily in the Soviet steppe or in rocky regions where competition is sparse.

As the climate warmed, various shrubs joined the herbs that had previously prevailed. The following communities are postulated, partly on the basis of comparisons with arctic and alpine vegetation today (Firbas 1949, p. 296; Berglund 1966): grass heaths, *Artemisia* communities, willow shrubs, stands of dwarf willow and dwarf birch, *Empetrum* heaths, juniper stands, *Hippophae* copses, tall herb meadows, and moss and lichen communities.

Temperate water plants appeared in the lakes, as a signal that the more slowly migrating forest vegetation was soon to come. By 12,500 years ago, tree birch had arrived in Denmark, marking the Bölling interstadial, which lasted for only 500 years. The birch forest was not continuous but rather parklike, and large areas of shrub tundra prevailed. The Bölling is identified as far southeast as central Poland, where an influx of tree birch, *Hippophae*, and temperate water plants into previously existing shrub tundra has three carbon dates within the time range of the Bölling of Denmark (Wasylikowa 1964).

After a temporary return to tundra conditions once again in Denmark for the interval 12,000 to 11,700 years ago (Older Dryas), a more substantial climatic warming occurred—the Alleröd interstadial, about 11,700 to 11,000 BP (Figure 4). Tree birch and aspen formed a forest cover over most of northwestern Europe, although the forest was open or parklike in the more northerly areas, with shrub heaths containing dwarf birch, willow, and juniper. Pine occurred in some areas, even as far north as southern Sweden, and farther to the southeast it was increasingly important. In central Poland, for example, pine shared dominance with birch (Wasylikowa 1964), although much of the landscape was still open enough to accommodate abundant juniper, *Hippophae, Artemisia*, and chenopods.

During the Alleröd the landscape became largely stabilized by its vegetation cover, and for the first time the lakes consistently received deposits of rich organic sediment rather than the silt previously supplied by frost action on unstable slopes. Temperate water plants like cattail give signs that the climate was even warmer than that suggested by the terrestrial vegetation, which lagged somewhat in its immigration from distant refuges.

The warming tendency was interrupted once again by another cold episode, lasting from about 11,000 to 10,300 years ago, generally called the Younger Dryas phase. In Denmark and southern Sweden the birch forest retreated or became restricted to groves. Dwarf shrub heath dominated. Frost action must have prevailed again on the hill slopes, because the Younger Dryas sediment once again consists of mineral-rich silt at many localities.

As a result of the cooling climate, the Scandinavian ice sheet at this time

JUTLAND, Denmark

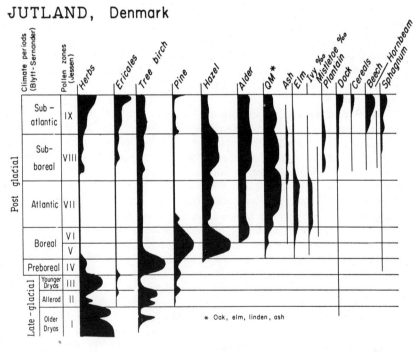

Figure 4 Generalized pollen diagram for the late-Weichselian and Holocene sequence in the Danish mainland. The Bölling interstadial is marked by the birch peak in the Older Dryas. From Iversen (1941).

apparently slowed down in its retreat across southern Norway, Sweden, and Finland and built the Ra-Central Swedish-Salpausselkä moraine from Oslo to Stockholm to Helsinki. Pollen diagrams to the south of the moraine show the record of the Alleröd interstadial, whereas those to the north do not (Fries 1965).

For the British Isles, Pennington (1975) makes a case for a much earlier re-vegetation of the polar desert than is the case in Denmark. A site in the Lake District of northwestern England shows that tundra with *Artemisia* developed as early as 14,300 BP and was succeeded about 13,000 BP by juniper and then birch woodland—500 years before birch arrived in Denmark. She attributes this earlier arrival of trees to the milder oceanic climate of Britain compared to Denmark. In Ireland, on the other hand, which is even more oceanic, the pollen sequence does not show this early development (Watts 1977), and evidence from the investigation of beetle remains in late-glacial sediments (Coope & Brophy 1972) suggests that the last word has not been said on the early vegetational and climatic succession in the British Isles following ice retreat.

The Alleröd/Younger Dryas fluctuation seems to be well shown in Britain, however. Perhaps the most striking record is in the English Lake District, where lakes affected by outwash from Younger Dryas valley-head glaciers show a layer of silt (even varved silt), representing outwash from glaciers reborn during this stadial interval, whereas lakes without such regrown glaciers have no such interruption (Pennington 1969).

Both the Younger Dryas ice-margin fluctuation and the vegetational shifts are usually attributed to a global or at least a continental climatic change, and they are often used as the basis for correlating events elsewhere in the world, not only in the Mediterranean area (van der Hammen et al 1971) but also in South America and Africa (van der Hammen & Vogel 1966). The problem has been reviewed by Mercer (1969), who shows that the correlation beyond northwestern Europe is not well supported by carbon dates, and he presents an opposing view that the climatic fluctuation, although real, was a secondary effect of melting of shelf ice. As the ice sheet retreated during the Alleröd, warm Atlantic water caused the warm-up and then the break-up of great ice shelves that must have covered most of the Arctic Ocean and the Norwegian Sea. Southward drift of ice masses from the break-up then cooled the surface waters of the North Atlantic enough to affect the climate of northwestern Europe north of the Alps.

HOLOCENE   When the climate finally warmed permanently about 10,300 years ago, a long chain of vegetational changes ensued. The elucidation of these changes and their causes has occupied the attention of European palynologists since the early days of 19th-century Scandinavian bog studies, when pine stumps were found consistently buried deep beneath the surface of treeless peatlands. The hundreds of pollen diagrams from all over Europe, now supported by great numbers of radiocarbon dates, provide a geographical and chronological framework for watching the forest change composition over time, and also force an evaluation of the relative importance of climatic, edaphic, biologic, and human factors in bringing about the changes.

The amount of information on the Holocene vegetational history is so vast that a few paragraphs cannot do justice to the geographic variations. The review that follows centers on the well-established study for Denmark, where the complexities have been succinctly described by Iversen (1973).

The climatic improvement was relatively abrupt. It is recorded first in Denmark by the appearance of temperate water plants, whose seeds can be so easily dispersed by birds. The first land plant to expand was juniper, which had previously been restricted to prostrate forms protected by snow in the winter. With warmer winters the juniper grew to tree size and left a conspicuous pollen record (Iversen 1973).

Aspen and birch, which had also been present during the Alleröd interstadial, soon overwhelmed the juniper, and then pine advanced from the southeast. The birch forest lasted for 1300 years in Denmark, and then the temperate hardwoods arrived (Figure 4). Hazel was first; it expanded in great profusion about 9000 years ago under the light shade of the pine-birch forest, and in turn the shade it produced hampered the reproduction of birch and pine. Thus hazel came to dominate, at least on the better soils; on sand plains and peaty soils pine could still compete.

Other hardwoods then entered the Danish scene in rapid succession, starting about 8500 years ago—elm, oak, alder, and lime. They did not arrive in the order of their temperature requirements (according to their modern distribution) or in the same order in which they had appeared during earlier interglacials, so one can only postulate differential rates of dispersal from different refuges—presumably in various parts of the Balkan peninsula or perhaps in Spain and Italy (Iversen 1973). Pollen diagrams from southern Europe provide clues about the refuges and about the early progress of northward migration, but no one has yet attempted to produce maps to show the rates of migration for the various trees.

These hardwood trees can reproduce in the shade of hazel, so they could gradually invade what may have been a continuous forest. Once established, they overtopped the hazel, which is a smaller tree and is also intolerant of heavy shade. By about 7000 years ago the climate had reached its maximum warmth for Holocene time, and the forest was luxuriant and stable—a mosaic dominated by different hardwood trees that depended on soil moisture and nutrients as well as on the accidents of invasion of blowdowns and other temporary openings in the forest. The climatic situation is shown especially by the occurrence of ivy and mistletoe, which are sensitive to low temperature and do not now occur in Denmark near localities where they are found as fossils. It is significant that they immigrated before alder and lime, even though these trees are less sensitive to low temperatures and should have come during the early stages of climatic warming—an indication that the sequence of immigration is not a direct reflection of climatic change.

The climate was also moister at this time in Denmark than it had been before. Low-lying sand plains, where pine had resisted the invasion of hardwoods because of the acid soils, were converted to raised bogs as a result of the increased humidity, and lake levels rose. The moister climate is usually attributed not to regional change in air-mass circulation but to geographic changes in land-sea relations. Melting of the ice sheets caused the inundation of southern Scandinavia and the expansion of the North Sea, and Denmark's climate shifted from continental to oceanic.

About 5000 years ago the relatively stable hardwood forest was abruptly disrupted. and a number of changes set in, roughly coincident with the arrival of agricultural man on the European scene. But not all forest changes are attributable to man alone, for the other three principal forces shaping the character of the vegetation also played a role—climatic change, progressive leaching of the soil, and delayed immigration of dominant tree species.

The most conspicuous change in the pollen sequence is the abrupt decrease in elm (Figures 4 and 5). This decrease is so sudden that it is natural to attribute it to the introduction of a disease, such as the Dutch elm disease in America today, but it is very difficult to find independent evidence for such an event. Ivy pollen decreases simultaneously, so perhaps sudden climatic cooling brought about the changes. Where studied in detail, however, the elm decline is followed immediately by the first appearance of pollen of plants associated with some form of agriculture—either the weeds resulting from land clearance or the cereal grains themselves. A sensible sequence of cultural events has been posited to explain the detailed forms

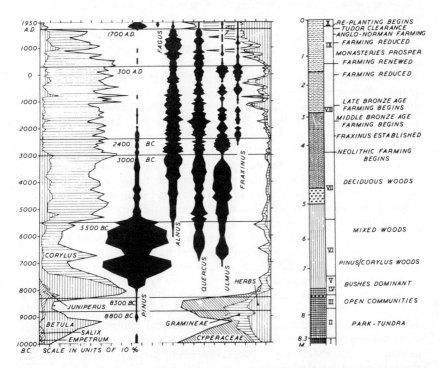

Figure 5 Holocene pollen diagram for Littleton Bog in south-central Ireland, showing possible correlations between pollen curves and cultural phases in the area beginning with the elm decline at the time of Neolithic farming 3000 BC. The gradual increase in herb and grass pollen from this time represents the inexorable influence of man on the landscape. The spurt of pine near the top of the sequence records forestry plantations. From Mitchell (1965).

of the pollen curves, and experiments in forest clearance and crop planting have been carried out to test the hypothetical reconstructions (Iversen 1973).

In this view the elm decline is attributed to felling the elm trees by girdling the bark to encourage the sprouting of young elm shoots as well as shrubs that can supply leaf fodder for the domestic animals introduced by the first farmers. The felling of big elm trees has the additional advantage of opening up the forest canopy and stimulating the growth of nearby oak trees, which supply acorns for pig food. The farming settlements as well as the soil disturbance by pigs and cattle lead to the proliferation of weeds like plantain and sorrel, which appear for the first time in quantity in the pollen profiles. Finally, when the forest clearance became more intensive, aided by burning of felled trees and brush, the cultivated fields yielded pollen of cereal grains as well as of weeds. Thereafter the abrupt minor fluctuations in the pollen curves for various trees, shrubs, and herbs can be attributed to successive abandonment and renewed clearance, and to various forms of herding and cultivation. The attendant disturbance of the soil, interruption of natural forest succession, and even introduction of plant species must be considered as strong factors in the forest development during all of subsequent time. It goes without saying that the western European forests of 5000 years ago will never return to their natural condition, for so much has happened to the environment since their first disturbance by man.

Among the natural environmental changes that have occurred after man's first significant influence on the vegetation is the gradual return to a somewhat cooler climate, as seen particularly with the decrease in ivy in the Danish pollen diagrams. Also the nutrient-rich soils that had supported the forests of lime and other hardwoods became progressively leached, and some areas were taken over by acid birch or oak forests. In moist locations the acid conditions led locally to reduction in soil organisms and thus in the decomposition of plant material; peat began to accumulate locally, isolating plant roots from the mineral substratum. Sphagnum moss took over the bog surface, and raised bogs expanded into once-forested areas.

About this time a new tree, the beech, appeared on the scene. It invaded rapidly from the south and took over much of the land previously occupied by lime that had been disturbed by man. Whether beech fruits were actually spread northward by man at a rate faster than would have occurred naturally can never be demonstrated, but certainly the disruption of the stable lime forest by man's activities set the stage for this new forest element.

Continued farming activities in successive cultural periods—Bronze Age, Iron Age, Roman period, the Viking period, and modern time—have resulted in deforestation of most of western Europe. Much of the land that is not actually cultivated now has been transformed into heathland after its cultivation several millennia ago. The few existing forest areas are mostly plantations or are intensively managed by selective wood-cutting.

This story of Holocene forest history for Denmark can be matched in its general form for other parts of western Europe. For example, in Sweden north of the modern limit of temperate deciduous trees, the most pronounced event was the temporary northward invasion of alder, hazel, oak, and elm during the midpostglacial warm period into areas previously dominated by birch and pine (Fries 1965). As in the south, this interval was interrupted by the elm decline and the subsequent irregularities resulting from forest clearance and agriculture.

Then the scene was changed by the immigration of beech and hornbeam in the south and especially by spruce in the north. Spruce came from the east from a Pleistocene refuge presumably in southeastern Europe (Figure 6). It was present in northwestern Europe along with pine during early Weischselian interstadials, and in the southeast during middle Weichselian interstadials, but in the early Holocene it did not accompany pine to the north. It migrated from Karelia (5000 BP) through Finland and reached Sweden about 2000 BP, spreading both around the northern end of the Gulf of Bothnia and across the archipelago (Moe 1970, Tallantire 1972). It is now found in Norway and is apparently still expanding. Various writers have discussed whether this phenomenon represents 1. the contemporaneous response to cooling climate, 2. delayed migration not controlled by climate, or 3. disturbance of the forest by man and his animals, making it easier for spruce to colonize gaps in the forest floor.

Eastern North America

EARLY AND MIDDLE WISCONSIN   The major North American areas with dated stratigraphic sequences referable to Early and Middle Wisconsin are in the Lake

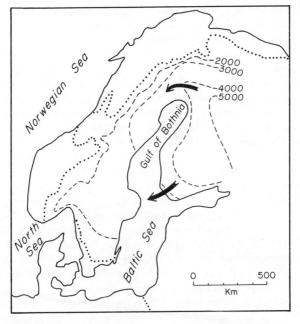

Figure 6   Map showing westward migration of spruce through Fennoscandia. Contours in years before present. Dotted line shows present range limit. Generalized from Moe (1970).

Erie area and in Illinois, Missouri, and the Southeast. Following Dreimanis & Karrow (1972), Early Wisconsin is considered to encompass the time from the end of the Sangamon interglacial to the beginning of the Port Talbot interstadial (about 55,000 BP in the Toronto area); Middle Wisconsin covers the time from then until the advance of the Late Wisconsin ice over the area (about 22,500 BP).

The Early Wisconsin of the Lake Erie area is beyond the range of accurate radiocarbon dating, and its few fossiliferous beds have been little studied. Pollen analysis of Middle Wisconsin sediments by Berti (1975) reveals a sequence starting with a brief interval (perhaps 55,000 to 50,000 BP) of pine, oak, and herbs, implying open vegetation and a climate warmer and drier than today's. The balance of the sequence, up until the Late Wisconsin till, is a monotonous assemblage of pine and spruce, even though it is interrupted twice by layers of Middle Wisconsin till. Present at three levels, however, are macrofossils of tundra plants; Berti considers that an open forest-tundra is the most likely reconstruction.

But throughout this interval pine pollen maintains a steady average of more than 60%. Here is the dilemma so commonly reached in modern American pollen studies—no surface samples have been found to provide an appropriate analogue for 60% pine pollen along with a strong record of tundra herbs. Today in the tundra of central Canada pine does not reach the tree line, which is formed of black spruce or in some cases white spruce or larch (Rowe 1972). Because of its high production and ease of dispersal, pine pollen reaches the tundra in quantity, but in no samples does it exceed 30% of the total pollen. Farther east it is much less, because pine does not occur in the boreal forest of Labrador to provide even a distant source. An alternative explanation is that in Middle Wisconsin time jack pine, along with spruce, did reach the northern tree line. The periglacial climatic effects of the ice sheet, which may have been near to the north all during this time, cannot be predicted. The vegetation that persisted in this area for so many thousand years with relatively little change, even after interruption by glaciation, implies long-range environmental stability and little periglacial effect of ice-sheet advance, as well as limited immigration of other trees into the area. This situation contrasts strongly with that which prevailed during the interglacial episodes such as the Holocene, when the composition of the temperate forest went through many changes as dominant trees immigrated at differential rates.

In southern Illinois, Early and Middle Wisconsin vegetation is recorded for several sites in the Vandalia area (Grüger 1972a, 1972b), supplemented by several short carbon-dated sections elsewhere in Illinois (Grüger 1972b). The Pittsburg site, which is a recently drained lake on Illinoian drift—a rare feature anywhere—starts with a late-Illinoian spruce-pine assemblage and then covers what must be most of the Sangamon interglacial (Figure 7). This is followed by an herb zone lasting until 35,000 years ago; it strongly resembles the pollen assemblage from modern prairie—grasses, sedges, *Ambrosia*, *Artemisia*, and chenopods, as well as some less common but diagnostic prairie herbs. The 10–20% oak pollen implies an oak savanna of the prairie border, but the *Artemisia* values suggest dry prairie farther from forest. Pine pollen holds values of 5–10% throughout the pollen zone—not enough to indicate the presence of pine trees in the area. In northern

Illinois at this time, a similar herb assemblage dominated, but with less oak and more pine, as was the case also on the north side of Lake Erie until 50,000 years ago (see above). The ice sheet during at least part of this time reached as far as northern Illinois, as indicated by the occurrence of Altonian till dated as younger than the Sangamon interglacial and older than 41,000 BP.

These are unexpected results: they can only mean that the Wisconsin climate in the Middle West up until about 35,000 years ago was drier than that of today, and not much colder.

About 35,000 years ago pine, birch, and alder apparently expanded into the Vandalia area at the expense of prairie, and oak maintained its presence (Grüger 1972b). Meanwhile in northern Illinois spruce and pine expanded into the prairie, and oak diminished in importance. This is the time of the latest Altonian ice advance in northern Illinois, dated as between 33,000 and 27,000 BP (Kempton &

PITTSBURG BASIN, Southern Illinois

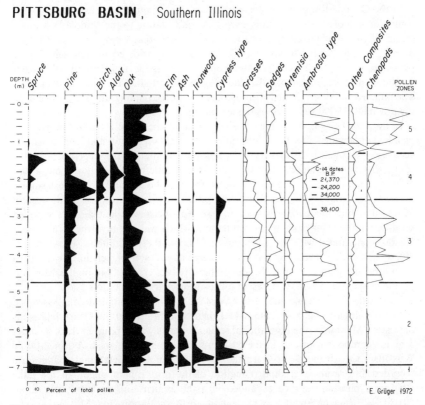

Figure 7 Summary pollen diagram from Pittsburg basin, southern Illinois. Zone 2 is correlated with the Sangamon interglacial, zone 3 with the Middle Wisconsin, zone 4 with the late Middle Wisconsin (Farmdalian) and the Late Wisconsin, and zone 5 with the Holocene. Redrawn from Grüger (1972a).

Hackett 1968). The expansion of pine and spruce had occurred much earlier in the Lake Erie region.

Spruce then expanded southward, reaching eastern Kansas by 25,000 BP, when pine diminished to 15% and oak was negligible (Grüger 1973). Spruce reached southern Illinois about 21,000 years ago. The Farmdalian interstadial is defined in Illinois as 28,000 to 22,000 years ago, largely on evidence from soils in loess, but at none of these sites does the pollen sequence give evidence for an interval of warmer climate. It appears instead that pine and its boreal associates and then spruce advanced southward during the entire time from 35,000 to 21,000 years ago to close what must have been a very broad prairie peninsula in Kansas, grading east to oak savanna in southern Illinois, with earlier extensions eastward to the Lake Erie area. Although the prairie was largely eliminated by this invasion of boreal forest, oak persisted in southern Illinois during the time of the spruce maximum.

In the Ozark Mountains of western Missouri, at about the same latitude as the Vandalia area of southern Illinois, spring deposits dating to the Middle Wisconsin (King 1973) provide a glimpse at the former vegetation. A pine-herb–pollen zone dating from 40,000 BP to about 20–25,000 BP is considered to represent a pine parkland, similar to the contemporaneous assemblage in southern Illinois except for the smaller proportion of oak. Macrofossils indicate that the pine was jack pine. The pine parkland terminated with the Late Wisconsin immigration of spruce from the north 20–25,000 years ago as the ice sheet advanced.

In the Ozarks as in other areas we are faced for the Middle Wisconsin with the problem of lacking suitable analogues by which the vegetation and climate can be accurately reconstructed. Today oak and pine dominate the vegetation in the southern Ozark highlands, but the pine is a southern type (yellow pine, *Pinus echinata*), and spruce is far away. King states that the assemblage is most similar to that of surface samples from the pine-hardwood forest of northwestern Minnesota, but the comparison is not so apt, because the major pine-pollen producer in the Minnesota area is white pine, which was not involved in Missouri. He also compares it with the aspen parkland of southern Manitoba. The principal differences are that the spruce-pollen percentages are higher in the modern Manitoba samples, and that the major herb-pollen type by far in the Missouri profile is sedge, which may represent local spring-marsh vegetation.

Thus in the three areas of the Middle West for which substantial Middle Wisconsin pollen profiles are available—the Lake Erie area, Illinois, and the Ozark highlands of western Missouri—the vegetational sequence starts with a prairie, which includes a major component of oak in the east and jack pine in the west. As the Middle Wisconsin ice advanced and retreated, no synchronous vegetational changes are recorded, but rather a gradual reduction of the prairie component. In the Lake Erie area, pine and spruce expanded into the prairie and replaced oak about 50,000 BP. Pine parkland was present in northern Illinois and Missouri 35,000 BP, with oak savanna in southern Illinois. Spruce then expanded in northern Illinois, and pine into southern. Spruce reached eastern Kansas 25,000 years ago, the Ozarks soon thereafter, and southern Illinois by 21,000 years ago. Pine was squeezed out completely soon after this. By this time the prairie openings were closed with forest, but some oak persisted in southern Illinois in the spruce forest. In the Southeast, several recent diagrams covering the time range of the Middle Wisconsin indicate a vegetation and climate significantly different from the Late Wisconsin. In the piedmont of northern Georgia an oak-hickory-cypress assemblage characterizes the interval from 30,000 to 24,000 BP (Watts 1973a). The component of cypress at these sites implies a climate as warm as that of today. On the Carolina coastal plain, a pollen zone with cedar, birch, oak, alder, and several herb types has carbon dates of 36,000 to 25,000 BP, preceding the Late Wisconsin development of boreal forest types (Whitehead 1973). The high proportion of herb pollen suggests a climate dry enough to allow the development of treeless areas. Meanwhile in south-central Florida the aridity was more strongly expressed, as pollen of herbs and shrubs dominated in an area of old sand dunes 37,000 years ago (Watts 1975).

LATE WISCONSIN   During the time of the Late Wisconsin glaciation, lasting from about 20,000 to 11,000 BP, a fringe of tundra locally bordered the ice sheet. The longest and most detailed record is at Wolf Creek in central Minnesota (Birks 1976), where an herb-pollen zone extends from the base of the section at 20,500 BP up to 13,600 BP. All during this time the area was practically surrounded by fluctuating ice lobes. The pollen flora at Wolf Creek includes many types now restricted to tundra regions. During the last 1000 years of the interval dwarf birch and other dwarf shrubs are represented: apparently the climate warmed enough to permit a change from herb tundra to shrub tundra. The spruce forest that then invaded the area prevailed until about 10,000 BP.

Elsewhere a Late Wisconsin herb-pollen assemblage has been found only in the northern part of the Great Lakes region and in New England, with an extension into Pennsylvania and down the Appalachian Mountains at least as far as Maryland (Maxwell & Davis 1972). This is not say that tundra did not occur elsewhere in the Middle West at this time. In contrast to Wolf Creek, which is located in a swale between two drumlins, most pollen sites in the region are in ice-block depressions that became lakes only after buried ice melted—perhaps many thousands of years after active ice left the region (Florin & Wright 1969). In northwestern Minnesota, for example, two sites in the Itasca moraine have basal carbon dates of about 11,000 years ago, yet regional evidence indicates that the moraine was formed more than 20,000 years ago. During the intervening 9000 years the climate must have been cold enough to permit the persistence of the buried ice—and the formation of permafrost as well, at least locally. Few geomorphic or stratigraphic indicators of past permafrost, like ice-wedge casts or involutions, have been found in Minnesota, but a major occurrence of ice-wedge casts in nearby Wisconsin has been attributed by Black (1965) to climatic conditions characterized by mean annual air temperature of − 5°C or less.

It seems more likely that a fringe of periglacial tundra did not in fact exist from Ohio to South Dakota during the time of maximal ice advance. Here the ice lobes penetrated far to the south because of the low elevation. They may well have invaded the forest, which perhaps could persist because of the southerly latitude.

Some evidence for forest—or at least for trees—comes from the logs in Wisconsin till in Ohio dated from 23,000 BP in the north to 17,000 in the south, recording the advance of the ice to its distant terminus (Goldthwait 1958, Burns 1958).

Today the boundary between tundra and forest is broadly transitional in central Canada. Pollen counts reflect the transition in a very general way, and maps and numerical analyses have been prepared to show the regional variations in the pollen rain (Davis & Webb 1975, Webb & McAndrews 1976, Birks et al 1975). A compilation of 102 analyses of lake muds from central Canada, when combined with vegetation maps prepared from forest-inventory data and air photographs, gives a picture of the representativity of various plants in the pollen rain (Lichti-Federovich & Ritchie 1968). The results (Table 2) show that spruce pollen is blown into the tundra from the forest-tundra and forest to the south; pine arrives in even greater quantities from areas even farther south, for it is a better pollen producer and disperser than spruce. Otherwise the pollen of dwarf birch, alder, sedges, and heath plants is conspicuous. The pollen production from the local vegetation is generally less than 100 grains cm^{-2} yr^{-1}, however, according to pollen-trap collections (Ritchie & Lichti-Federovich 1967).

In the forest-tundra and open coniferous forest the spruce (and some larch) is scattered or in groves along river courses or lower slopes, becoming denser southward, where paper birch, aspen, and jack pine occur locally. Many of the open areas are taken by lichens, which leave no pollen record, and by heath plants, which produce little pollen. The pollen assemblage for these transitional areas is therefore much the same as that of the tundra, except for smaller amounts of sedge pollen. The total pollen influx is much greater, however, averaging about 1000 grains cm^{-2} yr^{-1} in the forest tundra and 6000 in the open coniferous forest.

A closed coniferous forest is dominated by spruce and especially pine, both trees and pollen. Pollen of paper birch and alder is common, but pollen of sedge and heath plants is of minor importance.

These pollen percentages, shown in Table 2, should be applied to the interpretation of fossil pollen assemblages only with caution, because the composition of the ancient vegetation was significantly different in many cases. For example, the modern Canadian samples include appreciable percentages of pine pollen, even though jack

Table 2   Average pollen percentages for surface-sediment samples from 102 lakes in central Canada (from data of Lichti-Federovich & Ritchie 1968)

	Tundra	Forest/ tundra	Open coniferous forest	Closed coniferous forest A	Closed coniferous forest B	Closed coniferous forest C	Mixed forest (upland)	Mixed forest (lowland)
Picea	14	31	30	11	23	32	23	20
Pinus	22	20	33	64	49	38	28	32
Betula	26	13	11	5	13	8	16	10
Alnus	12	17	16	5	9	8	10	7
Artemisia	1.5	1.4	1.1	1.9	1.7	1.4	5.9	2.3
Cyperaceae	15	10	3	—	0.5	1.7	1.8	8.3
Ericaceae	1.7	0.8	0.6	—	0.1	0.1	—	—
Gramineae	—	—	—	—	—	—	3.1	8.6
Number of sites	14	10	14	8	22	22	12	14

pine occurs only in the southern part of the closed coniferous forest. The Late Wisconsin forests of the Middle West, however, contained no jack pine. If the contribution of long-distance pine is subtracted from the totals, the influx would be appreciably reduced, for the space in the forest now taken by jack pine would have been occupied by a more modest pollen producer.

The Late Wisconsin border between tundra and forest, even though it can only be vaguely defined from pollen diagrams, moved northward during the time of fluctuating ice retreat. It passed Wolf Creek in central Minnesota 13,600 BP (Birks 1976), Kotiranta Lake west of Duluth in northeastern Minnesota 11,500 BP (Wright & Watts 1969), and Lake of the Clouds next to Canada 10,000 BP. The border passed north of Lake Superior in Ontario 10,000 BP (Saarnisto 1974), but farther north the ice was apparently overtaken by the spruce forest.

In eastern North America (Figure 8) the tundra/forest border left the Maryland mountains 12,500 BP (Maxwell & Davis 1972), northeastern Pennsylvania 13,500 BP (W. A. Watts unpublished), Connecticut 12,000 BP (Davis 1967b), northern New York 12,000 BP (Miller 1973), Quebec City area 10,000–7000 BP (Richard 1971, Richard & Poulin 1976), and interior Labrador as late as 5500 BP (Morrison

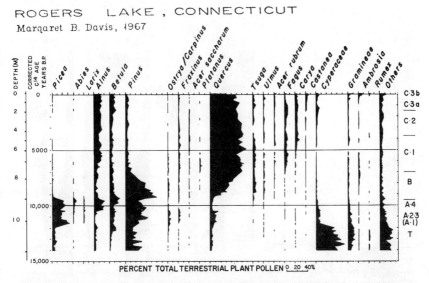

Figure 8 Pollen diagram for Rogers Lake, Connecticut, representative for northeastern United States. The high percentages of pine pollen in the herb-pollen zone (*T*) represent distant transport of pine pollen into tundra from forested regions to the south. The high percentages of *Quercus, Fraxinus,* and *Ostrya-Carpinus* pollen in the lower part of the spruce zone (*A*-2, 3) may reflect higher amounts of oak, ash, and ironwood trees in a spruce woodland than can be found today in Canada, but the relative decrease in the upper part (*A*-4) results from increase in influx of spruce, pine, and alder pollen, as shown from separate data on pollen influx. Zones B and C are dominated by oak pollen, but they show the progressive immigration of hemlock, beech, hickory, and chestnut. From Davis (1967b).

1970). Here apparently a fringe of tundra continued to separate the wasting ice from the closed boreal forest to the south.

The widespread spruce forest that characterized the area south of the tundra fringe was even more extensive than the modern spruce forest of Canada. Today spruce extends across the continent and from the northern tree line to the central Great Lakes area and New England, making a belt with an average width of 1000 km from north to south. White and black spruce have about the same distribution, although white spruce is less common near the northern tree line, and black spruce is confined to wetlands near the southern limit of its range (Rowe 1972). Before 14,000 BP the Late Wisconsin spruce forest extended from the central Great Lakes region far to the south—at least to eastern Kansas, where spruce pollen values average 75% (Grüger 1973), central Missouri up to 92% (King 1973), and southern Illinois 20–80% (Grüger 1972b).

The southern limit of the spruce forest is unknown, for no stratified sites have been found between Illnois and the Gulf of Mexico. The report of Brown (1938)

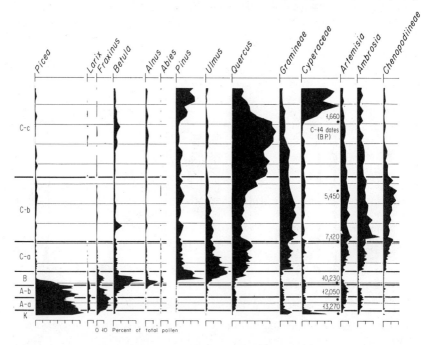

Figure 9   Pollen diagram for Kirchner Marsh, southeastern Minnesota. The herb zone (*K*) at the base probably represents tundra, terminating about 13,000 BP. The spruce zone (*A*) has no pine but has high proportions of oak and especially ash pollen in the lower part, unlike any spruce forest today. The spruce zone ends about 10,000 BP. The Holocene sequence illustrates the replacement of mixed deciduous forest (zone *C-a*) by prairie and oak savanna (zone *C-b*, starting about 7000 BP), followed by reversion to oak forest (zone *C-c*, starting about 5500 BP). Redrawn from Wright et al (1963).

concerning cones of white spruce in alluvial sediment in the blufflands of northern Louisiana is largely confirmed by a new find with 26% spruce pollen, as well as needles and a twig, demonstrating at least the local occurrence of spruce 1500 km south of the present limit of white spruce in Michigan (Delcourt & Delcourt 1977). A site in east Texas shows a few percent spruce pollen (Graham & Heimsch 1960). The western limit of the spruce forest is also unknown. A site at the northern edge of the Sandhills in western Nebraska in the heart of the modern prairie shows dominant spruce pollen 12,600 years ago (Watts & Wright 1966). In west Texas, also in prairie country today, pine pollen dominated 22,000 to 11,000 years ago, with a little spruce as an eastward extension of the Rocky Mountain woodland (Hafsten 1964). No prairie is recorded for the Late Wisconsin at any of these sites.

In the east spruce extended as far south as northern Georgia in the Appalachian piedmont (Watts 1970). Even on the Carolina coastal plain spruce-pollen values reach 30% (Whitehead 1973). Today red spruce occurs above 1300 m elevation in the southern Appalachians. This may have been the species that moved down to the piedmont and plains.

Even though the spruce forest of the Late Wisconsin formed a belt at least as broad as that of today, its composition was far from the same. Today it contains larch throughout its extent—entirely in lowlands in the southern part of its range, but reaching tree line on the north, where it succeeds rapidly after fires. Jack pine does not occur in the easternmost part (Labrador) and the western extremity (Alaska and western Yukon) of the boreal forest, nor is it found in the northern 300 km across central Canada (Halliday & Brown 1943). It is also uncommon in the southern fringe against the aspen parkland from Saskatchewan to Alberta. Balsam fir is found in most of the boreal forest in the east, especially around the Gulf of St. Lawrence, but in west-central Canada it is largely restricted to the southern part, and, unlike the other conifers, it does not reach the Rocky Mountains in Alberta. Northern white cedar occurs in the southern part of the boreal forest. Trees other than conifers include paper birch, aspen, and balsam poplar; they are not common close to the tree line, however. Black ash and American elm and other temperate deciduous trees extend locally into the boreal forest on favorable sites.

Overlapping the southern part of the range of spruce, jack pine, and fir from Minnesota eastwards are several other conifers, which with the temperate hardwoods from the south are important enough to define the transitional Great Lakes–St. Lawrence formation—white pine, red pine, and eastern hemlock.

In contrast to the modern boreal forest, the Late Wisconsin spruce forest in the Middle West lacked pine, although jack pine had been present in the Middle Wisconsin (Figure 9). The principal other types in the spruce pollen zone are birch, aspen, larch, fir, black ash, elm, and oak, along with sedges and *Artemisia*. This assemblage cannot be matched with the same general values from any part of the modern boreal forest, not only because of the lack of pine pollen but because of the relatively high values for pollen of temperate hardwoods. Black ash in particular seems anomalous: it reaches more than 15% of total pollen at some sites, yet today it is usually less than 1%, even when percentages are recalculated after subtraction of pine pollen. Black ash grows primarily in lowlands, although it also occurs on uplands with rich moist soils in parts of the transitional Great Lakes Forest. In the boreal forest proper, however, it is confined to lowlands and is uncommon. It might be postulated that in the Late Wisconsin it grew on unleached upland soils kept moist by late-lasting ground frost or even permafrost.

The 2–3% each of elm and oak pollen are likewise anomalous. Distant transport from the south seems unlikely because values do not increase systematically southward, at least not until southern Illinois (Grüger 1972b), which seems too far away to supply significant oak pollen to northern Minnesota 1500 km away. Elm pollen, which is dispersed less easily than oak, had no known major source in the south at this time. It seems likely that oak and elm, as well as black ash, were distributed locally on favorable sites throughout the spruce forest. American elm, like black ash, favors moist rich soils. The oak species involved is not known; of the several present today in the Great Lakes region, northern red oak has the most northerly distribution.

The several percent each of sedge and *Artemisia* pollen in the spruce zone at most sites also finds no match in surface samples from the modern spruce forest (even when pine is deleted from the pollen sum). Sedge pollen is diagnostic of forest-tundra sites today but not of a closed spruce forest. *Artemisia* species require sun and usually dry soils. Certain species occur in the tundra today but they are uncommon, probably because of the ubiquitous mat of moist turf. Favorable open, dry conditions for *Artemisia* growth in the Late Wisconsin spruce forest may have been provided by immature soils on ridge crests, lack of a turf mat, and occurrence of drying winds induced by the ice sheet nearby.

In the Appalachian region and the Atlantic coastal plain the Late Wisconsin spruce forest did contain pine, identified as jack pine on the basis of pollen and needle morphology (Watts 1970). Otherwise, the proportions of oak and other hardwoods are as low as they are in the Middle West. No sites have been investigated in the heart of the mixed mesophytic forest, made famous by the studies of Lucy Braun (1950), who postulated that the rich mixture of south-temperate hardwoods had been there for literally millions of years. It seems more likely, however, on the basis of the regional pollen rain at other sites in the southern Appalachians, that the mixed mesophytic forest that is now so extensive in the area did not occur in its present distribution. That is not to say that the trees involved were not present somewhere in the southern Appalachians in favorable sites, but they were not necessarily in the same combinations as today. That the Late Wisconsin climate was distinctly different from that of today, however, is demonstrated by the apparently widespread occurrence of spruce, jack pine, certain northern forest herbs, and northern aquatic plants (Watts 1970). In south-central Florida aridity persisted from the Middle Wisconsin (Watts 1975).

In conclusion it must be acknowledged that no satisfactory modern analogue for the Late Wisconsin spruce forest has been identified. The absence of pine in the midwestern diagrams may be accommodated by selecting southern Labrador as the analogue, but the minor pollen types cause problems. More northerly portions of the modern spruce forest may be open enough to show higher values of sedge pollen, but then the values for temperate hardwoods are less than ever. The lack of suitable analogues brings us back to the special conditions of Late Wisconsin time: 1. a great ice sheet to the north, providing climatic conditions (especially winds to maintain forest openings) that nowhere prevail today; 2. vast areas of unleached glacial deposits, with immature soils that favor the rapid immigration of spruce and also certain hardwoods like ash and elm, and 3. prior emigration of jack pine to refuges so distant from the Middle West that return was somewhat delayed after climatic warming. The Middle Wisconsin spruce forest of the Lake Erie area contained pine, and in northern Illinois pine was the dominant conifer of the prairie border. As the late-Altonian ice advanced into Illinois the pine moved south to fill the prairie and oak savanna, and spruce began to replace the pine. The trend continued during the Farmdalian interstadial in Illinois, and when the Late Wisconsin ice reached its maximal extent 20,000 years ago pine was practically gone from the Middle West.

One of the keen problems in Lake Wisconsin climatic history is the relation of the pollen sequence to the fluctuations in the ice-sheet margin. In most of the Great Lakes region the icelobes are well delineated and record in their moraines and stratigraphic relations a complex history of advance and retreat. In Minnesota, for example, at least 4 phases of ice advance are recognized 21,000 to 11,000 years ago. Ice-front shifts amounting to several hundred kilometers occurred. Although it is generally assumed that such major ice movements reflect regional changes in climate, the situation is complicated because the ice advances are not always synchronous from lobe to lobe. If the advance of an ice front results from a decrease in summer temperature (i.e. summer melting), then the response can be essentially immediate. But if the advance results from increase in winter snowfall far back in the source area of the glacier, the reaction at the margin can be long delayed. Thus climatic change may not be registered by immediate and synchronous advances of all ice lobes. A further possibility is that certain ice lobes with favorable morphology and thermal profiles may advance by surging—a phenomenon that may not be related to climate at all.

Unlike the relations at the margin of the Scandinavian ice sheet, where the Central Swedish moraine can clearly be correlated with the Younger Dryas phase of tundra development in the pollen sequence, in the Middle West there is no clear relation between the several ice-lobe fluctuations and the vegetational history, despite diligent search. In fact, not only has the Two Creeks/Valders fluctuation of the Lake Michigan lobe failed to leave a contemporaneous mark in the pollen record, but it proves not to be synchronous with the Alleröd/Younger Dryas oscillation (thereby weakening the case for global synchroneity of minor climatic changes), and in addition it is more complex than previously assumed (Mickelson & Evenson 1975).

Vegetational shifts and thus perhaps climatic change can best be detected in a pollen diagram if a major ecotone migrated past the site. Thus the Bölling and Alleröd interstadials in Europe are identified because the changes in proportions of herb pollen and tree pollen indicate that the tundra/forest border moved back and forth. No such fluctuation can be found in any of the American diagrams in which a tundra/forest border can be inferred. The best that can be said is that the border moved northward during Late Wisconsin time. At Wolf Creek in central Minnesota, for example, where the pollen record starts at 20,500 BP and covers the time of at least three close approaches of ice lobes, tundra changed to forest at 13,600 years ago and no reversals are evident.

The spruce-pollen zone in the Minnesota area does have a consistent subdivision, however, but its climatic implications are uncertain. A lower subzone is characterized by slightly more ash pollen and the upper subzone by more birch (Figure 9). *Artemisia* also has a maximum, primarily in the birch subzone. The change is dated about 12,250 BP. Ash certainly has a more temperate distribution than birch, and *Artemisia* seems to have higher values in the more northerly sites. These relations imply that a warming (ash subzone) and then a cooling (birch-*Artemisia* subzone) may have occurred. But the rise in birch may instead reflect delayed immigration, and the *Artemisia* maximum may indicate either expansion of the prairie or increased soil aridity related to winds.

HOLOCENE   The spruce forest that characterized the Middle West in the Late Wisconsin disappeared rapidly as the ice accelerated its retreat across the Great Lakes area about 11,000 years ago, and this event is taken to mark the Pleistocene/ Holocene boundary. The time ranges from about 12,000 years ago in the present prairie area to about 10,000 years ago in the modern coniferous forest area. The pollen record of the transformation is generally abrupt at any locality, occurring over just a few hundred years, although in the north, where spruce still occurs, the change is slower. In most cases it does not seem to be foreshadowed by any trends either in the spruce pollen curve or any other curves.

The trees that replaced the spruce forest may have depended more on the seed sources available and on other factors than on the subtleties of climatic change. Here began the same kind of grand progression as witnessed in western Europe at the end of the Younger Dryas. The sequence is less easy to summarize, however, because it is complicated by more pronounced climatic trends at the prairie border, and because there are too many major gaps in the geographic coverage of sites. Some of the main aspects of the migrational succession have been summarized by Davis (1976), who has prepared preliminary migration maps for some of the major trees. Other aspects of stability vs. invasion by new dominants are considered by Watts (1973b).

The early-Holocene climatic trend towards increased warmth and dryness (Hypsithermal interval) is most easily followed at the prairie border in Minnesota, where migrational problems are not severe. The climatic trends are independently indicated by evidence for shallowing or drying of lakes (Watts & Winter 1966). When the climate became too warm for regeneration of spruce, the space in the forest was taken largely by expansion of alder and birch, which were already at hand (Figure 9). Elm and oak soon followed. They may also have been present locally in the spruce forest, and oak at least had a major refuge in southern Illinois in the Late Wisconsin. Elm preceded oak in its expansion in some cases, perhaps because it is a better pioneer on moist unleached soils of the lower slopes. Its relatively large expansion even into northern Minnesota at this time is note-worthy (Amundson & Wright 1977).

The oak-elm forest, which also included ironwood, maple, ash, and basswood, soon changed character, as oak became predominant and prairie herbs expanded. Pollen curves for grasses, sedges, *Ambrosia, Artemisia,* other composites, and chenopods and other prairie herbs all increased. This transformation to prairie was probably completed in Nebraska and Kansas by 10,000 years ago, with only a very brief interval of deciduous woodland before development of essentially complete prairie. Farther to the northeast the transformation was not completed until about 7000 years ago, when prairie and oak savanna reached more than 100 km east of its present limit in southern and western Minnesota. Similar changes farther northeast can be detected, for example by oak expansion into coniferous forest.

Meanwhile in eastern Minnesota the vegetational sequence was modified by the involvement of pine. Jack pine was presumably the first to immigrate, for it has the most northerly distribution of the eastern pines today. Its principal Late Wisconsin refuge was the Appalachian highlands. Its immigration to Minnesota may have been delayed by the barrier provided by the western Great Lakes and the ice lobes persisting just to the north (Wright 1968). Red pine, which has the same pollen type, probably immigrated soon after.

Pine reached eastern Minnesota about 10,000 years ago. At the more southerly sites the spruce forest had already started its decline, and a brief interval of hardwoods ensued before the pine took over the landscape (Amundson & Wright 1977). In the center of the state (e.g. Wolf Creek), pine arrived precisely at the time of the spruce decline, and the complete transformation took only a few score years. In northeasternmost Minnesota, the spruce forest was still invading tundra when pine arrived, and for a few hundred years a pollen assemblage of spruce, pine, elm, and northern white cedar prevailed, at which point red pine may have arrived (Amundson & Wright 1977).

The continued climatic warming eventually brought about the replacement of jack and/or red pine in central Minnesota by temperate hardwoods, with elm being a more significant component than it is today in the so-called maple-basswood forest. Meanwhile, white pine had been migrating westward from its Late Wisconsin refuge in the Appalachian Mountains, and it reached eastern Minnesota just after the culmination of the northeastern expansion of the hardwoods 7000 years ago at the height of the prairie period. This led to a halt in westward expansion of white pine, and it remained in eastern Minnesota until the end of the prairie period, at which time it began an accelerated migration westward. The white pine and other northern conifers are apparently still expanding into the hardwood forest, which in turn is moving into the prairie (Jacobson 1975).

For this Holocene sequence in the Middle West, where migrational problems are not paramount and where modern analogues can be found for most pollen assemblages, the pollen diagram can be converted to a climatic diagram by a numerical procedure based on the relation of the modern pollen rain to modern vergetation and climate (Webb & Bryson 1972). Pollen analyses are made of surface samples collected in areas of known vegetation, where various climatic parameters are known from weather records. The pollen data are then transformed directly to climatic data, and curves for summer temperature, frequency of different air masses, or other parameters can then be graphed. Such climatic curves quantify the climatic conclusions worked out by qualitative comparisons of the same data. They work only as long as proper analogues can be found among the surface-sample analyses from various regions. Trouble thus comes in the Late Wisconsin of the Middle West, where the pollen assemblages are not matched anywhere (e.g. the absence of pine), and in the Holocene of the East, where migration of major types may not be determined by contemporary climate.

The Holocene vegetation history in the Appalachian Highlands and the Northeast is dominated by differential migration rates of major tree genera. With the initial warming of the climate at the end of the Pleistocene, the spruce-pine forest in New England was replaced immediately by pine-birch, including white pine, and this in turn by oak (Figure 8). The minor percentages of other hardwood types originally formed the basis for a subdivision with climatic interpretation—oak-hemlock, oak-hickory (warmest time), and oak-chestnut. With the more recent availability of pollen diagrams throughout the East, it now seems that this sequence reflects differential migration rates from Pleistocene refuges located in various parts of the Appalachians or coastal plain south of the glacial border. Besides the three genera already mentioned, beech, sweetgum, tupelo, pitch pine, and, farther south, some of the species of southern pines may be involved. The full story will come only after many additional detailed and carbon-dated pollen diagrams have been completed, but the preliminary maps of Davis (1976) give an idea of the patterns that can be expected. Migration directions as well as rates differed (Figure 10). Hemlock and beech, for example, are still progressing across northern Wisconsin and Michigan. Each new invasion caused readjustment of the forest as a result of mutual interactions related to shade and moisture tolerance and thus competition for regeneration. Other factors may also have been important: for example, the abrupt decrease in the hemlock throughout much of the East about 5000 years ago may record the introduction of a pathogen into the system.

This is not to say that climate may not also have been a factor, for in the mountains of northern New England the downward movement of conifers during the last few thousand years must reflect a return to cooler climate after the Hypsithermal (Likens & Davis 1975). In this case no migrational lag can be implicated, because the spruce was already close at hand.

CONCLUSIONS

Systematic studies of Quaternary vegetation history in eastern North America through pollen analyses have their roots in the European work of Iversen, Firbas, and many others, who stimulated ecologists like Sears and Deevey to investigate some of the more tractable problems of the late Pleistocene and Holocene in the 1930s and 1940s. A certain preoccupation with detailed trans-Atlantic paleoclimatic

---

$\longrightarrow$

Figure 10   Time lines (in thousands of years) showing migration of major trees from refuges in southeastern United States. From Davis (1976). Dashed line shows present range.

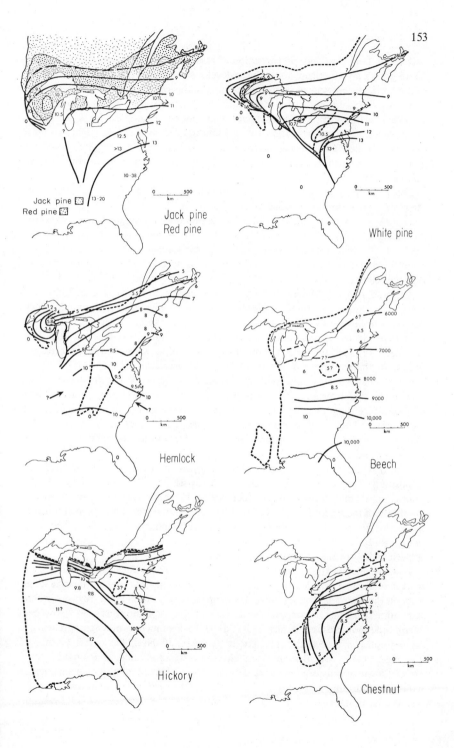

Jack pine
Red pine

Jack pine ▦ 13-20
Red pine ▦

White pine

Hemlock

Beech

Hickory

Chestnut

correlation tended to bias interpretation of results, and it has been only in the last 10–15 years that American studies have developed along independent lines suitable to the particular phytogeography and climate of this continent.

Among the more productive approaches in America has been the extensive use of pollen-surface samples, in efforts to find modern analogues for fossil-pollen assemblages, so that speculation on vegetational reconstructions might be more closely controlled. This effort has been tentatively extended to the next step—making quantitative paleoclimatic reconstructions based on climatic data for the surface-sample sites.

The surface-sample approach—never extensively applied in Europe because of the almost universal disturbance of the vegetation—has been successful mostly in a negative way. That is, the apparent absence of modern analogues for fossil-pollen assemblages, not only for the late Pleistocene but also for parts of the Holocene, has led to a realization that modern forest associations in many areas gradually or abruptly change in composition in response to several different factors besides regional climatic change, especially differential migration rates from distance Pleistocene refuges. Much research is now directed toward documenting the migration routes and rates by preparation of detailed and well-dated pollen diagrams in key areas. It is perhaps ironic that this aspect of American research has independently led to an interpretation of vegetational history that has been long accepted by European workers without the benefit of the surface-sample strategy—successive immigration of dominant trees from Pleistocene refuges.

A second approach emphasized by American workers involves the development of the method of determining absolute pollen influx to avoid the constraints imposed by percentage diagrams. This method depends primarily on the abundant use of radiocarbon dates, and it may make possible the distinction between tundra and open-forest vegetation—often a difficult problem in the interpretation of late-Pleistocene percentage diagrams.

A third aspect of American studies has developed because of the phytogeographic disposition of the continent—specifically the existence of sharp ecotones between prairie, hardwood forest, and conifer forest in some regions. The Holocene migration of these ecotones is strong evidence for climatic change, especially where it is backed up by paleolimnological data. Such a boundary exists in Soviet Europe, but its study has not been integrated with work in the heart of western Europe because of logistical and administrative difficulties.

While American studies have developed along the above lines, modern European work has reached new levels of detail on old problems and opened new fields as well. Surface-sample studies have been productively applied in regions where vegetation is less disturbed than normal, for example Scotland (Birks 1973), Sweden (Berglund 1966), and Greenland (Fredskild 1973), and numerous other comprehensive monographs have recently appeared in Norway, Finland, Denmark, Germany, The Netherlands, and Britain. Meticulous investigations at archaeological sites, always a forté of European workers, continue to provide a wealth of paleoecological data of a scale that is simply not possible in America, where significant human disturbance of the vegetation is only a century or so old. Iversen's (1973) beautiful account of the interplay of man and nature in Danish prehistory is an elegant example.

But perhaps the greatest overall contribution of European pollen studies in comparison with minuscule American results is the wealth of stratigraphic documentation of the various interglacial sediments, as well as particular investigations of the Pliocene-Pleistocene transition. This work may provide the best terrestrial complement to the long-scale Quaternary studies of deep-sea cores. It is at least in part an outgrowth of the practical geotechnical importance of geological stratigraphy in the region of the Rhine delta in The Netherlands, and in the region of the brown-coal deposits of nearby Germany. But interglacial deposits all over southern England, Ireland, and on the mainland beyond the limits of the last glaciation have been investigated, and some of the regional variations in interglacial vegetational history can now be worked out. Such studies have barely started in America, and there seems no prospect of much increase in the near future.

Knowledge of international developments in any scientific field is difficult to absorb, because of the unfamiliar geography, terminology, and language. The parallel yet independent approaches in the study of vegetational history in America and the countries of western Europe illustrate the value of scientific exchange. The areas are unified by certain climatic and geographic similarities and Quaternary geologic history. Understanding of the reaction of different phytogeographic provinces to nature's paleoclimatic experiments leads to special insights into the stability and variability of ecosystems.

Literature Cited

Amundson, D. C., Wright, H. E. Jr. 1977. Forest changes in Minnesota at the end of the Pleistocene. Unpublished.

Andersen, S. T. 1961. Vegetation and its environment in Denmark in the Early Weichselian Glacial (Last glacial). *Danmarks Geol. Under.* Ser. 2, No. 75. 175 pp.

Andersen, S. T. 1966. Interglacial vegetational succession and lake development in Denmark. *Paleobotanist* 15:117–27

Berglund, B. E. 1966. Late-Quaternary vegetation in eastern Blekinge, southeastern Sweden. A pollen-analytical study. I. Late-glacial time. *Bot. Not.,* Suppl. Ser., Vol. 12, No. 1

Bernabo, J. C., Webb, T. III. 1977. Cartographic analysis of the changing time-space patterns in the Holocene pollen record of northeastern North America. *Quat. Res. NY,* Vol. 7

Berti, A. A. 1975. Paleobotany of Wisconsinan interstadials, eastern Great Lakes region, North America. *Quat. Res. NY.* 5:591–620

Birks, H. J. B. 1973. Modern pollen rain studies in some arctic and alpine environments. In *Quarternary Plant Ecology*, ed.

H. J. B. Birks, R. G. West, pp. 143–68. Oxford: Blackwells. 317 pp.

Birks, H. J. B. 1976. Late-Wisconsin vegetational history at Wolf Creek, central Minnesota. *Ecol. Monogr.* Vol. 76, No. 4

Birks, H. J. B., Webb, T. III, Berti, A. A. 1975. Numerical analysis of surface pollen samples from central Canada: a comparison of methods. *Rev. Palaeobot. Palynol.* 20:133–70

Black, R. F. 1965. Ice-wedge casts of Wisconsin. *Wis. Acad. Sci. Arts Lett. Trans.* 54:187–222

Braun, E. L. 1950. *Deciduous forests of eastern North America.* New York: Blakiston. 596 pp.

Brown, C. A. 1938. The flora of Pleistocene deposits in the western Florida parishes, West Feliciana Parish, and East Baton Rouge Parish, Louisiana. In *Contributions to the Pleistocene history of the Florida Parishes of Louisiana. La Geol. Survey Geol. Bull.* 12:59–96

Burns, G. W. 1958. Wisconsin Age forests in western Ohio. II. Vegetation and burial conditions. *Ohio J. Sci.* 58:220–30

Coope, G. R., Brophy, J. A. 1972. Late Glacial environmental changes indicated by a coleopteran succession from North Wales. *Boreas* 1:97–142

Davis, M. B. 1967a. Late-glacial climate in northern United States: a comparison of New England and the Great Lakes region. *Quaternary Paleoecology*, ed. E. J. Cushing, H. E. Wright Jr., pp. 11–44. New Haven: Yale Univ. Press. 433 pp.

Davis, M. B. 1967b. Pollen accumulation rates at Rogers Lake, Connecticut, during late- and post-glacial time. *Rev. Palaeobot. Palynol.* 2:219–30

Davis, M. B. 1976. Pleistocene biogeography of temperate deciduous forests. In *Ecology of the Pleistocene*, ed. R. C. West, W. G. Hoag, La. State Univ., *Geoscience and Man*, 13:13–26. 76 pp.

Davis, M. B., Brubaker, L. B., Webb, T. III. 1973. Calibration of absolute pollen influx. In *Quaternary Plant Ecology*, ed. H. J. B. Birks, R. G. West, pp. 9–25. Oxford: Blackwell. 317 p.

Davis, R. B., Webb, T. III. 1975. The contemporary distribution of pollen in eastern North America: a comparison with the vegetation. *Quat. Res. NY* 5:395–434

Delcourt, P. A., Delcourt, H. R. 1977. The Tunica Hills, Louisiana-Mississippi: late-glacial locality for spruce and deciduous forest species. *Quat. Res. NY,* Vol. 7

Dreimanis, A., Karrow, P. F. 1972. Glacial history of the Great Lakes-St. Lawrence region, the classification of the Wisconsin(an) Stage, and its correlatives. *Int. Geol. Congr., 24th.* 12:5–15

Florin, M.-B., Wright, H. E. Jr. 1969. Diatom evidence for the persistence of stagnant glacial ice in Minnesota. *Geol. Soc. Am. Bull.* 80:695–704

Firbas, F. 1949. *Spät- und nacheiszeitliche Waldgeschichte Mitteleuropas nördlich der Alpen.*, Vol. 1. Jena: Fischer. 480 pp.

Firbas, F. 1958. Über das Fagus-Vorkommen im "interglazial" von Wasserburg am Inn (Oberbayern). *Veröff. Geobot. Inst. Eidg. Tech. Hochsch. Stift. Rübel, Zürich.* 33:81–90

Fredskild, B. 1973. Studies in the vegetational history of Greenland. *Medd. Grønl.,* Vol. 198. No. 4. 245 pp.

Frenzel, B. 1964. Zur Pollenanalyse von Lössen. *Eiszeitalter Ggw* 15:5–39

Fries, M. 1965. The Late-Quaternary vegetation of Sweden, in *The plant cover of Sweden. Acta Phytogeogr. Suec.* 50:269–80

Goldthwait, R. P. 1958. Wisconsin Age forests in western Ohio. I. Age and glacial events. *Ohio J. Sci.* 58:209–19

Graham, A., Heimsch, C. 1960. Pollen studies of some Texas peat deposits. *Ecology* 41:785–90

Grüger, E. 1972a. Late Quaternary vegetation development in south-central Illinois. *Quat. Res. NY* 2:217–31

Grüger, E. 1972b. Pollen and seed studies of Wisconsinan vegetation in Illinois, USA. *Geol. Soc. Am. Bull.* 83:2715–34

Grüger, J. 1973. Studies on the late-Quaternary vegetation history of northeastern Kansas. *Geol. Soc. Am. Bull.* 84:239–50

Hafsten, U. 1964. A standard pollen diagram for the southern High Plains, USA, covering the period back to the early Wisconsin glaciation. *Int. Quat. Congr. 6th.* 2:407–20

Halliday, W. E. D., Brown, A. W. A. 1943. The distribution of some important forest trees in Canada. *Ecology* 24:353–73

Iversen, J. 1941. Land occupation in Denmark's Stone Age. *Danmarks Geol. Under.,* Ser. 2, No. 66. 68 pp.

Iversen, J. 1964. Retrogressive vegetational succession in the Post-Glacial. *J. Ecol.* 52: Suppl. 59–70

Iversen, J. 1973. The development of Denmark's nature since the Last Glacial. *Danmarks Geol. Under.,* Ser. 5, No. 7-c. 126 pp.

Jacobson, G. L. 1975. A palynological study of the history and ecology of white pine in Minnesota. PhD thesis. Univ. Minnesota. 140 pp.

Janssen, C. R. 1967. A comparison between the recent regional pollen rain and the subrecent vegetation in four major vegetational types in Minnesota. *Rev. Palaeobot. Palynol.* 2:331–42

Janssen, C. R. 1973. Local and regional pollen deposition. In *Quaternary Plant Ecology*, ed. H. J. B. Birks, R. G. West, pp. 31–42. Oxford: Blackwell. 326 pp.

Kempton, J. P., Hackett, J. E. 1968. Stratigraphy of the Woodfordian and Altonian drifts of central northern Illinois. In *The Quaternary of Illinois*, ed. R. E. Bergstrom, pp. 27–34. *Univ. Illinois Coll. Agric., Spec. Publ. 14.* 179 pp.

King, J. E. 1973. Late Pleistocene palynology and biogeography of the western Ozarks. *Ecol. Monogr.* 43:539–65

Lichti-Federovich, S., Ritchie, J. C. 1968. Recent pollen assemblage from the western Interior of Canada. *Rev. Palaeobot. Palynol.* 7:297–344

Likens, G. E., Davis, M. B. 1975. Postglacial history of Mirror Lake and its watershed in New Hampshire, USA: an initial report. *Verh. Int. Ver. Limnol.* 19:982–93

Mamakowa, K. 1968. Flora from the Paudorf Interstadial at Lazek near Zaklikow (SE Poland). *Acta Palaeobot.* 9:29–44

Mamakowa, K., Mook, W. G., Srodon, A.

1975. Late Pleistocene flora at Katy (Pieniny Mts., West Carpathians). *Acta Palaeobot.* 16:147–72

Maxwell, J. A., Davis, M. B. 1972. Pollen evidence of Pleistocene and Holocene vegetation on the Allegheny Plateau, Maryland. *Quat. Res.* 2:506–30

McAndrews, J. H. 1976. Fossil history of man's impact on the Canadian flora: an example from southern Ontario. In *Man's Impact on the Canadian Flora,* ed. J. K. Morton. *Can. Bot. Assoc. Bull. Suppl.* 9:1–6

Menke, B., Behre, K. E. 1973. History of vegetation and biostratigraphy. In *State of research on the Quaternary of the Federal Republic of Germany. Eiszeitalter Ggw.* 23/24:251–67

Mercer, J. H. 1969. The Alleröd oscillation: a European climatic anomaly? *Arct. Alp. Res.* 1:227–34

Mickelson, D. M., Evenson, E. B. 1975. Pre-Twocreekan age of the type Valders till, Wisconsin. *Geology* 3:587–90

Miller, N. G. 1973. Late-glacial and post-glacial vegetation change in southwestern New York state. *NY State Mus. Sci. Ser. Bull.* 420. 102 pp.

Mitchell, G. F. 1965. Littleton Bog, Tipperary: an Irish vegetational record. *Geol. Soc. Am. Spec. Pap.* 84:1–16

Moe, D. 1970. The post-glacial migration of *Picea abies* into Fennoscandia. *Bot. Not.* 123:61–66

Morrison, A. 1970. Pollen diagrams from interior Labrador. *Can. J. Bot.* 48:1957–75

Pennington, W. 1969. *The history of British vegetation.* London: English Univ. Press. 152 pp.

Pennington, W. 1975. A chronostratigraphic comparison of Late-Weichselian and Late-Devensian subdivisions, illustrated by two radiocarbon-dated profiles from western Britain. *Boreas.* 4:157–171

Richard, P. 1971. Two pollen diagrams from the Quebec City area, Canada. *Pollen Spores* 13:523–60

Richard, P., Poulin, P. 1976. Un diagramme pollinique au Mont des Eboulements, region de Charlevoix, Quebec. *Can. J. Earth Sci.* 13:145–56

Ritchie, J. C., Lichti-Federovich, S. 1967. Pollen dispersal phenomena in arctic-sub-arctic Canada. *Rev. Palaeobot. Palynol.* 3:255–66

Rowe, J. S. 1972. Forest regions of Canada. *Can. For. Serv., Publ.* 1300. 172 pp.

Saarnisto, M. 1974. The deglaciation history of the Lake Superior region and its climatic implications. *Quat. Res. NY* 4:316–39

Simpson, I. M., West, R. G. 1958. On the

stratigraphy and paleobotany of a late-Pleistocene organic deposit at Chelford, Cheshire. *New Phytol.* 57:239–50

Tallantire, P. A. 1972. The regional spread of spruce (*Picea abies* (L.) Karst.) within Fennoscandia. *Norw. J. Bot.* 19:1–16

van der Hammen, T., Maarleveld, G. C., Vogel, J. C., Zagwijn, W. H. 1967. Stratigraphy, climatic succession and radiocarbon dating of the Last Glacial in the Netherlands. *Geol. Mijnbouw.* 46:79–95

van der Hammen, T., Vogel, J. C. 1966. The Susacan-interstadial and the subdivision of the Late-glacial. *Geol. Mijnbouw* 45:33–35

van der Hammen, T., Wijmstra, T. A., Zagwijn, W. H. 1971. The floral record of the late Cenozoic of Europe. In *The Late Cenozoic Glacial Ages,* ed. K. K. Turekian, pp. 391–424. New Haven: Yale Univ. Press. 606 pp.

von Post, L. 1930. Problems and working lines in the post-Arctic forest history of Europe. *Proc. Int. Bot. Congr. 5th*

Wasylikowa, K. 1964. Vegetation and climate of the late-glacial in central Poland, based on investigations made at Witow, near Leczyea. *Biul. Peryglacjalny* 13:262–417

Watts, W. A. 1970. The full-glacial vegetation of northwestern Georgia. *Ecology* 51:17–33

Watts, W. A. 1973a. The vegetation record of a mid-Wisconsin interstadial in north-west Georgia. *Quat. Res. NY* 3:257–68

Watts, W. A. 1973b. Rates of change and stability in vegetation in the perspective of long periods of time. In *Quaternary Plant Ecology,* ed. H. J. B. Birks, R. G. West. pp. 194–206. Oxford: Blackwell. 317 pp.

Watts, W. A. 1975. A late Quaternary record of vegetation from Lake Annie, south-central Florida. *Geology* 3:344–46

Watts, W. A. 1977. The Late Devensian vegetation of Ireland. *Phil. Trans. R. Soc. London, Ser. B.* In press

Watts, W. A., Winter, T. C. 1966. Plant macrofossils from Kirchner Marsh, Minnesota—a paleoecological study. *Geol. Soc. Am. Bull.* 77:1339–59

Watts, W. A., Wright, H. E. Jr. 1966. Late-Wisconsin pollen and seed analysis from the Nebraska sandhills. *Ecology* 47:202–10

Webb, T. III. 1974. Corresponding patterns of pollen and vegetation in lower Michigan: a comparison of quantitative data. *Ecology* 55:17–28

Webb, T. III, Bryson, R. A. 1972. Late- and postglacial climatic change in the northern Midwest, USA: quantitative estimates derived from fossil pollen spectra by multivariate statistical analysis. *Quat. Res. NY* 2:70–115

Webb, T. III, McAndrews, J. H. 1976. Corresponding patterns of contemporary pollen and vegetation in central North America. *Geol. Soc. Am. Mem.* 145:267–300

West, R. G. 1961. Interglacial and interstadial vegetation in England. *Proc. Linn. Soc. London* 172:81–89

Whitehead, D. R. 1973. Late-Wisconsin vegetational changes in unglaciated eastern North America. *Quat. Res. NY* 3:621–31

Wright, H. E. Jr. 1967. The use of surface samples in Quaternary pollen analysis. *Rev. Palaeobot. Palynol.* 2:321–30

Wright, H. E. Jr. 1968. The roles of pine and spruce in the forest history of Minnesota and adjacent areas. *Ecology* 49:937–55

Wright, H. E. Jr. 1972. Interglacial and postglacial climates: the pollen record. *Quat. Res. NY* 2:274–82

Wright, H. E. Jr., Watts, W. A. 1969. Glacial and vegetational history of northeastern Minnesota. *Minn. Geol. Surv., Spec. Publ. Ser.* 11. 50–59 pp.

Wright, H. E. Jr., Winter, T. C., Patten, H. L. 1963. Two pollen diagrams from southeastern Minnesota: problems in the regional late-glacial and postglacial vegetational history. *Geol. Soc. Am. Bull.* 74:1371–96

Zagwijn, W. H. 1960. Aspects of the Pliocene and Early Pleistocene vegetation in the Netherlands. *Meded. Geol. Sticht., Ser. C-*III-1, No. 5. 78 pp.

Zagwijn, W. H. 1974. The Pliocene-Pleistocene boundary in western and southern Europe. *Boreas* 3:75–98

Zagwijn, W. H. 1975. Variations in climate as shown by pollen analysis, especially in the Lower Pleistocene of Europe, in *Ice ages: ancient and modern,* ed. A. E. Wright, F. Moseley, pp. 137–52. *Geol. J.. Spec. Issue 6*

Ann. Rev. Earth Planet. Sci. 1977. 5: 159–78
Copyright © 1977 by Annual Reviews Inc. All rights reserved

PALEOECOLOGICAL TRANSFER FUNCTIONS[1]

✖10070

Harvey Maurice Sachs[2]
CLIMAP, Department of Earth Sciences, Case Western Reserve University, Cleveland, Ohio 44106

T. Webb III
CLIMAP, Department of Geological Sciences, Brown University, Providence, Rhode Island 02912

D. R. Clark
Center for Climatic Research, University of Wisconsin, Madison, Wisconsin 53706

INTRODUCTION

The reconstruction of ancient environments has long been a major goal of paleontology. For older rocks whose fossils lack conspecific living representatives, the argument by analogy is difficult. Conversely, for younger sediments with fossil assemblages dominated by extant forms, stronger inferences can be drawn. For the late Pleistocene, approaches utilizing *paleoecological transfer functions* have been introduced, and allow computation of calibrated estimates of past conditions. By definition *paleoecological transfer functions* are empirically derived equations for calculating quantitative estimates of past atmospheric or oceanic conditions from paleontological data. The relationships represented in these functions are based on the spatial correlations between modern climatic data and data for tree rings, pollen, Foraminifera, diatoms, coccoliths, or Radiolaria, which are used as *calibration sets*.

Although the adoption of multivariate methods for calculating these functions is recent, many earlier studies produced qualitative estimates of late Pleistocene conditions. Schott (1935) apparently initiated the study of fossil planktonic Foraminifera from deep-sea cores. In a succession of papers, Ericson & Wollin (1968, and papers cited therein) attempted to reconstruct past conditions in the Atlantic and Pacific (1970, cf. Morin et al 1970) from the varying abundances of particular

---

[1] Case Western Reserve University Contribution #119.

[2] Current address: Department of Geological and Geophysical Sciences, Guyot Hall, Princeton University, Princeton, NJ 08540.

159

species. Early techniques for the climatic interpretation of tree ring patterns are reviewed in Fritts and others (1971), and Webb & Bryson (1972) review earlier estimates of past climate from pollen spectra. Most of the early studies were qualitative or semiquantitative, and attempts to estimate numerical values of past climate parameters were few.

The use of numerical techniques in micropaleontology, including transfer functions, became feasible with advances in computer technology and with the assembly of large data sets. In 1971 and 1972, three groups independently introduced techniques that enabled quantitative estimates of past conditions to be made. These techniques involved study of planktonic Foraminera in deep-sea sediments (Imbrie & Kipp 1971), variations in tree ring widths (Fritts et al 1971), and changes in pollen assemblages in lake sediments (Webb & Bryson 1972). Since then multivariate methods have been applied to diatoms (Cooke-Poferl et al 1975, Maynard 1976, Sancetta 1977), Radiolaria (Lozano & Hays 1976, Moore 1973, Sachs 1973a, b), and coccoliths (Geitzenauer et al 1976), and in other studies (Cline & Hays 1976). Other methods have been used by Hecht (1973), Williams & Johnson (1975), and Lynts & Judd (1971) with Foraminifera. Paleoecological transfer functions share the following characteristics:

1. They produce calibrated quantitative estimates of some parameter of a past environment, such as seasonal or monthly air or ocean surface temperature, or air mass durations.
2. They utilize explicit algebraic methods to formulate these estimates.
3. The algorithms rely on multivariate computational techniques to analyze multi-component fossil (or tree ring width) data.
4. The functions are calibrated from an adequate sample of modern distributional data.
5. The calibrated function is then applied to older samples to estimate environmental parameters for past times.

In this way, paleontological data are quantitatively transformed into meteorological or oceanographic variables. Using paleoecological transfer functions to estimate conditions, the CLIMAP group (CLIMAP 1976) has presented maps of ocean surface temperature on the 18,000 yr BP Glacial Maximum datum. Because paleontological data are presented at temperature fields, these maps can be used as input data to calibrate general circulation models (Gates 1976) whose output can be checked against independent estimates of land climate derived from pollen transfer functions. Interest generated by this interaction has spurred continued development of this approach.

In the papers that introduced paleoecological transfer functions, several techniques were used. Each paper also used different algebraic notation, making comparisons among the techniques difficult. Following Webb & Clark (1977), we attempt to remedy the situation by using a common notation in this review. We also review the methods and assumptions of the transfer function approach, and discuss its limitations and accomplishments.

Basic Assumptions

Several major assumptions underlie the use of paleoecological transfer functions, including the following:

1. We assume that a multivariate approach is preferable to estimating past conditions from fluctuations of a single species or other variable. Although this is reasonable for most ecosystems, it has not been rigorously demonstrated. Multivariate estimates may be buffered from some types of depositional or ecological anomalies that selectively affect single species (Sachs 1973c), and the use of these methods has allowed the discovery of such anomalies (Lozano & Hays, 1976, Robertson 1975, Sachs 1973c).

2. We require that biological responses be systematically related to the physical attributes of the environment, and that the physical parameters of interest (e.g. temperature) either be important controls or linearly related to important controls on the biota (Imbrie & Kipp 1971). Noncorrelated parameters are readily detected with all transfer function approaches used. However, if an indirectly linked relationship suffers a change in the linkage through time, errors may be introduced that cannot be internally detected.

3. We assume that the following simple relationships exist between modern conditions and those in the time of interest: (a) Evolutionary change of the fossil species has been negligible, so ecological roles are constant. (b) Climatic or oceanographic conditions fall within the range of the modern calibration data. (c) Preservational conditions have been fairly stable through time. Testing of these assumptions is a potentially fruitful area of research, and one which already has provided important results.

Mathematical Formulation of the Basic Model

Given the data in hand and the above assumptions, the general form of the basic model may be represented by a response function Θ, which relates the set of biological responses X to a set of climatic factors C and to a set of nonclimatic factors D. These latter variables might include measurements of competition, secondary succession, nutrient availability, soil texture, dissolution, and anthropogenic disturbance. The following equation expresses this model:

$$X = \Theta(C, D), \tag{1}$$

in which Θ represents an equation, or system of equations, on the climatic and nonclimatic variables. If climate, C, exerts the only significant control on the biological responses, or if nonclimatic factors can be eliminated, a less formidable *climatic response function*, Θ_c might be developed:

$$X = \Theta_c(C). \tag{2}$$

The inverse of this model, Φ, would permit unknown values of the climatic variables to be calculated from known biological responses:

$$C = \Phi(X). \tag{3}$$

This equation thus fulfills the paleoclimatic goal of yielding a means for "...processing one time-varying signal (or set of signals) in a core, to yield another signal consisting of paleotemperature estimates" (Imbrie et al 1973, p. 11), and thus Φ acts as a *paleoecological transfer function.*

THE LINEAR EMPIRICAL MODEL   The basic model introduced by Imbrie & Kipp (1971) Webb & Bryson (1972), and Fritts et al (1971) involves linear algebraic equations employed as empirical transfer functions. Using the notation of Table 1 in Webb & Clark (1977), we can define the following matrices: Let $_n\mathbf{C}_q$ be the set of climatic data, comprising n observations of q climatic variables. Then $_n\mathbf{X}_p$ is the analogous matrix of paleontological data determined at the same n locations for p variables. [The data for each variable in $_n\mathbf{C}_q$ and $_n\mathbf{X}_q$ are standardized. See Equation A.1 in Webb & Bryson (1972).] Then there may exist a matrix of *calibration functions,* $_p\boldsymbol{\beta}_q$, such that

$$_n\mathbf{C}_q = {_n\mathbf{X}_p}\boldsymbol{\beta}_q + {_n\boldsymbol{\delta}_q}, \tag{4}$$

where $_n\boldsymbol{\delta}_q$ is the matrix of random noise. If $_p\boldsymbol{\beta}_q$ is known, any observation on the p biological variables may immediately be converted into estimates of the corresponding q climatic variables. The problem, now, is to estimate $_p\boldsymbol{\beta}_q$.

Despite the variations in statistical procedure employed, all paleontological transfer functions evaluated to date can be expressed as a simple linear transformation, symbolized by Equations 3 or 4. In each technique, the calibration matrices of spatial arrays of modern biota and climate, $_n\mathbf{X}_p$ and $_n\mathbf{C}_q$, are input to a statistical procedure that determines $_p\mathbf{B}_q$, an estimate of $_p\boldsymbol{\beta}_q$ in Equation 4. The calibration function is then applied to each observation in the application data matrix, $_1\mathbf{X}_p^f$, in order to estimate the associated climate $_1\mathbf{C}_q^f$. $_1\mathbf{X}_p^f$ is usually a temporal array of fossil data from a core of lake or deep-sea sediment.

THE ALGEBRAIC METHODS

In work on marine plankton and terrestrial pollen data, several algebraic methods have been applied for the calculations in Equation 4. Several methods are outlined in this section, including multiple regression, stepwise multiple regression, two forms of principal components analysis with multiple regression, canonical correlation analysis, temperature indices with regression and a weighted-average technique. The end product of each method is a linear combination of the biological variables in $\mathbf{X}$ that "best" predicts the variables in $\mathbf{C}$.

Method 1: Multiple Regression Analysis

Linear least-squares multiple regression provides a straightforward technique for solving Equation 4. The least-squares method and theory are well known (Draper & Smith 1968), and are widely used for fitting equations to data. Computer programs for its usage are generally available and well documented (Nie et al 1975). Using standard variables, the regression equation and solution for $\mathbf{B}$ is:

$$_n\mathbf{C}_q = {_n\mathbf{X}_p}\mathbf{B}_q + {_n\boldsymbol{\varepsilon}_q}, \tag{5}$$

where

$$_p\mathbf{B}_q = (_p\mathbf{X}_n\mathbf{X}_p)^{-1}{}_p\mathbf{X}_n\mathbf{C}_q. \tag{6}$$

The matrix of residuals, $_n\varepsilon_q$, contains the differences between observed and predicted climatic values. As a result of the least-squares procedure, the residuals for a given climatic variable have zero correlation with that climatic variable and the included independent variables, and the average of the residuals is zero. In a successful analysis, the residuals are small and represent random variations and observational errors in the data.

Each column of $\mathbf{B}$ contains the estimated coefficients of a calibration function for one climatic variable. The least-squares solution for $\mathbf{B}$ (Equation 6) gives the most precise, unbiased estimates of the coefficients obtainable by a linear-algebraic function of the data. Further, the sum of squared residuals for the regression (diagonal elements of $_q\varepsilon'_n\varepsilon_q$) is the minimum obtainable by a linear-algebraic estimator. These properties rely on several assumptions about the regression equation and the data:

1. The model is correct. The right independent (biological) variables have been chosen, so that a linear equation explains all but the random variations ($_n\delta_q$) in the dependent (climatic) variable.

2. The data are typical. Additional observations would not greatly change the estimates.

3. The observations are statistically independent. The random component in one observation does not affect the random component in any other observation.

4. The random (unexplainable) component in each climatic observation comes from a normal distribution with mean zero and constant (although unknown) variance for that climatic variable.

5. The independent variables have no random components.

In addition to satisfying the "best estimator" criteria, these assumptions are necessary for the validity of t and F significance tests and confidence intervals for the coefficients.

These assumptions are seldom met exactly, but the least-squares estimator is robust enough that useful results are usually obtained. Experience with transfer functions shows that the first two assumptions, which require 1. the right equation and the right variables, and 2. good data, are the most crucial (Bryson & Kutzbach 1974, Fritts et al 1971, Imbrie & Kipp 1971, Webb & Bryson 1972). The other assumptions remain secondary until the large uncertainties about 1 and 2 are resolved.

The several other methods that have been employed for estimating $\mathbf{B}$ have been developed in an effort to deal with the problem of choosing the form of the calibration equation and the variables to be used. In Methods 2–5, variables are selectively dropped, transformed, or recombined to express better the climatic variables in terms of the biological variables. The resulting number of variables r is usually less than the original number p, a useful feature for taxonomic groups with high species diversity (e.g. diatoms and Radiolaria).

Method 2: Stepwise Multiple Regression Analysis

The second technique is a common variation of multiple regression which provides an objective method for selecting variables. The procedure progressively "improves" the regression equation by adding and/or dropping biological variables according to their ability to reduce the residual sum of squares for an individual climatic variable (Draper & Smith 1968). By choosing the desired level of stepwise "improvement," the analyst is able to combine scientific knowledge with objective analysis in selecting the number of variables $r_i \leq p$ to use. The resulting equation appears identical to Equation 5:

$$_nC_q = {}_nX_rB_q^* + {}_n\varepsilon_q, \tag{7}$$

but each column i of $\mathbf{B}^*$ contains least-squares coefficients computed using only $r_i < p$ biological variables, with zero coefficients for any unused variables. If all variables are employed ($r_i = p$ for all i), this procedure reverts to Method 1 and $\mathbf{B}^* = \mathbf{B}$.

Stepwise regression is particularly useful when some of the biological variables are highly correlated (e.g. pollen from tree types that are commonly found growing together). For a linear model, these variables may represent primarily redundant information, and may only give small and possibly meaningless improvement in the equation.

Method 3: Principal Components Analysis Plus Multiple Regression

Use of the third method stems from the concept that regression against the original biological variables (Methods 1 and 2) may not be the best way to extract climatic information (Fritts 1974). For certain biological systems, assemblages of biological variables may be better predictors (Imbrie & Kipp 1971, p. 81). Principal components analysis (PCA) is an objective method that combines the original variables into linear combinations (eigenvectors) and concentrates within the first few components the primary patterns of variation among all the variables and leaves the least coherent aspects ("noise") for the last few components. The eigenvectors are mutually orthogonal and thus as transformed variables (component scores) are uncorrelated.

PCA often groups ecologically similar species into assemblages that can be biologically interpreted (Birks et al 1975, Imbrie & Kipp 1971, Webb 1974). The eigenvariables can then be regressed against the climatic variables using Method 1 or 2, as shown by:

$$_nC_q = {}_nX_pE_r'B_q^{**} + {}_n\varepsilon_q \tag{8}$$

and

$$_rB_q^{**} = {}_r(E_pX_n'X_pE')_rE_pX_n'C_q, \tag{9}$$

where the eigenvector matrix $\mathbf{E}$ is derived from the correlation matrix $\mathbf{R}$.

Because $_nX_pE_r'$ is the principal components matrix, Equation 8 represents a multiple regression of uncorrelated variables, the principal components, against each of the climatic variables as opposed to fitting the original variables X to C as in Equation 5. Normally, when Method 3 is used, r is less than p and the r eigenvectors retained are those with biological meaning (Imbrie & Kipp 1971, Webb 1974). The eigenvectors are often rotated by a varimax routine (Kaiser 1958), which reorients the eigenvectors so each is closely aligned with clusters of the original variables (Imbrie & Kipp 1971, Webb & McAndrews 1976). This step enhances interpretability of the assemblages.

When all eigenvectors are used in the regression (i.e. $r = p$), the results of Method 3 are exactly the same as those from Method 1. This equality is apparent when the two steps in this procedure—the calculation of principal components and their regression against the climatic variables—are combined. Thus,

$$_nC_q = {_nX_p}B_q^{***} + {_n\varepsilon_q}, \tag{10}$$

with

$$_pB_q^{***} = {_pE_r'}B_q^{**}. \tag{11}$$

Hence, if all eigenvectors are retained,

$$_pB_q^{**} = ({_pX_n'}X_p)X_n'C_q = {_pB_q}, \tag{12}$$

because

$$_pE_p'E_p = I$$

in Equation 9. As p-r increases, $_pE_r'B_q^{**}$ departs from $_pB_q$, and therefore the results of Method 3 differ from the results of Method 1.

VARIATIONS OF METHOD 3   PCA is a scale-dependent technique since the eigenvectors produced (and thus the assemblages represented) depend on the numerical measures of similarity between the variables. Of the many possible similarity measures (Orloci 1972), only row-normalized cross-products and Pearson moment–correlation coefficients have been in common use. The row-normalized cross-products matrix (cos Θ matrix) has been favored for marine data. This matrix gives maximum weight to large-valued variables and little weight to small-valued variables. Its use is probably appropriate for foraminiferal data, which tend to be dominated by a few species, but W. H. Hutson (personal communication) has suggested that calculating eigenvectors from the variance-covariance or correlation matrices would include more information from the less frequent species. Testing of these procedures is in progress.

The main measure of similarity used for pollen data has been the common correlation coefficient (Pearson product-moment), which weights variables equally. As with the row-normalization procedure initially employed by Imbrie & Kipp (1971), there is some danger that this technique will amplify the random errors associated with counts of the least frequent species.

Imbrie and Kipp: a variation   This technique is widely used for marine data (e.g. CLIMAP 1976). The eigenvector matrix (E^*) is derived from the row-normalized cross-products matrix (cos Θ matrix) and a varimax rotation of the eigenvectors is always used. As discussed above, this method tends to yield assemblages dominated by one or a few large-valued variables.

In the initial use of this method, Imbrie & Kipp (1971) rescaled each variable as a percent of its range. This initial rescaling by the range, however, is no longer done (Imbrie & Kipp 1971, p. 80, Imbrie et al 1973, Kipp 1976), because the resulting estimates are in general less accurate, tend to amplify counting errors, and on occasion may lead to unreasonable values.

Imbrie & Kipp (1971) used a curvilinear regression by including cross-products and squares of the principal components in addition to the components themselves, thus providing $[r + r(r+1)/2]$ independent variables. The justification for this type of regression is primarily empirical, based on observing an increase in accuracy when the results of a regression using the squared terms were compared with the results of the equation using only linear terms. Due allowance was made for the degrees of freedom lost in this empirical test. These results were found in sea-bed calibration studies with Foraminifera (Kipp 1976), with Radiolaria (Lozano & Hays 1976, Moore 1973, Sachs 1973a), and with coccoliths Geitzenauer et al (1976). For this reason, many workers using this method have continued to employ the curvilinear terms in their regression equations. In addition, applications to downcore marine studies have shown that the curvilinear equations yield a lower proportion of unreasonable estimates.

Method 4: Canonical Correlation Analysis

Canonical correlation analysis was used by Webb & Bryson (1972) and extends the approach of regression of biological "assemblages" against single climatic variables to an approach of finding linear combinations of the biological data $(_nX_pH_q)$ that are maximally correlated with linear combinations of the climatic variables $(_nC_qG_q)$. These linear combinations are chosen such that

$$_qH_p'X_n'C_qG_q = _q\Lambda_q,   \tag{13}$$

where $_q\Lambda_q$ is a diagonal matrix of canonical correlation coefficients, and

$$_qH_p'X_n'X_pH_q = _qI_q,   \tag{14}$$

$$_qG_q'C_n'C_qG_q = _qI_q.   \tag{15}$$

Equations 14 and 15 state that the linear combinations (canonical variates) are orthonormal, i.e. standardized and uncorrelated. From Equations 10 and 11, canonical regression coefficients (****) can be calculated by

$$_pB_q^{****} = _pH_r\Lambda_rG_q^{-1},   \tag{16}$$

$$_nB_q^{****} \cong _pB_q,   \tag{17}$$

but for $r < q$, which is achieved by setting some of the canonical correlations (λ's) equal to zero, the form of equation 5 becomes

$$_nC_q = _nX_pB_q^{****} + _n\varepsilon_q.   \tag{18}$$

In the analysis of one set of pollen data, Bryson & Kutzbach (1974) showed that $_p\mathbf{B}_q^{****} = {_p}\mathbf{B}_q$ even when the final four out of eight λ's are set equal to zero. They also showed that although the eight canonical variates explained 90% of the climatic variance, only 50% of the pollen variance was explained. This implies that at least half of the biological variance in this system arises either from such non-macroclimatic causes as edaphic, cultural disturbance, and pollen-depositional effects, or from nonlinear interactions within the pollen data or between the pollen and climatic data. Although rough calculations of this sort can be made when Method 3 is used, canonical correlation analysis gives this information directly.

In the initial use of canonical correlation for finding paleoecological transfer functions, Webb & Bryson (1972) included site factors (such as latitude) with their pollen data. The validity of this approach as a covariance adjustment is discussed by Webb & Clark (1977), who chose not to employ this variation in their comparison of transfer function algorithms.

Method 5: Temperature-Indices-with-Regression

This method, like Method 3, first transforms the biological data before regressing these data against $\mathbf{C}$. The transformation, however, is more severe than that of Method 3, because $_n\mathbf{X}_p$ is reduced to a column vector $_n\mathbf{D}_1$.

In the form of this method developed by Hecht (1973), each element (d_i) of $_n\mathbf{D}_1$ is an algebraic measure of the distance between the ith sample $(_1\mathbf{X}_i)$ and a sample designated as a standard $(_1\mathbf{X}_t)$. For regression against temperature, the standard sample is chosen to be at a temperature extreme within the data set. The distance (d_i) is therefore likely to be related to the temperature difference between a given sample and the standard. This distance thus may serve as a temperature index that the regression step recalibrates in degrees Celsius. Hecht (1973) used Parks distance coefficient,

$$d_i = p^{-1/2}\left[\sum_{j=1}^{p}(x_{ij} - x_{tj})^2\right]^{1/2}, \tag{19}$$

where x_{ij} and x_{tj} are the elements of $_1X_i$ and $_1X_t$ respectively, as a measure of the distance between each sample and the standard, but conceivably other distance measures or similarity coefficients could be used to measure this distance (Orloci 1972).

Hecht & Kipp (1974) have generalized Method 4 in two ways. First, they have expanded the column vector $_n\mathbf{D}_1$ to a matrix $_n\mathbf{D}_r$, in which the algebraic distances to several standard samples are entered. Second, they include the option of augmenting this matrix by adding squares and cross products $(_n\mathbf{D}_r)^2$. In this form, Method 5 gives results that are similar to those of the Imbrie and Kipp version of Method 3.

Williams & Johnson (1975) introduced a technique formally analogous to Hecht's distance index with regression. Instead of computing the "distance" from a standard sample of the fauna at each station, Williams and Johnson computed the Shannon-Wiener and Brillouin diversity indices and used these in the regression against temperature. Since diversity of core top assemblages (by either measure) is highly correlated with temperature in this area of the southern Indian Ocean, the diversity index regression technique yielded reasonable temperature estimates. As the authors point out, however, many factors other than temperature govern diversity in the world ocean. This technique should be used only in restricted regions.

A VARIATION: WEIGHTED AVERAGES   The weighted average technique of Jones (1964, cited in Hutson 1976), which was used by Lynts & Judd (1971), Berger (1969) and Berger & Gardner (1975), is formally similar to Hecht's distance index with regression or Williams and Johnson's diversity index with regression. In its simplest form, estimates of climatic variables are calculated from weighted-average formula

$$t^f = \left(\sum p_i C_i\right)/\left(\sum p_i\right), \tag{20}$$

where t^f is the estimated temperature for the fossil assemblage, p_i is the proportion of the ith species, and e_i is the optimum temperature for that form. The values for e are selected by the investigator from available distributional data for species in plankton (Lynts & Judd 1971) or sediments (Berger & Gardner 1975). Berger & Gardner improved their results by using a regression relationship to calibrate t^f against observed temperatures:

$$t^{f'} = at^f + K \tag{21}$$

or

$$t^{f'} = a\left[\left(\sum p_i e_i\right)/\sum p_i\right] + k, \tag{22}$$

where a and k are respectively the empirically fitted slope and intercept. Recasting this in matrix notation, we have

$$C_1^f = (a)_n X_p E_1 + K, \tag{23}$$

where C_1^f is the vector of estimated temperatures or other climatic variables, $_n X_p$ is the standardized matrix of occurrences of species in samples, and $_n E_1$ is the vector of climatic preferences for species.

Although the method has been used only for estimation of a single parameter, temperature, it could be used for any climatic variables $(_n C_q)$ for which there is information on species preferences $(_p E_q)$.

EVALUATION OF PALEOECOLOGICAL TRANSFER FUNCTIONS

The approaches discussed above have been compared by Webb & Clark (in press). Using a single calibration set of modern pollen and climate data and one profile of fossil data from Kirchner Marsh, they applied multiple regression, stepwise multiple regression, principal components analysis with multiple regression and canonical correlation analysis. In each case, they restricted themselves to a linear model.

To the extent that each of these models approaches a regression technique, the results will be similar. The first major conclusion of their study is that all methods give approximately the same results (within analytical precision) for samples less than about 10,000 years old. Thus, when the assumptions for calculating paleoecological transfer functions are met, and the fossil system is directly comparable to the calibration data set, the quality of estimates from all multivariate techniques is comparable.

If the basic assumptions are met, therefore, the techniques largely converge and yield similar results. In this case the investigator may choose a method for its other advantages. Regression and stepwise regression offer conceptual simplicity and the ability of standard computer programs to handle large data sets efficiently. The temperature indices with regression method also offer conceptual simplicity, but only one environmental parameter can be estimated. Principal components with regression and canonical correlation both produce summary variables ("assemblages"), which may be useful in interpreting relationships among the paleontological variables—a consideration beyond their use in calibration functions. Bryson & Kutzbach (1974) showed that canonical correlation has the additional advantage of directly evaluating the relationship of the paleontological data to the climatic or oceanographic data.

The No-Analogue Problem

The second major observation by Webb and Clark is that the various estimates are discordant for the time interval from about 10,000 to 13,000 yr BP. Examination of the pollen spectra in that part of their core indicates that the pollen assemblages are unlike those found in the calibration set. Thus, under "no-analogue" conditions, estimates made by different transfer functions will often differ. In fact, it has been proposed (Hutson 1976) that the existence of divergent estimates is prima facie evidence for no-analogue conditions.

Several "no-analogue" situations have been described: As one example, the latest Glacial Maximum of North Atlantic had an extensive low-salinity surface water layer north of about 40°N (van Donk 1970, Sancetta et al 1973, Sachs 1976). Because comparable conditions are not found in the Atlantic today, the calibration set cannot reflect this particular case of paleoclimatic interest (Assumption 3b).

In a second example, in a series of studies of Antarctic (Hays et al 1976) and North Pacific (Robertson 1975, Sachs 1973b) Radiolaria, the abundances of particular forms downcore greatly exceeded their abundance in modern sediments. The form identified (incorrectly) as *Pterocanium korotnevi* by Sachs and as *Lychnocanium grande* by Robertson has a widespread abundance peak in the North Pacific at about 60,000 yr BP. Similarly, *Cycladophora davisiana* in Antarctic (Hays et al 1976) and Subarctic Pacific sediments (Robertson 1975, Sachs, in progress) has a widespread anomalous abundance peak close to 18,000 yr BP. In the modern ocean, high abundance of this form has been found only in the Okhotsk Sea (Robertson 1975), and interpretation of the 18,000 yr BP abundance maximum in Antarctic and other waters remains speculative. Although Robertson (1975) included the species in his transfer function other workers have not, feeling that it responded in a manner not analogous to its behavior in recent times. A similar problem caused

Webb & Bryson (1972) to exclude *Fraxcinus, Ulmus,* and *Larix* from their paleo-ecological transfer functions.

Dissolution of the tests of Foraminifera or coccoliths is well documented (Berger 1971, Berger & Roth 1975, Sliter et al 1975), and may cause a third type of problem. If the dissolution regime at past times differed from that at present, due to changed bottom water circulation or other factors, changes in the fossil assemblages present would be expected and might lead to spurious results. As discussed below under Quality Control, such a change was simulated by Berger & Gardner (1975).

Evolution of single species or the community also may produce no-analogue situations. Sachs (1973b) found substantial downcore abundances of species now extinct, and excluded them from his analysis. This action implicitly assumes that these species did not interact significantly with those included in the study. Imbrie & Kipp (1971) and Sachs (1973c) found that the communalities (sums of squares of factor estimates) decreased in their oldest samples. As suggested by Imbrie and Kipp, this decrease is probably due to subtle evolutionary changes in community structure which decrease the climatic resolution of these ancient assemblages in terms of modern factors.

Hutson (1976) has theoretically and empirically studied some possible types of no-analogue situations. A recognizable no-analogue situation will be characterized by downcore abundances for at least one species that are significantly higher than the maximum abundance in surface (calibration) samples, or by analogously defined anomalous abundance ratios between two species downcore and in surface sediments. Hutson recognized two classes of no-analogue conditions: environmental and biological. Environmental no-analogue situations may include the latest Glacial Maximum North Atlantic and artifact assemblages produced by dissolution. Biological no-analogue situations may arise from incomplete sampling of modern environments, that is, from a calibration set not extensive enough to include conditions encountered downcore. Evolution may also cause biological no-analogue situations.

After experimenting with the behavior of several paleoecological transfer function methods under no-analogue conditions, Hutson (1976) concluded that the weighted-average technique offers an advantage over other methods: If the no-analogue condition is caused by downcore abundance in excess of the abundance of a particular species in the calibration set, then the weighted-average technique will estimate a temperature approaching the optimum for the species involved as its abundance increases. Hutson refers to this as interpolative behavior of the model, and contrasts it with other transfer function approaches, which tend to respond to this type of no-analogue condition by extrapolating to estimated values beyond the range of those in the calibration data set.

Hutson's point requires the assumption that temperature preferences of the species involved did not change because of the physical changes that led to the formation of the no-analogue assemblage. However, if this assumption can be made, and if there is sufficiently good statistical information to assign values of temperature preferences (e_i in Equation 22), then the weighted-average technique should be used.

Quality Control

The precision of paleoecological transfer functions is conventionally estimated from the caliber of estimates for test samples withheld from the calibration set during development of the function (Webb & Bryson 1972, Sachs 1973a). The confidence intervals of estimates are generally ± 5–10% of the observed range of the variable in the calibration set (80% confidence interval). In studies using the Imbie and Kipp principal components with regression technique, six foraminiferal transfer functions had an average standard error of estimate of $1.5°C$ (range 1.24–$1.90°$); six Radiolarian functions had an average s.e.e. of $1.7°$ (range 0.92–$2.7°$) (CLIMAP 1976). These figures are for transfer functions in various stages of development, and the range of temperatures covered varied from function to function. Significant improvement is expected in most cases.

Estimating the accuracy of transfer functions is much more difficult, as there is rarely an independent source of information on ancient temperatures. Isotopic paleotemperature determinations have large uncertainties because a value for the ice-volume effect in the Quaternary has not been accepted by all workers, and because glacial/interglacial changes in regional evaporation/precipitation patterns can markedly affect the isotopic signal in particular regions (reviewed by Imbrie et al 1973, Berger & Gardner 1975). In view of these uncertainties, it is not now possible to calibrate oceanic transfer functions from isotopic data.

In the absence of independent downcore checks, Berger & Gardner (1975) attempted to utilize a transfer function calibrated for North Atlantic Foraminifera on a set of surface samples from the South Pacific. They found that the quality of their estimates was lower than the precision of the original transfer function, and that the scatter was nonrandom. The dissolution regime in the South Pacific is different from the North Atlantic (Berger 1970), and there are indications that some species may have different temperature optima in different regions (Bé & Tolderlund 1971, Morin et al 1970). This comparison, therefore, must be considered to involve a mild "no-analogue" situation. However, such situations may be good approximations of the net change in conditions between glacial and interglacial conditions, and estimates of ancient conditions may not be as accurate as the precision of modern calibrations.

Another test of paleoecological transfer function accuracy is the comparison of parameter estimates based on different fossil groups counted in the same core samples. Where this comparison has been used, discrepancies between estimates usually have been reasonable (T.C. Moore Jr., unpublished data).

Lacking adequate external tests, internal checks have been used by various workers. In the Imbrie and Kipp technique, the communality is computed for each unknown sample. This figure, the sum of squares of the factor estimates for each sample, measures the success of the varimax factors in accounting for the composition of the particular sample. It is clear that a large decrease in communality indicates a failure of the transfer function (Hutson 1976), but it has not yet been shown that all no-analogue situations will lead to communality decreases.

Thus, Hutson (1976) urges that multiple approaches be used: If possible, transfer functions should be developed for more than one taxonomic group. Comparison of results may show divergent estimates indicative of anomalous conditions. If only one group is available, both regression-based and nonregression-based transfer functions should be used. For example, divergent estimates from weighted-average estimates and principal components with regression estimates would indicate the existence of a no-analogue situation. The divergence of estimates in Webb & Clark (in press) for samples more than about 10,000 yr old clearly shows this effect.

NEAR MISSES

Paleoecological transfer functions use explicit multivariate algebraic methods to analyze adequate multicomponent fossil data sets and calibrate equations for estimating conditions at past times. For contrast, we note several "near misses" which lacked either calibration or an adequate number of samples.

Kanaya & Koizumi (1966) published a benchmark study of diatom assemblages from the North Pacific. Working with 118 samples from the upper parts of cores, they identified some 220 taxa which were grouped into seven assemblages by a recurrent group analysis (Fager 1957). From this information, they constructed a diatom temperature index defined as

$$Td = [Xw/(Xc + Xw)] \times 100 \tag{24}$$

where Xw is the frequency of warm water forms and Xc is the frequency of cold water forms. The species are assigned to the warm water or cold water groups by their affinities, as indicated in the recurrent group analysis. Values of Td were calculated for their surface sediment set and for deep-sea core V20-130. From their figures, it is clear that Td usefully reflects temperature changes in the Transitional Water. Had Kanaya and Koizumi performed a regression to calibrate Td in terms of surface temperature, their study would have met all the criteria of a transfer function study. Without this step, it is only possible to estimate relative fluctuations in surface water temperature.

The method of Kanaya & Koizumi (1966) was applied to other North Pacific cores by Donahue (1970). Nigrini (1970) applied this method to the estimation of past environmental conditions from Radiolarian faunas from North Pacific cores. Like the preceding studies, Nigrini's index (called Tr) was not calibrated, but it was explicitly defined and based on analysis of a large number of surface sediment samples. A similar study in the equatorial Pacific was published by Johnson & Knoll (1974).

Another type of uncalibrated quantitative analysis was presented by Blackman & Somayajulu (1966). They did a factor analysis of the foraminiferal assemblages from 56 Southeast Pacific surface sediment samples to identify major faunal groups in the region. The importance of these groups was then determined for two deep-sea cores. Blackman and Somayajulu were thus able to note changing faunal patterns in the region, but they did not attempt to estimate past ocean temperatures or other objective environmental parameters.

Luz (1973) extended the work of Blackman & Somayajulu (1966). Confronted with a small calibration set and extensive dissolution of Foraminifera in the tropical Southeast Pacific, Luz factor–analyzed his calibration and downcore data together. He then developed a regression equation with the surface data and applied it to the downcore samples. Luz did not separately treat the calibration and ancient samples, but the method otherwise qualifies under the present definition of paleo-ecological transfer functions.

An even smaller set of samples was used by Bandy et al (1971) and Casey (1971) for Radiolaria-based paleotemperature estimates. They used seven Pacific Basin Holocene samples from five areas extending from near the equator to the Antarctic. Eight taxonomic groups were utilized as temperature indices: The presence of collosphaerids was taken to indicate average summer temperatures greater than 20°C; temperatures about 10°C were indicated by the "fairly common" presence of three well-defined species. For lower temperatures, an abundance ratio of two Antarctic to two subantarctic species was used. This method suffers several drawbacks: 1. Only a small fraction of the taxa available are used. If Assumption 1 is correct, this system may not be well protected from distributional anomalies. 2. The species used belong to rather different taxonomic groups, and those of Casey's indices that were studied by Johnson (1975) show greatly varying suscepti-bility to dissolution. Given the nature of the index used, differential dissolution may yield spurious "temperature" signals. 3. It does not seem possible to reduce Casey's indices to an explicit equation yielding a continuous range of temperature estimates in response to faunal variations. 4. The surface sample set used to establish the estimators was too small.

DISCUSSION

Paleoecological transfer functions transform fossil abundance data into estimates of meteorological or oceanographic variables. When the assumptions of the system are met, these estimates are of high quality and have value for many other studies. Because the results are expressed in terms of physical variables, it is possible to compare transfer function results from different fossil groups in the same region and study the quality of the estimates. It also is possible to contrast results from different regions. In this way, we can estimate changes in regional gradients in temperature or other parameters through time.

Regional comparisons can be expanded to world maps of estimated ocean surface temperatures at particular times, such as the map for the 18,000 yr BP Glacial Maximum (CLIMAP 1976). This map can serve as proxy data for the evaluation of general circulation models of the atmosphere (Gates 1976). The predictions of the models can be checked against transfer function results from continents (Webb & Bryson 1972) or other data such as quartz accumulation rates in deep-sea cores, which give information on changes in wind patterns (Heath and others 1973).

These data are also available for evaluation of other models of climatic change. For example, Sachs (1976) found that data complied by the CLIMAP group and others supports the depiction of glacial climate presented by Weyl (1968), but not his model for the inception or termination of a glacial epoch.

Studies of dated deep-sea cores yield time series whose study complements map studies in the search for causes of climatic change. In this case, estimates derived from transfer functions serve as summary variables, reducing temporal fluctuations of many species abundances to fluctuations of a single parameter with oceanographic or meteorological significance. These fluctuations can be studied by digital frequency techniques to isolate the characteristic periodicites of climatic change. Imbrie & Kipp (1971) found periodicities resembling the "Milankovitch" cycles of insolation variations due to perturbations in the earth's orbit. This work has been extended to nonbiological parameters by Pisias (1976), who suggests important cyclicities approximating 23,000 and 100,000 yr. Hays, Imbrie and Shackleton (1976) have found further evidence by applying elegant statistical techniques to the records of two Antarctic cores. From these studies, it seems that perturbations of the earth's orbit are an important driving mechanism of climatic change.

The use of these methods relies on choosing records with appropriate resolution and adequate length. For the various fossil groups, there seem to be characteristic values for these variables. For tree ring data, the maximum time scale is probably several hundred years, but with resolution of individual years. For pollen data, the Holocene record is very good, but the robustness of pre-Holocene transfer functions is not yet well-established. Resolution of pollen studies is typically 10^1–10^2 years (Kutzbach 1975). For marine plankton, time scales on the order of 10^5 years pose no problems for most fossil groups, but satisfactory results for samples older than 10^6 years have not been obtained. Resolution varies from decadal, for exceptional regions such as the Santa Barbara Basin (Kipp & Towner 1975), to several thousand years in regions of slow deposition such as most of the Pacific Ocean.

Limitations

It is clear that one problem facing world-wide reconstruction of past climate from paleoecological transfer functions is a lack of data overlap. Most marine studies have inadequate resolution for fine-scale studies of Holocene changes. On a global basis, there is little hope for relief from this situation, since bioturbation probably limits ultimate resolution in most deep-sea regions (Berger & Heath 1968). On the other hand, no-analogue situations have made the application of pollen transfer functions to pre-Holocene samples difficult, and the latest Glacial Maximum seems well beyond the reach of tree ring studies.

A major limitation of the paleoecological transfer function approach is that it is based on a linear empirical model. In the case of plankton data, the relationship between sea-bed assemblages and surface oceanographic parameters is considered, but the mechanisms controlling the distribution of the plankton and the conversion of living assemblages to fossil assemblages are ignored. If there were sufficient knowledge of the plankton, deterministic models could be developed which might circumvent some "no-analogue" problems, but their development does not seem imminent.

The assumption of constancy of ecosystem response has limited paleoecological transfer functions to the most recent geological past. The extent of this limitation is not yet completely clear, but it is probable that the marine record can be used for most of the Pleistocene for Foraminifera, Radiolaria, and diatoms. At appropriate sites, pollen analysts may have similar success. It is hoped that reanalysis of the data assembled for transfer function studies will shed light on the evolution and changes of communities through time, and on other problems as well. Analysis of the basic assumptions of the method, as begun by Hutson (1976), is leading to better understanding of the models and their behavior.

Prospects

Over the next five years or so, we would anticipate the following developments in paleoecological transfer function work:

1. There will be a general consolidation and improvement of results. In the cases of many plankton studies, taxonomic schemes are still being optimized, calibration sample coverage is inadequate, and the best possible oceanographic data have not yet been assembled. This consolidation will include much testing of the limitations of transfer functions posed by dissolution or other no-analogue phenomena (Kipp 1976, Hutson 1976).

Ledbetter & Ellwood (1976) have proposed guidelines for selecting sampling intervals in deep-sea cores, and these will be tested. We anticipate that analytical and empirical solutions will be sought to the analogous question of the optimum number of calibration samples, which is a function of the order of curvature of the surface that adequately fits the distribution of the physical parameter modelled, and of a number of other variables.

2. In the marine realm, the rapid spread of the transfer function approach makes it possible to estimate conditions at the same point with transfer functions based on more than one microfossil group. If the groups involved have different ecological controls, their different transfer functions may yield estimates of changes in upper water mass properties. This approach is only one example of possible opportunities for improved interpretation of transfer function studies, and we anticipate much work along these lines.

3. Clearly, there will be a great deal of sophisticated analysis of results obtained from both spatial arrays and time series. For time series from selected intervals it may become possible, given optimal conditions, to detect phase differences in the regional response to climatic changes (Pisias 1976) and thus to find areas that are particularly sensitive indicators of incipient climatic change.

4. Work is beginning to extend the transfer function approach to older time intervals. This may involve "bootstrapping" techniques which allow the estimation of the preferences of species which are now extinct, and which therefore allow their inclusion in the transfer function. Such an approach, if feasible, would overcome a major no-analogue limitation.

SUMMARY AND CONCLUSIONS

1. Given adequate samples of fossils in modern sediments, and good data on the distribution of environmental parameters under modern conditions, it is possible to construct transfer functions objectively, relating the distribution of fossils to the distribution of the parameter of interest.

2. For appropriate older samples, it is possible to use this relationship to estimate

the values of the environmental parameters under ancient conditions. This operation is crucially dependent on a strong and stable correlation between parameter variations and fossil distribution, and on choosing fossil samples young enough that significant evolution has not occurred.

3. If these constraints are satisfied, most algorithms that have been proposed yield comparable results.

4. Hutson (1976) has pointed out that failure of the transfer function due to "no-analogue" conditions may be cryptic, but that chances for discovery are increased if more than one transfer function algorithm is applied to a given group of samples.

ACKNOWLEDGEMENTS

National Science Foundation grants from the Climate Dynamics Program (NSF OCD-14934 to Brown University and NSF ATM74-23041 A01 to the Center for Climatic Research, University of Wisconsin-Madison) and from the International Decade of Ocean Exploration (NSF ID075-23032 to Case Western Reserve University and NSF ID076-00398 to Brown University) supported the writing of this article. We wish to thank S. Howe for detailed comments on the Assumptions and Algebraic Methods sections, Dr. W. H. Hutson for preprints and stimulating ideas, and R. M. Cline for providing abstract preprints of articles in G.S.A. Memoir 145.

Literature Cited

Bandy, O. L., Casey, R. E., Wright, R. C. 1971. Late Neogene Planktonic zonation, magnetic reversals, and radiometric dates, Antarctic to the Tropics. In *Antarctic Oceanology*, ed. J. L. Reid, 1:1–26. *Am. Geophys. Union Res. Ser. 15*

Bé, A. W. H., Tolderlund, D. S. 1971. Distribution and ecology of living planktonic foraminifera in surface waters of the Atlantic and Indian Oceans. In *Micropalaeontology of Oceans*, ed. B. M. Funnell, W. R. Riedel, pp. 105–149. Cambridge University Press, London.

Berger, W. H. 1969. Ecological patterns of living planktonic foraminifera. *Deep-Sea Res.* 16:1–24

Berger, W. H. 1970. Biogenous deep-sea sediments: Fractionation by deep-sea circulation. *Geol. Soc. Am. Bull.* 81:1385–1402

Berger, W. H. 1971. Sedimentation of planktonic foraminifera. *Mar. Geol.* 11:325–58

Berger, W. H., Gardner, J. V. 1975. On the determination of Pleistocene temperatures from planktonic foraminifera. *J. Foraminiferal Res.* 5:102–13

Berger, W. H., Heath, G. R. 1968. Vertical mixing in pelagic sediments. *J. Mar. Res.* 26:134–43

Berger, W. H., Roth, P. H. 1975. Oceanic micropaleontology: Progress and prospect. *Rev. Geophys. Space Phys.* 13:561–85

Birks, H. J. B., Webb, T. III, Berti, A. A. 1975. Numerical analysis of surface pollen samples from central Canada: a comparison of methods. *Rev. Palaeobot. Palynol.* 20:133–69

Blackman, A., Somayajulu, B. L. K. 1966. Pacific Pleistocene cores: faunal analyses and geochronology. *Science* 154:886–89

Bryson, R. A., Kutzbach, J. E., 1974. On the analysis of pollen climate canonical transfer functions. *Quat. Res.* 4:162–74

Casey, R. E. 1971. Neogene radiolarian biostratigraphy and paleotemperatures: Southern California, the experimental Mohole, Antarctic core E14-8. *Palaeogeogr. Palaeoclimatol. Palaeoecol.* 12:115–30

CLIMAP, 1976. The surface of the ice-age earth. *Science.* 191:1131–37

Cline, R. M., Hays, J. D., eds. 1976. Investigation of Late Quaternary Paleo-Oceanography and Paleo-Climatology. *Geol. Soc. Am. Mem.* 145. 464 pp.

Cooke-Poferl, K., Burckle, L. H., Riley, S. 1975. Diatom evidence bearing on Late

Pleistocene climatic changes in the Equatorial Pacific. *Geol. Soc. Am. Abstr. Programs* 7:1038

Donahue, J. G. 1970. Pleistocene diatoms as climatic indicators in North Pacific Sediments. In *Geological Investigations of the North Pacific*, ed. J. D. Hays. *Geol. Soc. Am. Mem.* 126:121–38

Draper, N. R., Smith, H. 1968. *Applied Regression Analysis*. New York: Wiley. 407 pp.

Ericson, D. B., Wollin, G. 1968. Pleistocene climates and chronology in deep-sea sediments. *Science* 162:1227–34.

Ericson, D. B., Wollin, G. 1970. Pleistocene climates in the Atlantic and Pacific Oceans: a comparison based on deep-sea sediments. *Science* 167:1483–85

Fager, E. W. 1957. Determination and analysis of recurrent groups. *Ecology* 38:586–95.

Fritts, H. C. 1974. Some quantitative methods for calibrating ring widths with variables of climate. *Am. Quat. Assoc. Abstr.* pp. 2–5

Fritts, H. C., Blasing, T. J., Hayden, B. P., Kutzbach, J. E. 1971. Multivariate techniques for specifying tree-growth and climate relationships and for reconstructing anomalies in paleoclimate. *J. Appl. Meteorol.* 10:845–64

Gates, W. L., 1976. Modeling the ice-age climate. *Science* 191:1138–44

Geitzenauer, K. R., Roche, M. McIntyre, A., 1976. Modern Pacific coccolith assemblages: derivation and application to late Pleistocene paleotemperature analysis. See Cline & Hays 1976, pp. 423–48

Hays, J. D., Imbrie, J., Shackleton, N. J. 1976. Variations in the earth's orbit: Pacemaker of the Ice Ages. *Science* 194:1121–32

Hays, J. D., Lozano, J. A., Shackleton, N., Irving, G. 1976. Reconstruction of the Atlantic and Western Indian Ocean sectors of the 18,000 BP Antarctic Ocean. See Cline & Hays 1976, pp. 337–72

Heath, G. R., Dauphin, J. P., Opdyke, J. P., Moore, T. C. Jr. 1973. Distribution of quartz, opal, calcium carbonate and organic carbon in Holocene, 600,000 and Brunhes/Matuyama age sediments of the North Pacific. *Geol. Soc. Am. Abstr. Programs.* 5:662

Hecht, A. D. 1973. A model for determining Pleistocene paleotemperatures from planktonic foraminiferal assemblages. *Micropaleontology* 19:68–77

Hecht, A. D., Kipp, N. G. 1974. Experiments in transfer function paleoecology: application to Pleistocene paleotemperatures.

Geol. Soc. Am. Abstr. 6:787

Hutson, W. H. 1976. Transfer functions under no-analogue conditions. PhD. thesis. Brown Univ., Providence, RI.

Imbrie, J., Kipp, N. G. 1971. A new micropaleontological method for quantitative paleoclimatology: application to a late Pleistocene Caribbean core. *The Late Cenozoic Glacial Ages*, ed. K. Turekian, pp. 71–181. New Haven, Yale Univ. Press. 606 pp.

Imbrie, J., Van Donk, J., Kipp, N. G. 1973. Paleoclimatic investigation of a late-Pleistocene Caribbean deep-sea core: comparison of isotopic and faunal methods. *Quat. Res.* 3:10–38

Johnson, D. A., Knoll, A. H. 1974. Radiolaria as Paleoclimatic indicators: Pleistocene climatic fluctuations in the Equatorial Pacific Ocean. *Quat. Res.* 4:206–16

Johnson, T. C., 1975. *The dissolution of siliceous microfossils in deep sea sediments.* PhD thesis. Univ. Calif., San Diego. 163 pp.

Kaiser, H. F. 1958. The varimax criterion for analytic rotation in factor analysis. *Psychometrika* 23:187–200

Kanaya, T., Koizumi, I. 1966. Interpretation of diatom thanatocoenoses from the North Pacific applied to a study of core V20-130. (Studies of deep-sea core V20-130. Part IV.) *Sci. Rep. Tohoku Univ. Jpn. Ser. 2 (Geol.)* 37:89–130

Kipp, N. G. 1976. New transfer function for estimating past sea-surface conditions from sea-bed distribution of planktonic foraminiferal assemblages in the North Atlantic. See Cline & Hays 1976, pp. 3–41

Kipp, N. G., Towner, D. P., 1975. The last Millenium of climate: Foraminiferal record from coastal basin sediments. *Proc. WMO/IAMAP Symp. Long-term Clim. Fluctuations. W.M.O. Publ. 421*

Kutzbach, J. E., 1975. Diagnostic studies of past climates. In *The Physical Basis of Climate and Climate Modelling, GARP Publ. Ser. No. 60, ICSU/WMO,* pp. 119–26. Geneva: WMO. 256 pp.

Ledbetter, M. T., Ellwood, B. B., 1976. Selection of sample intervals in deep-sea sedimentary cores. *Geology* 4:303–4

Lozano, J., Hays, J. D. 1976. Relationship of radiolarian assemblages to sediment types and physical oceanography in the Atlantic and Western Indian Ocean sectors of the Antarctic Ocean. See Cline & Hays 1976

Luz, B. 1973. Stratigraphic and paleoclimatic analysis of late Pleistocene tropical Southeast Pacific cores. *Quat. Res.* 3:56–72

Lynts, G. W., Judd, J. B. 1971. Late Pleistocene paleotemperature at Tongue of the Ocean, Bahamas. *Science.* 171:1143–44

Maynard, N. G. 1976. Relationship between diatoms in surface sediments of the Atlantic Ocean and the biological and physical oceanography of overlying waters. *Paleobiology* 2:99–121

Moore, T. C. Jr. 1973. Late Pleistocene-Holocene oceanographic changes in the northeastern Pacific. *Quat. Res.* 3:99–109

Morin, R. W., Theyer, F., Vincent, E. 1970. Pleistocene climates in the Atlantic and Pacific Oceans: a reevaluated comparison based on deep-sea sediments. *Science* 169:365–66

Nie, N. H., Hull, C. H., Jenkins, J. G., Streinbrenner, K. Bent, D. H., 1975. *Statistical package for the Social Sciences,* 2nd ed. New York: McGraw-Hill. 675 pp.

Nigrini, C. A. 1970. Radiolarian assemblages in the North Pacific and their application to a study of Quaternary sediments in core V20-130. In *Geological investigations of the North Pacific.* ed. J. D. Hays. *Geol. Soc. Am. Mem.* 126:139–83

Orloci, L. 1972. On objective functions of phytosociological resemblance. *Am. Midl. Nat.* 88:28–55

Pisias, N. G. 1976. Late Quaternary sediment of the Panama Basin: sedimentation rates, periodicities, and controls of carbonate and opal accumulation. See Cline & Hays 1976, pp. 375–91

Robertson, J. H., 1975. *Glacial to Interglacial Oceanographic Changes in the Northwest Pacific, including a continuous Record of the last 400,000 years.* PhD thesis. Columbia Univ., New York. 355 pp.

Sachs, H. M. 1973a. North Pacific radiolarian assemblages and their relationship to oceanographic parameters. *Quat. Res.* 3:73–88

Sachs, H. M. 1973b. Late Pleistocene history of the North Pacific: evidence from a quantitative study of Radiolaria in core V21-173. *Quat. Res.* 3:89–98

Sachs, H. M. 1973c. Quantitative radiolarian-based Paleo-Oceanography in late Pleistocene Subarctic Pacific Sediments. PhD thesis, Brown Univ. 208 pp.

Sachs, H. M. 1976. Evidence for the role of the oceans in climatic change: tests of Weyl's theory of Ice Ages. *J. Geophys. Res.* 81:3141–50

Sancetta, C. A. 1977. *Oceanography of the North Pacific during the Last 18,000 Years Derived from Fossil Diatoms.* PhD thesis. Oregon State Univ., Corvallis. 148 pp

Sancetta, C., Imbrie, J., Kipp, N. G. 1973. Climatic record of the past 130,000 years in North Atlantic deep-sea core V23-82: Correlation with the terrestrial record. *Quat. Res.* 3:110–16

Schott, W. 1935. Die Foraminiferen in den Aquatorialen Teil des Atlantischen Ozeans. *Dtsch. Atl. Exped.* 3:43–134

Sliter, W. V., Bé, A. W. H., Berger, W. H. (eds.) 1975. *Dissolution of Deep-Sea Carbonates. Cushman Found. Foraminiferal Res. Spec. Publ. 13,* 159 pp.

van Donk, J. 1970. *The oxygen isotope record in deep-sea sediments.* PhD thesis. Columbia Univ. New York. 136 pp.

Webb, T. III. 1974. Corresponding patterns of pollen and vegetation in lower Michigan: a comparison of quantitative data. *Ecology.* 55:17–28

Webb, T. III, Bryson, R. A. 1972. Late and postglacial climatic change in the northern Midwest, U.S.A.: quantitative estimates derived from fossil pollen spectra by multivariate statistical analysis. *Quat. Res.* 2:70–115

Webb, T. III, Clark, D. R. 1977. Calibrating micropaleontological data in climatic terms: a critical review. *Ann. NY Acad. Sci.* (In press)

Webb, T. III, McAndrews, J. H. 1976. Corresponding patterns of contemporary pollen and vegetation in central North America. Cline & Hays 1976, pp. 267–99

Weyl, P. K. 1968. The role of the oceans in climatic change: a theory of the ice ages. *Meteorol. Monogr.* 8:37–62

Williams, D. F., Johnson, W. C. 1975. Diversity of Recent planktonic foraminifera in the southern Indian Ocean and Late Pleistocene paleotemperatures. *Quat. Res.* 5:237–50

Ann. Rev. Earth Planet. Sci. 1977. 5:179–202
Copyright © 1977 by Annual Reviews Inc. All rights reserved

COMPOSITION OF ✖10071
THE MANTLE AND CORE[1]

Don L. Anderson
Seismological Laboratory, California Institute of Technology, Pasadena,
California 91125

INTRODUCTION

Free oscillations, surface wave dispersion, travel times, and high-resolution body wave data, combined with laboratory data, provide constraints on the composition and state of the various regions of the earth. Anderson & Hart (1976) have recently presented an earth model based on a large body of free oscillation and travel time data. This model, designated C2, incorporates structure in both the mantle and core found using special high-resolution body wave techniques. Much of the structure is beyond the resolving power of such gross earth data as free oscillation periods and differential travel times. This model, shown in Figure 1, provides an appropriate basis for a discussion of the composition of the various regions of the earth. Model 1066A of Gilbert & Dziewonski (1975), based on more normal mode data but less body wave data, is also shown.

These models, like most models based on free oscillation periods, assume that the elastic properties of the earth are independent of frequency. It is well known that absorption decreases the intrinsic elastic properties and causes velocity dispersion (Zener 1948). This means that laboratory, body wave, surface wave, and free oscillation data cannot be directly compared unless the absorption-dispersion correction is applied. The method for correction of seismic data is given in Liu et al (1976) and consistent earth models have been determined from these considerations by Hart et al (1976). These models have slightly higher velocities in the upper mantle than models based on uncorrected free oscillation data (Figure 2). The physical dispersion accompanying absorption also complicates comparisons of laboratory and seismic data since they are taken at quite different frequencies. These complications should be taken into account when one attempts to infer the composition of the earth by comparing laboratory and seismic data.

The uppermost 700 km of the mantle can be split into five distinct regions separated by regions of transition. These represent average conditions and are

[1] Contribution No. 2804, Seismological Laboratory, California Institute of Technology, Pasadena, California 91125.

weighted most heavily by the properties of the oceanic mantle. These regions include a 40-km-thick lid, a 180-km-thick low-velocity zone terminating at about 250 km depth, and relatively homogeneous regions between 250–375, 425–500, and 550–650 km. The intervening regions have high velocity gradients and probably

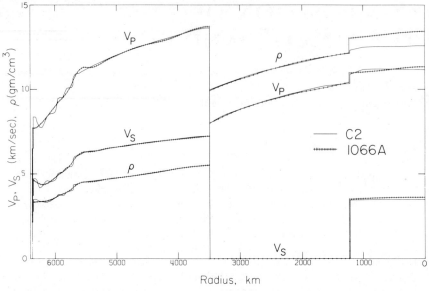

Figure 1   Two recent earth models: C2 (Anderson & Hart 1976) and 1066A (Gilbert & Dziewonski 1975).

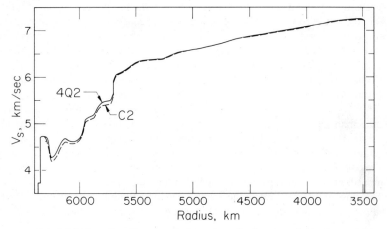

Figure 2   Model C2 based on frequency-independent elastic properties and a model based on absorption-corrected normal mode periods (Hart et al 1976).

represent transitions between lower and higher pressure phases. The mantle is also mildly inhomogeneous between 675–1245 and 1550–1770 km, indicating possible phase changes or compositional gradients in the lower mantle. The lowermost 350 km of the mantle has abnormally low velocity gradients, suggestive of a high temperature gradient in this region. The only clearly homogeneous regions of the lower mantle are between 1250–1550 and 1770–2520 km depth. The upper and lower regions of the outer core also appear to be inhomogeneous. A homogeneous region occurs between depths of 3170 and 4170 km. A velocity and density jump occurs at the inner core–outer core boundary.

THE UPPER MANTLE AND TRANSITION REGION

We define the upper mantle as the region between the base of the crust and the seismic discontinuity near 370 km. The top layer is the mantle part of the lithosphere or the "lid" of the low-velocity zone. Using Bullen's (1967) notation, as extended by Anderson (1967a), this is region B'. It is variable in thickness but averages about 50–70 km in oceanic regions (Anderson 1965, Kanamori & Press 1970) and may be as thick as 150–200 km under older shield areas (Brune & Dorman 1963).

The low-velocity zone (LVZ), region B'', is a pronounced feature under oceanic and tectonic regions, but evidence for its existence under continental shields is weak. It is a required property of the average earth as defined by free oscillation and great circle surface wave dispersion data. The base of the LVZ is at roughly 250 km.

Between 250 and 370 km region B''', the mantle is relatively homogeneous. No important phase changes have been found in the laboratory that correspond to this depth interval, and normal mantle mineral assemblages (olivine, pyroxene, garnet) can be expected. Partial solubility of pyroxene into the garnet structure, however, may occur near the base of this region (Ringwood 1975).

The uppermost mantle, or lid, is characterized by relatively high density and seismic velocities ($\rho = 3.5$ g cm^{-3}, $V_p = 8.38$ km sec^{-1}, and $V_s = 4.71$ km sec^{-1}). These values are greater than can be obtained in a rock composed solely of olivine and pyroxene, and the presence of substantial amounts of garnet is implied. The lid falls well within the experimental field of eclogites, but not on any reasonable extrapolation of the mantle below the LVZ. It should be pointed out that the resolving power for density in such a small region is poor.

Below 250 km and extending to the 375 km discontinuity, the velocities and densities are consistent with a mixture of olivine and pyroxene, the presumed major constituents in this region.

Between 375 and 425 km the velocities increase by 11%. This corresponds to a 12–13% increase at standard conditions. This can be compared with the increases of 13–17% for V_p and 8–16% for V_s involved in the olivine–γ-spinel reaction (Liebermann 1975). Whitcomb & Anderson (1970) and Burdick & Anderson (1975) have interpreted the region of the mantle between 425 and 500 km as being composed predominantly of β-spinel, a phase some 7% denser than olivine and several percentage points less dense than γ-spinel. The density jump across this region cannot be well resolved, but in model C2 is 7% at normal conditions, consistent with the olivine–β-spinel reaction. This depth range appears to have the same composition as the olivine region above.

The velocity jumps associated with the 500 km discontinuity are 4% and 6% for V_p and V_s respectively, giving a total increase through the two discontinuities of 16% and 19%, respectively. The V_p jump is within Liebermann's range, but the V_s jump is slightly greater than the maximum value of 16% reported by Liebermann for the β-γ transition in Mg_2GeO_4. The total density jump, 13%, is also slightly greater than expected for the olivine-spinel reaction but it should be remembered that the density resolution is poor.

The pyroxene component of the mantle begins to dissolve in garnet somewhere in this depth range (Ringwood 1975), leading to about 40% garnet solid solution that has a density increase of about 10% over that of pyroxene. This is about the density increase associated with the olivine-spinel reaction and the velocity increases might be expected to be comparable. However, if this garnet has the lower velocities typical of other garnets, then other reactions in the pyroxene-garnet component of the mantle must be invoked. One possibility is the transformation to ilmenite structure (Ringwood 1975). The pyroxene-ilmenite transformation leads to density increases of $\sim 12\%$ for V_p and 6% for V_s (Liebermann 1974).

Another approach to the problem of crystal structure in the upper mantle has recently been suggested (Anderson 1976). The bulk modulus is relatively insensitive to composition but is a good diagnostic of crystal structure. The bulk modulus increases in the order pyroxene, olivine, β-spinel, $(Mg, Fe)O$ (rock salt), garnet, ilmenite, γ-spinel and $(Mg, Fe)_2SiO_4$ (mixed oxides), $(Mg, Fe)SiO_3$ (mixed oxides), and perovskite. This approach suggests that the phases in the upper mantle are olivine-pyroxene-garnet (< 350 km), β-spinel plus garnet-pyroxene solid solution (350–500 km), $(\beta + \gamma)$-spinel plus garnet (500–670 km), and $(\gamma + \beta)$-spinel plus mixed oxides or perovskite (> 670 km).

The Lithosphere

Uppermost mantle velocities, P_n, are typically 8.0 to 8.2 km sec^{-1}, and the spread is about 7.9–8.6 km sec^{-1}. Some long refraction profiles give evidence for a deeper layer in the lithosphere having a velocity of 8.6 km sec^{-1} (Kind 1974). The lithosphere appears to contain at least two layers.

The lid of the low-velocity zone is too restricted a region to expect good resolving power using gross earth data alone. Nevertheless, the average values of ρ, V_p, and V_s, when corrected to standard conditions, are 3.55 g cm^{-3}, 8.72 km sec^{-1}, and 4.99 km sec^{-1}, respectively. For this calculation we have assumed a temperature of 950°C at 40 km and derivatives appropriate for olivine. The corrections for temperature amount to 0.5 and 0.3 km sec^{-1}, respectively, for V_p and V_s. The pressure corrections are much smaller. Short-period surface wave data has better resolving power for V_s in the lid. Applying the same corrections to Forsyth's (1973) surface wave data we obtain 4.48–4.55 km sec^{-1} and 4.51–4.64 km sec^{-1} for 5 m.y. old and 25 m.y. old oceanic lithosphere. Presumably, velocities can be expected to increase further for older regions. A value for V_p of 8.6 km sec^{-1} is commonly observed near 40 km depth in the oceans (Kosminskaya et al 1972). This corresponds to about 8.87 km sec^{-1} at standard conditions. These values can be compared with 8.48 and 4.93 km sec^{-1} calculated for olivine aggregates (Simmons & Wang 1971). Eclogites have V_p and V_s as large as 8.8 and 4.9 km sec^{-1} in certain directions and as high as 8.61 and 4.86 km sec^{-1} as average values (Manghnani et al 1974) (see Figure 3). Pyrope garnet has estimated velocities of 9.6 and 5.4 km sec^{-1}, respectively (Wang & Simmons 1974).

All of the above suggests that corrected velocities of at least 8.6 and 4.8 km sec^{-1}, for V_p and V_s, respectively, occur in the lower lithosphere and this requires substantial amounts of garnet at relatively shallow depths. The estimate of the density of the lithosphere, although more uncertain, is also consistent with a large proportion of garnet. At least 26% garnet is required to satisfy the compressional velocities. The density of such an assemblage is about 3.4 g cm^{-3}. If one is to honor the higher seismic velocities and the upper mantle density, even greater proportions of garnet are required.

Most refraction profiles, particularly at sea, sample only the uppermost lithosphere. P_n velocities of 8.0–8.2 km sec^{-1} are consistent with peridotite or harzburgite, the refractory residual after basalt removal. Anisotropies are also appropriate for olivine-pyroxene assemblages. The sequence of layers, at least in oceanic regions, seems to be basalt, peridotite, eclogite.

Anisotropy of the upper mantle is a potentially useful petrological constraint (Anderson 1965, Ringwood 1975) although it can also be caused by organized heterogeneity, such as laminations or parallel dikes and sills or aligned partial melt zones, and stress fields. Under oceans the uppermost mantle, the P_n region, exhibits an anisotropy of 7% (Morris et al 1969). The fast direction is in the direction of spreading, and the magnitude of the anisotropy and the high P_n velocities suggest that oriented olivine crystals control the elastic properties, as originally suggested by Anderson (1965). Pyroxene exhibits a similar anisotropy, whereas garnet is isotropic. The preferred orientation is presumably due either to the temperature gradient away from the ridge or to higher stresses in this direction. A peridotite layer at the top of the oceanic mantle is consistent with the observations.

The average anisotropy of the upper mantle is much less than the values given above. Forsyth (1973) studied the dispersion of surface waves and found shear wave anisotropies, averaged over the upper mantle, of 2%. Shear velocities in the lid vary from 4.26 to 4.46 km sec^{-1}, increasing with age; the higher values correspond to a lithosphere 10–50 m.y. old. This can be compared with shear wave velocities of 4.30–4.86 km sec^{-1} and anisotropies of 1–4.7% found in relatively unaltered eclogites (Manghnani et al 1974). The compressional velocity range in the same samples is 7.90–8.61 km sec^{-1}, reflecting the large amounts of garnet.

On the basis of data available at the time, Ringwood (1975) suggested that the V_p/V_s ratio of the lithosphere was smaller than measured on eclogites. He therefore ruled out eclogite as an important constituent of the lithosphere. The eclogite-peridotite controversy regarding the composition of the suboceanic mantle is long standing and still unresolved (Wyllie 1971). Newer and more complete data on the

V_p/V_s ratio in eclogites is summarized in Figure 3. The data are grouped by composition. The high-V_p/V_s eclogites are generally low density and contain alteration products, such as plagioclase or olivine. The higher-density eclogites are consistent with the properties of the lower lithosphere. Also shown schematically in Figure 3 are lines for orthopyroxenes, olivines, and garnets. The elastic properties of clinopyroxene are not well known, but the eclogite data suggest that they serve to increase the V_p/V_s ratio. Figure 3 indicates that garnet and clinopyroxene may be important components of the lithosphere. A lithosphere composed primarily of olivine and $(Mg, Fe)SiO_3$, i.e. pyrolite or refractory phases remaining after basalt removal, does not satisfy the seismic data for the bulk of the lithosphere.

I suggest that oceanic crustal basalts represent only part of the basaltic fraction of the upper mantle. The peridotite layer represents the depleted materal complementary to the extruded basalts, but more basaltic material exists at depth, intruded from the low-velocity layer. It is likely that the low-velocity zone is layered with a higher volatile and melt fraction concentrated toward its top. As the slab cools, this basaltic material is incorporated at the base of the plate and as the plate thickens it eventually transforms to eclogite, yielding the observed velocities and increasing the thickness and mean density of the plate. Eventually the plate becomes denser than the underlying asthenosphere and conditions become appropriate for subduction.

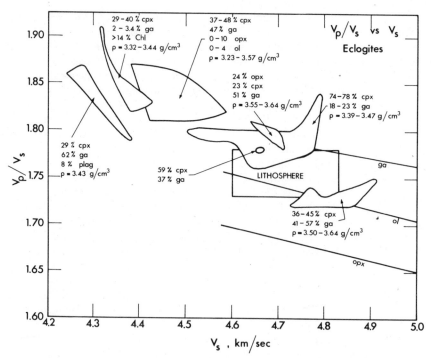

Figure 3 V_p/V_s versus V_s for eclogites and for the lithosphere.

O'Hara (1970) has argued that erupted lavas are not the original liquids produced by partial melting of the upper mantle, but are residual liquids from processes that have left behind complementary eclogite accumulates in the upper mantle. Such a model is consistent with the seismic observations of high P_n velocities at midlithospheric depths and with the propensity of oceanic lithosphere to plunge into the asthenosphere. The latter observation suggests that the average density of the lithosphere is greater than that of the asthenosphere.

The Low-Velocity Zone (LVZ)

Anderson & Sammis (1970) demonstrated that the geophysical data (seismic velocities, attenuation, heat flow) could be explained if the LVZ were a region of partial melt. This explanation, in turn, demanded the presence of volatiles in order to suitably depress the solidus of mantle materials. The beginning of the LVZ marks the crossing of the geotherm with the wet solidus of peridotite. Its termination would be due to (a) a crossing in the opposite sense of the geotherm and the solidus, (b) the absence of water, or (c) the removal of water into high-pressure hydrous or hydroxylated phases. In all cases the boundaries of the LVZ would be expected to be sharp. Later petrological studies showed that undersaturated magmas could only be partial melts of pyrolite provided there was some water present ($\sim 0.4\%$ H_2O) in the LVZ (Green 1973). Anderson et al (1971) showed that small amounts of melt ($\sim 1\%$) could explain the velocity reduction if the melt occurred as thin grain boundary films. There is therefore a good convergence of geophysical and petrological views regarding the physical nature of the low-velocity zone. Such a zone also makes it easier to understand the extreme mobility of lithospheric plates and the thickening with time of the lithosphere (Parker & Oldenburg 1973). Any attempt to determine the composition of the LVZ from seismic data is futile because of the strong dependence on melt zone geometry. Hydrous pyrolite satisfies the geophysical and petrological constraints.

Partial melting is not the only explanation of the LVZ. High attenuation is associated with rapid changes in velocity (Zener 1948) and a subsolidus explanation of the LVZ and the high-attenuation zone is conceivable. However, laboratory and geological data indicate that partial melting should and does occur in the upper mantle.

Region B'''

The velocities and densities of the 300–400-km depth region of the mantle extrapolated to standard conditions using the method of Burdick & Anderson (1975) are given in Table 1 for model C2 and the Helmberger-Engen model. Also given are parameters estimated from the mineral assemblage of pyrolite (8% FeO). The velocities in pyrolite and dunite are substantially higher than observed. If the FeO content of pyrolite is arbitrarily raised to 17% then the velocities are acceptable but the density is probably too high. Eclogites also satisfy the velocities, but again their densities are high. The density, however, is only a weak constraint because of resolution limitations of the geophysical data.

The velocities between 300 and 400 km can also be explained on the pyrolite model with the addition of hydrous phases or small amounts of partial melt. We

Table 1   Properties of upper mantle (300–400 km) and possible upper mantle mineral assemblages (STP)

V_p (km sec^{-1})	V_s (km sec^{-1})	ρ (g cm^{-3})	V_p/V_s	Model or mineral assemblage
8.29	4.67	3.33	1.72	model C2
8.19	4.69	—	1.75	Helmberger-Engen
8.37	4.87	3.38	1.72	pyrolite 8% FeO
8.19	4.71	3.48	1.74	pyrolite 17% FeO
8.16	4.70	3.35	1.74	pyrolite 0.4% H$_2$O

estimate that about 5% hydrous phase will lower the velocities of pyrolite by the required amount. The amount of hydrous phase depends on its composition, but the above estimate can be accomplished with about 1% H$_2$O. When absorption is taken into account the velocities in this region increase and pyrolite may prove acceptable.

The Transition Region

The transition region, Bullen's region C, has received much attention from seismologists and petrologists. It now appears to consist of three regions of relatively rapid velocity increase separated by more homogeneous regions. The regions of high velocity gradients are 370–430, 500–570, and 670–700 km (Helmberger & Wiggins 1971, Helmberger & Engen 1974). There is some suggestion of moderate additional structure down to about 1200 km although the mantle is relatively homogeneous below about 900 km. The 670 km discontinuity is the best reflector of seismic energy (Engdahl & Flinn 1969, Whitcomb & Anderson 1970) indicating that at least part of the rapid increase in velocity is discontinuous, although a large part of the increase must be spread out to satisfy direct arrival amplitudes. Lateral heterogeneity, however, may also be involved.

The inhomogeneous nature of the transition region can be well understood in terms of the sequence of phase changes in a peridotite mantle. The three major discontinuities are due primarily to the α-olivine, β-spinel, γ-spinel, and post-spinel (mixed oxides, strontium plumbate, or perovskite plus ferropericlase structures) in the olivine component of the mantle and the pyroxene, garnet, ilmenite, spinel plus stishovite, perovskite, and mixed oxides transformations in the pyroxene component of the mantle. The 670 km discontinuity is primarily due to pyroxene-garnet solid solution transforming to mixed oxides or perovskite (Anderson 1976).

The major density jumps in mantle minerals are associated with the transitions from olivine to β-phase (8.3%), spinel to mixed oxides or Sr$_2$PbO$_4$ structure (8.5%), pyroxene to spinel plus stishovite (17.3%), pyroxene to garnet (10%), garnet to mixed oxides (11%), and ilmenite to mixed oxides or perovskite (4.2–7.0%). The transitions from β-phase to spinel, spinel plus stishovite to ilmenite, and mixed oxides to perovskite-bearing assemblages (or vice versa) all involve less than 2.8% changes in density. From the seismic equation of state (Anderson 1967b) we estimate that the percentage velocity jumps are about 1.5 times that of the density jumps for the phase changes that do not involve coordination changes. When a coordination increase is involved the cation-anion distances increase and a smaller velocity increase can be expected (Liebermann & Ringwood 1973). For example spinels have similar elastic properties to the mixed oxides.

Olivine-spinel solid solution equilibria in the system Mg_2SiO_4-Fe_2SiO_4 have been studied at several laboratories over the pressure range 40–200 kbar at 800 and 1000°C. Ringwood & Major (1966) discovered the β-phase of spinel in compositions close to pure Mg_2SiO_4. This was verified by Akimoto & Fujisawa (1966) and Akimoto (1972) who constructed a phase diagram for the olivine–β-spinel–γ-spinel portion of the Mg_2SiO_4-Fe_2SiO_4 system. The existence of β-spinel in the Mg_2SiO_4-rich assemblages introduces an additional loop in the phase diagram and indicates that the transition region will be complex, even for a purely olivine mantle, and that a first-order discontinuity may occur in this region (Whitcomb & Anderson 1970). The stability field of β-spinel is extensive at high temperature. It is not yet completely clear whether the transformation of β- to γ-spinel occurs in the mantle before the intervention of other, more dense phases. Present evidence suggests that the olivine component of the mantle is still in the mixed $\beta + \gamma$ field below 670 km (Anderson 1976).

$(Mg, Fe)SiO_3$ pyroxene remains stable to substantially higher pressures than required to cause olivines with a similar Mg/Fe ratio to transform to spinels (Akimoto 1972). This means that the major phases in the "spinel" region of an olivine-pyroxene mantle are β- or γ-spinel and untransformed pyroxene. Garnet is stable until about 200 kbar (> 600 km) at which point it transforms to a garnet-ilmenite solid solution (Ringwood & Major 1967).

However, at high pressures, enstatite is capable of dissolving extensively in the garnet structure (Ringwood 1975). The transformation is rapid at 90 kbar and complete by 150 kbar. This indicates that the pyroxene component of the mantle, in the presence of garnet or Al_2O_3, disappears between 270 and 450 km. This transformation involves a density increase of about 10%. There is probably adequate Al_2O_3 in the mantle for this transformation to proceed to completion. The lower part of region B''' (250–375 km) is therefore olivine and pyroxene with increasing amounts of garnet with depth, and the β-spinel region of the mantle (425–500 km) is mainly β-spinel and garnet.

LOWER MANTLE

On a gross scale the earth is chemically and gravitationally zoned, with the crust, mantle, and core the main subdivisions. The crust itself is chemically stratified. Although it is often assumed that the mantle is homogeneous in composition, this seems unlikely. The gravitational potential increases with depth, providing a driving force for downward concentration of denser material. If the whole of the mantle was ever in chemical equilibrium there would be a rapid increase of iron content with depth. The fact that a large iron gradient cannot be accommodated by the seismic data indicates that the whole mantle has never been in chemical equilibrium or that iron-bearing phases are distinct from the magnesium-bearing phases. Nevertheless it would not be surprising if the iron content of the lower mantle were greater than that of the upper mantle.

The $(FeO + Fe_2O_3)/MgO$ ratio of basalts, the liquid fraction of the upper mantle, is about 1.2, much greater than the complementary refractory crystals, 0.23. Chondritic values of this ratio are near 0.44. In an initially chondritic earth the above ratios can be satisfied with 32% basaltic fraction. Incomplete basalt extraction from the upper mantle could explain the values inferred for pyrolite. It is of interest that 200 m.y. of sea floor spreading with the present rate of crustal generation would yield a layer of basalt 44 km thick over the surface of the earth. The lower mantle would be enriched in iron, relative to the upper mantle, if it were more primitive, i.e. not having been subjected to basalt removal, or if the ultimate fate of the iron-rich downgoing slabs is their assimilation in the lower mantle.

Possible Phase Assemblages in the Lower Mantle

Birch (1952) suggested that mantle minerals may disproportionate to simple oxides in the lower mantle. The constitution of the lower mantle would therefore be primarily MgO (periclase) and a then hypothetical rutile form of SiO_2, now known as stishovite.

Until recently, speculations regarding the phase assemblages likely to be present in the lower mantle were based on shock wave data and on behavior of analog compounds. There is now abundant direct static experimental data on transformations in such pertinent mantle compositions as $(Mg, Fe)_2SiO_4$, $(Mg, Fe)SiO_3$, $CaSiO_3$, and $Mg_3Al_2SiO_3O_{12}$ (Liu 1974, 1975). Most of the data has been obtained by quenching experiments from ~ 300 kbar and $\sim 1800°C$. Although there is still some disagreement regarding the results it appears that complete disproportionation to the mixed oxides or to rock salt plus perovskite structure occurs via a series of intermediate phase changes and partial disproportionations involving spinel, garnet, ilmenite, rock salt, rutile, and perovskite structures. Reconnaissance experiments suggest that perovskite is a higher pressure phase than the oxides, stishovite plus magnesiowustite, but the complete phase diagrams have not yet been worked out. At lower mantle pressures these two phase assemblages have very similar densities.

Other postulated phase changes in A_2BO_4 compounds such as Sr_2PbO_4 and K_2NiF_4 (Ringwood 1975) have not been found in experiments on $(Mg, Fe)_2SiO_4$. Sr_2PbO_4 structures occur in oxides having large cation/anion ratios (Sr_2PbO_4, Ca_2PbO_4, Ca_2SnO_4, Cd_2SnO_4, Mn_2GeO_4) and such structures have nearly the same densities as isochemical rock salt plus rutile mixtures. K_2NiF_4-type structures are also exhibited only by compounds with large cations. They average 4% denser than the mixed oxides, although they are less dense than the rock salt plus perovskite assemblage.

Ringwood (1975) pointed out that where perovskites are formed between rock salt and rutile-type oxides, their densities are characteristically between 5 and 10% greater than the isochemical mixed oxides and average about 7.5% denser. However, the perovskites considered by Ringwood all had large cation radii, between 1.31 and 1.60 Å for the A cation and 0.61 to 1.18 Å for the B cation, compared with 1.40 Å for the oxygen ion, in the formula ABO_3. For comparison the cationic radii for $MgSiO_3$ are 0.89 and 0.40 Å for coordinations appropriate for the orthorhombic perovskite structure. There is a tendency for the volume difference between the mixed oxides and perovskite to decrease with cation radii.

$MgSiO_3$ (perovskite), recently measured by Liu (1974, 1975), is only 2.8% denser than the mixed oxides. The assemblage $MgSiO_3$ (perovskite) plus MgO (olivine stoichiometry) is only 1.9% denser than the oxides. Whether perovskite or an oxide mixture is the denser assemblage at lower mantle pressures and temperatures depends critically on their respective equations of state and, in particular, their bulk moduli and pressure derivatives. As is discussed later, the similarity in density between mixed oxides and assemblages involving perovskite makes it necessary to increase the FeO content of the lower mantle regardless of the actual phase assemblage. The possibility of phases denser than the mixed oxides has been suggested to account for the high density of the lower mantle (Ringwood 1975), which is denser than the mixed oxide assemblage of pyrolite composition. Alternatively, Anderson et al (1971) have suggested that the lower mantle contains more FeO and SiO_2 than pyrolite and is therefore closer to chondritic silicate abundances.

To test the alternatives it is necessary to estimate the compressibility or bulk modulus (K) of $MgSiO_3$ (perovskite). The bulk modulus of MgO plus SiO_2 (stishovite) is about 2.29 Mbar. Using the perovskite data tabulated by Jones & Liebermann (1974) and the relation proposed by Anderson & Anderson (1970), we estimate that the bulk modulus of $MgSiO_3$ (hypothetical cubic perovskite) is 12 to 24% greater than the mixed oxides. Using the relation proposed by Davies (1974) between the densities and bulk sound velocities of silicates and their oxide mixture we estimate K (perovskite) to be 10% greater than the oxide mixture. Because of the greater incompressibility of the perovskite phase the density difference will decrease with pressure. Since the perovskites used in the above analysis involve twelvefold coordination of silicon, while $MgSiO_3$ (perovskite) involves eightfold coordination we may have underestimated the modulus of mantle perovskite since the Si-O distance is smaller than in the analog compounds. We therefore adopted the higher value for K (perovskite) and calculated density-pressure trajectories for both the mixed oxides and $MgSiO_3$ (perovskite) with several assumptions about dK/dP.

Using the Birch-Murnaghan equation with $K'_0 = 4$ the trajectories cross at about 1 Mbar pressure. Above 1 Mbar the mixed oxides are denser than the perovskite phase. The seismic parameter Φ, however, is about 20% greater for perovskite than for mixed oxides throughout the mantle. At 400 kbar the density difference is less than 1%. The density difference between Mg_2SiO_4 (mixed oxides) and $MgSiO_3$ (perovskite) plus MgO (periclase) is 1.9% at zero pressure and also rapidly decreases with pressure. Whether the density trajectories cross or not in the mantle it is clear that perovskite phases at lower mantle pressures will not serve to substantially increase the density. The same argument can be used for other high-pressure phases. However, the seismic velocities will be higher for the initially denser phases because of their greater bulk moduli. In order to satisfy both the densities and the elastic parameters in the lower mantle it seems necessary to adjust the iron content.

The structure of the perovskite phase of $(Mg, Fe)SiO_3$ is similar to that of the orthorhombic rare earth orthoferrites with 8-8-6 coordination rather than an "ideal" perovskite with 12-6-6 coordination. This results in shorter bond lengths, smaller cell volumes, and increased repulsion between ions. The bulk modulus is therefore likely to be even greater than estimated here. This has two important implications in interpreting shock wave and seismic data. First, the density difference between mixed oxides and perovskite structures will decrease even more rapidly with pressure than given here and, second, the seismic velocities will be even higher. There is an increased possibility that mixed oxides become denser than perovskite at moderate pressure.

Seismic Data for the Lower Mantle

The most direct method for inferring the composition of the lower mantle involves comparing lower mantle densities with shock wave data on various rocks and minerals. Shock wave data from summaries in Al'tshuler & Sharipdzhanov (1971, 1972) and Clark (1966) on bronzitites and dunites containing 7–11% FeO are approximately 3–7% less dense, when corrected for temperature, than lower mantle densities. The lower mantle therefore has more FeO ($>11.4\%$) or SiO_2 ($>42\%$) than, for example, the Twin Sisters dunite. About 18% FeO (weight percent) would be consistent with the dunite data. Above about 1 Mbar pressure the lower mantle is apparently even more enriched in one or both of these components, or consists of a phase assemblage more dense by about 3% than the mixed oxides, or the densest phase exhibited by the shocked dunites. Pyroxene with about 15% FeO would satisfy the data for the upper half of the lower mantle. More FeO or a denser phase than is achieved in shock compression is required in the deeper mantle. The deep mantle can have as much as 18–21% FeO. It has been suggested earlier (Anderson et al 1971, Burdick & Anderson 1975) that the lower mantle is enriched in both FeO and SiO_2 relative to the upper mantle and may have a near pyroxene stoichiometry.

Some previous interpretations of shock wave data on olivine- and pyroxene-rich rocks have suggested that some of these materials, notably bronzitite, iron-rich dunite, and fayalite have transformed to phases denser than the mixed oxides. For example, Davies & Gaffney (1973) infer that bronzitite has transformed to a phase with a zero-pressure density of 4.41 g cm^{-3} or 7% denser than the mixed oxides. They interpreted this as perovskite, but this is much greater than the density of $(Mg, Fe)SiO_3$ (perovskite). Also, they assumed a bulk modulus for stishovite some 11% higher and for FeO some 20% higher than current estimates. When these corrections are made, the high-pressure bronzitite data is consistent with an oxide mixture. Their interpretations for Twin Sisters dunite, Mooihoek (iron-rich) dunite, and fayalite are, respectively, about 1, 4, and 5% denser than the mixed oxides or, equivalently, the Sr_2PbO_4 structure. These estimates are all greater than those obtained using different techniques (Anderson & Kanamori 1968 and Ahrens et al 1969).

It is possible to interpret the shock wave data on dunites and bronzitites as mixed oxides simply by adopting the preferred values for K_0 of stishovite and wustite. Using the values for K_0 of 3000 and 1480 kbar for SiO_2 (stishovite) and FeO, respectively, given by Kalinin & Pan'kov (1974) and Al'tshuler & Sharipdzhanov

(1972) we have recalculated the hugoniots for oxide mixtures corresponding to the composition of the Twin Sisters and Mooihoek dunites, fayalite, and the Stillwater and Bushfeld bronzitites. In all cases the shock wave data is consistent with mixed oxides or, equivalently, the Sr_2PbO_4 structure for the dunites and fayalite. There is no need to invoke denser structures such as perovskite for the bronzitites or K_2NiF_4 or $CaFe_2O_4$ for the olivines nor is it necessary to invoke the low-spin form for Fe^{2+}.

With these reinterpretations there is no basis, from shock wave data, for assuming that any silicate transforms to a denser structure than the mixed oxides. Silicates are now known to transform to the perovskite structure under static compression and intense heating (Liu 1974, 1975). It is probable that the densities of the mixed oxides and perovskite are nearly coincident at high pressure, a result of the lower compressibility of the denser phase. In the latter case it is immaterial, at least for inferring composition from density, whether the lower mantle is mixed oxides or contains perovskite. However, perovskite should have higher velocities than the mixed oxides.

In addition to the density of the lower mantle we also have, as additional constraints, the seismic velocities. Unfortunately, it is difficult to determine velocities in shock wave experiments. However, we can estimate the combination $K_s/\rho = V_p^2 - (4/3)V_s^2 = (\partial P/\partial \rho)_s$ from shock wave data. Anderson & Kanamori (1968) have shown that shock wave data on many rocks are consistent with the properties of the mixed oxides. Since we have good equation of state data for MgO and SiO_2 (stishovite), the major components of the mantle, and some data for FeO, it is possible to determine both the density and bulk sound speed of oxide mixtures with fair precision.

Figure 4 gives the variation of density, ρ, and bulk sound speed, $c = \Phi^{1/2}$, as a function of pressure in the lower mantle. Two models, C2 and PEM (Dziewonski et al 1975), are shown. Also shown are the mixed oxides calculations of Al'tshuler & Sharipdzhavov (1971) for four different compositions. Curves 1 and 2 are for chondritic SiO_2 contents. Calculations are given for FeO contents of 21, 14, and 9%. The best agreement for density is the chondritic curve with 14% FeO. Pyrolite with 9% FeO is about 4% underdense. There is a suggestion of iron enrichment with depth in the lower mantle, especially for the velocity, which can be better resolved than the density. Pyrolite has approximately the right velocity at the top of the lower mantle but becomes increasingly too fast with depth. At the base of the mantle the inferred FeO content approaches 21%.

Ringwood (1975) has suggested that the density deficit can be accounted for with the pyrolite composition if phases denser than the mixed oxides occur in the lower mantle. We have therefore estimated the seismic velocity associated with the required density increase by use of a relation derived from data tabulated in Davies (1974), $\Delta c/c_0 = 1.33 \, \Delta \rho/\rho_0$, where Δc is the velocity increase associated with a $\Delta \rho$ density increase and c_0 and ρ_0 are, respectively, the velocity and density of the mixed oxides assemblage. This leads to velocities in the hypothetical denser phase that are greatly in excess of those observed. We conclude that the properties of the lower mantle are consistent with a mixed oxides assemblage with an average lower

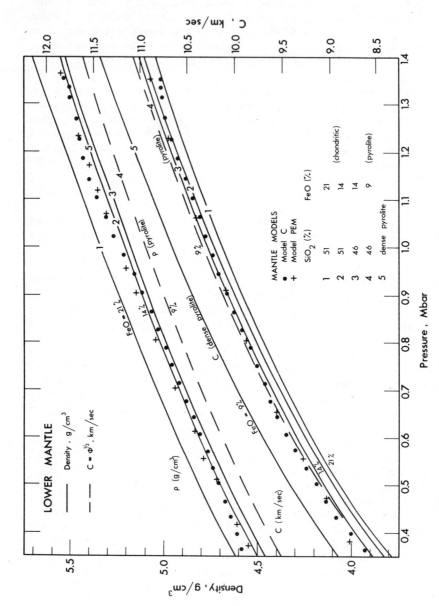

Figure 4 Density and hydrodynamic sound speeds for the lower mantle and for several mineral assemblages.

mantle FeO content of 14% with a suggestion that the FeO content increases from about 12 to 21% through the lower mantle. These conclusions are all consistent with those drawn from a direct comparison of lower mantle densities with shock wave data on dunites and bronzitites.

It is significant that the same iron content satisfies both the density and bulk sound speed for the temperatures and phase assemblage chosen. A denser phase assemblage would increase both the density and velocity, as would a decrease in temperature. To decrease the iron content and maintain consistency it would be necessary to invoke further phase changes and an increase in temperature.

The temperatures used in the lower mantle for the present calculations are those

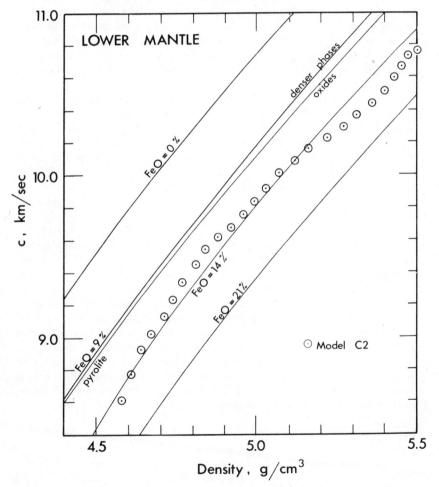

Figure 5   Birch plots for lower mantle and several possible lower mantle compositions.

Table 2 Inferred zero-pressure properties of the lower mantle ($T \approx 1600°K$). From Butler & Anderson (in preparation)

Radius (km)	ρ_0 (g cm^{-3})	K_0 (kbar)	V_{p_0} (km sec^{-1})	V_{s_0} (km sec^{-1})	$(\partial K/\partial P)_0$
4825–5125	3.97	2.07	9.84	5.80	4.08
3850–4600	3.97	2.18	10.01	5.83	3.77

of Reynolds & Summers (1969), which represented a synthesis of available estimates. These are slightly ($\sim 0.4°K$ km^{-1}) superadiabatic. Using estimates of upper mantle temperatures and an adiabatic gradient in the lower mantle leads to estimates of temperature at the base of the mantle that are about 800°K less than the Reynolds-Sumner geotherm. The effect is to decrease the theoretical estimates of ρ and c at the base of the mantle by about 0.04 g cm^{-3} and 0.043 km sec^{-1}, respectively, without affecting values at the top of the lower mantle. The Birch c-ρ plot (Figure 5) is hardly affected.

Use of shock wave data for inferring the composition of the lower mantle is the most direct method, but it is limited to a comparison of density and the seismic parameter Φ. It would be useful if the seismic velocities, V_p and V_s, could be used directly. For this purpose Sammis et al (1970) extended the finite strain formalism to allow extrapolation of V_p and V_s as well as ρ and ϕ, or K. This method is also limited because of the large extrapolations involved and the lack of reliable data on all the pertinent high-pressure phases, such as ilmenite and perovskite. Even the values for stishovite are uncertain. Nevertheless, values extrapolated from the mantle are given in Table 2, and Figures 6 and 7.

The solid lines are lines of constant mean atomic weight and the dashed lines are isostructural joins. Note the offsets for the spinel–mixed oxides reaction. This is probably a feature to be expected for any phase change that involves an increase in coordination and a consequent increase in the ionic spacing. This is a basic limitation in the use of velocity-density systematics to infer composition. An offset to higher density does not necessarily mean an increase in iron content or mean atomic weight except in isostructural compounds. Systematics involving mean atomic distances or molar volumes (Anderson & Anderson 1970) should be more useful in this regard. Liebermann & Ringwood (1973) were the first to recognize the coordination effect. The high inferred values for V_{p_0} and V_{s_0} for the lower mantle imply high SiO_2 (stishovite) contents or SiO_2-rich compounds with Si in 6-coordination with oxygen. Figure 8 gives extrapolated bulk moduli and density of the various regions of the mantle (Anderson 1976).

THE CORE

The earth's core can be divided into three regions: 1. the outermost core, $3485 > r > 3200$ km; 2. a relatively homogeneous region, $3200 > r > 2200$ km; 3. the transition region, $2200 > r > 1215$ km; and 4. the solid inner core, $4 > 1215$ km.

The subdivisions of the outer core are based on equation of state fits (R. Butler & D. Anderson, in preparation). The outer core has low, or vanishing, rigidity and therefore is either completely or almost completely molten. The transition region has been controversial but now appears to be a region of low velocity gradient

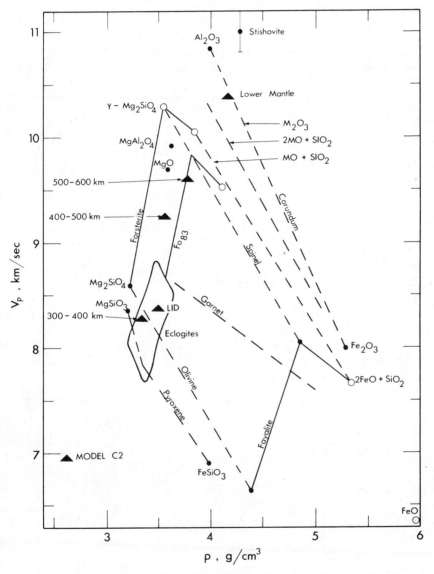

Figure 6 V_p versus ρ for various regions of the lower mantle and various minerals.

without discontinuities or velocity reversals. The outer part of the inner core appears to have a high velocity gradient.

The outer core is generally presumed to be completely molten because of the absence of transmitted shear waves and the reflection coefficients of ScS and PcP waves. However, the outer core could contain up to about 30% solids in suspension and still exhibit the liquid-like behavior required by the geophysical data. Since a light alloying element is required to explain the density of the core it is likely that the core has a melting interval rather than a well-defined melting temperature. The solidus, or eutectic temperature, of the core can be quite a bit lower than the melting point of pure iron or the liquidus temperature of the alloy. If core

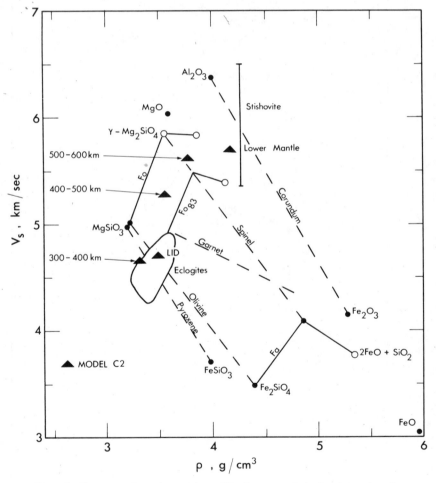

Figure 7 V_s versus ρ for various regions of the lower mantle and various minerals.

temperatures are below the liquidus then solid iron would coexist with an iron-rich melt solution. The solid iron could either be held in suspension by turbulent convection or settle to the center of the core to form the solid inner core. In the latter case the inner core would grow with time and the temperatures in the outer core would lie on the liquidus. Since the liquidus is pressure dependent, the composition of the core would vary with depth.

Figure 9 is a plot of density vs pressure for the core, and estimates of the density of iron and nickel from shock wave data, corrected to temperatures thought to be appropriate for the core. The effect of a light alloying element (silicon) combined with iron is also shown. The outer core is about 9% less dense than iron at core conditions. Interpolating between Fe and $Fe_{80}Si_{20}$ the mean atomic weight of the outer core is about 49.

The most plausible candidate for the light alloying element in the core is sulfur (Murthy & Hall 1972, Anderson et al 1971).

The sulfur content of the core can be estimated using the shock wave results for FeS (King & Ahrens 1973) and FeS_2 (Al'tshuler & Kormer 1961). On the basis of cosmic abundances we assume that the nickel content of the core is 10% of the iron content. FeS or FeS_2 is then added until the core density is satisfied. The results using FeS_2 and one of the extrapolations for FeS give Fe-Ni increasing from 89% at the top of the core to 92% at the base of the outer core. Another extrapolation of the FeS results gives a constant 88% for the Fe-Ni content of the core. The sulfur content is 7.6 to 12.4% with 11% being most consistent with the majority of the data. This gives a mean atomic weight of 51.9 for the core. The FeS content of

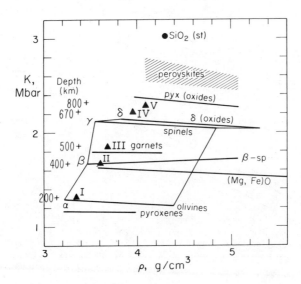

Figure 8 Bulk modulus versus density for various minerals and the various regions of the mantle (Anderson 1976).

the core, consistent with the shock wave data, is 34–21%, which can be compared with the average of 31% for carbonaceous chondrites.

The seismic properties of iron and iron alloys at core pressures are much harder to estimate. Calculations for iron and nickel (Al'tshuler & Kormer 1961), molten iron (Zharkov 1960), and an Fe-Si alloy (Balchan & Cowen 1966) are given in Figure 10. The velocities in the outer core are 4–7% greater than calculated for liquid iron. The two curves shown for the inner core are $V_p = \{[K + (4/3)\mu]/\rho\}^{1/2}$ and $V_\phi = (K/\rho)^{1/2}$. The bulk sound speed, V_ϕ, except for a sharp drop at the inner core–outer core boundary, is nearly continuous. Both the bulk modulus and the density of a solid are slightly greater than for a melt of the same composition and it appears that the inner core may be roughly the same composition as the outer core. There is evidence for inhomogeneity on both sides of the boundary, which may be due to differing amounts of fluid and solid phase. The outermost inner core may be partially molten. The lowermost outer core has a smaller velocity gradient than the rest of the outer core perhaps due to increasing amounts of iron with depth.

The high Poisson's ratio (0.45) of the inner core has been used to argue that it cannot be a crystalline solid. However, Al'tshuler & Sharipdzhanov (1971) have

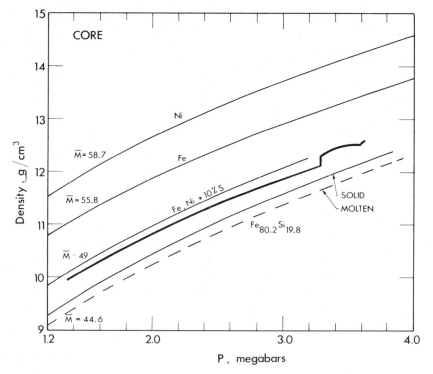

Figure 9 Density of the core.

reported on shock wave measurements of the Poisson's ratio (σ) of copper and iron to pressures of 1 and 2 Mbar. They obtained 0.43 at about 2 Mbar and the data extrapolate smoothly to the value found for the inner core. Unlike other elastic properties, Poisson's ratio increases with both temperature and pressure. Some metals have extremely high Poisson's ratios even at low temperatures and pressures, e.g. gold has a σ of 0.412 at room temperature increasing to 0.427 at 800°K, indium has a σ of 0.444 at room temperature, and a copper-aluminum alloy has been measured with a σ of 0.456 (Simmons & Wang 1971). It does not seem necessary, therefore, to invoke a residual fluid character or an electronic phase transition to explain the properties of the inner core. The rapid increase of V_p at the outer part of the inner core, however, may indicate the elimination of a melt fraction with depth in this region.

The results of this paper reopen the chondritic controversy regarding the composition of the earth. Although the earth is enriched in iron relative to chondrites, the composition of the mantle from the seismic and shock wave data is consistent with the silicate phase of chondrites. Similarly, the Fe/S ratio in the core is in the chondritic range.

The possibility that the outer core is an iron-sulfur melt with iron in suspension presents an interesting dynamic problem. The iron particles will tend to settle out unless held in suspension by turbulent convection. If the composition of the core is such that it is always on the iron-rich side of the eutectic composition the iron

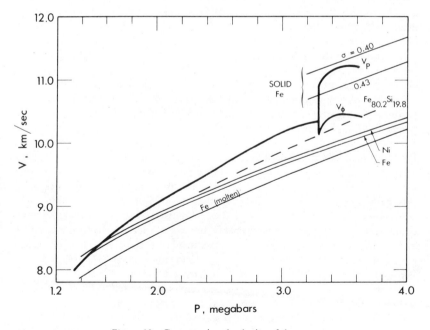

Figure 10 Compressional velocity of the core.

will settle to the inner core–outer core boundary and increase the size of the solid inner core. Otherwise it will melt at a certain depth in the core. The end result is an outer core that is chemically inhomogeneous and on the liquidus throughout.

SUMMARY

Gross earth data such as free oscillation periods, great circle surface wave dispersion, and differential travel times are useful for establishing average values for velocity and density in the various depth regions of the earth. This data must be supplemented by travel time, $dt/d\Delta$, and amplitude studies to determine fine structure of the mantle and core. Lateral variations are clearly important and it is necessary to establish their extent, both laterally and radially, in the next generation of gross earth modeling.

The lithosphere appears to be layered. The high velocity, >8.6 km sec^{-1}, found at depths greater than 40 km demands the presence of substantial amounts of garnet. Eclogite has the required properties and may have formed from basalt intrusion at depth. This layer is overlain by material having the properties of peridotite. It is interpreted as the residual crystals left behind upon basalt removal. The low-velocity zone is probably partially molten hydrous pyrolite. Subsolidus explanations for the low-velocity, low-Q zone in a homogeneous upper mantle can be ruled out if the boundaries are sharp. The region below the LVZ has velocities slightly lower than pyrolite, although uncertain temperature and pressure corrections are involved in the calculation. Taken at face value the velocities imply a small degree of partial melting, a greater FeO content than pyrolite, or a small amount of hydrous phase. The high attenuation in this region suggests that the temperature derivatives of velocity may also be high and we may therefore have underestimated the room temperature values.

The structure of the upper mantle transition region is due to the successive collapse of olivine to the β-phase and then to spinel, and the transformation of pyroxene to the garnet structure. Immediately above 670 km the major phases are β- plus γ-spinel and pyroxene-garnet solid solution. The vicinity of the 670 km discontinuity is likely to be complex. A first-order phase change is required by the P′P′ precursor reflection data and high velocity gradients are required by P and S wave amplitude data. Possible post-spinel phase assemblages for the olivine component of the mantle are $(Mg, Fe)SiO_3$ (ilmenite) plus $(Mg, Fe)O$ (rock salt), $(Mg, Fe)SiO_3$ (perovskite) plus $(Mg, Fe)O$ (rock salt), $(Mg, Fe)O$ (rock salt) plus SiO_2 (stishovite), and $(Mg, Fe)_2SiO_4$ (strontium plumbate). The garnet and pyroxene components of the mantle will probably transform successively to the complex garnet phase, ilmenite, perovskite, and mixed oxide structures. It is not completely clear whether the ultimate phase assemblage in the lower mantle is mixed oxides or perovskite plus $(Mg, Fe)O$. The density difference between these two assemblages is less than 1% at lower mantle pressures. Perovskite therefore cannot be invoked to explain the density difference between the lower mantle and pyrolite.

An FeO content of 14% satisfies the seismic data for the lower mantle. This is close to chondritic abundances. The FeO content may vary slightly with depth, particularly in the lowermost mantle.

The properties of the core are consistent with Fe-Ni in cosmic proportions mixed with 11% S. The FeS/Fe ratio is in the chondritic range. The total iron content of the earth, however, is greater than chondritic. The regions on both sides of the mantle core and outer core–inner core boundaries are inhomogeneous. These could be due to abnormal chemical or thermal gradients.

ACKNOWLEDGMENTS

This research was partially supported by the Advanced Research Projects Agency of the Department of Defense monitored by the Air Force Office of Scientific Research under Contract No. F44620-72-C-0078. The author was a Fulbright-Hays Fellow under the auspices of the Australian-American Educational Foundation at the Research School of Earth Sciences of the Australian National University in Canberra, Australia in 1975. I wish to acknowledge useful discussions with A. Hales, R. Liebermann, A. E. Ringwood, D. Green, L. Liu, R. Butler, H. Liu, H. Kanamori, and R. Hart.

Literature Cited

Ahrens, T. J., Anderson, D. L., Ringwood, A. E. 1969. Equations of state and crustal structures of high-pressure phases of shocked silicates and oxides. *Rev. Geophys.* 7:667–707

Akimoto, S. 1972. The system MgO-FeO-SiO$_2$ at high pressures and temperatures — phase equilibria and elastic properties. *Tectonophysics* 13:161–87

Akimoto, S., Fujisawa, H. 1966. Olivine-spinel transition in system Mg$_2$SiO$_2$-Fe$_2$SiO$_4$ at 500°C. *Earth Planet. Sci. Lett.* 1:237–40

Al'tshuler, L., Kormer, S. 1961. On the internal structure of the earth. *Izv. Acad. Sci. USSR Phys. Solid Earth* 1:33–36

Al'tshuler, L., Sharipdzhanov, L. 1971. Distribution of iron in the earth and its chemical differentiations. *Izv. Acad. Sci. USSR Phys. Solid Earth* 4:3–16

Al'tshuler, L., Sharipdzhanov, I. 1972. Additive equation of state of silicates at high pressure. *Izv. Acad. Sci. USSR Phys. Solid Earth* 3:167–77

Anderson, D. 1965. *Phys. Chem. Earth* 6:1–31

Anderson, D. 1967a. Phase changes in the upper mantle. *Science* 157:1165–75

Anderson, D. 1967b. A seismic equation of state. *Geophys. J. R. Astron. Soc.* 13:9–30

Anderson, D. 1976. The 650 km mantle discontinuity. *Geophys. Res. Lett.* 3:347–49

Anderson, D., Anderson, O. 1970. The bulk modulus-volume relationship for oxides. *J. Geophys. Res.* 75:3494–3500

Anderson, D., Hart, R. 1976. An earth model based on free oscillations and body waves. *J. Geophys. Res.* 81:1461–75

Anderson, D., Kanamori, H. 1968. Shock-wave equations of state for rocks and minerals. *J. Geophys. Res.* 73:6477–6502

Anderson, D., Sammis, C. 1970. Partial melting in the upper mantle. *Phys. Earth Planet. Inter.* 3:41–50

Anderson, D., Sammis, C., Jordan, T. 1971. Composition and evolution of the mantle and core. *Science* 171:1103–12

Balchan, A., Cowen, I. 1966. Shock compression of two iron-silicon alloys to 2.7 megabars. *J. Geophys. Res.* 71:3577–88

Birch, F. 1952. Elasticity and constitution of the Earth's interior. *J. Geophys. Res.* 57:227–86

Brune, J., Dorman, J. 1963. Seismic waves and earth structure in the Canadian shield. *Bull. Seismol. Soc. Am.* 53:167–210

Bullen, K. 1967. Basic evidence for Earth division. In *The Earth's Mantle*, ed. T. Gaskell, pp. 1–40. London/New York: Academic. 509 pp.

Burdick, L., Anderson, D. 1975. Interpretation of velocity profiles of the mantle. *J. Geophys. Res.* 80:1070–74

Clark, S. P. Jr. 1966. Handbook of physical constants. *Geol. Soc. Am. Mem.* 97. 587 pp.

Davies, G. 1974. Elasticity, crystal structure and phase transitions. *Earth Planet. Sci. Lett.* 622:339–46

Davies, G., Gaffney, E. 1973. Identification of phases of rocks and minerals from Hugoniot data. *Geophys. J.* 33:165–83

Dziewonski, A., Hales, A., Lapwood, E. 1975. Parametrically simple earth models consistent with geophysical data. *Phys. Earth Planet. Inter.* 10:12–48

Engdahl, E., Flinn, E. 1969. Seismic waves

reflected from discontinuities within the upper mantle. *Science* 163:177–79

Forsyth, D. 1973. Anisotropy and the structural evolution of the oceanic upper mantle. PhD thesis. Mass. Inst. Technol., Cambridge, Mass. 255 pp.

Gilbert, F., Dziewonski, A. 1975. An application of normal mode theory to retrieval of structural parameters and source mechanisms from seismic spectra. *Philos. Trans. R. Soc. London A* 278:187–269

Green, D. 1973. Experimental melting studies on model upper mantle compositions at high pressure under both water-saturated and water-undersaturated conditions. *Earth Planet. Sci. Lett.* 19:37–53

Hart, R., Anderson, D., Kanamori, H. 1976. Shear velocity and density of an attenuating Earth. *Earth Planet. Sci. Lett.* 32: 25–30

Helmberger, D., Engen, G. 1974. Upper mantle shear structure. *J. Geophys. Res.* 79:4017–28

Helmberger, D., Wiggins, R. 1971. Upper mantle structure of the midwestern United States. *J. Geophys. Res.* 76:3229–45

Jones, S., Liebermann, R. 1974. Elastic and thermal properties of fluoride and oxide analogues in the rock salt, fluorite, rutile and perovskite structures. *Phys. Earth Planet. Inter.* 9:101–7

Kalinin, V., Pan'kov, V. 1974. Equations of state of rocks. *Izv. Earth Phys.* 7:10–20

Kanamori, H., Press, F. 1970. How thick is the lithosphere? *Nature* 226:330–31

Kind, R. 1974. Long range propagation of seismic energy in the lower lithosphere. *J. Geophys.* 40:189–202

King, D., Ahrens, T. 1973. Shock compression of iron sulphide and possible sulfur content of the earth's core. *Trans. Am. Geophys. Union* 54:476

Kosminskaya, I., Puzyrev, N., Alekseyer, A. 1972. Exploration seismology: Its past, present, and future. *Tectonophysics* 13: 309–23

Liebermann, R. 1974. Elasticity of pyroxene-garnet and pyroxene-ilmenite phase transformations in germanates. *Phys. Earth Planet. Inter.* 8:361–74

Liebermann, R. 1975. Elasticity of olivine (γ), Beta (β), and Spinel (α) polymorphs of germanates and silicates. *Geophys. J.* 42:899–929

Liebermann, R., Ringwood, A. 1973. Birch's law and polymorphic phase transformations. *J. Geophys. Res.* 78:6926–32

Liu, H., Anderson, D., Kanamori, H. 1976. Velocity dispersion due to anelasticity; Implications for seismology and mantle composition. *Geophys. J.* 47:41–58

Liu, L. 1974. Silicate perovskite from phase transformations of pyrope-garnet at high pressure and temperature. *Geophys. Res. Lett.* 1:277–80

Liu, L. 1975. Post-oxide phases of forsterite and enstatite. *Geophys. Res. Lett.* 2:417–19

Manghnani, M., Ramananantoandro, R., Clark, S. 1974. Compressional and shear wave velocities in granulate facies rocks and eclogites to 10 kbar. *J. Geophys. Res.* 79:5424–46

Morris, G., Raitt, R., Shor, G. 1969. Velocity anisotropy and delay-time maps of the mantle near Hawaii. *J. Geophys. Res.* 74: 4300–16

Murthy, V., Hall, H. 1972. The origin and composition of the Earth's core. *Phys. Earth Planet. Inter.* 6:125–30

O'Hara, M. 1970. Upper mantle composition inferred from laboratory experiments and observations of volcanic products. *Phys. Earth Planet. Inter.* 3:236–45

Parker, R., Oldenburg, D. 1973. Thermal model of ocean ridges. *Nature* 242:137–39

Reynolds, R., Summers, A. 1969. Calculations on the composition of the terrestrial planets. *J. Geophys. Res.* 74:2494–2500

Ringwood, A. 1975. *Composition and Petrology of the Earth's Mantle.* New York: McGraw-Hill. 618 pp.

Ringwood, A., Major, A. 1966. Synthesis of Mg_2SiO_4-Fe_2SiO_4 solid solutions. *Earth Planet. Sci. Lett.* 1:241–45

Ringwood, A., Major, A. 1967. Some high pressure transformations of geophysical interest. *Earth Planet. Sci. Lett.* 2:106–10

Sammis, C., Anderson, D., Jordan, T. 1970. Application of finite strain theory to ultrasonic and seismological data. *J. Geophys. Res.* 75:4478–80

Simmons, G., Wang, H. 1971. *Single Crystal Elastic Constants and Calculated Aggregate Properties: A Handbook.* Cambridge, Mass: MIT Press. 370 pp. 2nd ed.

Wang, H., Simmons, G. 1974. Elasticity of some mantle crystal structures, 3, spessaritic-almandine garnet. *J. Geophys. Res.* 79:2607–14

Whitcomb, J., Anderson, D. 1970. Reflection of P'P' seismic waves from discontinuities in the mantle. *J. Geophys. Res.* 75:5713–28

Wyllie, P. 1971. *The Dynamic Earth: Textbook in Geosciences.* New York: Wiley. 416 pp.

Zener, C. 1948. *Elasticity and Anelasticity of Metals.* Chicago: Chicago Univ. Press. 170 pp.

Zharkov, V. 1960. The physics of the Earth's core. Thermodynamic properties, I. *Izv. Acad. Sci. USSR Phys. Solid Earth* 3: 1417–25

Ann. Rev. Earth Planet. Sci. 1977. 5: 203–26

TRANSMISSION ELECTRON MICROSCOPY IN EARTH SCIENCE

×10072

P. E. Champness

Department of Geology, Manchester University, Manchester M13 9PL, Great Britain

INTRODUCTION

This review is concerned mainly with conventional transmission electron microscopy (TEM) and its application to mineralogy and geology. However, I indulge in some speculation at the end about new developments and how they might be used to advantage by the mineralogist.

Although TEM became a routine tool of the physical metallurgist in the 1960s and the theory of image formation from crystalline material was well established by then, it was not until the present decade that it was adopted to any great extent by mineralogists. The main reason for the long delay was that there was no reliable method for preparing thin foils of nonmetallic materials; studies were restricted to cleavage fragments of layer minerals or to powdered fragments, and the latter technique only allowed examination of microstructural features smaller than about one micron.

The advent of reliable ion-thinning devices in the early 1970s (Barber 1970, Gillespie et al 1971, Heuer et al 1971) solved the problem of specimen preparation. Foils in which hundreds of square microns are transparent to the electron beam can now be prepared routinely. Discs are drilled from optical thin sections and then thinned by sputtering both surfaces with a beam of energetic argon ions.

The electron microscope has not only extended the optical microscope into magnifications of several tens or hundreds of thousand times, but it also provides diffraction information. Thus the high-resolution image can be correlated in situ with the crystallography of the specimen. The high-voltage (1000 kV) electron microscope is of particular value for mineralogical studies because the penetration is increased approximately tenfold compared with 100 kV, and ionization damage, a serious problem with framework silicates at 100 kV, is virtually eliminated (Lorimer & Champness 1973a, 1974). Analytical electron microscopy is also proving to be extremely useful. This technique, which is discussed in a later section, has improved the spatial resolution for chemical analysis by an order of magnitude over that obtainable in the microprobe analyser.

RESOLUTION

The basic reason for the use of the electron microscope is the high resolution that results from the small wavelength of electrons (0.037 Å for an accelerating voltage of 100 kV). Like their optical counterparts magnetic lenses are subject to chromatic and spherical aberrations. Chromatic aberration, which arises when electrons of different wavelengths are brought to focus at different image planes, limits the usable thickness of foils to about 0.2 μm at 100 kV. But it is spherical aberration that limits the ultimate resolution of the electron microscope. It is inherent in magnetic lenses that the outer regions focus more strongly than the inner ones, with the result that the semi-angle of the lens, α, must be limited by an objective aperture if the aberration is to be reduced. However, diffraction effects limit the resolution to a value of $(0.61\lambda)/\alpha$ (where λ is the wavelength of the electrons). Thus there is an optimum value for α that balances the effects of diffraction and spherical aberration. The resolution is then given by:

$$R = B\lambda^{3/4}C_s^{1/4}, \tag{1}$$

where C_s is the coefficient of spherical aberration and $B \sim 0.5$. In a modern electron microscope operating at 100 kV with a C_s value of 1 mm for the objective lens, a resolution of about 2 Å can be attained. For routine analysis the resolution is reduced to about 5 Å, mainly because of the chromatic aberration introduced by the specimen. As the accelerating voltage is increased the wavelength decreases, and thus the resolution should increase. However, the high-voltage microscope has not yet achieved a better resolution than the 100-kV microscope due to problems of mechanical and electrical stability. (See Cosslett 1968 for a more detailed discussion.)

TRANSMISSION ELECTRON MICROSCOPY

In a review of this nature I can only attempt the barest outline of the theory of diffraction and image formation. For a more thorough treatment the reader is referred to articles especially written for mineralogists by Van de Biest & Thomas (1976), Gard (1976), and Amelinckx & Van Landuyt (1976). Loretto & Smallman (1975) have recently published a concise and practical account of the quantitative interpretation of electron micrographs.

Contrast

In most applications of TEM to crystalline materials the image is formed either by removing all the diffracted beams (bright-field image) or by allowing only one strongly diffracted beam to pass through the aperture in the back focal (diffraction) plane (dark-field image) (see Figure 1). *Diffraction contrast* arises because electrons are very strongly diffracted by crystals. Thus if a uniformly thick crystal is bent as in Figure 2, the area in which the lattice planes are inclined to the electron beam at the Bragg angle θ_B will appear darker than its surroundings in the bright-field

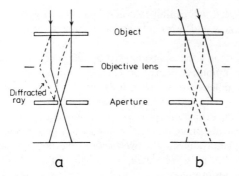

Figure 1   Schematic ray diagram illustrating the formation of (*a*) a bright-field image and (*b*) a dark-field image. In the latter the incident beam has been tilted so that the diffracted beam travels along the optic axis and spherical aberration is minimized.

image and will be brighter than its surroundings in the dark-field image formed with that reflection. The image is said to exhibit a *bend contour*.

Dynamical interaction between the transmitted and diffracted beams leads to an oscillation of the transmitted and diffracted intensities with depth as shown in Figure 3. For the exact Bragg condition the depth periodicity is known as the *extinction distance*, ξ_g (**g** is the reciprocal lattice vector), and is inversely proportional to the structure factor for the reflection; as the orientation deviates further from the exact Bragg angle the periodicity of the oscillations decreases. Consequently

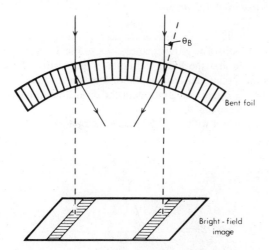

Figure 2   Diagrammatic representation of the contrast from a bent foil. Regions where the lattice planes are inclined at the Bragg angle θ_B to the electron beam will appear dark in the bright-field image because the electrons will be diffracted outside the objective aperture.

a wedge-shaped crystal will exhibit *thickness contours*. In general, of course, foils vary both in thickness and in orientation and show *extinction contours* (see for instance Figures 7 and 8*a*).

Any inhomogeneity in a foil can be imaged by diffraction contrast if it alters the local diffracting conditions. A precipitate can show a number of different types of contrast. If it has a different structure factor from the matrix, either because it has a different structure or because it has a different composition (and thus different scattering factors), the extinction distance will be different in the two phases. Thus, when the foil is near the Bragg orientation for matrix or precipitate, the intensity in the image will generally differ in the two regions (Figure 3). Examples of *structure factor contrast* are shown by the lamellae of Figure 6*a* and *b*, the lamellae *B* of Figure 7, and the voids in Figure 8*a*.

If the precipitate has a different orientation from the matrix it will be possible to orient the foil so that one phase is diffracting strongly while the other is not. An example of *orientation contrast* is shown by the augite lamellae, *A*, in Figure 7.

Coherent precipitates (those for which the lattice is continuous across the interface with the matrix) show *matrix contrast* if their volume is different from the matrix. The contrast arises because the strain causes the lattice planes in the matrix to bend, thus changing the local diffracting conditions; consequently small precipitates will appear larger than they really are. An example is shown by the platelets near the extinction contour of Figure 7.

Another type of contrast is shown by inclined precipitates, stacking faults, grain boundaries, and twins. These features interrupt the regular oscillation of the diffracted and transmitted intensities of Figure 3 and show *displacement fringes* at their interfaces (see inset, Figure 7).

A dislocation produces contrast in the TEM in much the same way as the coherent precipitate (in fact it is difficult to distinguish between a small precipitate and a dislocation loop). The lattice planes close to the dislocation core are highly distorted (Figure 4), and the Bragg condition is only satisfied locally. Thus the image records the strain field around the dislocation and the image is generally about two orders of magnitude larger than the dislocation. Images of dislocations can be seen in Figures 8 and 9.

For any general defect with a displacement vector $\mathbf{R}$ there will be certain lattice

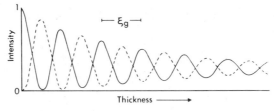

Figure 3   Theoretical profiles of thickness fringes in bright field (full line) and dark field (dashed line). The damping of the fringes is due to absorption. ξ_g is the extinction distance. (After Hashimoto et al 1960.)

planes that are *not* distorted, i.e. those containing the vector **R**. (In vector notation this situation can be described by the equation $\mathbf{g} \cdot \mathbf{R} = 0$.) This is the basis of defect analysis; images are formed in bright or dark fields with only one reflection at the exact Bragg condition in each case (*two-beam conditions*). If the defect is invisible in two of the images and the **g** vectors of the operating reflections are noncollinear, then **R** can be found by cross multiplication of the two **g** vectors (i.e. it must be perpendicular to both of them). This analysis applies exactly to the case of the screw dislocation where **R** is the Burgers vector **b** (which is parallel to the dislocation line **u**). But for an edge dislocation the displacement has two components, one parallel to **b** and the other, a smaller one, parallel to the vector product of **b** and **u** (**u** is perpendicular to **b**; see Figure 4). Consequently the dislocation may not be completely out of contrast when $\mathbf{g} \cdot \mathbf{b} = 0$; there will be some residual contrast unless $\mathbf{g} \cdot \mathbf{b} \times \mathbf{u} = 0$ also. However, the $\mathbf{g} \cdot \mathbf{b}$ criterion can still be used to find the direction of **b** by finding those **g** vectors for which contrast is much weaker than in other reflections. In materials with complex structures, such as minerals, the image contrast will also be affected by the elastic anisotropy, and the isotropic invisibility criteria described above may not apply. However, contrast experiments and calculations of image contrast suggest that for quartz (Ardell et al 1974), pyroxenes (Kirby 1976), and olivine (Boland et al 1971, Boland 1976), the isotropic criteria may be employed, though it may be wise to compute the images expected from different Burgers vectors and compare them with the observed image (McCormick 1976, Boland 1976).

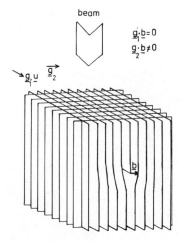

Figure 4 Diagram of an edge dislocation with Burgers vector **b**; **u** is parallel to the dislocation line. The lattice planes whose normal is $\mathbf{g}_1$ are not distorted by the dislocation ($\mathbf{g}_1 \cdot \mathbf{b} = 0$ and $\mathbf{g}_1 \cdot \mathbf{b} \times \mathbf{u} = 0$), but the lattice planes whose normal is $\mathbf{g}_2$ are distorted ($\mathbf{g}_2 \cdot \mathbf{b} \neq 0$). The dislocation will thus be visible in images formed when reflection $\mathbf{g}_2$ is at the exact Bragg condition, but invisible when reflection $\mathbf{g}_1$ is at the exact Bragg condition.

Electron Diffraction

From what has already been said it is clear that knowledge of the exact orientation of the crystal with respect to the electron beam is crucial to the interpretation of image contrast. This information can be gained from the diffraction pattern and fortunately, because of the strong diffraction suffered by electrons, one can view the pattern directly on the screen of the instrument while the specimen is tilted into the desired orientation (modern microscopes allow tilting up to $\pm 60°$ about any axis). Also, because the radius of the Ewald sphere is very large (due to the short wavelength of electrons) and because the diffracted intensity is not confined to the reciprocal lattice point, but is a spike perpendicular to the foil surface (due to the limited thickness of the crystal), the pattern is, to a good approximation, an undistorted layer of the reciprocal lattice and is readily interpreted.

The minimum area from which the electron diffraction pattern can be recorded is limited by the spherical aberration of the objective lens. The error in the specimen plane is equal to $C_s \alpha^3$; for a microscope operating at 100 kV with a C_s value of 3 mm and $\alpha = 0.02$ radians (for a low-order reflection), it is about 0.025 μm, and the area from which the diffraction pattern can be reliably recorded is about 1 μm in diameter. Because C_s varies approximately as λ^{-1} and $\alpha \simeq \lambda \mathbf{g}$, the size of the selected area at 1 MeV should be improved by about 20 times (λ at 1 MeV is 0.0087 Å). However, this cannot usually be achieved because of the practical difficulty of making an aperture small enough to isolate the desired area in the first image plane.

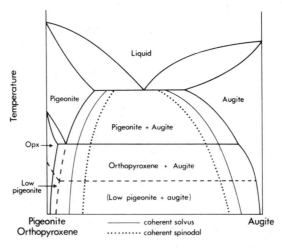

Figure 5 Schematic phase diagram for the clinopyroxenes. In quickly cooled pigeonites the transformation to orthopyroxene does not occur, but the structure transforms to low pigeonite, space group $P2_1/c$. As this structure is thought to be metastable (Papike et al 1973), the transformation is indicated by a dashed line in the orthopyroxene+augite stability field. The coherent solvus is that within which homogeneous nucleation can take place. Spinodal decomposition can take place within the coherent spinodal.

As electron diffraction patterns can be recorded from such small areas they have obvious advantages over X-ray diffraction patterns in the determination of the orientation and nature of small precipitates.

There are a number of reasons why one would not normally attempt to derive crystallographic information from electron diffraction patterns. Firstly, the manipulation of the crystal into different orientations is much more difficult than for X-ray diffraction and one is limited to a tilt of $\pm 60°$ with a single foil. Secondly, the magnification of the diffraction pattern depends upon the height of the specimen. Although this drawback can be overcome to some extent by evaporating a standard such as aluminium onto the specimen, electron diffraction cannot rival powder X-ray diffraction for the accurate measurement of cell parameters. The third reason is that the intensities of the diffracted beams cannot normally be used to calculate structure factors because of the strong dynamical interactions between the diffracted beams. Fourthly, double diffraction leads to the appearance of reflections forbidden by space-group symmetry elements such as screw axes (but not by the lattice type), with the consequence that the space group cannot be determined from the diffraction pattern.[1] However, some minerals and many synthetic phases occur as fine powders that are too small for analysis by single crystal X-ray diffraction, and electron diffraction must be used. Gard (1976) has reviewed the use of electron diffraction in conjunction with X-ray powder diffraction for such problems.

EXSOLUTION

One of the most fruitful applications of TEM in mineralogy has been in the study of phase transformations. Optical microscopy, such as the pioneering work of Hess (1941), established the nature and morphology of exsolution lamellae in plutonic pyroxenes and other minerals, and later X-ray diffraction studies by Bown & Gay (1960) and others characterised exsolution products that were below the resolution of the optical microscope. However, X-ray diffraction cannot detect small-volume fractions of precipitates and, as it provides no morphological information, X-rays cannot be used to determine the *mechanism* of exsolution reactions. This is where TEM has come into its own.

I have chosen the pyroxene group of silicates as an illustration of the "state of the art" as it illustrates so well the variations in phase distributions that result from different cooling rates. A schematic phase diagram for the clinopyroxenes is shown in Figure 5.

The first pyroxenes to be examined by TEM were from lunar basalts. Typically they showed coherent modulations on a scale of about 0.01 μm with diffuse interfaces. Often the finest structures showed a "tweed" texture with modulations approximately parallel to (001) and (100) (Figure 6a), while coarser ones were parallel to (001)

---

[1] However, J. W. Steeds and co-authors have developed techniques whereby forbidden reflections can be identified from the careful observation of extinction contours. The presence or absence of a center of symmetry can also be inferred—something that cannot be done with normal X-ray techniques. For a review see Steeds et al (1973).

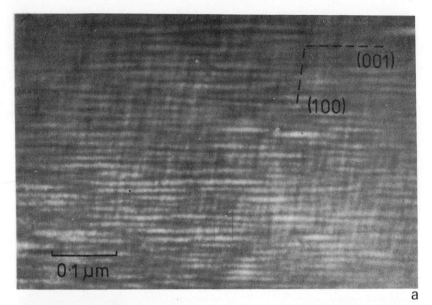

a

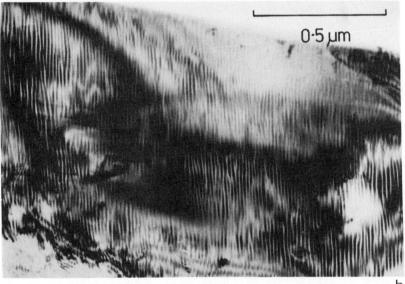

b

Figure 6 Microstructures of clinopyroxenes. (*a*) Tweed structure approximately parallel to (001) and (100) of augite from Apollo 12 basalt 12052. (*b*) Modulated structure parallel to (001) of a synthetic sample of composition $En_{45.9}Di_{54.1}$ [McCallister & Yund (1975), reproduced by permission]. (*c*) (001), *A*, and (100), *B*, lamellae of pigeonite in an augite from the Whin Sill, Northern England. Between the large lamellae is a tweed structure of which the (001) component is the more prominent. Note the absence of the tweed adjacent to the lamellae and the antiphase boundaries in the pigeonite (formed during the C → P

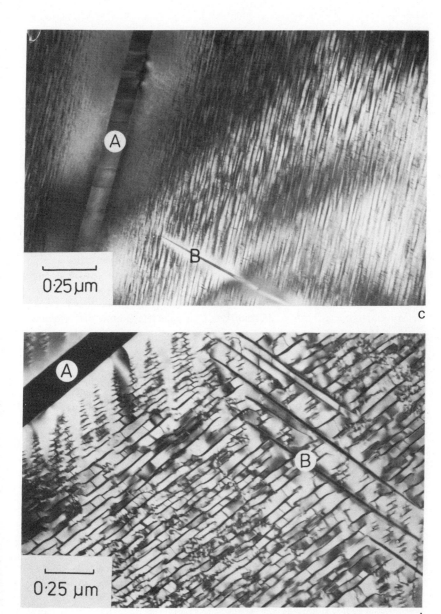

c

d

transition. P. P. K. Smith, unpublished). (*d*) (001), *A*, and (100), *B*, lamellae of augite in pigeonite from the same rock as (*c*). Small augite platelets close to the (001) lamellae have nucleated on the *APBs*, but those further from the lamellae probably nucleated homogeneously. Dark field $\mathbf{g} = \bar{1}02$. The APBs are in contrast because $\mathbf{R} = \frac{1}{2}\langle \mathbf{a} + \mathbf{b}\rangle$, so that $\mathbf{g} \cdot \mathbf{R} \neq 0$. The augite appears dark as it is not diffracting into the objective aperture. (P. P. K. Smith, unpublished).

only as in Figure 6b. These textures have generally been interpreted as resulting from spinodal decomposition (Champness & Lorimer 1971, 1972, Christie et al 1971, Lally et al 1975), that is the evolution of sinusoidal composition waves with progressively increasing wavelength and amplitude in which the overall symmetry reflects that of the parent structure (for a review of spinodal decomposition see Cahn 1968). Fast cooling is expected to favor this mechanism, assuming that it is kinetically feasible, because any nucleation and growth process has a nucleation barrier and requires long-range diffusion (Champness & Lorimer 1976). Modulated structures attributed to spinodal decomposition have recently been reproduced in synthetic samples by McCallister & Yund (1975). An example is shown in Figure 6b.

Plutonic clinopyroxenes show exsolution microstructures that are very different from those described above. Often, (001) and (100) lamellae can be clearly seen in the optical microscope, but the electron microscope reveals other, smaller, lamellae; all of them are distributed heterogeneously (see for instance Champness & Copley 1976), and we conclude that they nucleated heterogeneously (i.e. on defects such as grain boundaries). Slow cooling favors heterogeneous nucleation and growth because, although only a small number of nuclei form, the diffusivities of the Mg, Fe, and Ca ions are sufficiently high and the time available is sufficiently long that the transformation can keep up with equilibrium. By the time the temperature of the coherent solvus (that at which homogeneous nucleation can occur randomly in the matrix) or the spinodal is reached (Figure 5), all the solute has been drained from the matrix into the precipitates.

Intermediate cooling rates, such as occur in hypabyssal rocks, also produce heterogeneously nucleated lamellae, but equilibrium is not maintained. Solute is "stranded" between the lamellae, the concentration being higher midway between the lamellae than adjacent to them, and a number of secondary exsolution structures can form. "Tweed" textures (Figure 6c) have been found between the (001) lamellae in some clinopyroxenes from lunar basalts (Lally et al 1975, Nord et al 1976) and in augite from the center of the Whin Sill (P. P. K. Smith and P. A. Copley, in preparation) and probably formed when the temperature reached that of the coherent spinodal (Figure 5). Small (001) augite platelets have been found in a lunar pigeonite (Nord et al 1973) and in a pigeonite from the center of the Whin Sill (Smith and Copley, in preparation). In Figure 6d evidence of the calcium profile can be seen in the distribution of the augite platelets. The platelets close to the lamellae nucleated on the antiphase boundaries that formed during the inversion from $C2/c$ to $P2_1/c$ symmetry, but those further away, where the supersaturation was higher, appear to have nucleated homogeneously.

The exsolution of clinopyroxene from orthopyroxene (or vice versa) cannot take place by the spinodal mechanism because the two phases have different structures. Exsolution is a more difficult process than in the clinopyroxenes; even orthopyroxenes from the Bushveld and other plutonic complexes contain metastable (transition) phases, despite the very slow cooling they have experienced (Champness & Lorimer 1973, 1974). Figure 7 shows an orthopyroxene from the Stillwater complex. The augite lamellae, A, nucleated on the grain boundary. They were followed at a lower temperature by the intermediate phase B, which nucleated on

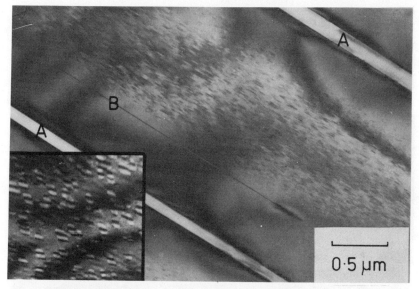

Figure 7 Microstructure of an orthopyroxene from the Stillwater Complex showing large (100) augite lamellae *A*, intermediate phase *B*, and a fine distribution of platelets that nucleated homogeneously. The inset shows the platelets inclined to the electron beam. They display displacement fringe contrast. From Champness & Lorimer (1973), reproduced by permission of Chapman & Hall Ltd.

defects such as subgrain boundaries. Finally, a second intermediate phase nucleated homogeneously. This consists of (100) Ca-rich platelets only 18 Å wide. Evidence of the Ca profile left by the growth of the augite lamellae can be clearly seen in the absence of the platelets adjacent to the augite lamellae. The existence of the profile was verified directly by Lorimer & Champness (1973b) by analysis in a combined electron microscope and microanalyser.

DEFORMATION STRUCTURES

The study of minerals deformed in the earth or in the laboratory has attracted much interest recently; the ultimate aim is to make deductions about the tectonic history of rocks from their microstructures. It is now known that under most conditions the deformation of minerals, like that of metals and other crystalline materials, is controlled by the generation and movement of dislocations (line defects), and thus the TEM is an ideal tool for such studies. (For a recent review see Christie & Ardell 1976).

There are, however, a number of important differences between the deformation of metals and minerals. Firstly, experiments at slow rates of deformation must be conducted under a confining pressure of several kilobars in order to suppress fracture and to increase ductility (this also simulates the overburden in the earth).

Secondly, the effect of water dissolved in the structure is dramatic. Quartz, which shows the most pronounced effects, exhibits a strength close to the theoretical one when dry crystals are deformed at 500–700°C, but wet quartz is very weak. Griggs & Blacic (1965) have suggested that the role of water is to hydrolyse the strong Si-O-Si bonds at the dislocation lines to silanol groups (Si-OH.HO-Si), which can then be more easily broken and thus permit migration of the dislocations. Water

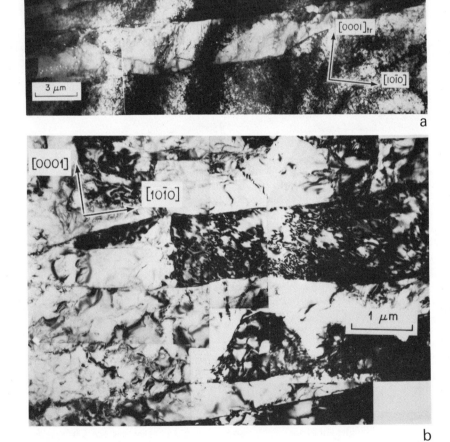

a

b

Figure 8 (a) Sub-basal lamella in a naturally deformed meta-quartzite. In the center is a recovered region which has the appearance of a subgrain. Note the bubbles that contain water precipitated during recovery. (b) Elongated subgrains in a quartzite experimentally deformed at 800°C and 10^{-7} sec^{-1}. From Ardell et al (1976). Reproduced by permission of the Institute of Physics, London.

also enhances the climb of edge dislocations (the movement of the dislocation from one slip plane to another by the absorption of point defects).

At low temperatures and/or high strain rates (shock deformation) the framework and chain silicates also behave differently from metals. Quartz deformed by shock either by meteoritic impact or in the laboratory contains submicroscopic glassy layers and lenses between regions of crystalline quartz: there is no evidence that dislocations play any part in the deformation (Müller 1969, Christie & Ardell 1976). Under similar conditions pyroxenes suffer extensive mechanical twinning and plastic deformation (which are both common in shocked metals) in addition to vitrification.

The conditions of deformation that are of most interest to geologists are the slow strain rates responsible for mountain building and the flow of the asthenosphere; these are thought to be 10^{-10} to 10^{-15} sec^{-1} at differential stresses less than a kilobar. Unfortunately these flow rates are unattainable in the laboratory and the aim of experiments must be to determine the flow laws and deformation mechanisms of rocks over a sufficiently wide range of conditions that extrapolation can be made to geological conditions. However, the microstructures of quartzites, and to a lesser extent peridotites, deformed at the highest temperatures and the slowest strain rates, resemble those of natural tectonites (Figures 8 and 9). These structures contain elongated subgrains formed as a result of dislocations climbing out of their slip plane into low-energy arrays; recrystallization may also occur in quartz at the original grain boundaries (Ardell et al 1973, Kohlstedt & Goetze 1974). The dislocation densities in natural quartzites and peridotites are generally lower than in their experimental counterparts and they show more extensive recovery, even though they were deformed at lower temperatures. This is consistent with the lower strain rates in the earth.

The correlation of microstructures in slowly deformed and experimental rocks had led to the conclusion that they are the result of a process of solid-state creep that is controlled by dislocation climb, i.e. deformation under constant stress in which dislocation production is balanced by recovery. Weertman (1970) and Stocker & Ashby (1973) also concluded from theoretical considerations that dislocation creep is the dominant deformation mechanism in the upper mantle.

Much progress has been made in identifying the Burgers vectors, **b**, of the dislocations responsible for plastic deformation in minerals, although the experiments are more difficult than in metals due to the difficulty of obtaining two-beam conditions. As explained in the introduction, it may also sometimes be impossible to achieve true invisibility of the dislocation for the two values of **g** that are needed to define **b** uniquely, and some subjectivity is involved in deciding whether a dislocation is invisible or not.

HIGH RESOLUTION

In the applications of TEM that have been described so far only one beam (usually the direct one) was allowed to pass through the objective aperture, and hence the image contained no information about the lattice or the atomic arrangement within the unit cell. But the resolution of the electron microscope is now such

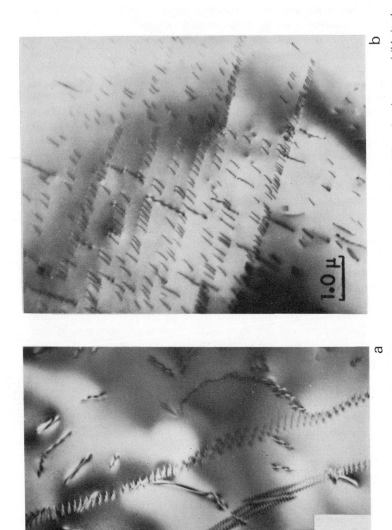

Figure 9 (*a*) Sub-boundaries approximately parallel to (100) in a naturally deformed olivine from a dunite. The subgrains were visible in the optical microscope as fine striations. From Boland et al (1971), reproduced by permission of Springer-Verlag. (*b*) Olivine deformed at 1535°C and 1.1×10^{-3} sec^{-1}. The dislocations are parallel to [010] and the Burgers vector is [100]. Some of the dislocations have lined up forming a tilt boundary approximately parallel to [100]. From Kohlstedt & Goetze (1974), copyrighted by American Geophysical Union.

that, if enough diffracted beams are allowed to interfere in the image plane, a two-dimensional projection of the electron density can be recorded with a resolution of about 3 Å. In order to minimize the dynamical interactions between the diffracted beams, the crystal must be very thin, typically less than about 10 nm. If the image is under-focused by about 1000 Å it corresponds approximately to the projected charge distribution, i.e. regions of the structure with a high concentration of super-posed heavy atoms appear as dark areas (Allpress & Sanders 1973). This technique was first applied to the ternary Nb oxides, for which it was shown that what was once referred to as a "solid-solution" actually consists of a series of closely related parent phases that can intergrow to produce more or less well-ordered phases with the appropriate stoichiometry (see Allpress & Sanders 1973 for a review). The first structure that was solved using electron microscopic images was published by Iijima & Allpress (1974), who showed that the structure previously proposed for $2Nb_2O_5 \cdot 7WO_3$ from limited X-ray evidence was incorrect in certain details. Tourmaline was the first mineral whose structure was imaged in the TEM (Iijima et al 1973), and a number of other minerals were studied by Buseck & Iijima (1974). The images all showed close correspondence with the structures determined using X rays. An example, an image of idocrase, is shown in Figure 10.

Because the resolution of structure images is limited to about 3 Å, i.e. only gross features within unit cells are visible, and because the image must always be a projec-

Figure 10   Structure image of idocrase showing a section perpendicular to [001]. A unit cell is outlined and the insert shows several unit cells at the same scale as the image. The channels and areas of high electron density can be distinguished. From Buseck & Iijima (1974).

tion of the structure, the technique has not been used to solve a completely unknown structure, nor is it likely to provide such information in the future. Direct imaging of perfect mineral structures is likely only to be applied to problems with which X rays are unable to deal, for instance to very small crystals or to precipitates (particularly transition phases) that occur in low-volume fractions. The basic features of the structures will almost certainly have to be known for such studies to be successful.

On the other hand, the study of defects in minerals has wide potential and a number of important problems have already been solved. It also has the advantage that it is often only necessary to form a one-dimensional (lattice) image, and in this case the thickness criterion is less rigid: crystals several hundreds of Å thick can be used (Allpress & Sanders 1973). Yada (1967, 1971) was a pioneer in this field. He examined transverse sections of chrysotile asbestos and confirmed predictions that both concentric and spiral lattices occur (as well as rare, more complex structures) and that most of the cylinders were hollow.

More recently Pierce & Buseck (1974) and Morimoto et al (1974) examined pyrrhotites, $Fe_{1-x}S$, with different stoichiometries. The structure of the compound Fe_7S_8, in which the iron vacancies are ordered, had already been determined by X-ray diffraction. The correlation of the structure images obtained from it with the known structure gave the authors confidence in interpreting the images of the nC

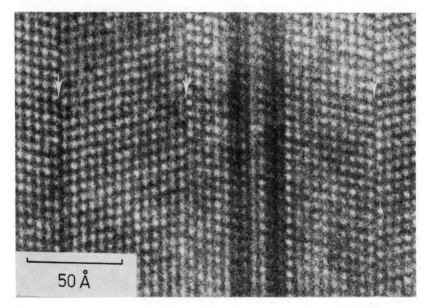

Figure 11   High-resolution electron micrograph of amosite asbestos showing a section perpendicular to [010]. (200) and (001) lattice fringes are resolved and clearly show the (100) twinning. Notice the $\pm c/2$ displacement in the twin boundary (arrowed). From Hutchison et al (1975).

type pyrrhotite (composition from about Fe_9S_{10} to $Fe_{11}S_{12}$), whose distribution of vacancies was unknown. It was found that the nonintegral value of the c parameter for these minerals arose from a statistical distribution of vacancies rather than from intergrowths of different superstructures.

Nakajima et al (1975) imaged the structure of the nonstoichiometric mineral mullite, $xAl_2O_3 . SiO_2$ (where x varies from 1.5 to 2). As in the case of nC pyrrhotite, the *average* structure had been determined by X-ray diffraction, but the *real* arrangement of the oxygen vacancies was unknown. Nakajima et al convincingly showed that the arrangement of the vacancies was partly statistical and partly periodic in the (100) plane. The appearance of nonintegral superstructure reflections in the diffraction patterns of mullites was attributed to the out-of-step displacement of "blocks" of the ordered structure parallel to c. The diffuseness of certain reflections depends upon the variation in the size of the blocks.

In an important study of fibrous monoclinic amphiboles, Hutchison et al (1975) were able to show that the anomalous features found in electron diffraction patterns by Chisholm (1973) were due to the presence of frequent (100) twins (Figure 11). They also confirmed the suggestion by Chisholm that both monoclinic and orthorhombic amphiboles contained Wadsley defects ("mistakes" in the stacking sequence of the Si-O chains that result in a local change in composition).

High-resolution TEM has great potential for solving many other problems in mineralogy that involve defect structures for which X-ray diffraction only gives an "average" picture.

CHEMICAL ANALYSIS

With most current models of transmission electron microscopes it is possible to use the characteristic X rays produced by the interaction of the electron beam with the specimen to provide a chemical analysis. Usually the X rays are intercepted by an energy-dispersive X-ray detector, but X-ray spectrometers may also be provided, as in the analytical electron microscope EMMA-4 (Cooke & Openshaw 1970, Cliff & Lorimer 1972). The obvious advantage of providing X-ray spectrometry within the TEM is that it allows the direct correlation of chemical data with microstructural and diffraction information.

As the specimen must be thin enough to transmit electrons, typically 0.1–0.2 μm at 100 kV, very little diffusion of the electron beam occurs within the sample: the bell-shaped region from which the majority of X rays are produced during conventional electron probe microanalysis is absent (Lorimer & Cliff 1976). Thus, with a sample of thickness 0.1 μm and a probe diameter of 0.1 μm the analysed area at 100 kV is roughly that covered by the probe; the spatial resolution is at least an order of magnitude better than in a microprobe analyser. The limitation of the technique is that the activated volume is small compared to a thick sample and X-ray counting rates are low. The potential accuracy of analysis is therefore inferior to that produced using the microprobe analyser.

If the specimen is thin enough to transmit 100 kV electrons it is also transparent to most of the primary X rays produced by the beam and, to a first approximation,

X-ray absorption and fluorescence can be neglected. Thus, although the absolute intensity of the characteristic X rays for an element will be a function of specimen thickness, the *ratio* of the X-ray intensities for any two elements will be independent of thickness (Cliff & Lorimer 1972), i.e.:

$$I_1/I_2 = k_{12}(C_1/C_2), \tag{2}$$

where I_1 and I_2 are the observed characteristic intensities, C_1 and C_2 are the weight fractions of the elements, and k_{12} is a factor that can be calculated or determined experimentally from a thin-film standard.

The crystal spectrometers have better wavelength discrimination and better peak-to-background characteristics than energy-dispersive detectors and are thus superior both for detecting adjacent elements in the Periodic Table and for detecting trace elements. However, the energy-dispersive detector has two distinct advantages over the crystal spectrometers. Firstly, all elements above $Z = 11$ can be recorded simultaneously. Secondly, because the detector is extremely stable, the factor k_{12} in Equation 1 is a constant for a given instrument, voltage and detector. Thus once the k values have been determined from suitable thin standards, analyses can be performed without further recourse to standards (Cliff & Lorimer 1975). Thus analysis of thin films, particularly with the energy-dispersive detector, is rapid and extremely simple.

Apart from the inferior accuracy of the analysis of thin films, the major limitation of the technique compared with electron probe analysis is that no absolute weight fractions are obtained, and consequently an independent determination or estimation of light elements such as water must be made. However, a number of useful analyses of minerals have already been published; analyses that could not have been obtained in any other way.

The first mineralogical application of analytical electron microscopy was by Lorimer & Champness (1973b) who showed, by monitoring I_{Ca}/I_{Si}, that the presence of the precipitate-free zone in the Stillwater orthopyroxene of Figure 6 could be attributed to a depletion of calcium adjacent to the augite lamellae. They also showed that the Ca content of the clinopyroxene corresponded to an augite and not to a subcalcic augite, as had been suggested by Boyd & Brown (1969) from electron probe analysis.

Analytical electron microscopy has obvious potential in the analysis of exsolved phases that are too small for the microprobe to tackle. Gittos et al (1976) have used it to analyse the coexisting phases in exsolved orthoamphiboles, while Cliff et al (1976) analysed the two phases in persterites, labradorites, and bytownites. Buchanan (1976) investigated the exsolved phases in an augite and a coexisting inverted pigeonite from the Bushveld complex. From these compositions and those of the bulk crystals (obtained from microprobe analysis), he was able to calculate the temperatures of primary crystallization and of exsolution by using the method suggested by Wood & Banno (1973).

The analytical microscope is also of great value in the analysis of small crystals, either natural or synthetic, which are intimately mixed with other material. In 1966 Carpenter et al described a new mineral named jennite from Crestmore, California and reported that, from wet chemical analysis, it had the formula $Na_2Ca_8Si_5O_{30}H_{22}$. Re-examination of the original mineral and another specimen from Israel by analytical electron microscopy failed to detect any sodium (Gard et al 1977) and gave a formula of $9CaO . 6SiO_2 . 11H_2O$ (water was assumed to be the same as found by Carpenter et al). Gard et al conclude that the original specimen was probably contaminated.

In our laboratory B. J. Wood and co-workers are using the analytical microscope to analyze coexisting phases from experimental charges. The aim is to apply the data to the chemical analyses of coexisting natural minerals, as obtained by microprobe, and thus to determine the conditions of their crystallization.

Another fruitful field that is now being exploited is the identification of mineral particles in environmental samples (Rubin & Maggiore 1974, Beaman & File 1976, Kramer 1976). The energy-dispersive detector is ideal for such studies because the "fingerprint" from a particle can be displayed and recorded in seconds and compared with the spectra from standards. Figure 12 shows the spectra from the four common varieties of asbestos, which are easily distinguished from one another. Often the

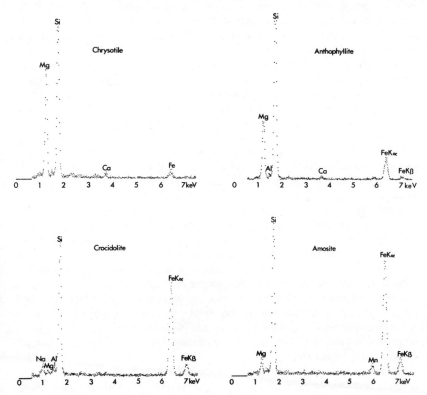

Figure 12   The X-ray "fingerprint" obtained using an energy-dispersive detector from the four common varieties of asbestos.

spectrum is sufficient to identify the particle, for instance when its source is known and its chemistry has been defined. In such cases the calculation of the concentration of particles is fairly rapid. In other cases, however, it may be necessary to record and measure a diffraction pattern or perform a quantitative analysis in order to obtain an unequivocal identification (Champness et al 1977). Both processes are time-consuming when many particles have to be examined.

CONCLUSION

Even from this brief review it should be clear that the use of the transmission electron microscope has had a similar impact on the science of mineralogy in this decade to that which the advent of the electron microprobe analyser had in the 1960s. In fact the newer techniques of thin-film analysis and high resolution complement and extend the use of the electron probe and X-ray diffraction (which must be the technique that has had most impact on mineralogy in this century).

What of the future? X-ray spectrometry of thin films could obviously be put to great advantage in tackling many more mineralogical problems, particularly in conjunction with experimental petrology. TEM is more sensitive than X-ray diffraction for determining the onset of solution and exsolution in experimental charges and thus could be used to determine phase diagrams.

I also expect to see TEM, particularly high-resolution TEM, used more often to solve specific mineralogical problems that X rays cannot tackle. The present limit of resolution for structure images may well be reduced to 1.5 or 2 Å or less for high-voltage instruments if technical problems can be overcome (Cowley & Iijima 1976). There is little hope of improving the resolution of 100 kV instruments significantly because of the dependence of R on the fourth root of C_s (Equation 1). High resolution in the high-voltage electron microscope would have the added bonus for mineralogists that the chromatic aberration introduced by the damaged surface of ion-thinned foils (which can be up to 0.05 μm thick) would be considerably reduced.

In-situ experiments in the high-voltage electron microscope are providing new information about reactions in metals under conditions similar to those that occur in service. Because of the high penetrating power of 1000 kV electrons specimens several microns thick can be used and the surface effects that occur with thin ($< 0.2 \mu m$) foils can be minimized. Gases such as oxygen or water vapor can be introduced into the environmental cell (which either has thin windows or is differentially pumped through small apertures) and the specimen can be heated (Tighe 1976). Although many mineralogical reactions occur very slowly, there are a number of oxidation and hydration reactions that could be studied using an environmental cell; some polymorphic transitions have already been investigated in heating stages (see for instance Heuer et al 1976).

The newest technique, whose potential still remains to be fully evaluated, is the scanning transmission electron microscope, or STEM (for an introductory review see Crewe 1971). This instrument has a construction similar to a conventional scanning electron microscope except that the samples are thin films viewed in transmission rather than being bulk samples viewed in reflection. STEM attachments are available for most new TEMs and greatly add to the instrument's versatility. The sole purpose of the electron-optical system is to form a small spot on the specimen; there are no lenses after the sample. The advantage of STEM over TEM is that the electrons that have been elastically scattered, inelastically scattered, or have not interacted with the sample can be separated and used to form an image either separately or in combination.

It is frequently claimed by the manufacturers that, as chromatic aberration is eliminated (because of the absence of image-forming lenses) much thicker samples can be examined in STEM than in TEM at the same voltage. Fraser et al (1976) have shown that, provided both instruments are operated so as to optimize the resolution, TEM and STEM are comparable for quantitative microscopy (such as the determination of the nature of stacking faults). In fact STEM is unlikely to replace TEM for routine quantitative microscopy for the simple reason that the image is instantaneously visible on the screen of the TEM, whereas in STEM the image is produced on a scanning raster.

The great practical advantage of STEM is the provision of an extremely small electron probe that can be used to obtain electron diffraction patterns or to carry out "micro-microprobe" analysis from very small regions of a specimen. The maximum current I that can be focussed into probe of diameter d at a given voltage is given by

$$I_{max} \propto \beta d^{8/3}/C_s^{2/3}, \tag{3}$$

where β is the brightness of the electron source. In a modern TEM equipped with a conventional thermionic tungsten filament and fitted with a STEM attachment it is possible to obtain a current of 10^{-9} amps (the minimum if quantitative X-ray spectrometry is to be carried out) in a probe 250 to 400 Å in diameter. If the same instrument is fitted with a cold tungsten field-emission gun, β is approximately 10^3 to 10^4 times that of the thermionic filament, and a current of 10^{-9} amps could be focussed into a probe about 10 to 20 Å in diameter. This would provide a remarkable analysis facility, but it is to be expected that specimen damage would be severe.

ACKNOWLEDGEMENTS

I am extremely grateful to G. W. Lorimer for many helpful discussions and for critically reading the manuscript. The following kindly supplied micrographs from their work: R. H. McCallister and R. A. Yund, P. P. K. Smith, J. M. Christie, J. N. Boland and A. C. McLaren, D. L. Kohlstedt, P. R. Buseck and J. L. Hutchison.

Literature Cited

Allpress, J. G., Sanders, J. V. 1973. The direct observation of the structure of real crystals by lattice imaging. *J. Appl. Crystallogr.* 6:165–90

Amelinckx, S., Van Landuyt, J. 1976.

Contrast effects at planar interfaces. In *Electron Microscopy in Mineralogy*, ed. H.-R. Wenk et al, pp. 68–112. Berlin: Springer. 564 pp.

Ardell, A. J., Christie, J. M., Kirby, S. H.,

McCormick, J. W. 1976. Electron microscopy of deformation structures in quartz. In *Developments in Electron Microscopy and Analysis. Proc. Electron Microsc. Anal. Group 75, Bristol, 1975,* pp. 153–6. London: Academic

Ardell, A. J., Christie, J. M., McCormick, J. W. 1974. Dislocation images in quartz and the determination of Burgers vectors. *Philos. Mag.* 29: 1399–411

Ardell, A. J., Christie, J. M., Tullis, J. A. 1973. Dislocation substructures in deformed quartz rocks. *Cryst. Lattice Defects.* 4: 275–85

Barber, D. J. 1970. Thin foils of non-metals made for electron microscopy by sputter-etching. *J. Mater. Sci.* 5: 1–8

Beaman, D. R., File, D. M. 1976. Quantitative determination of asbestos fiber concentrations. *Anal. Chem.* 48: 101–10

Boland, J. N. 1976. An electron microscope study of dislocation image contrast in olivine (Mg, Fe)$_2$SiO$_4$. *Phys. Status Solidi* 34A: 361–7

Boland, J. N., McLaren, A. C., Hobbs, B. E. 1971. Dislocations associated with optical features in naturally-deformed olivine. *Contrib. Mineral. Petrol.* 30: 53–63

Bown, M. G., Gay, P. 1960. An X-ray study of exsolution phenomena in the Skaergaard pyroxenes. *Mineral. Mag.* 32: 379–88

Boyd, F. R., Brown, G. M. 1969. Electron probe study of pyroxene exsolution. *Mineral. Soc. Am. Spec. Pap.* 2: 211–16

Buchanan, D. L. 1976. The quantitative analysis of Bushveld pyroxenes. See Ardell et al 1976, pp. 477–80

Buseck, P. R., Iijima, S. 1974. High resolution electron microscopy of silicates. *Am. Mineral.* 59: 1–21

Cahn, J. W. 1968. Spinodal decomposition. *Trans. AIME* 242: 166–80

Carpenter, A. B., Chalmers, R. A., Gard, J. A., Speakman, K., Taylor, H. F. W. 1966. Jennite, a new mineral. *Am. Mineral.* 51: 56–74

Champness, P. E., Cliff, G., Lorimer, G. W. 1977. The identification of asbestos. *J. Microsc.* In press

Champness, P. E., Copley, P. A. 1976. The transformation of pigeonite to orthopyroxene. See Amelinckx & Van Landuyt 1976, pp. 228–33

Champness, P. E., Lorimer, G. W. 1971. An electron microscopic study of a lunar pyroxene. *Contrib. Mineral. Petrol.* 33: 171–83

Champness, P. E., Lorimer, G. W. 1972. Electron microscopic studies of some lunar and terrestrial pyroxenes. *Proc. Int. Mater.*

Symp., 5th, 1971, pp. 1245–55. Berkeley: Univ. Calif. Press

Champness, P. E., Lorimer, G. W. 1973. Precipitation (exsolution) in an orthopyroxene. *J. Mater. Sci.* 8: 467–74

Champness, P. E., Lorimer, G. W. 1974. A direct lattice-resolution study of precipitation (exsolution) in orthopyroxene. *Philos. Mag.* 30: 357–65

Champness, P. E., Lorimer, G. W. 1976. Exsolution in silicates. See Amelinckx & Van Landuyt 1976, pp. 174–204

Chisholm, J. E. 1973. Planar defects in fibrous amphiboles. *J. Mater. Sci.* 8: 475–83

Christie, J. M., Ardell, A. J. 1976. Deformation structures in minerals. See Amelinckx & Van Landuyt 1976, pp. 374–403

Christie, J. M., Lally, J. S., Heuer, A. H., Fisher, R. M., Griggs, D. T., Radcliffe, S. V. 1971. Comparative electron petrography of Apollo 11, Apollo 12 and terrestrial rocks. *Proc. Lunar Sci. Conf., 2nd, Geochim. Cosmochim. Acta Suppl.* 2, 1: 69–89

Cliff, G., Champness, P. E. Nissen, H.-U., Lorimer, G. W. 1976. Analytical electron microscopy of exsolution lamellae in plagioclase feldspars. See Amelinckx & Van Landuyt 1976, pp. 258–65

Cliff, G., Lorimer, G. W. 1972. Quantitative analysis of thin metal foils using EMMA-4 —the ratio technique. *Proc. Eur. Congr. Electron Microsc., 5th.* 140–41

Cliff, G., Lorimer, G. W. 1975. The quantitative analysis of thin specimens. *J. Microsc.* 103: 203–7

Cooke, C. J., Openshaw, I. K. 1970. Combined high resolution electron microscope and X-ray microanalysis. *Proc. Ann. Electron Microsc. Soc. Am. Meet., 28th,* Baton Rouge, pp. 552–53

Cosslett, V. E. 1968. The high voltage electron microscope. *Contemp. Phys.* 9: 333–54

Cowley, J. M., Iijima, S. 1976. The direct imaging of crystal structures. See Amerlinckx & Van Landuyt 1976, pp. 123–36

Crewe, A. V. 1971. A high-resolution scanning electron microscope. *Sci. Am.* 224: 26–35

Fraser, H. L., Jones, I. P., Loretto, M. H. 1976. Limiting factors in specimen thickness in conventional and scanning transmission electron microscopy. *Philos. Mag.* In press

Gard, J. A. 1976. Interpretation of electron diffraction patterns. See Amerlinckx & Van Landuyt 1976, pp. 52–67

Gard, J. A., Taylor, H. F. W., Cliff, G., Lorimer, G. W. 1977. A re-examination of jennite. *Am. Mineral.* In press

Gillespie, P., McLaren, A. C., Boland, J. N. 1971. Operating characteristics of an ion bombardment apparatus for thinning non-metals for transmission electron microscopy. *J. Mater. Sci.* 6:87–89

Gittos, M. F., Lorimer, G. W., Champness, P. E. 1976. The phase distributions in some exsolved amphiboles. See Amelinckx & Van Landuyt 1976, pp. 238–47

Griggs, D. T., Blacic, J. D. 1965. Quartz: anomalous weakness of synthetic crystals. *Science.* 147:292–95

Hashimoto, H., Howie, A., Whelan, M. J. 1960. Anomalous electron absorption effects in metal foils. *Philos. Mag.* 5:967–74

Hess, H. H. 1941. Pyroxenes of common mafic magmas. *Am. Mineral.* 26:515–35, 573–94

Heuer, A. H., Firestone, R. F., Snow, J. D., Green, H. W., Howe, R. G. Christie, J. M. 1971. An improved ion-thinning apparatus. *Rev. Sci. Inst.* 42:1177–84

Heuer, A. H., Nord, G. L., Lally, J. S., Christie, J. M. 1976. Origin of the (c) domains in anorthite. See Amelinckx & Van Landuyt 1976, pp. 345–53

Hutchison, J. L., Irusteta, M. C., Whittaker, E. J. W. 1975. High-resolution electron microscopy and diffraction studies of fibrous amphiboles. *Acta Crystallogr.* A31:794–801

Iijima, S., Allpress, J. G. 1974. Structural studies by high resolution electron microscopy: tetragonal tungsten bronze type of structures in the system $Nb_2O_5-WO_3$. *Acta Crystallogr.* A30:22–9

Iijima, S., Cowley, J. M., Donnay, G. 1973. High resolution microscopy of tourmaline. *Tschermaks Mineral. Petrogr. Mitt.* 20:216–24

Kirby, S. H. 1976. The role of crystal defects in the shear-induced transformation of orthoenstatite to clinoeustatite. See Amelinckx & Van Landuyt 1976, pp. 465–72

Kohlstedt, D. L., Goetze, C. 1974. Low-stress, high-temperature creep in olivine single crystals. *J. Geophys. Res.* 79:2045–51

Kramer, J. R. 1976. Fibrous cummingtonite in Lake Superior. *Can. Mineral.* 14:91–98

Lally, J. S., Heuer, A. H., Nord, G. L., Christie, J. M. 1975. Subsolidus reactions in lunar pyroxenes: an electron petrographic study. *Contrib. Mineral. Petrol.* 51:263–81

Loretto, M. H., Smallman, R. E. 1975. *Defect Analysis in Electron Microscopy.* London: Chapman & Hall. 134 pp.

Lorimer, G. W., Champness, P. E. 1973a. Mineralogical applications of HVEM.

In *High Voltage Electron Microscopy*, eds. P. R. Swann, C. J. Humphreys, N. J. Goringe, pp. 301–11. London: Academic. 475 pp.

Lorimer, G. W., Champness, P. E. 1973b. Combined electron microscopy and analysis of an orthopyroxene. *Am. Mineral.* 58:243–48

Lorimer, G. W., Champness, P. E. 1974. The origin of the phase distribution in two perthitic alkali feldspars. *Philos. Mag.* 28:1391–1403

Lorimer, G. W., Cliff, G. 1976. Analytical electron microscopy of minerals. See Amelinckx & Van Landuyt 1976, pp. 506–19

McCallister, R. H., Yund, R. A. 1975. Kinetics and microstructure of pyroxene exsolution. *Carnegie Inst. Washington Yearb.* 74:433–46

McCormick, J. W. 1976. Computer simulation of dislocation images in quartz. See Amelinckx & Van Landuyt 1976, pp. 113–22

Morimoto, N., Nakazawa, H., Watanabe, E. 1974. Direct observation of metal vacancy in pyrrhotite, $Fe_{1-x}S$ by means of an electron microscope. *Proc. Jpn. Acad.* 50:756–59

Müller, W. F. 1969. Elektronenmikroskopischer Nachweis amorpher Bereiche in stosswellenbeanspruchten Quarz. *Naturwiss.* 56:279–80

Nakajima, Y., Morimoto, N., Watanabe, E. 1975. Direct observation of oxygen vacancy in mullite, $1.89 Al_2O_3 . SiO_2$ by high resolution electron microscopy. *Proc. Jpn. Acad.* 51:173–78

Nord, G. L., Heuer, A. H., Lally, J. S. 1976. Pigeonite exsolution from augite. See Amelinckx & Van Landuyt 1976, pp. 220–27

Nord, G. L., Lally, J. S., Heuer, A. H., Christie, J. M., Radcliffe, S. V., Griggs, D. T., Fisher, R. M. 1973. Petrologic study of igneous and metaigneous rocks from Apollo 15 and 16 using high voltage transmission electron microscopy. *Proc. Lunar Sci. Conf. 4th, Geochim. Cosmochim. Acta, Suppl. 4,* 1:953–70

Papike, J. J., Prewitt, C. T., Sueno, S., Cameron, M. 1973. Pyroxenes: comparisons of real and ideal structural topologies. *Z. Kristallogr.* 138:254–73

Pierce, L., Buseck, P. R., 1974. Electron imaging of pyrrhotite superstructures. *Science.* 186:1209–12

Rubin, I. B., Maggiore, C. J. 1974. Elemental analysis of asbestos fibers by means of electron probe techniques. *Environ. Health Perspect.* 9:81–94

Steeds, J. W., Tatlock, G. J., Hampson, J. 1973. Real space crystallography. *Nature Phys. Sci.* 246:126–28

Stocker, R. L., Ashby, M. F. 1973. On the rheology of the upper mantle. *Rev. Geophys.* 11:391–426

Tighe, N. J. 1976. Experimental techniques. See Amelinckx & Van Landuyt 1976, pp. 144–73

Van de Biest, O., Thomas, G. 1976. Fundamentals of electron microscopy. See Amelinckx & Van Landuyt 1976, pp. 18–51

Weertman, J. 1970. The creep strength of the earth's mantle. *Rev. Geophys.* 8:145–68

Wood, B. J., Banno, S. 1973. Garnet-orthopyroxene and orthopyroxene-clinopyroxene relationships in simple and complex systems. *Contrib. Mineral. Petrol.* 42:109–24

Yada, K. 1967. Study of chrysotile asbestos by a high resolution electron microscope. *Acta Crystallogr.* 23:704–7

Yada, K. 1971. Study of microstructure of chrysotile asbestos by high resolution electron microscopy. *Acta Crystallogr.* A27:659–64

Ann. Rev. Earth Planet. Sci. 1977. 5: 227–55

GEOCHEMISTRY OF ATMOSPHERIC RADON AND RADON PRODUCTS

×10073

Karl K. Turekian,[1] *Y. Nozaki, and Larry K. Benninger*

Department of Geology and Geophysics, Yale University, New Haven, Connecticut 06520

INTRODUCTION

The ^{222}Rn in the atmosphere comes from the earth's surface. Once in the atmosphere, this rare-gas daughter of ^{238}U decays with a half-life of 3.8 days to produce daughter activities according to the scheme shown in Figure 1. The charged daughter ions thus produced are chemically reactive and soon become irreversibly associated

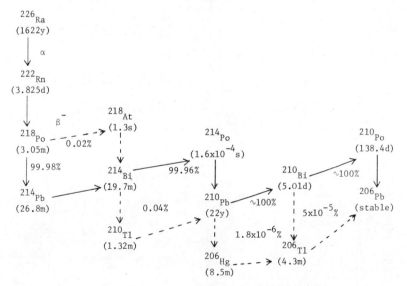

Figure 1 ^{222}Rn and its decay chain (after Moore, Poet & Martell 1973).

[1] This research supported by U.S. Energy Research and Development Administration Grant E(11-1)3573.

227

with aerosol particles. Thereafter their behavior is simply the behavior of the aerosol particles, with respect to growth and transport in the atmosphere and to removal from the atmosphere. Thus the relative activities of the long-lived ^{222}Rn daughters, ^{210}Pb, ^{210}Bi, and ^{210}Po, are set by the rates of production of each, their rates of decay, and their common rate of removal from the atmosphere. In principle the abundances of ^{222}Rn and its daughters in the atmosphere should contain information about the mean residence time of aerosols, relative to removal to the earth's surface.

The principle of continuity (what goes up must come down again) means that on the global scale there should be an atomic balance between ^{222}Rn flux from the earth's surface and the removal of the longest-lived daughter ^{210}Pb (22-yr half-life). Thus if the atmospheric ^{210}Pb flux were well-known geographically, it would uniquely determine the global ^{222}Rn flux. It would also provide a constraint on either the aerosol mean residence time or the mean air transit time, if one or the other is known. By considering ^{222}Rn profiles in addition, it is possible to fix both the mean transit time and the aerosol mean residence time. In this review we have attempted to evaluate the various estimates of ^{222}Rn flux from the earth's

Table 1 ^{222}Rn flux by direct measurement

Location	Mean ^{222}Rn flux (atom cm^{-2} sec^{-1})[a]	Reference
Dublin, Ireland (53°20′ N, 6°15′ W)	1.3	Smyth (1912)
Saclay, France (48°42′ N, 2°12′ E)	0.7	Servant (1964) (cited Wilkening, Clements & Stanley 1975)
Aachen, W. Germany (50°46′ N, 6°06′ E)	0.8	Israël & Horbert (1970)
Innsbruck, Austria (47°17′ N, 11°25′ E)	0.41	Zupancic (1934)
	0.77	Zeilinger (1935)
Graz, Austria (47°05′ N, 15°22′ E)	0.89	Kosmath (1935)
Warsaw, Poland (52°15′ N, 21°00′ E)	0.34	Pensko, Wardaszko & Wochna (1968)
USSR (European)	0.33	Sisigina (1967)
USSR (Kazakhstan)	0.24	Sisigina (1967)
Manila, Philippines (14°36′ N, 120°59′ E)	0.52	Wright & Smith (1915)
Osaka, Japan (34°40′ N, 135°30′ E)	0.5	Megumi & Mamuro (1973)
Wellington, New Zealand (41°17′ S, 174°47′ E)	0.004	Rosen (1957)
Hawaii (19°30′ N, 155°30′ W)		
lava fields	0.012	Wilkening (1974)
thin soil	0.08	Wilkening (1974)
deep soil	1.5	Wilkening (1974)
sea surface, shallow water	0.15	Wilkening & Clements (1975)
sea surface, deep water	0.0074	Wilkening & Clements (1975)
Continental United States		
Fairbanks, Alaska (64°50′ N, 147°50′ W)	0.23	Wilkening, Clements & Stanley (1975)
Yucca Flats, Nevada (37°00′ N, 116°03′ W)	1.0	Kraner, Schroeder & Evans (1964)
Socorro, New Mexico (34°04′ N, 106°55′ W)	1.6	Wilkening & Hand (1960)
	1.8	Pearson & Jones (1966)
	0.52	Clements & Wilkening (1974)
Texas		
plains	0.7	Wilkening, Clements & Stanley (1975)
central and coastal	0.4	Wilkening, Clements & Stanley (1975)
Illinois		
Champaign County (40°07′ N, 88°14′ W)	2.5	Pearson & Jones (1966)
Argonne (41°32′ N, 88°05′ W)	1.0	Pearson & Jones (1966)
Lincoln, Massachusetts (42°24′ N, 71°18′ W)	1.7	Kraner, Schroeder & Evans (1964)

[a] 1 atom ^{222}Rn cm^{-2} sec^{-1} = 1.865 dpm ^{210}Pb cm^{-2} yr^{-1};
1 dpm ^{210}Pb cm^{-2} yr^{-1} = 0.54 atom ^{222}Rn cm^{-2} sec^{-1}.

surface, the ^{210}Pb flux from the atmosphere, and the mean residence time of aerosols. We also propose a total global model for ^{222}Rn and its daughters.

The knowledge of both the ^{222}Rn flux and the ^{210}Pb flux distributions may have a bearing on some current problems involving man's welfare. The ^{222}Rn flux has been related to the development of atmospheric electricity because of the production of charged species as the result of radioactive decay (see Israël, 1951). Recently it has been proposed as a monitor, in wells and thermal springs, of earthquake activity (see Scholz, Sykes & Aggarwal 1973). If a reliable method for measuring fluxes of Rn from the ground itself is ever developed for routine operations, the ambient flux becomes important in identifying anomalous stations or times.

The ^{210}Pb residence time bears directly on the expected behavior of man-made aerosols with similar properties. A knowledge of the worldwide mean residence time as well as the local variants is fundamental in identifying the downwind extent of influence of a polluting source.

MEASUREMENT OF THE RADON-222 FLUX TO THE ATMOSPHERE

Two methods of estimating the ^{222}Rn flux from the earth's surface to the atmosphere have been used. The first is the collection of ^{222}Rn emitted during a known time into a vessel covering a known area of the soil. This is designated as the direct measurement technique. The second is the integration of ^{222}Rn profiles in the atmosphere and estimating the flux by assuming balance between ^{222}Rn emanation and radioactive decay. The results by the direct measurement technique are listed in Table 1 and by the atmospheric profile integration method in Table 2.

Neither of these methods is entirely satisfactory. Direct measurement has the disadvantage that convenient, though not necessarily representative, areas of relatively small size are actually measured for ^{222}Rn flux. Probably the most serious omission, as the result of this sampling method, is terrain covered by

Table 2 ^{222}Rn flux from atmospheric ^{222}Rn profile

Location	^{222}Rn flux (atom cm^{-2} sec^{-1})	Reference
Saclay, France (48°42′ N, 2°12′ E)	0.65	Servant (1964) cited Wilkening, Clements & Stanley (1975)
USSR		
Central European (55° N, 35° E)	0.35	Kirichenko (1970)
North (60° N, 40° E)	0.18	Kirichenko (1970)
Caucasus (42° N, 45° E)	0.53	Kirichenko (1970)
South Urals (55° N, 60° E)	0.53	Kirichenko (1970)
South central desert (40° N, 60° E)	0.62	Kirichenko (1970)
South central mountains (40° N, 70° E)	0.88	Kirichenko (1970)
United States		
Alaska		
Yukon basin (65° N, 150° W)	0.33	Anderson & Larson (1974)
	0.39	Wilkening, Clements & Stanley (1975)
Tanana basin (65° N, 147° W)	0.37	Wilkening, Clements & Stanley (1975)
Socorro, New Mexico (34°04′ N, 106°55′ W)	0.91	Wilkening, Clements & Stanley (1975)

large vegetation. As pointed out by Pearson & Jones (1965), soil water should be rich in dissolved radon, and transpiration should deliver ^{222}Rn to the atmosphere. They report on a single experiment done in summer at midday in which the radon emanation per unit area from the leaves of field corn (*Zea mays L.*) was 2.6 times that from adjacent soil. This result has been criticized by Mattsson (1970), whose data imply a much smaller ^{222}Rn flux due to transpiration by birch trees.

^{222}Rn fluxes based upon standing crops of ^{222}Rn in the atmosphere suffer from the assumption of a local steady state between emanation and decay. This problem was recognized by Moore, Poet & Martell (1973), who criticized earlier workers for applying a steady-state model to essentially maritime air masses. Even over continents and with constant emanation, however, the half-life for ^{222}Rn build-up in the atmosphere is 3.8 days. As this time scale is of the same order as that for large-scale circulation in the troposphere, the general validity of the steady-state assumption in continental air masses is also open to question. This subject is treated quantitatively below.

It is thus quite likely that measurements like those presented in Tables 1 and 2 can only give underestimates of the average flux of ^{222}Rn from the continents. The integrated atmospheric profile measurements will only approach the steady-state value after the air mass has traversed a continent for ten days or more (see last section) so that the estimated fluxes will generally be systematically lower than the true continental flux. By comparing data in Table 1 with those in Table 2, it appears that ^{222}Rn fluxes from atmospheric profiles are as high as, or higher than, those measured directly. Thus, if the atmospheric profile fluxes are lower than the true steady-state values, then the flux estimates from direct measurements are low as well. This suggests that a true average ^{222}Rn flux probably exceeds the commonly used values of 0.7 atom cm^{-2} sec^{-1} (Israël 1951) or 0.75 atom cm^{-2} sec^{-1} (Wilkening, Clements & Stanley 1975). In the last section we attempt to quantify this number on a global basis.

MEASUREMENT OF THE LEAD-210 ATMOSPHERIC FLUX

The measurement of the ^{210}Pb flux from the atmosphere to the earth's surface can be made by using either man-made or natural collectors. In the former case wet and dry atmospheric precipitation is collected in a suitable fashion over a long enough period of time to accommodate seasonal and episodic variations. In the latter case natural repositories such as snow fields, lake sediments, and soils are used to integrate over decades if not centuries. A third method of estimating flux has been suggested, namely, the measurement of the concentration of ^{210}Pb in surface air and the transformation of this value into a flux through the deposition velocity (estimated or derived from nearby measurements).

One difficulty suggested for long-term repository estimates is the possibility of contamination by ^{210}Pb produced by nuclear devices and injected into the air. The evaluation of the importance of this process becomes important in using long-term repositories for flux measurements.

Lead-210 Flux Estimates Based on Measurements on Atmospheric Precipitation

LEAD-210 IN PRECIPITATION   The ^{210}Pb concentration in a sample representing a single precipitation event will clearly depend on a complex array of variables, including the history of the air mass and the particular characteristics of the storm (intensity, duration, height of cloud cover, etc). It is therefore not surprising that ^{210}Pb concentrations in precipitation vary considerably. In 48 rainfall samples collected at Fayetteville, Arkansas (mid-continent, USA) Gavini, Beck & Kuroda (1974) found a concentration range of 1.7–24.8 dpm ^{210}Pb l^{-1}. Except by chance, no one rain sample would provide an accurate estimate of the weighted average ^{210}Pb concentration in precipitation at Fayetteville. The problem is a universal one.

Increasing the number of samples improves the estimate of the average ^{210}Pb concentration in precipitation. The simple numerical average ($\pm$ one standard deviation) concentration for the 48 rains analyzed by Gavini, Beck & Kuroda (1974) is 8.0 ($\pm$4.4) dpm ^{210}Pb l^{-1}. But, for purposes of computing the annual wet deposition of ^{210}Pb, such an average will generally be in error (in the Fayetteville case it would be too high) since the computation involves the product of the intensive (^{210}Pb concentration) and extensive (precipitation volume) parameters, and these are not strictly independent. Generally, light rains (or early fractions of heavier ones) tend to have higher concentrations of ^{210}Pb than heavy rains (or bulk samples). A weighted average ^{210}Pb concentration is the correct method of computing the annual wet deposition of ^{210}Pb. That is:

$$^{210}\text{Pb flux (dpm cm}^{-2}\text{ yr}^{-1}) = \lambda_{Pb}\Sigma(r_i \times [^{210}\text{Pb}]_i)/f, \tag{1}$$

where, for an individual precipitation event i, r_i is the volume of precipitation per cm^2 of earth surface, and $[^{210}\text{Pb}]_i$ is the atomic concentration of ^{210}Pb; λ_{Pb} is the decay constant for ^{210}Pb; and f is the fraction of annual precipitation represented by Σr_i.

TOTAL LEAD-210 DEPOSITION   Atmospheric aerosols may be removed either by precipitation-scavenging or by settling under the influence of gravitational or electrostatic forces. The relative contributions of the two mechanisms have been determined at only a few locations and for only limited periods of time. It would appear that no universally applicable rule exists to assess their relative importance. For any site and sampling period, however, for which dry deposition contributes appreciably to the total atmospheric ^{210}Pb deposition, estimates of deposition based solely on episodic collections of wet precipitation, no matter how complete, will be low. Continuously exposed integrated samples may be expected to be more representative of the true ^{210}Pb flux.

The atmospheric ^{210}Pb flux determinations in Table 3 are based upon such integrated collections. At New Haven, Connecticut, comparison of the flux determined from artificial collectors with that determined from the standing crop of excess ^{210}Pb in nearby salt-marsh sediment and soil profiles (see below, Table 5) suggests that the artificial collectors behave like vegetated surface terrain in this locality.

A plot of total monthly ^{210}Pb deposition in the collector at New Haven vs monthly precipitation is shown in Figure 2. Plots of the same parameters from

Table 3 ^{210}Pb flux in precipitation + dry fallout

Location	^{210}Pb flux (dpm cm^{-2} yr^{-1})	Reference
Milford Haven, UK (51°44′ N, 5°02′ W)	0.51 ± 0.08	Peirson, Cambray & Spicer (1966)
Moscow, USSR (55°45′ N, 37°42′ E)	0.69 ± 0.06	Baranov & Vilenskii (1965)
India		Joshi, Rangarajan & Gopalakrishnan
Bombay (18°57′ N, 72°55′ E)	1.50 ± 0.36	(1969)
Srinagar (34°06′ N, 74°55′ E)	1.09 ± 0.29	
Ootacamund (11°23′ N, 76°40′ E)	0.52 ± 0.16	
Delhi (28°45′ N, 77°20′ E)	0.80 ± 0.26	
Bangalore (12°57′ N, 77°30′ E)	0.49 ± 0.19	
Nagpur (21°12′ N, 79°04′ E)	0.61 ± 0.21	
Gangtok (27°12′ N, 88°23′ E)	> 1.38 ± 0.76	
Calcutta (22°34′ N, 88°25′ E)	0.61 ± 0.29	
Hokkaido, Japan (41°50′ N, 140°25′ E)	2.2 ± 0.6	Fukuda & Tsunogai (1975)
New Haven, USA (41°18′ N, 72°55′ W)	0.92 ± 0.09	Benninger (1976) and unpublished data
Australia		Bonnyman & Molina-Ramos (1971)
Perth (31°58′ S, 115°49′ E)	0.26 ± 0.07	
Wokalup (33°07′ S, 115°52′ E)	0.28 ± 0.11	
Port Hedland (20°24′ S, 118°36′ E)	0.19 ± 0.06	
Darwin (12°23′ S, 130°44′ E)	0.57 ± 0.15	
Alice Springs (23°42′ S, 133°52′ E)	0.34 ± 0.06	
Adelaide (34°55′ S, 138°35′ E)	0.32 ± 0.07	
Meadows (35°11′ S, 138°45′ E)	0.35 ± 0.08	
Townsville (19°13′ S, 146°48′ E)	0.23 ± 0.07	
Brisbane (27°30′ S, 153°00′ E)	0.39 ± 0.05	
Samford (27°22′ S, 152°53′ E)	0.37 ± 0.07	
Sydney (33°55′ S, 151°10′ E)	0.32 ± 0.09	
Berry (34°48′S, 150°41′ E)	0.42 ± 0.11	
Melbourne (37°45′ S, 144°58′ E)	0.30 ± 0.07	
Warragul (38°11′ S, 145°55′ E)	0.40 ± 0.08	
Hadspen (41°30′ S, 147°05′ E)	0.32 ± 0.05	
Hobart (42°54′ S, 147°18′ E)	0.18 ± 0.02	
New Zealand		L.P. Gregory, New Zealand National
Kaitaia (35°08′S, 173°18′ E)	0.37 ± 0.10	Radiation Laboratory, personal
Auckland (36°55′ S, 174°45′ E)	0.30 ± 0.07	communication. Also NRL, 1968–
New Plymouth (39°03′ S, 174°04′ E)	0.44 ± 0.09	1975.
Greymouth (42°28′ S, 171°12′ E)	0.75 ± 0.07	
Invercargill (46°26′ S, 168°21′ E)	0.21 ± 0.07	
Dunedin (45°52′ S, 170°30′ E)	0.18 ± 0.05	
Christchurch (43°33′ S, 172°40′ E)	0.14 ± 0.01	
Havelock North (39°40′ S, 176°53′ E)	0.18 ± 0.01	
Suva, Fiji (18°08′ S, 178°25′ E)	0.48 ± 0.10	
Rarotonga (21°15′ S, 159°45′ W)	0.31	

collectors in Moscow, USSR (Baranov & Vilenskii 1965), and Hokkaido, Japan (Fukuda & Tsunogai 1975) yield more or less similar distributions. A review of these data and those from the remaining stations of Table 3 reveals the following relationships.

Considering the accumulated monthly data for any one location, total ^{210}Pb deposition is always positively correlated with precipitation. Although the sample correlation coefficients all differ significantly from zero, most are quite low (range 0.19–0.73), and most are not significantly increased by removal of a few extreme points from the fit. Thus not more than about half of the sample variance is explained by variations in the intensity of precipitation of constant concentration. The average monthly ^{210}Pb concentration in precipitation is quite variable so that little smoothing is achieved by averaging over this time period.

The presence of a strong seasonal effect in ^{210}Pb concentration in precipitation would, of course, increase the variance in the pooled data. Fukuda & Tsunogai (1975) reported such an effect for Hokkaido, Japan. Summer (May to August) and winter (November to February) samples define separate linear relationships between ^{210}Pb deposition and rainfall, for which the concentration (= slope) is 3.5 times higher in winter than in summer. This is due to the strong north-west monsoon in winter. Baranov & Vilenskii (1965) found no seasonal pattern in ^{210}Pb deposition vs precipitation at Moscow, however, and this latter relationship appears to be the more general case. No seasonal effects, for example, are evident in any of the

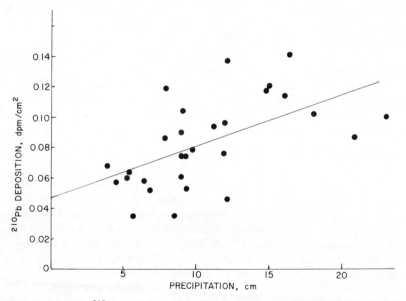

Figure 2   Monthly ^{210}Pb total deposition plotted against monthly precipitation at New Haven, Connecticut (August 1973–January 1976). Data from Benninger (1976) and unpublished data.

Table 4   Estimation of "dry" fallout from the intercept of the least-squares line

	^{210}Pb annual deposition (dpm cm^{-2} yr^{-1})		
	total	"dry"	dry/total
Kaitaia	0.37 ± 0.10	0.20 ± 0.04	0.54 ± 0.18
Darwin	0.54 ± 0.15	0.29 ± 0.04	0.54 ± 0.17
Milford Haven	0.51 ± 0.08	0.17 ± 0.05	0.33 ± 0.11
New Haven	0.92 ± 0.09	0.56 ± 0.13	0.61 ± 0.15

data from New Zealand; Milford Haven, England; or New Haven, Connecticut. The generally more arid Australian stations show a more pronounced seasonality in precipitation than any of the above stations, yet no seasonal effect in slope or intercept in plots of ^{210}Pb deposition vs precipitation is discernible.

In the idealized linear model for ^{210}Pb deposition vs precipitation, the intercept at zero precipitation for a regression line has been used as an estimate of dry deposition. The intercepts are shown in Table 4 for a few selected areas from Table 3. The least-squares line, in every case, is a significantly better fit than is a line through $(0,0)$ and $(\bar{x}, \bar{y})$, but it must be recognized that the reliability of such estimates of dry deposition depends upon the adequacy of the linear model.

In summary, total deposition data are of greater value in estimating atmospheric ^{210}Pb deposition at a given locality than are data on individual precipitation events since dry deposition is included. Even with this constraint, one-month collection periods are not adequate for the approximation of annual fluxes, especially where strong seasonal differences exist. Even annual deposition at most stations is not as highly correlated with annual precipitation as would be desired for a good representation of the flux. The basis for extrapolation of short records in arriving at an annual flux estimate is so weak that atmospheric ^{210}Pb fluxes calculated from fewer than six to twelve months' data may be uncertain by a factor of two for many places, and several years' data may fix the average flux to no better than 10–50% (see Table 3).

Deposition Velocities and the Lead-210 Flux

There are many more data on the ^{210}Pb content of surface air than actual flux measurements. There have been attempts at using such concentration data to obtain fluxes on the premise that the flux must be related to the concentration through a "deposition velocity." Originally the deposition velocity was defined to relate dry deposition on an exposed surface to concentration in air at some reference point above the surface (Chamberlain 1960). Indeed, the classic paper by Lambert & Nezami (1965b) estimating the zonal fluxes of ^{210}Pb uses this as one component in estimating the total ^{210}Pb flux (see below). The deposition velocity concept has been extended to relating the surface-air concentration of ^{210}Pb (and other nuclides, stable and radioactive) to total fluxes, including deposition by all processes. This is not a strictly accurate transformation. Figure 3 shows the plot of ^{210}Pb flux against surface-air concentration for Australia, England, and the New York–Connecticut area for which long-range data (over several years) are available. Although the "average" line yields a "deposition velocity" of about 0.85 cm sec^{-1}, the range is from 1.33 to 0.43 cm sec^{-1}. A similar calculation for data from Antarctica yields a deposition velocity of 0.12 cm sec^{-1}. Clearly one universal value for the deposition velocity should be used with full cognizance of the large errors possible when transforming concentrations to fluxes for any nuclide. Indeed, short-term measurements of either flux or atmospheric concentration indicate that too short a record of concentration can compound the error. In light of this, we have not calculated any ^{210}Pb fluxes based on an average deposition velocity and surface-air concentrations.

Lead-210 Fluxes Based on Natural Repository Analyses

In addition to the direct measurement of ^{210}Pb fluxes by collecting atmospheric precipitation in artificial collectors it is possible, in principle, to use natural collection systems. These range from lichens on rocks to the large continental ice sheets. The reliability of each type of "collector" or reservoir as an indicator of the ^{210}Pb flux varies, even though each may record a stratigraphy of ^{210}Pb concentrations useful for chronological purposes. For example the ^{210}Pb flux can be enhanced in the sediments of some parts of standing bodies of water by bottom transport of

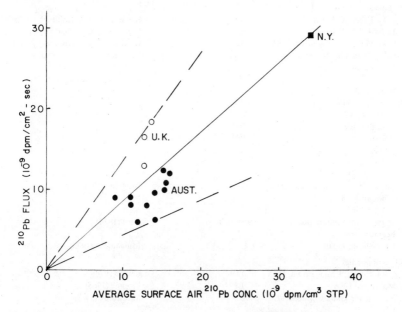

Figure 3 ^{210}Pb annual flux plotted against mean annual ^{210}Pb concentration in surface air at stations in Australia, England, and the United States. See sources in Table 3 for Australia and England. United States data based on Benninger (1976) and ERDA-HASL 306 (July 1976).

Table 5 ^{210}Pb flux in natural collectors and repositories

	^{210}Pb flux (dpm cm^{-2} yr^{-1})	Snow accumulation (g cm^{-2} yr^{-1})	Reference
I. Glaciers			
Greenland, Camp Century (77°10′ N, 61°00′ W, 1885 m)	0.091	32	Crozaz & Langway (1966)
Greenland, Site 2 (76°59′ N, 56°04′ W, 2000 m)	0.11	63	Windom (1969)
Antarctica (82°07′ S, 55°06′ E, 3715 m)	0.006	3.1	Picciotto et al (1968)
Antarctica (80°00′ S, 120°00′ W, 1200 m)	0.005	9.75	Windom (1969)
St. Elias Range, Yukon Territory, Canada (60°45′ N, 139°30′ W, 2600 m)	0.039	39	Windom (1969)
Mt. Olympus, Washington, USA (47°52′ N, 123°36′ W, 2000 m)	0.16	65	Windom (1969)
Mt. Orizaba, Mexico (19°01′ N, 97°15′ W, 5400 m)	0.04	11	Windom (1969)
Mt. Popocatepetl, Mexico (19°01′ N, 98°37′ W, 4700 m)	0.009	5.9	Windom (1969)
II. Lakes			
Lake Tahoe, California, USA	0.5	—	Koide, Soutar & Goldberg (1972)
Trout Lake, Wisconsin, USA	6.4	—	Koide, Soutar & Goldberg (1972)
Lake Mendota, Wisconsin, USA	3.2	—	Koide, Soutar & Goldberg (1972)
Lake Leman, Switzerland	0.43	—	Krishnaswami et al (1971)
Lake Pavin, France	1.3	—	Krishnaswami et al (1971)
Lake Michigan, USA (from North to South)			Robbins & Edgington (1975)
Station 100A	0.84	—	
Station 103	0.71	—	
Station 105	0.60	—	
Station 11	0.38	—	
Station 17	0.40	—	
Station 29	1.44	—	
Station 31	0.31	—	
Linsley Pond, Connecticut, USA			Brugam (1975)
Before 1938	1.6	—	
After 1961	2.3	—	
Mirwart aquaculture pond, Belgium	0.95, 0.78	—	Petit (1974)
Lake Shinji, Japan	0.9, 1.8, 2.2, 1.5	—	Matsumoto (1975)
III. Soil profiles			
Cook Forest, Pennsylvania, USA	1.0	—	Lewis (1976)
Maryland, USA	1.2	—	Fisenne (1968)
Farm River knoll, Connecticut, USA	0.8	—	McCaffrey (1977)
Boulder, Colorado, USA	1.6	—	Moore & Poet (1976)
IV. Salt marsh profiles			
Farm River salt marsh	1.0	—	McCaffrey & Thomson (1974)

sediments or by differences in sediment type and thus in ^{210}Pb scavenging efficiency (see Benninger 1976 for a discussion of this problem). The types of natural collectors or repositories that have been used to assay ^{210}Pb fluxes are the following: (a) glaciers and ice caps, (b) lichens on rock, (c) natural and artificial lake sediments, (d) soils, and (e) salt marsh sediments.

GLACIERS AND ICE CAPS   Goldberg (1963) was the first to show the possibility of using ^{210}Pb to date cores from glacial accumulations. Profiles of ^{210}Pb in glacial ice also provide a direct estimate of ^{210}Pb flux at that site. Since then, techniques have been developed for measuring ^{210}Pb profiles in both the large ice caps of Greenland and Antarctica and the smaller montane permanent snow fields. The major problem in the montane snow fields is the presence of ^{210}Pb associated with dust rather than from purely atmospheric production from radon. If this effect is found to be large, corrections must be made for the supported ^{210}Pb from ^{226}Ra measurements, on the assumption of secular equilibrium.

The most recent flux measurements in glaciers and ice caps are listed in Table 5. The ice caps receive low fluxes of ^{210}Pb, with Antarctica about a factor of 20 lower than Greenland. This can be ascribed to the fact that there are no continental areas releasing radon at these latitudes. It could also reflect the fact that aerosols are scavenged very efficiently prior to arrival of the air masses at these elevations (an orographic effect). This is most strikingly seen in the permanent snow fields of western North America. The ^{210}Pb fluxes are a factor of two to fifty smaller than expected at sea level in the same area.

LICHENS   Lichens frequently form a continuous cover on rocks at high latitudes and act as efficient collectors of ions from atmospheric precipitation. Persson (1970) assayed lichen patches of known area in northern Sweden between 1961 and 1969 and found no variation in ^{210}Pb concentration. On the basis of unit area accumulation and estimated atmospheric ^{210}Pb flux for that region, he estimated that the mean retention time of ^{210}Pb by lichens is about seven years. The ^{210}Pb concentrations from his Swedish lichens are essentially the same as those from lichens taken in a traverse from 60°N to 70°N in Finland (Kauranen & Miettinen 1969), which indicates that the atmospheric flux of ^{210}Pb in northern Europe is constant north of 60°N.

NATURAL AND ARTIFICIAL LAKE DEPOSITS   Lake sediments were among the first continental repositories to which ^{210}Pb chronology was applied, and ^{210}Pb fluxes were a by-product. The major problem about using this type of data as a true measure of the atmospheric flux of ^{210}Pb is that there are redistributions of ^{210}Pb within a lake by specific chemical associations with certain sediment types and by physical transport of sediments. In addition, supply of ^{210}Pb-enriched top soil can influence the standing crop in small lakes. Consequently most natural lakes do not provide a reliable estimate of the atmospheric ^{210}Pb flux. Table 5 shows examples of these perturbations with time and location. Very few direct measurements of ^{210}Pb flux from atmospheric precipitation near the lakes studied are available. It is possible that shallow, man-made reservoirs that can be periodically drained for sampling purposes will be more reliable as ^{210}Pb flux monitors, as in the aquaculture ponds in Belgium studied by Petit (1974). Even here, however, there are complications. There has evidently been physical or biological mixing to depths as great as 30 cm to yield the observed ^{210}Pb distribution with depth in the core. The integrated ^{210}Pb in the cores yields the fluxes given in Table 5. These are higher than the atmospheric ^{210}Pb flux of Peirson, Cambray & Spicer (1966) for England (see Table 3), but not incompatible with the value predicted by the global model discussed below.

SOIL PROFILES   Benninger, Lewis & Turekian (1975) showed that undisturbed soil profiles can be treated as good repositories for atmospherically supplied ^{210}Pb. Most of the ^{210}Pb is trapped and preserved with a residence time of thousands of years in the organic-rich top soil. Very little percolates into the ground water regime or flows off into stream channels. On this basis it is evident that a soil profile measured for ^{210}Pb and ^{226}Ra should show the standing crop of excess ^{210}Pb. This was first done by Fisenne (1968) for an undisturbed soil profile in Maryland. In addition to ^{210}Pb supply, there is ^{222}Rn escape from the soil so that the actual level of ^{210}Pb support is less than the measured ^{226}Ra activity. For soils of the northeastern USA this causes the atmospheric ^{210}Pb flux to be under-estimated by only about 10%. It is possible to obtain both the radon flux out of the soil and the true ^{210}Pb atmospheric flux if the radon-release properties of the soil are known or estimated. The data listed in Table 5 are the only ones to date, but a good regional distribution pattern of ^{210}Pb atmospheric flux can clearly be obtained by this method to complement the more tedious, continuous, direct measurement of the ^{210}Pb concentration in atmospheric precipitation.

SALT MARSHES   Most of the ^{210}Pb in salt marshes is derived from the atmosphere (to which it is exposed most of the time), as it has been shown that neither coastal sea water (Benninger 1976) nor streams (Rama, Koide & Goldberg 1961, Lewis 1976) have significant ^{210}Pb to contribute. A marsh should therefore differ little from an upland meadow as a collector of atmospheric ^{210}Pb, and salt-marsh sediments should provide a good reservoir for estimating the atmospheric flux. McCaffrey & Thomson (1974) have done this in a Connecticut salt marsh (Table 5). Comparison of their flux value with flux estimates from an adjacent soil profile (McCaffrey 1977) and a nearby precipitation collector (Benninger 1976) indicates that salt marshes are indeed good detectors for ^{210}Pb atmospheric flux.

Is There a Man-Made Lead-210 Flux?

If a fission or thermonuclear device contains lead as a structural material, the intense neutron flux on detonation will produce ^{210}Pb by the reaction ^{208}Pb$(2n, \gamma)^{210}$Pb (Peirson, Cambray & Spicer 1966). That at least some nuclear devices contain lead can probably be inferred from the fact that James & Fleming (1966) ranked ^{210}Pb and its daughters only behind ^{90}Sr as a threat to human health should nuclear explosives be used in excavating a new Panama Canal. Thus it is probable, a priori, that some of the nuclear tests introduced artificial ^{210}Pb into the atmosphere. The magnitude and location of such injections has been debated.

Stebbins (1961) suggested that some of the ^{210}Pb in the stratosphere was injected by equatorial bomb tests. Peirson, Cambray & Spicer (1966) invoked a 30–50% contribution of stratospherically aged artificial ^{210}Pb to surface-air concentrations at Chilton, England to explain ^{210}Po/^{210}Pb activity ratios of about 0.3 in the early months of 1962 and 1963. Jaworowski (1966) suggested that temporal increases in ^{210}Pb concentrations in deer antlers, lichens, mountain-glacier ice, and human bones between the 1950s and the 1960s might be due to ^{210}Pb from bombs, the artificial contribution having reached, by his calculation, 150% of the natural in 1962. Krey (1967) interpreted nearly simultaneous peaks in ^{90}Sr and ^{210}Pb concentrations in the upper stratosphere of the Northern Hemisphere during 1966 as probably having resulted from the 1961–1962 bomb test-series; the ^{210}Pb concentrations were as much as an order of magnitude above levels considered normal. Baltakmens (1969) ascribed high concentrations of ^{210}Pb in milk produced in New Zealand during July–August 1965 to combined natural and artificial ^{210}Pb fallout.

Against these indications of artificial ^{210}Pb production in nuclear testing, Bhandari, Lal & Rama (1966) interpreted the absence of high concentrations of ^{210}Pb in the stratosphere at high and mid-latitudes during 1959–1961 as implying no appreciable bomb production during this period. The Project Stardust analyses of Feely et al (1965) should no longer be taken as corroborative of artificial ^{210}Pb in the stratosphere, as Feely & Seitz (1970) concluded, upon re-analysis of some of the samples in 1968, that the earlier results were high because of incomplete separation of fission products. The results of the redeterminations were similar to levels considered natural. Crozaz (1967) found no correlation between ^{90}Sr and ^{210}Pb in recent glacier ice from Antarctica and the Tyrolean Alps. Persson (1970) found no evidence of enhanced ^{210}Pb deposition in lichens collected at 62°N in Sweden during 1961–1969. The data of Dodge & Thomson (1974) on ^{210}Pb in annual layers in a coral sample from Bermuda show no effect from the major atmospheric test series. Since the mean residence time of ^{210}Pb in ocean surface waters is less than two years (Nozaki, Thomson & Turekian 1976), the corals probably should have detected any significant change in the atmospheric ^{210}Pb flux.

Analysis of environmental samples from test sites should establish whether particular test series produced significant ^{210}Pb. Beasley (1969) found concentrations of ^{210}Pb in sediments and soils from Bikini and Eniwetok atolls that did not exceed concentrations expected naturally, although he interpreted the presence of ^{207}Bi [produced ^{207}Pb$(p, n)^{207}$Bi or ^{206}Pb$(p, \gamma)^{207}$Bi] as evidence of the presence of stable lead at the test site. Noshkin et al (1975) have since shown, by analysis of annual bands from a coral sampled in Bikini lagoon, that small amounts of artificial ^{210}Pb probably were produced in the nuclear tests of 1954, 1956, and 1958. While these observations suggest that nuclear testing at Bikini and Eniwetok did not inject massive quantities of artificial ^{210}Pb into the atmosphere, Beasley (1969) correctly limits this conclusion to those test programs. He points out, however, that the absence of reported ^{207}Bi in worldwide fallout makes significant ^{210}Pb production in bomb tests doubtful.

THE MEAN RESIDENCE TIME OF AEROSOLS

Radon as a noble gas remains part of the gaseous atmosphere until it decays radioactively. The radioactive daughters are chemically reactive and are quickly associated with the aerosol burden of the atmosphere. As part of the aerosol burden they are subject to transfer to the earth's surface by rain and snow and dry deposition. Since this all is going on as the radioactive daughters themselves are undergoing radioactive decay, the degree of disequilibrium among the daughters provides a measure of the survival time of the particles (with which the radio-nuclides are associated) in the air column relative to removal to the surface.

There are several major assumptions in this approach:

1. The short-lived immediate progeny of ^{222}Rn (^{218}Po, ^{214}Pb, ^{214}Bi) have half-lives of the order of minutes and can be presumed to be predominantly converted to ^{210}Pb before any major removal occurs. Disequilibria amongst these nuclides have actually been observed in air close to the ground (Shapiro & Forbes-Resha 1975), but generally their activities are in equilibrium with ^{222}Rn throughout most of the atmosphere.
2. The long-lived daughters, ^{210}Pb, ^{210}Bi, and ^{210}Po, produced from atmospheric ^{222}Rn are associated with the same types of aerosols, and these aerosols are typical of the predominance of aerosol particles in the air. Martell & Moore (1974) report that 90% of the ^{210}Pb and ^{210}Bi activity in aerosols is associated with particles having diameters equal to or less than 0.3 μm. Residence time calculations based on ^{210}Pb and its daughters are thus longer than for chemical species associated with larger-diameter particles.
3. The source of ^{210}Pb, ^{210}Bi, and ^{210}Po is atmospheric radon and represents a single air mass without mixing with other air masses. If any of these nuclides is supplied by an independent source it must be accounted for. If more than one air mass is involved, a method of isolating each must be available.
4. The removal follows a first-order law. That is, the rate of removal is proportional to the number of atoms per unit volume of air. This assumes that all the processes of association of the nuclides being generated by radioactive decay with aerosols act to simulate a first-order process. The actual mechanisms must be much more complex than this simple law implies.

The Box Model

The simplest way of treating the data obtained by analyzing air samples is to assume that an analysis or set of analyses is a representative sample of a well-mixed box into which radon is supplied continuously. If the radon flux into the box is sustained it is possible to use the ratio of radon (or its short lived progeny) to ^{210}Pb to obtain a residence time by the relation:

$$\lambda_{Rn} N_{Rn} = \lambda_{Pb} N_{Pb} + \lambda_R N_{Pb}. \qquad (2)$$

The mean residence time then becomes:

$$\tau_R = \frac{1}{\lambda_R} = \frac{1}{\lambda_{Pb}} \frac{A_{Pb}/A_{Rn}}{(1 - A_{Pb}/A_{Rn})},$$ (3)

where τ_R is the mean residence time of ^{210}Pb relative to removal from the atmosphere to the ground, and A_{Pb} and A_{Rn} are the activities of ^{210}Pb and ^{222}Rn respectively in a unit volume of atmosphere. Some of the earliest estimates of τ_R were made this way and yielded values ranging from about one to nine days (Blifford, Lockhart & Rosenstock 1952, Haxel & Schumann 1955, Lehmann & Sittkus 1959, Rangarajan et al 1975). Such a calculation has the intrinsic limitation that a fairly constant radon flux must be assumed.

Much effort has been expended on measuring the activity ratios ^{210}Po/^{210}Pb or ^{210}Bi/^{210}Pb, or both, in the atmosphere or in rains to arrive at residence times. In principle this should be less dependent on short time-scale changes in radon flux.

The advantage of using the ^{210}Pb-^{210}Po pair is that ^{210}Po has a sufficiently long half-life to be less sensitive to lag times between collection and analysis, thus permitting measurements to be made more easily in remote areas. The advantage of the ^{210}Pb-^{210}Bi pair is that with the five-day half-life of ^{210}Bi, it is less sensitive to extraneous sources than is ^{210}Po.

Most of the current debate about aerosol residence times, aerosols supplied from the earth's surface, and mixing of air masses (including stratospheric) has arisen because of the discrepancies that have been observed between residence times obtained from the ^{210}Pb-^{210}Po couple and the ^{210}Pb-^{210}Bi couple. Where such a discrepancy exists, the ^{210}Pb-^{210}Bi residence time is always lower than the ^{210}Pb-^{210}Po residence time.

It is not tractable to consider the general equation accounting for all of these components. Instead we will proceed to increasing levels of complexity with an evaluation of the methods of simplification used by the various workers in the field.

THE LEAD-210: BISMUTH-210 COUPLE   The simplest relationship is based on the direct couple ^{210}Pb-^{210}Bi. This was first done by Fry & Menon (1962), who arrived at an aerosol mean residence time of 6.6 days in Arkansas. For this couple in a single air mass assuming steady state we have

$$\tau_R \, (\text{Bi-Pb}) = \frac{1}{\lambda_{Bi}} \times \frac{A_{Bi}/A_{Pb}}{(1 - A_{Bi}/A_{Pb})}.$$ (4)

There are two basic problems associated with the use of a direct measure of the activity ratio of ^{210}Bi/^{210}Pb found in a surface-air sample. The first is that if the air mass has suddenly arrived at the surface from some other environment, the measurement indicates the residence time in the recently evacuated reservoir and not the tropospheric box that is expected from ambient conditions. Such a situation can exist when stratospheric air penetrates and dominates the normal tropospheric column as a result of folding of the tropopause or at a storm front. A long residence time determined from the ^{210}Bi/^{210}Pb activity ratio, then, is due to the stratospheric residence time and not the tropospheric residence time. In principle such an event could be identified by the increased ^{90}Sr content of the air with a characteristic ^{90}Sr/^{210}Pb activity ratio of 130 (Krey & Krajewski 1969–1971). Poet, Moore & Martell (1972) use this ratio and the assumption of a sufficiently long stratospheric residence time to permit secular equilibrium to be established for ^{210}Bi, and correct their data accordingly.

The second problem is the supply of ^{210}Pb and ^{210}Bi from the earth's surface. By this is meant any aerosol supplied by the elevation of soil material and fine components of regoliths, and by the emanations from plants. Even the oceanic surface has been suggested as a source of ^{210}Pb, ^{210}Bi, and ^{210}Po.

Rates of supply of ^{210}Pb and its progeny from the earth's surface are difficult to assess. One method has been to use the ^{210}Pb/^{226}Ra activity ratio as the monitor. Unfortunately, different parts of a soil profile have different activity ratios of ^{210}Pb/^{226}Ra, and it would be important to know which part is the dominant supplier of aerosols locally. We have virtually no information on the ratio of ^{210}Po and ^{210}Bi supply relative to ^{226}Ra, thus making the appropriate correction difficult.

As we shall see, there is good evidence that most of the ^{210}Po is from sources on the earth's surface. The effect on the residence time calculated from the ^{210}Pb-^{210}Bi couple is not greatly affected even if most of ^{210}Po, in equilibrium with ^{210}Pb and ^{210}Bi, is from the earth's surface.

THE LEAD-210: POLONIUM-210 COUPLE   Because of the strong time constraints in the analysis of the relatively short-lived ^{210}Bi (five-day half-life) the use of the ^{210}Pb-^{210}Po couple has been suggested as an alternative way of arriving at residence times (Lehmann & Sittkus 1959; Burton & Stewart 1960, Lambert & Nezami 1965a; Peirson, Cambray & Spicer 1966, Francis, Chesters & Haskin 1970, Gavini, Beck & Kuroda 1974).

The equation for the Po-Pb couple is:

$$\tau_R \,(\text{Po-Pb}) = [-b+(b^2-4ac)^{1/2}]/2a, \tag{5}$$

where $a = A_{\text{Pb}} - A_{\text{Po}}$; $b = -A_{\text{Po}}(1/\lambda_{\text{Bi}} + 1/\lambda_{\text{Po}})$; and $c = -A_{\text{Po}}/\lambda_{\text{Bi}}\lambda_{\text{Po}}$.

It should be noted that although τ_R (Po-Pb) accounts for the presence of ^{210}Bi between ^{210}Pb and ^{210}Po in the decay chain, it does not refer in any way to a measured value of ^{210}Bi activity.

COMPARISONS OF RESIDENCE TIMES   Comparison of τ_R (Bi-Pb) and τ_R (Po-Pb) from the same samples of air demonstrates that they are commonly discordant (Table 6), with τ_R (Bi-Pb) shorter than τ_R (Po-Pb). Where agreement is found between the two, the "true" aerosol residence time of the air mass is indicated. This does not mean, however, even if the sample is taken at the base of the troposphere, that the air mass necessarily has its most recent origin in the troposphere only. Indeed the ideal test for the origin of the air mass is the ratio of ^{90}Sr/^{210}Pb. A value of about 130 is indicative of stratospheric air, and a value of about 1.5 is indicative of tropospheric air (Moore, Poet & Martell 1973). Unfortunately, many of the

Table 6   Aerosol residence times based on ^{210}Pb, ^{210}Bi, and ^{210}Po in air and rain samples

Collection date & time	Mean residence time (days)		
	τ_R(Bi-Pb)	τ_R(Po-Pb)	τ'_R

I. Boulder, Colorado surface air (Poet, Moore & Martell 1972)

Sept. 5, 1967	4.7 ± 0.2	44 ± 3	2.9 ± 0.3
Sept. 6, 1967	4.7 ± 0.4	19 ± 1	4.0 ± 0.6
Jan. 9, 1968	5.4 ± 0.2	24 ± 3	4.5 ± 0.3
April 2, 1969	10.9 ± 0.5	77 ± 8	6.5 ± 0.8
June 4 to 6, 1969			
1400–2000	4.1 ± 0.4	17 ± 2	3.5 ± 0.6
2000–0200	3.7 ± 0.3	14 ± 2	3.2 ± 0.5
0200–0800	3.0 ± 0.4	11 ± 2	2.7 ± 0.7
0800–1400	5.1 ± 0.4	13 ± 1	4.8 ± 0.5
1400–2000	7.4 ± 0.6	20 ± 2	4.4 ± 0.6
2000–0200	6.3 ± 0.6	24 ± 2	5.4 ± 0.7
0200–0800	13.0 ± 2.0	30 ± 2	12.0 ± 2.0
0800–1400	4.6 ± 0.5	33 ± 3	3.3 ± 0.7
Nov. 3, 1970	8.6 ± 0.6	15 ± 1	8.2 ± 0.5
Nov. 16, 1970	3.3 ± 0.8	15 ± 1	2.8 ± 1.2
Dec. 14, 1970	4.1 ± 0.1	38 ± 3	2.6 ± 0.2
Feb. 13, 1971	3.3 ± 0.1	11 ± 1	3.0 ± 0.2
Feb. 16, 1971	5.0 ± 0.4	17 ± 1	4.4 ± 0.5
May 17, 1971	4.8 ± 0.6	19 ± 1	4.1 ± 0.8
Dec. 27, 1971	1.6 ± 0.1	11 ± 1	1.3 ± 0.3
Jan. 2, 1972	4.6 ± 0.9	27 ± 1	3.6 ± 1.1

II. Fayetteville, Arkansas 1973 rain (Gavini, Beck & Kuroda 1974)

April 9	$26 \rightarrow \infty$	35 ± 5	—
April 16	3^{+2}_{-1}	29 ± 4	—
April 25	6^{+4}_{-3}	40 ± 6	—
May 1	$53 \rightarrow \infty$	29^{+2}_{-3}	—
May 5, 6	$34 \rightarrow \infty$	31^{+3}_{-2}	—
May 11	$36^{+\infty}_{-17}$	48^{+7}_{-6}	—
May 21	20^{+10}_{-5}	29^{+2}_{-3}	—
May 23	27^{+68}_{-13}	52 ± 9	—
May 24	$29 \rightarrow \infty$	29^{+2}_{-3}	

Table 6   (*continued*)

Collection date & time	Mean residence time (days)		
	τ_R(Bi-Pb)	τ_R(Po-Pb)	τ'_R
May 26	$15\,^{+15}_{-6}$	16 ± 3	—
May 28	$28 \to \infty$	$31\,^{+6}_{-5}$	—
May 31	5 ± 1	26 ± 3	—
June 1	$53\,^{+240}_{-27}$	$61\,^{+11}_{-9}$	—
June 5	6 ± 2	$16\,^{+2}_{-3}$	—
June 12	$240 \to \infty$	43 ± 6	—
June 13	$7\,^{+3}_{-2}$	26 ± 3	—
June 16	$95\,^{+\infty}_{-62}$	40 ± 6	—

samples do not have this information, thus leaving open the question of whether the conformable residence times are true tropospheric residence times or represent stratospheric air masses with longer residence times.

We can simplify the problem of the meaning of old τ_R (Po-Pb) in tropospheric air by considering either the stratospheric mixing effect or the effect of earth-surface supply of ^{210}Po. We have already discussed the method of assessing the role of aged (mainly stratospheric) air in yielding an old concordant residence time. We now have to consider the effect of earth-surface-derived ^{210}Po accompanied by ^{210}Pb, ^{210}Bi, and the higher members of the uranium decay series, including uranium itself.

Vilenskii (1970) was the first to observe that ^{210}Po derived from ground sources could seriously alter the ^{210}Po/^{210}Pb activity ratio in the air and thus the calculated τ_R (Po-Pb). He argued that dust from the ground was the most likely source of this perturbing fraction and that ^{226}Ra in atmospheric samples might be used to monitor it.

If we choose this method of correcting for the addition of ^{210}Pb and ^{210}Po to the atmosphere, then we must make two assumptions. First, we must assume that ^{210}Po is in equilibrium with ^{210}Pb in the source and, second, that we know the ^{210}Pb/^{226}Ra activity of the source. As we shall see, neither of these assumptions is firmly grounded.

Table 7 shows the ^{210}Pb/^{226}Ra in several soil profiles. The highest values occur in the top soil (or A-1 horizon) and the value approaches secular equilibrium with depth. It is not possible to determine what the actual material supplied to the atmosphere is. Indeed one would expect it to vary with location and season.

Table 7 ^{210}Pb/^{226}Ra activity ratios in some soil profiles

	% organic matter	^{210}Pb/^{226}Ra activity ratio
Hubbard Brook, New Hampshire, USA (Kharkar et al 1974)		
A$_1$ Horizon (#1)	71	69
A$_1$ Horizon (#2)	75	68
A$_2$ Horizon	0	0.76
B Horizon	14	1.0
Cook Forest State Park, Pennsylvania, USA (Lewis 1976)		
	88	88
	58	39
	17.5	9.4
	3.8	1.9
Emporium, Pennsylvania, USA (Lewis 1976)		
	87	50
	45	25
	7.5	2.8
	2.8	1.3
	3.9	1.0
Moscow, USSR (Vilenskii 1970)		
Sandy (0–4 cm)	—	25
Loamy (0–5 cm)	—	3.8
Loamy (0–5 cm)	—	3.1
"Surface" soils (0–1 cm) in the western US (Moore & Poet 1976)		
Grand Junction, Colorado	—	2.7
Cripple Creek, Colorado	—	2.6
Goodland, Kansas	—	5.8
Loveland, Colorado	—	2.2
Brighton, Colorado	—	1.7
Golden, Colorado	—	2.6
Ralston Reservoir, Colorado	—	4.6
Boulder, Colorado	—	3.6
Big Spring, Texas	—	4.2

Although the method of correction using ^{226}Ra seems attractive, it clearly is not as feasible a method as was thought earlier.

The method used by Poet, Moore & Martell (1972) does not use the ^{226}Ra tag method but does assume that ^{210}Pb, ^{210}Bi, and ^{210}Po are in secular equilibrium in the material supplied to the atmosphere from the ground. The relevant equations are:

$$A_{Pb} + R_{Bi} = A_{Bi}\left(1 + \frac{1}{\tau'_R \lambda_{Bi}}\right) \tag{6}$$

and

$$A_{Bi} + R_{Po} = A_{Po}\left(1 + \frac{1}{\tau'_R \lambda_{Po}}\right) \tag{7}$$

where R_{Bi} and R_{Po} are the earth-surface rates of supply of ^{210}Bi and ^{210}Po and τ'_R is the corrected mean residence time. If the ^{210}Bi and ^{210}Po are in secular equilibrium with ^{210}Pb, then

$$R_{Bi} = \frac{\lambda_{Po}}{\lambda_{Bi}} R_{Po}. \tag{8}$$

On this basis, the calculated mean residence time based on the measured activities of ^{210}Pb, ^{210}Bi, and ^{210}Po in the air becomes

$$\tau'_R = \frac{A_{Bi} - A_{Po}}{\lambda_{Bi}(A_{Pb} - A_{Bi}) - \lambda_{Po}(A_{Bi} - A_{Po})} \tag{9}$$

Table 6 lists the values of τ'_R for a number of air samples that show disparate $\tau_R(Bi-Pb)$ and $\tau_R(Po-Pb)$. Using τ'_R it is possible to partition the contributions of ^{210}Pb, ^{210}Bi, and ^{210}Po from atmospheric radon and earth-surface sources. They are modeled for an earth-surface component that is in equilibrium with ^{210}Pb. On this basis an average of about 87% of the ^{210}Po is assignable to the earth-surface component.

Of course, there is no certainty that the earth-surface component is in equilibrium or indeed if the $^{210}Pb/^{210}Bi/^{210}Po$ activity proportions are the same at various places on the earth. For example, the $^{210}Po/^{210}Pb$ activity ratio can be greater than one if certain marine sources are involved (Turekian, Kharkar & Thomson 1974, Lambert, Sanak & Ardouin 1974), or it can be less than one in films associated with vegetation (Moore, Martell & Poet 1976).

Since the ^{210}Bi-^{210}Pb couple is least subject to these uncertainties, it appears to provide the best estimate of aerosol residence times, as propounded by Martell and his co-workers, if steady state can be assumed. We will show in the next section that the steady-state condition is rarely met.

Vertical Diffusion Models

Under steady-state conditions where a constant flux of ^{222}Rn is supplied from the surface, it is possible to calculate the distribution of ^{222}Rn and its long-lived daughters with height in the troposphere. The equations and a discussion of the values of the appropriate diffusion constants have been detailed by Jacobi & Andre (1963). Moore, Poet & Martell (1973) have applied this model to a set of data from the central United States obtained by themselves and others for ^{222}Rn, and by themselves for the long-lived daughters of ^{222}Rn. The complication of stratospheric diffusion at the top of the troposphere and the supply from the earth's surface of a particulate burden with a characteristic ^{210}Pb, ^{210}Bi, ^{210}Po complexion, makes the fit of actual data less perfect than the simple model requires. The ^{210}Pb profile is best fit when a $\tau_R = 7.2$ day is used (Figure 4). This is a mean value and indicates the possibility of differences in residence time with height in the troposphere. The upper troposphere mean residence time appears to be less than a factor of three greater than the lower troposphere value (Moore, Poet & Martell 1973). In the next section, the question of whether the steady-state model is applicable in the central United States is discussed.

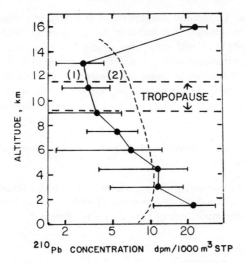

Figure 4   Observed (Moore, Poet & Martell 1973) ^{210}Pb concentrations (1) as a function of altitude over west-central United States (average of 12 profiles—error bars represent one standard deviation). The dashed line (2) is the distribution expected using the Jacobi & Andre (1963) model with an average ^{222}Rn vertical distribution and eddy diffusion coefficients variable with elevation.

GLOBAL MODELS FOR RADON AND ITS DAUGHTERS

From the previous section it can be seen that most mean residence time calculations using ^{222}Rn and its daughters ^{210}Pb, ^{210}Bi, and ^{210}Po are dependent on a steady-state assumption. As Moore, Poet & Martell (1973) point out, some of the disparities in mean residence time calculations can be ascribed to the extent to which steady state is approached at various points on the earth's surface.

The Lambert-Nezami Model

The steady-state assumption is implicit in the global model of Lambert & Nezami (1965b), who attempted to estimate the zonal ^{210}Pb flux. Having very few reliable global ^{210}Pb flux data, the worldwide zonal values were deduced in the following way. The total ^{210}Pb flux and the dry deposition component were determined for Paris. It was then assumed: (*a*) that ^{210}Pb deposition by dry deposition (found to account for 50% of the total flux at Paris), contributes elsewhere in proportion to the local concentration of ^{210}Pb in ground-level air, in effect using an average dry deposition velocity; and (*b*) that the concentration of ^{210}Pb in rainfall has the same distribution over latitude as do the concentrations of artificial radioactivity from nuclear testing.

Regarding the first assumption, the dry deposition value at any location is a complex function of vegetation and climate and thus must have a large uncertainty associated with it. The second assumption is evidently very weak, inasmuch as fission products are globally distributed in the stratosphere, while ^{222}Rn and its daughters are chiefly circulated in the troposphere. A consequence of these assumptions is that the maximum deposition of ^{210}Pb is expected at 50°N, where Lambert & Nezami estimated a flux of about 0.5 dpm cm^{-2} yr^{-1} on the basis of their measurements in Paris. As we have seen, this low value is almost completely restricted to western Europe. The model of Lambert & Nezami, however, was probably the best for its time.

A Model Based on Large-Scale Atmospheric Circulation and Aerosol Residence Times

We propose a self-consistent global model for ^{222}Rn and its long-lived daughters, taking into account the sharp discontinuity of the ^{222}Rn flux between continents and oceans, and the importance of both air transit time and mean residence time of aerosols.

In order to proceed, we assume the following:

1. The measured ^{210}Pb fluxes between about 15° and 55° in latitude for both hemispheres are the primary data.
2. The air movement between these latitudes is predominantly from west to east and the importance of diffusion is negligibly small, relative to the advective term (Assaf 1969). It is further assumed that the transit time is an averaged one applicable to the one-dimensional model.
3. The mean residence time of aerosols is the same all over the earth.
4. The radon emissivity from the ocean surface is negligibly small (Table 1), and that from the continental surface is an average value independent of geographic or climatic variations.

Considering one-dimensional transport of ^{222}Rn and ^{210}Pb by westerlies, the following equations are then set up:

$$\frac{\partial N_{\text{Rn}}}{\partial t} = -U \frac{\partial N_{\text{Rn}}}{\partial x} + E - \lambda_{\text{Rn}} N_{\text{Rn}}; \tag{10}$$

$$\frac{\partial N_{\text{Pb}}}{\partial t} = -U \frac{\partial N_{\text{Pb}}}{\partial x} + \lambda_{\text{Rn}} N_{\text{Rn}} - (\lambda_{\text{Pb}} + \lambda_R) N_{\text{Pb}}; \tag{11}$$

$$\frac{\partial N_{\text{Bi}}}{\partial t} = -U \frac{\partial N_{\text{Bi}}}{\partial x} + \lambda_{\text{Pb}} N_{\text{Pb}} - (\lambda_{\text{Bi}} + \lambda_R) N_{\text{Bi}}; \tag{12}$$

where N is the amount of nuclide in the air column (atoms cm^{-2}), t is time, x is the distance, U is the mean wind velocity (cm sec^{-1}), E is the ^{222}Rn emanation rate (atom cm^{-2} sec^{-1}), λ_{Rn}, λ_{Pb}, and λ_{Bi} are the radioactive decay constants (sec^{-1}) and λ_R is the first-order removal rate constant (sec^{-1}) equivalent to the reciprocal of the aerosol mean residence time, τ_R. By replacement of x/U with the transit time of the air mass, T, and λN with the activity, A, we obtain the global steady-state solution of Equations 10, 11, and 12:

$$A_{\text{Rn}} = A_{\text{Rn}}^0 \exp\left(-\lambda_{\text{Rn}} T\right) + E\left[1 - \exp\left(-\lambda_{\text{Rn}} T\right)\right]; \tag{13}$$

$$A_{\text{Pb}} = A_{\text{Pb}}^0 \exp\left(-k_{\text{Pb}} T\right) + \frac{\lambda_{\text{Pb}}(A_{\text{Rn}}^0 - E)}{k_{\text{Pb}} - \lambda_{\text{Rn}}} \exp\left(-\lambda_{\text{Rn}} T\right) + \frac{\lambda_{\text{Pb}}(A_{\text{Rn}}^0 - E)}{\lambda_{\text{Rn}} - k_{\text{Pb}}} \exp\left(-k_{\text{Pb}} T\right)$$

$$+ \frac{\lambda_{\text{Pb}} E}{k_{\text{Pb}}}\left[1 - \exp\left(-k_{\text{Pb}} T\right)\right]; \tag{14}$$

and

$$A_{\text{Bi}} = A_{\text{Bi}}^0 \exp\left(-k_{\text{Bi}} T\right) + \frac{\lambda_{\text{Bi}} \lambda_{\text{Pb}}(A_{\text{Rn}}^0 - E)}{(k_{\text{Bi}} - \lambda_{\text{Rn}})(k_{\text{Pb}} - \lambda_{\text{Rn}})} \exp\left(-\lambda_{\text{Rn}} T\right)$$

$$+ \frac{\lambda_{\text{Bi}} \lambda_{\text{Pb}}(A_{\text{Rn}}^0 - E)}{(\lambda_{\text{Pb}} - \lambda_{\text{Bi}})(k_{\text{Pb}} - \lambda_{\text{Rn}})} \exp\left(-k_{\text{Pb}} T\right) + \frac{\lambda_{\text{Bi}} \lambda_{\text{Pb}}(A_{\text{Rn}}^0 - E)}{(\lambda_{\text{Bi}} - \lambda_{\text{Pb}})(k_{\text{Bi}} - \lambda_{\text{Rn}})} \exp\left(-k_{\text{Bi}} T\right)$$

$$+ \frac{\lambda_{\text{Bi}}(k_{\text{Pb}} A_{\text{Pb}}^0 - \lambda_{\text{Pb}} E)}{(\lambda_{\text{Bi}} - \lambda_{\text{Pb}})k_{\text{Pb}}}\left[\exp\left(-k_{\text{Pb}} T\right) - \exp\left(-k_{\text{Bi}} T\right)\right]$$

$$+ \frac{\lambda_{\text{Bi}} \lambda_{\text{Pb}} E}{k_{\text{Bi}} k_{\text{Pb}}}\left[1 - \exp\left(-k_{\text{Bi}} T\right)\right], \tag{15}$$

where $k_{\text{Pb}} = \lambda_{\text{Pb}} + \lambda_R$; $k_{\text{Bi}} = \lambda_{\text{Bi}} + \lambda_R$; and A_{Rn}^0, A_{Pb}^0, and A_{Bi}^0 are the activities of ^{222}Rn, ^{210}Pb, and ^{210}Bi at the boundaries (the coast lines), where the emissivity E of ^{222}Rn

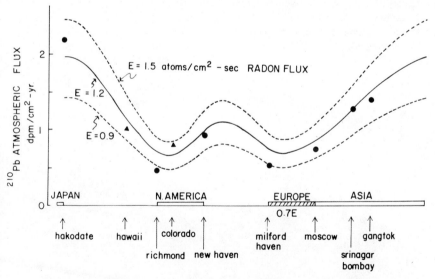

Figure 5   The longitudinal variation of ^{210}Pb flux in mid-Northern Hemisphere (15°–55° N). The curves were calculated from Equation 14 by setting $\Delta T = 12$ days (corresponding to $\tau = 5$ days). Circles and triangles show the primary data based on ^{210}Pb in long-term precipitation and air column, respectively. (Hawaii based on $\tau_R = 5$ days and Richmond using deposition velocity calculated from stable lead.)

changes sharply. Equations 13, 14, and 15 are applied for each continent or ocean mass proceeding from west to east. In the Northern Hemisphere they are the Pacific Ocean, North America, the Atlantic Ocean, Europe, and Asia. The emanation rate of ^{222}Rn from Europe is assumed to be 70% of the other continents because of the surrounding seas. In the Southern Hemisphere, both Australia and Africa are assigned a ^{222}Rn emanation rate 70% of the continental rate as an approximation to relative areas of land and sea between 15°S and 55°S. Since the westerlies circulate around the spherical earth, ^{222}Rn and its daughters must be continuous along the wind direction. Therefore, once the parameters, E, T, and λ_R are set, we can obtain A^0_{Rn}, A^0_{Pb}, and A^0_{Bi} at each boundary between continent and ocean and the distributions of Rn and its daughters simultaneously. On the basis of the systematics of radioactive decay, the radon emissivity must be balanced by the decay of ^{222}Rn in the atmosphere and ^{210}Pb deposition in the whole system. Consequently only the air mass transit time T and the aerosol mean residence time τ_R $(= 1/\lambda_R)$ are independent parameters in Equations 13, 14, and 15. In Figure 5, three modelled ^{210}Pb flux distributions are compared with the measured values. The best fit determines E, the specific continental emission rate of ^{222}Rn, to be 1.2 atoms cm^{-2} sec^{-1}. The value of ΔT, the transit time across the North Pacific Ocean (9000 km), and τ_R, the aerosol mean residence time, are coupled. In order to retain the ^{210}Pb flux distribution curve, any change in ΔT must be accompanied by a change in τ_R. To calculate the curves in Figure 5, a value of 12 days for ΔT was set, which seems reasonable on the basis of air trajectories determined by a variety of means; this corresponds to $\tau_R = 5$ days. If we double the transit time for the same distance to 24 days, then τ_R becomes 18 days. The application of the model to the Southern Hemisphere using the same constants

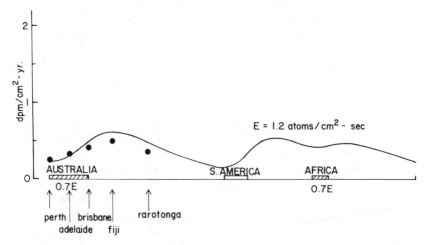

Figure 6 The longitudinal variation of ^{210}Pb flux in mid-Southern Hemisphere (15°–55° S). The curves were calculated by using the same time parameters as Figure 5 and $E = 1.2$ atoms cm^{-2} sec^{-1}. Circles are the primary data based on ^{210}Pb in long-term precipitation.

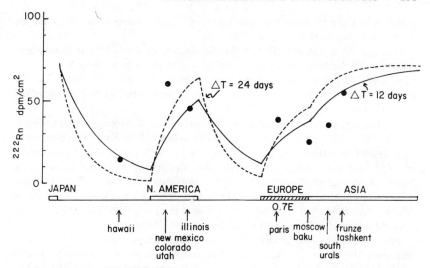

Figure 7   The distribution of ^{222}Rn in the atmosphere of the mid-Northern Hemisphere for $E = 1.2$ atoms cm^{-2} sec^{-1} (Equation 13): solid line, $\Delta T = 12$ days; dashed line, $\Delta T = 24$ days. Circles show observed data. (North American data summarized in Moore, Poet & Martell 1973).

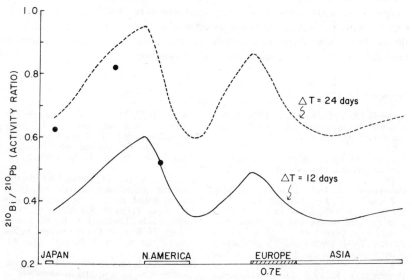

Figure 8   The longitudinal variation of ^{210}Bi/^{210}Pb activity ratio in the atmosphere of the mid-Northern Hemisphere. The solid line was calculated by using $\Delta T = 12$ days and $\tau = 5$ days and the dashed line by $\Delta T = 24$ days and $\tau_R = 18$ days. The data in the central United States (Moore, Poet & Martell 1973) in Hawaii (Moore et al 1974) and Japan (Tsunogai & Fukuda 1974) are shown by circles.

is shown in Figure 6. Since the distribution of ^{222}Rn is independent of λ_R, the air transit time may be better set by ^{222}Rn data.

Figure 7 shows the modelled ^{222}Rn distribution in the Northern Hemisphere together with the measured values. Although the fit seems to be better for $\Delta T = 12$ days than the longer values, there is some ambiguity, especially based on the measurements in the Rocky Mountains, which appear to be higher than any of the predicted values.

One consequence of the calculations is the prediction of the longitudinal variation of the ^{210}Bi/^{210}Pb activity ratio. This is shown in Figure 8 for both time parameters. Two important features of Figure 8 should be noted. The first is that even with a constant real aerosol mean residence time, the apparent mean residence time based on Equations 4 or 9 by a simple application of the ^{210}Bi/^{210}Pb ratio varies considerably. The second point is that the USA western plains data of Moore, Poet & Martell (1973) are compatible with $\tau_R = 5$ days, but their Hawaiian data (Moore et al 1974) seem to be consistent with a longer τ_R.

This one-dimensional model cannot adequately accommodate the true three-dimensional nature of the atmosphere, but we believe it provides a base for further work. Specifically the model suggests that, over time, good data for ^{222}Rn and its daughters in the air column on the east coasts of North America and Asia will most strongly constrain the parameters of a model. Secondly, the two-dimensional solution should be based on ^{210}Pb fluxes on the continents. It appears to us that the most conservative way of going about this is to measure old soil profiles. Although secondary problems involving vegetation effects will have to be assayed, soils do provide a unique long-term collector of both dry and wet material from the air. The yearly differences are thus accommodated, and the debate about the importance of dry and wet fallout can be short-circuited.

Literature Cited

Anderson, R. V., Larson, R. E. 1974. Atmospheric electricity and radon profiles over a closed basin and the open ocean. *J. Geophys. Res.* 79:3432–35

Assaf, G. 1969. Emanated products as a probe of atmospheric transport. *Tellus* 21:820–28

Baltakmens, T. 1969. Lead-210 in New Zealand milk. *Aust. J. Sci.* 31:111–14

Baranov, V. I., Vilenskii, V. D. 1965. Pb-210 in the atmosphere and in fallout. *Sov. At. Energy* 18:645–48

Beasley, T. M. 1969. Lead-210 production by nuclear devices: 1946–1958. *Nature* 224:573

Benninger, L. K. 1976. *The uranium-series radionuclides as tracers of geochemical processes in Long Island Sound.* PhD thesis. Yale Univ., New Haven, Conn. 151 pp.

Benninger, L. K., Lewis, D. M., Turekian,

K. K. 1975. The use of natural Pb-210 as a heavy metal tracer in the river-estuarine system. In *Marine Chemistry in the Coastal Environment, Am. Chem. Soc. Symp. Ser. 18,* ed. T. M. Church, pp. 201–10.

Bhandari, N., Lal, D., Rama. 1966. Stratospheric circulation studies based on natural and artificial radioactive tracer elements. *Tellus* 18:391–406

Blifford, I. H., Lockhart, L. B., Jr., Rosenstock, H. B. 1952. On the natural radioactivity in the air. *J. Geophys. Res.* 57:499–509

Bonnyman, J., Molina-Ramos, J. 1971. Concentrations of lead-210 in rainwater in Australia during the years 1964–1970. *Tech. Rep. CXRL/7, Commonw. X-Ray Radium Lab.,* Melbourne.

Brugam, R. B. 1975. *The human disturbance history of Linsley Pond, North Branford,*

Connecticut. PhD thesis. Yale Univ., New Haven, Conn. 184 pp.

Burton, W. M., Stewart, N. G. 1960. Use of long-lived natural radioactivity as an atmospheric tracer. *Nature* 186:584–89

Chamberlain, A. C. 1960. Aspects of the deposition of radioactive and other gases and particles. *Int. J. Air Pollut.* 3:63–88

Clements, W. E., Wilkening, M. H. 1974. Atmospheric pressure effects on ^{222}Rn transport across the earth-air interface. *J. Geophys. Res.* 79:5025–29

Crozaz, G. 1967. Datation des glaciers par le plomb-210. *Proc. Symp. Radioact. Dating and Methods of Low-level Counting, Monaco, 1967,* pp. 385–92. Vienna: Int. Atomic Energy Agency

Crozaz, G., Langway, C. C. Jr. 1966. Dating Greenland firn-ice cores with Pb-210. *Earth Planet. Sci. Lett.* 1:194–96

Dodge, R. E., Thomson, J. 1974. The natural radiochemical and growth records in contemporary hermatypic corals from the Atlantic and Caribbean. *Earth Planet. Sci. Lett.* 23:313–22

Feely, H. W., Friend, J. P., Krey, P. W., Russell, B. A. 1965. Flight data and results of radiochemical analyses of filter samples collected during 1961 and 1962. *HASL Tech. Rep. 153,* US At. Energy Comm. (now ERDA) Health and Safety Lab., N.Y.

Feely, H. W., Seitz, H. 1970. Use of lead-210 as a tracer of transport processes in the stratosphere. *J. Geophys. Res.* 75:2885–94

Fisenne, I. M. 1968. Distributions of lead-210 and radium-226 in soil. *Rep. UCRL-18140,* pp. 145–58. Washington, D. C.: US At. Energy Comm. (now ERDA)

Francis, C. W., Chesters, G., Haskin, L. A. 1970. Determination of ^{210}Pb mean residence time in the atmosphere. *Environ. Sci. Technol.* 4:586–89

Fry, L. M., Menon, K. K. 1962. Determination of the tropospheric residence time of lead-210. *Science* 137:994–95

Fukuda, K., Tsunogai, S. 1975. Pb-210 in precipitation in Japan and its implication for the transport of continental aerosols across the ocean. *Tellus* 27:514–21

Gavini, M. B., Beck, J. N., Kuroda, P. K. 1974. Mean residence times of the long-lived radon daughters in the atmosphere. *J. Geophys. Res.* 79:4447–52

Goldberg, E. D. 1963. Geochronology with lead-210. In *Radioactive Dating,* pp. 121–31. Vienna: Int. At. Energy Agency

Haxel, O., Schumann, G. 1955. Selbstreinigung der Atmosphäre. *Z. Physik* 142:127–32

Israël, H. 1951. Radioactivity of the atmo-

sphere. In *Compendium of Meteorology,* ed. T. F. Malone, pp. 155–61. Boston: Am. Meteorol. Soc. 1334 pp.

Israël, H., Horbert, M. 1970. Tracing atmospheric eddy mass transfer by means of natural radioactivity. *J. Geophys. Res.* 75:2291–97

Jacobi, W., Andre, K. 1963. The vertical distribution of radon 222, radon 220, and their decay products in the atmosphere. *J. Geophys. Res.* 68:3799–3814

James, R. A., Fleming, E. H. Jr. 1966. Relative significance index of radionuclides for canal studies. *Tech. Rep. UCRL-50050-1,* Lawrence Radiat. Lab., Univ. Calif., Livermore

Jaworowski, Z. 1966. Temporal and geographical distribution of radium D (lead-210). *Nature* 212:886–89

Joshi, L. U., Rangarajan, C., Gopalakrishnan, S. 1969. Measurement of lead-210 in surface air and precipitation. *Tellus* 21:107–12

Kauranen, P., Miettinen, J. K. 1969. ^{210}Po and ^{210}Pb in the arctic food chain and the natural radiation exposure of Lapps. *Health Phys.* 16:287–95

Kharkar, D. P., Thomson, J., Turekian, K. K., McCaffrey, R. J. 1974. The distribution of uranium and its decay series nuclides and trace metals in a Hubbard Brook, New Hampshire soil profile. In *AEC Tech. Rep. COO-3573-8.* Yale Univ., New Haven, Conn.

Kirichenko, L. V. 1970. Radon exhalation from vast areas according to vertical distribution of its short-lived decay products. *J. Geophys. Res.* 75:3539–49

Koide, M., Soutar, A., Goldberg, E. D. 1972. Marine geochronology with ^{210}Pb. *Earth Planet. Sci. Lett.* 14:442–46

Kosmath, W. 1935. Die Exhalation der Radiumemanation aus dem Erdboden und ihre Abhängigkeit von den meteorologischen Faktoren. *Beitr. Geophys.* 43:258–79

Kraner, H. W., Schroeder, G. L., Evans, R. D. 1964. Measurements of the effects of atmospheric variables on radon-222 flux and soil-gas concentrations. In *The Natural Radiation Environment,* ed. J. A. S. Adams, W. M. Lowder, pp. 191–215. Chicago: Univ. Chicago Press. 1069 pp.

Krey, P. W. 1967. Sr-90, Ce-144 and Pb-210 in the upper stratosphere. In *HASL-181.* US At. Energy Comm. (now ERDA) Health and Safety Lab., N.Y.

Krey, P. W., Krajewski, B. T. 1969–1971. In *HASL 207, 210, 217, 224, 239, 242.* US At. Energy Comm. (now ERDA) Health and Safety Lab., N.Y.

Krishnaswami, S., Lal, D., Martin, J. M., Meybeck, M. 1971. Geochronology of lake sediments. *Earth Planet. Sci. Lett.* 11: 407–14

Lambert, G., Nezami, M. 1965a. Determination of the mean residence time in the troposphere by measurement of the ratio between the concentration of lead 210 and polonium 210. *Nature* 206: 1343–44

Lambert, G., Nezami, M. 1965b. Importance des retombées sèches dans le bilan du plomb 210. *Ann. Géophys.* 21: 245–51

Lambert, G., Sanak, J., Ardouin, B. 1974. Origine marine des excès de polonium 210 dans la basse atmosphere Antarctique. *J. Rech. Atmos.* 8: 647–48

Lehmann, L., Sittkus, A. 1959. Bestimmung von Aerosolverweilzeiten aus dem RaD- und RaF—Gehalt der atmosphärischen Luft und des Niederschlags. *Naturwissenschaften* 46: 9–10

Lewis, D. M. 1976. *The geochemistry of manganese, iron, uranium, lead-210 and major ions in the Susquehanna River.* PhD thesis. Yale Univ. New Haven, Conn. 272 pp.

Martell, E. A., Moore, H. E. 1974. Tropospheric aerosol residence times: a critical review. *J. Rech. Atmos.* 8: 903–10

Matsumoto, E. 1975. ^{210}Pb geochronology of sediments from Lake Shinji. *Geochem. J.* 9: 167–72

Mattsson, R. 1970. Seasonal variation of short-lived radon progeny, Pb-210 and Po-210, in ground level air in Finland. *J. Geophys. Res.* 75: 1741–44

McCaffrey, R. J. 1977. *A record of the accumulation of sediment and trace metals in a Connecticut salt marsh.* PhD thesis. Yale Univ., New Haven, Conn. In preparation

McCaffrey, R. J., Thomson, J. 1974. A record of trace metal fluxes to a Connecticut salt marsh as determined by Pb^{210} dating. *AEC Tech. Rep. COO-3573-8* Yale Univ., New Haven, Conn.

Megumi, K., Mamuro, T. 1973. Radon and thoron exhalation from the ground. *J. Geophys. Res.* 78: 1804–8

Moore, H. E., Martell, E. A., Poet, S. E. 1976. Sources of polonium-210 in atmosphere. *Environ. Sci. Technol.* 10: 586–91

Moore, H. E., Poet, S. E. 1976. ^{210}Pb fluxes determined from ^{210}Pb and ^{226}Ra soil profiles. *J. Geophys. Res.* 81: 1056–58

Moore, H. E., Poet, S. E., Martell, E. A. 1973. ^{222}Rn, ^{210}Pb, ^{210}Bi, and ^{210}Po profiles and aerosol residence times versus altitude. *J. Geophys. Res.* 78: 7065–75

Moore, H. E., Poet, S. E., Martell, E. A., Wilkening, M. H. 1974. Origin of ^{222}Rn

and its long-lived daughters in air over Hawaii. *J. Geophys. Res.* 79: 5019–24

National Radiation Laboratory (New Zealand) 1968–1975. *Environ. Radioact. Ann. Rep., 1967–1974. Tech. Rep. NRL-F/ 28, NRL-F/33, NRL-F/38, NRL-F/43, NRL-F/48, NRL-F/50, NRL-F/52, NRL-F/54,* Nat. Radiat. Lab., Christchurch, New Zealand

Noshkin, V. E., Wong, K. M., Eagle, R. J., Gatrousis, C. 1975. Transuranics and other radionuclides in Bikini lagoon: concentration data retrieved from aged coral sections. *Limnol. Oceanogr.* 20: 729–42

Nozaki, Y., Thomson, J., Turekian, K. K. 1976. The distribution of ^{210}Pb and ^{210}Po in the surface waters of the Pacific Ocean. *Earth Planet. Sci. Lett.* 32: 304–12

Pearson, J. E., Jones, G. E. 1965. Emanation of radon-22 from soils and its use as a tracer. *J. Geophys. Res.* 70: 5279–90

Pearson, J. E., Jones, G. E. 1966. Soil concentrations of "emanating radium-226" and the emanation of radon-222 from soils and plants. *Tellus* 18: 655–61

Peirson, D. H., Cambray, R. S., Spicer, G. S. 1966. Lead-210 and polonium-210 in the atmosphere. *Tellus* 18: 427–33

Persson, B. R. 1970. ^{210}Pb-atmospheric deposition in lichen-carpets in northern Sweden during 1961–1969. *Tellus* 22: 564–71

Pensko, J., Wardaszko, T., Wochna, M. 1968. Natural atmospheric radioactivity and its dependence on some geophysical factors. *Atompraxis* 14: 255–58

Petit, D. 1974. ^{210}Pb et isotopes stables du plomb dans des sediments lacustres. *Earth Planet. Sci. Lett.* 23: 199–205

Picciotto, E., Cameron, R., Crozaz, G., Deutsch, S., Wilgain, S. 1968. Determination of the rate of snow accumulation at the pole of relative inaccessibility, eastern Antarctica: a comparison of glaciological and isotopic methods. *J. Glaciol.* 7: 273–87

Poet, S. E., Moore, H. E., Martell, E. A. 1972. Lead 210, bismuth 210, and polonium 210 in the atmosphere: accurate ratio measurement and application to aerosol residence time determination. *J. Geophys. Res.* 77: 6515–27

Rama, Koide, M., Goldberg, E. D. 1961. Lead-210 in natural waters. *Science* 134: 98–99

Rangarajan, C., Gopalakrishnan, S., Chandrasekaran, V. R., Eapen, C. D. 1975. The relative concentrations of radon daughter products in surface air and the significance of their ratios. *J. Geophys. Res.* 80: 845–48

Robbins, J. A., Edgington, D. M. 1975. Determination of recent sedimentation rates in Lake Michigan using Pb-210 and Cs-137. *Geochim. Cosmochim. Acta* 39: 285–304

Rosen, R. 1957. Note on some observations of radon and thoron exhalation from the ground. *N.Z. J. Sci. Technol.* 38: 644–54

Scholz, C., Sykes, L., Aggarwal, Y. 1973. Earthquake prediction: a physical basis. *Science* 181: 803–10

Servant, J. 1964. *Radon and its short-lived daughters in the lower atmosphere.* PhD. thesis. University of Paris.

Shapiro, M. H., Forbes-Resha, J. L. 1975. ^{214}Bi/^{214}Pb ratios in air at a height of 20 m. *J. Geophys. Res.* 80: 1605–13

Sisigina, T. I., 1967. Radon emanation from the surface of some types of soils of the European part of the USSR and Kazakhstan. In *Radioactive Isotopes in the Atmosphere and their Use in Meteorology,* ed. I. L. Karol et al, pp. 29–33. Jerusalem: Isr. Program Sci. Transl. AEC Rep. AEC-tr-6711.

Smyth, L. B. 1912. On the supply of radium emanation from the soil to the atmosphere. *Philos. Mag.* 24: 632–7

Stebbins, A. K. 1961. *Second special report on the high altitude sampling program.* US Dept. Def. Rep. DASA 539-B: 127–33, Defense Atomic Support Agency, Washington, D.C.

Tsunogai, S., Fukuda, K. 1974. Pb-210, Bi-210 and Po-210 in meteoric precipitation and the residence time of tropospheric aerosol. *Geochem. J.* 8: 141–52

Turekian, K. K., Kharkar, D. P. Thomson, J. 1974. The fates of Pb^{210} and Po^{210} in the ocean surface. *J. Rech. Atm.* 8: 639–46

Vilenskii, V. D. 1970. The influence of natural radioactive atmospheric dust in determining the mean stay time of lead-210 in the troposphere. *Izv., Acad. Sci., USSR, Atmos. Oceanic Phys.* 6: 307–10

Wilkening, M. H. 1974. Radon-222 from the island of Hawaii: deep soils are more important than lava fields or volcanoes. *Science* 183: 413–15

Wilkening, M. H., Clements, W. E. 1975. Radon-222 from the ocean surface. *J. Geophys. Res.* 80: 3828–30

Wilkening, M. H., Clements, W. E., Stanley, D. 1975. Radon-222 flux measurements in widely separated regions. In *The Natural Radiation Environment II,* ed. J. A. S. Adams, 2: 717–30. USERDA CONF-720805. 959 pp.

Wilkening, M. H., Hand, J. E. 1960. Radon flux at the earth-air interface. *J. Geophys. Res.* 65: 3367–70

Windom, H. L. 1969. Atmospheric dust records in permanent snowfields: implications to marine sedimentation. *Geol. Soc. Am. Bull.* 80: 761–82

Wright, J. R., Smith, O. F. 1915. The variation with meteorological conditions of the amount of radium emanation in the atmosphere, in the soil gas, and in the air exhaled from the surface of the ground, at Manila. *Phys. Rev.* 5: 459–82

Zeilinger, P. R. 1935. Über die Nachlieferung von Radiumemanation aus dem Erdboden. *Terr. Magn. Atmos. Electr.* 40: 281–94

Zupancic, P. R. 1934. Messungen der Exhalation von Radium-emanation aus dem Erdboden. *Terr. Magn. Atmos. Electr.* 39: 33–46

Ann. Rev. Earth Planet. Sci. 1977. 5:257–86

GEOCHRONOLOGY OF SOME ALKALIC ROCK PROVINCES IN EASTERN AND CENTRAL UNITED STATES[1]

×10074

Robert E. Zartman

U.S. Geological Survey, Denver Federal Center, Denver, Colorado 80225

INTRODUCTION

Recent efforts to understand the origin of igneous rocks have emphasized the association between petrographic suites and geologic environment. Plate tectonic theory has been particularly useful in relating certain types of calc-alkalic rocks to their position near active plate boundaries. In contrast, alkalic rocks are often emplaced after cessation of orogenic forces or in completely anorogenic settings of the craton. The role of plate tectonics in dictating such magmatism is not readily understandable. One obvious prerequisite to any effort at identification of underlying causes is the accurate dating of the igneous rocks. My objective in this paper is to review our present state of knowledge concerning a chronology for some alkalic rocks of eastern and central United States.

Although representing less than five percent of the volume of all igneous rocks, the alkalic clan includes a variety of interesting petrographic types, such as syenite, alkalic granite, alkalic gabbro, lamprophyre, and mica peridotite. These rocks typically occupy discordant stocks, plugs, dikes, sills, and diatremes at relatively shallow levels in the crust; and they may also include eruptive counterparts. The literature abounds with contributions regarding their structure, petrography, mineralogy, and geochemistry; and I do not stray too far from my intended objective into these areas.

My working definition of an alkalic rock corresponds essentially to that of Sorensen (1974), but I also include certain ultramafic rocks (Wyllie 1967) that are often found in petrologic and spatial association. Generally, rock names are accepted as the original authors used them without defending the specific genetic implication sometimes attached to the name.

This paper focuses upon several petrographically coherent groups of alkalic rocks, or provinces, that occur within the eastern and central United States. Although each province contains some felsic and some mafic members, two rather different

[1] Publication authorized by the Director, U.S. Geological Survey.

manifestations of igneous activity are recognized, depending on whether syenite-granite-monzonite-gabbro or mica peridotite-kimberlite-lamprophyre dominates. In the former case, represented by the New Hampshire and eastern Massachusetts occurrences, differentiated igneous rocks occupy batholiths, stocks, and ring dikes, which suggest a relatively slow intrusion of complexly evolved magma chambers into the upper reaches of the crust. In the latter case, represented by the western Appalachian Mountains and 38th-parallel-lineament occurrences, mafic to ultra-mafic magma has been explosively introduced into the crust in a highly fluid state and has formed dikes, sills, and diatremes. A more equal representation of the two kinds of emplacement features is found in the Gulf Coastal region occurrence, which includes both the felsic Magnet Cove Complex and the ultramafic Murfrees-boro kimberlite diatremes.

Isotopic age investigations, particularly of the past decade, have added a fascinating aspect to the study of alkalic rocks. Unexpected patterns in both time and space have frequently emerged. No longer is it safe to assume that rocks are always contemporaneous in age within a suite of alkalic rocks that is closely associated geographically or petrographically. Once conditions for the formation of alkalic rocks are established, conditions that are probably localized by zones of crustal weakness, repeated episodes of magma injection can span several geologic periods or even the entire Phanerozoic Eon! Evidently, these fracture systems attest to intraplate stresses, which, although not as obvious as those operating at sub-duction zones and ocean ridges, can periodically create conduits for ascending magma. Several recent attempts have been made to correlate alkalic magmatism with times of special plate activity. The onset of rifting, change in direction of plate motion, and migration of a plate over melting centers, or "hot-spots," in the mantle might activate such magmatism.

This is not to deny that some alkalic intrusive and eruptive rocks are triggered more directly by orogenic forces. Of the alkalic Cenozoic volcanic fields and intrusive centers scattered throughout the western cordillera of the United States, many cannot be discussed independently of a broader tectonic framework. By concentrating on those occurrences removed from concurrent orogenic activity, however, emphasis can be placed on some temporal and spatial peculiarities of rocks uniquely intruded into otherwise stable continental crust.

A variety of decay constants has been used in previous literature, which makes direct comparison of data difficult. Therefore, to establish uniformity throughout this paper, all ages have been converted according to the following values of the constants.

Potassium-40: $\lambda_\beta = 4.72 \times 10^{-10}$ yr^{-1}
$\lambda_e = 5.85 \times 10^{-11}$ yr^{-1}
$40_{K/K} = 0.0122$ weight percent

Rubidium-87: $\lambda_\beta = 1.39 \times 10^{-11}$ yr^{-1}
Uranium-238: $\lambda = 1.54 \times 10^{-10}$ yr^{-1}
Uranium-235: $\lambda = 9.72 \times 10^{-10}$ yr^{-1}
Thorium-232: $\lambda = 4.88 \times 10^{-11}$ yr^{-1}
Uranium-238 [fission]: $\lambda_f = 6.85 \times 10^{-17}$ yr^{-1}

In several instances, dating by one or more methods results in significant disagreement between radiometric ages. Only where important scientific controversy persists over these discrepancies do I elaborate and, in one case, present new data bearing on the matter. Otherwise, no attempt will be made to systematically review spurious results reflecting altered isotopic systems rather than primary ages.

CASE HISTORIES

New Hampshire

One of the most extensive occurrences of alkalic rocks in the United States, known collectively as the White Mountain Plutonic Series and related rocks (Billings 1956), is situated along a 200-km-long north-northwest trending belt in New Hampshire

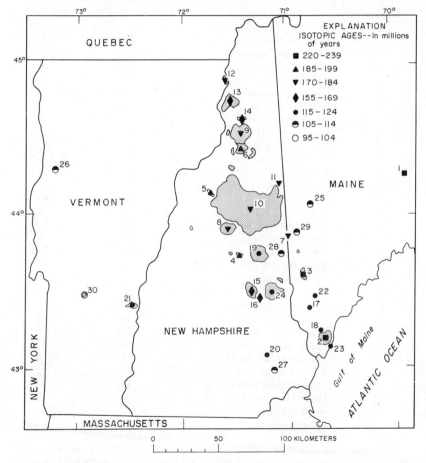

Figure 1 Distribution of isotopic ages for alkalic rocks of the New Hampshire province. Numbers refer to dated localities given in Table 1.

and adjacent states (Figure 1). Field relations clearly establish these alkalic rocks as postdating the Devonian Acadian orogeny, during which time the high-grade metamorphic and igneous country rock was formed. Although they are somewhat dwarfed by voluminous Paleozoic calc-alkalic intrusions, the rocks of the White Mountain Plutonic Series have become a classic for petrographic, structural, geochemical, and geophysical study (Chapman & Williams 1935, Billings 1945, Chapman 1976). Included in this suite of mildly alkaline rocks are granite, quartz syenite, syenite, and lesser amounts of intermediate and mafic units, which together

Table 1 Summary of radiometric ages for alkalic rocks from the New Hampshire province (White Mountain Plutonic Series) and related rocks

Number	Location	Age, millions of years	Reference[a]
1	Litchfield, Maine	238	1
2	Agamenticus, Maine	228	2, 3, 4
3	Abbott Mountain, Maine	221	4
4	Red Hill, N.H.	197	4, 5
5	Cannon Mountain, N.H.	194	3, 4
6	Pliny Range, N.H.	188	4
7	Whales Back, N.H.	184	4
8	Mad River, N.H.	182	3, 4
9	Pilot Range, N.H.	181	3, 4
10	White Mountain, N.H.	168–180	3, 4, 6, 7, 8
11	Baldface, N.H.	171	4
12	Mount Monadnock, Vt.	171	4
13	Gore Mountain, N.H.	168	3, 4
14	Percy Peaks, N.H.	165	4
15	Belknap Mountain, N.H.	158	3, 4
16	Pine Moutain, N.H.	156	4
17	Lebanon, Maine	125	4
18	Tatnic Hills, Maine	122	4
19	Ossipee Mountain, N.H.	121	3, 4
20	Mount Pawtuckaway, N.H.	121	3
21	Mount Ascutney, Vt.	120	3, 4, 9
22	Alfred, Maine	120	4
23	Cape Neddick, Maine	116	4
24	Merrymeeting, N.H.	116	3, 4
25	Pleasant Mountain, Maine	112	4
26	Barber Hill, Vt.	111	5
27	Little Rattlesnake Hill, N.H.	111	4
28	Green Mountain, N.H.	110	4
29	Burnt Meadow, Maine	108	4
30	Cuttingsville, Vt.	98	5, 10

[a] References: (1) Burke et al 1969; (2) Hoefs 1967; (3) Foland et al 1971; (4) Foland & Faul 1977; (5) Armstrong & Stump 1971; (6) Tilton et al 1957; (7) Aldrich et al 1958; (8) Hurley et al 1960; (9) Faul et al 1963; (10) Zen 1972.

occupy a batholith, several stocks, and related ring dikes. The Moat Volcanics, probably an extrusive manifestation of the series, are locally preserved inside some of the plutonic bodies associated with collapsed calderas.

Prior to the advent of modern radiometric dating methods, considerable uncertainty surrounded attempts to assign an age to the rocks of the White Mountain Plutonic Series. Their posttectonic character and relatively shallow emplacement depths required that these bodies be younger than the deformation that affected the country rock. Correlation with the alkalic rocks of eastern Massachusetts—to be discussed in the following section—had led some workers to favor an age between Middle Devonian and Pennsylvanian, preferring, generally, the Mississippian. Despite the lack of evidence supporting their view, few geologists seriously thought the rocks were younger than the Triassic sedimentary rocks of the nearby Connecticut Basin.

Lyons et al (1957; see also Jaffe et al 1959) obtained Pb-alpha ages of 177 to 195 m.y. on zircon from several bodies of these alkalic rocks. The geologic time scale then in use placed these ages in the Late Permian, but time-scale revisions soon showed that the ages were actually Early Jurassic. These Pb-alpha results were almost immediately confirmed by U-Pb zircon and by K-Ar and Rb-Sr biotite determinations on the granite from near Redstone, N.H. (Tilton et al 1957, Aldrich et al 1958, Hurley et al 1960). This discovery meant that despite their petrographic similarities, the alkalic rocks of New Hampshire could not be time correlatives of those from eastern Massachusetts (Toulmin 1961). Nevertheless, the different discrete bodies of the White Mountain Plutonic Series were otherwise to be considered essentially contemporaneous for another decade. Important to such an interpretation was the common appearance within many of the plutons of a distinctive rock type called Conway Granite. In retrospect, it can be noted that the only verification of the Early Jurassic age, aside from the Pb-alpha measurements, involved samples collected from a single locality of the White Mountain batholith. Some indication of future complications was given by two K-Ar biotite ages of 137 and 125 m.y. from the complex at Mount Ascutney in east-central Vermont (Faul et al 1963) and, later, in an Rb-Sr whole-rock isochron age of 227 m.y. from the complex at Agamenticus in southern Maine (Hoefs 1967).

A major redirection of thought concerning the petrogenetic processes involved in forming these alkalic rocks was necessitated by two companion papers (Foland et al 1971, Armstrong & Stump 1971) that firmly documented the true time span for emplacement of these plutons. These papers left no doubt that the belt of the White Mountain Plutonic Series contains bodies—often including Conway Granite—ranging in age from Early Triassic to Early Cretaceous.

Recently, Foland & Faul (1977) have expanded the sample coverage to include almost all known igneous bodies associated with the province. In their comprehensive study they are able to establish that (a) three rather broad pulses of magmatism at 220 to 235 m.y., 155 to 200 m.y., and 95 to 125 m.y. are discernible, and (b) no regular progression relating ages to locations occurs, as might accompany the migration of a single melting center. A compilation of all available data (Table 1 and Figure 1) reveals the dispersion of ages of the White Mountain Plutonic Series.

By contrast, individual complexes show almost no difference in age among lithologic units, except for the White Mountain batholith, which is itself a composite of several intrusive centers.

Although generally conceding an anorogenic setting for these alkalic rocks, some workers have attempted to relate the White Mountain Plutonic Series to modern tectonic theory. Regionally, these rocks lie almost in a line with a landward projection of the offshore New England seamount chain of Cretaceous age. Also, to the north, they merge with the Lower Cretaceous alkalic rocks of the Monteregian Hills, Quebec (Gold 1967). Together these features form a linear structure of Mesozoic igneous activity extending for more than 2000 km across the North Atlantic and adjacent North American continent. It has been argued that such a pattern could be caused by either (a) the trace of a mantle "hot-spot" under a moving crustal plate (Morgan 1972), or (b) emplacement along a fracture related to transform faulting in the North Atlantic (Ballard & Uchupi 1975, Foland & Faul 1977). The lack of any regular time-transgressive pattern of ages along the linear structure makes the second hypothesis more attractive. The recurrence of igneous activity scattered along the belt could then represent reactivation of the fracture zone, perhaps in response to early stages of Atlantic opening or subsequent changes in plate motion.

Eastern Massachusetts

The alkalic igneous rocks of eastern Massachusetts and northeastern Rhode Island (Figure 2) share renown with the White Mountain Plutonic Series; indeed they were once regarded as a part of it. Occurrences of granite, syenite, monzonite, and a variety of less common rock types constitute this important petrographic province (Warren 1913, Clapp 1921, Warren & McKinstry 1924, Toulmin 1964). The rocks mainly occupy simple to complex stocklike bodies, but later faulting has often obscured the original size and shape of the pluton. A number of concepts of modern igneous petrology have roots in the pioneer studies of these rocks, and the province continues to attract petrologists, mineralogists, and geochemists. It is somewhat ironic, then, that an area offering these early insights into the nature of igneous processes has been so slow in giving up the mysteries about its age.

Similar to the alkalic rocks of New Hampshire, those of eastern Massachusetts are definitely posttectonic with respect to the dynamothermally metamorphosed terrane that constitutes much of its country rock. A long-held interpretation that this metamorphism was produced by the Devonian Acadian orogeny that caused metamorphism in adjacent areas to the northwest necessitated a maximum Middle Devonian age for the alkalic rocks. The fact that unmetamorphosed, fossiliferous Lower and Middle Cambrian strata—intruded by alkalic granite at one locality—are also present and overlie the metamorphic rocks was explained as the local protection of favorable sites during deformation. A remarkably well-preserved volcanic unit, the Newbury Formation, of latest Silurian to earliest Devonian(?) age, which may be an extrusive phase of the alkalic suite, represents another "favorable-site" situation. Less equivocal is the occurrence of cobbles of alkalic rock in the Pennsylvanian Pondville Conglomerate, establishing a minimum age for the source of these cobbles.

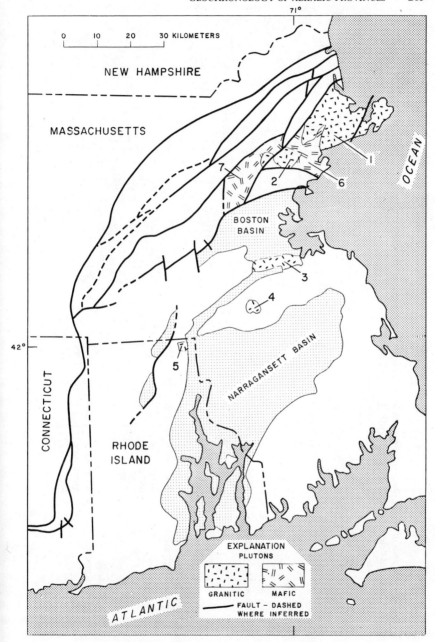

Figure 2 Simplified geologic map of the alkalic rocks of the eastern Massachusetts province. Granitic plutons: 1, Cape Ann; 2, Peabody; 3, Quincy; 4, Rattlesnake Hill; 5, Cumberland, R.I. Mafic plutons: 6, Salem; 7, Lexington.

Table 2 Radiometric ages of alkalic rocks from eastern Massachusetts and Rhode Island (where multiple samples from a rock unit have been analyzed, the number of determinations is indicated in [ ])

Number	Pluton	Mineral	K-Ar	Rb-Sr	Age, millions of years				Ref.[a]
					$\frac{206\text{Pb}}{238\text{U}}$	$\frac{207\text{Pb}}{235\text{U}}$	$\frac{207\text{Pb}}{206\text{Pb}}$	$\frac{208\text{Pb}}{232\text{Th}}$	
GRANITIC PLUTONS									
1	Cape Ann stock								
	Cape Ann granite	Whole-rock isochron [2]	—	415–435	—	—	—	—	1, 2
		Amphibole [8]	353–410	—	—	—	—	—	2, 3
		Biotite [2]	352–357	—	—	—	—	—	2, 3
		Zircon	—	—	417	421	452	402	2
	Wenham monzonite	Amphibole	352	—	—	—	—	—	3
2	Peabody stock								
	Peabody granite	Whole-rock isochron	—	367	—	—	—	—	2
		Amphibole [6]	350–403	—	—	—	—	—	2, 3
		Biotite [3]	356–367	—	—	—	—	—	2, 3
		Zircon [2]	—	—	369, 356	379, 366	445, 435	368, 351	2
3	Quincy stock								
	Quincy granite	Whole-rock isochron [2]	—	313–367	—	—	—	—	2, 4
		Amphibole [5]	430–458	—	—	—	—	—	2
		Pyroxene [2]	934–1005	—	—	—	—	—	2, 5
		Zircon	—	—	413	416	437	422	2

Locality / Rock	Material	Age	Whole-rock isochron	Reference[a]
Blue Hill granite				
Porphyry	Whole-rock isochron	—	282	4
	Amphibole	301	—	2
4 Rattlesnake Hill stock				
Alkalic granite	Amphibole [6]	343–375	—	5
	Biotite [2]	254–308	—	5
5 Cumberland stock				
Alkalic granite	Amphibole [3]	188–236	—	5
	Biotite	230	—	3

MAFIC PLUTONS

Locality / Rock	Material	Age	Whole-rock isochron	Reference[a]
6 Salem stock				
Gabbro-diorite	Biotite [3]	414–489	460–493	2
	Pyroxene	1780	—	6
Gabbro-diorite horn-felsed by Peabody granite	Biotite	356	—	3
Essexite	Biotite [2]	395–410	—	2, 3
7 Lexington stock				
Granodiorite	Amphibole	346	—	6
Diorite	Amphibole [3]	417–570	—	6

[a] References: (1) Bottino & Fullagar 1968; (2) Zartman & Marvin 1971; (3) Zartman et al 1970; (4) Bottino et al 1970; (5) Lyons & Krueger 1976; (6) Appendix, this work.

In recent years it has been recognized that the geology of southeastern New England forms an upper Precambrian eastern basement to the tectonically active Paleozoic Appalachian orogen (Fairbairn et al 1967, Naylor 1975). Within this eastern basement, an extensive system of faults hampers attempts at paleo-geographic reconstruction and accentuates problems of correlation into the orogen. Thus, without extrapolating beyond the coastal area, one really cannot define, from field relations, the age of the alkalic rocks more closely than post-Middle Cambrian and pre-Pennsylvanian.

An extensive Pb-alpha zircon study on several granitic rocks, including the Cape Ann and Peabody Granites, was the first major effort to decipher the geochronology of the region (Webber et al 1956; see also Quinn et al 1957). Together with a K-Ar whole-rock age for the Quincy Granite (Hurley et al 1960), these determinations supported a Mississippian age of 230–280 m.y., which was in agreement with geologic interpretation then in vogue (such ages are considered Permian by currently accepted time scales).

Subsequently, a number of workers have applied K-Ar, Rb-Sr, and U-Th-Pb dating techniques in an attempt to better define the emplacement history of the eastern Massachusetts alkalic rocks (Bottino & Fullagar 1968, Bottino et al 1970, Zartman et al 1970, Zartman & Marvin 1971, Lyons & Krueger 1976). Although these efforts led to a general consensus that the alkalic granites were at least as old as Devonian, the emerging pattern of ages was still ambiguous (Table 2). Zartman & Marvin (1971) placed great emphasis on the U-Th-Pb zircon analyses in assigning a Late Ordovician age to the Quincy, Cape Ann, and Peabody Granites. They explained some distinctly younger K-Ar amphibole and biotite and Rb-Sr whole-rock ages as the product of later metamorphism. Lyons & Krueger (1976) pointed out the possibility of older, relict zircons in the granites, which might result in anomalously old $^{207}Pb/^{206}Pb$ ages. Likewise, suspicion arose over the possibility of excess radiogenic argon in amphiboles, particularly where the phenomenon occurred in coexisting pyroxenes.

Two other findings related to this debate have prompted further investigation of the eastern Massachusetts alkalic rocks. Firstly, radiometric age studies on the similar rocks of the White Mountain Plutonic series in New Hampshire demonstrated that intrusion extended over more than 100 m.y. (Foland et al 1971, Armstrong & Stump 1971, Foland & Faul 1977). Separate plutons containing almost identical igneous rock types are not all contemporaneous, but rather they attest to a remarkably long period of very similar magmatism. Secondly, Bell & Dennen (1972) and Dennen (1972) have reported on a chemical and petrographic study that relates the alkalic granites of eastern Massachusetts with an appreciable volume of gabbroic, dioritic, and granodioritic rocks. Field relations show that the more mafic members of the alkalic series both intrude and are intruded by the granites, thereby signifying more than a single cycle of magmatic differentiation. Consequently, the underlying assumption that all of the alkalic rocks are cogenetic, which was made by Zartman & Marvin (1971) in assigning them a unique age of 450 ± 25 m.y., must now be considered obsolete.

In addition to the published results for the alkalic granitic rocks, published and

Table 3 New uranium-thorium-lead isotopic ages of zircon from alkalic granite of eastern Massachusetts

Sample number and mesh size	Concentration (ppm)			Isotopic composition of lead (atom percent)				Age, millions of years			
	U	Th	Pb	^{204}Pb	^{206}Pb	^{207}Pb	^{208}Pb	$\dfrac{^{206}\text{Pb}}{^{238}\text{U}}$	$\dfrac{^{207}\text{Pb}}{^{235}\text{U}}$	$\dfrac{^{207}\text{Pb}}{^{206}\text{Pb}}$	$\dfrac{^{208}\text{Pb}}{^{232}\text{Th}}$
Cape Ann granite											
163 (−200+250)	886.4	400.5	62.85	0.033	81.93	5.045	12.99	422±4	427±4	451±6	420±4
170A (−100+150)	158.6	67.9	10.98	0.038	82.24	5.130	12.59	413±4	418±4	446±6	411±4
170A (−200+270)	203.6	86.6	14.41	0.010	83.28	4.766	11.94	430±4	432±4	442±5	439±4
Wenham monzonite											
175 (−100+150)	207.5	135.0	14.58	0.109	74.71	5.667	19.51	375±4	378±6	398±10	378±5
175 (−200+325)	245.2	157.1	16.39	0.022	78.11	4.562	17.31	381±4	383±4	393±5	392±4
Peabody granite [light-greenish-gray variety]											
172 (−100+150)	327.8	147.3	20.97	0.039	81.65	5.007	13.31	380±4	382±4	392±6	384±4
172 (−200+325)	420.4	191.0	25.84	0.020	82.31	4.774	12.89	370±4	373±4	392±5	374±4
Quincy granite											
98 (−100+150)	839.4	370.1	56.43	0.191	75.93	6.997	16.88	358±4	370±8	436±16	335±6
210 (−200+270)	767.9	321.3	53.52	0.319	70.59	8.570	20.52	333±5	345±12	433±26	320±8

The isotopic composition of the common lead used to correct for nonradiogenic lead in the zircon is ^{206}Pb/^{204}Pb = 17.90, ^{207}Pb/^{204}Pb = 15.65, and ^{208}Pb/^{204}Pb = 38.20.

new analyses on the more mafic members of the suite (Table 2), together with new U-Th-Pb zircon analyses on the granitic rocks (Table 3), give further insight into the problem. A concordia plot of all existing zircon analyses is shown in Figure 3. Examined in this light, the evidence becomes convincing that within the eastern Massachusetts province igneous activity extended from Early Ordovician until Late Devonian time. Based on presently available data, a "best estimate" for the age of the main alkalic plutons is given in Figure 4, although many problems of interpretation remain.

Specifically, these problems are as follows:

1. A Late Ordovician or Early Silurian age for the Quincy Granite and the eastern part of the Cape Ann Granite is reconfirmed by the new zircon data. As

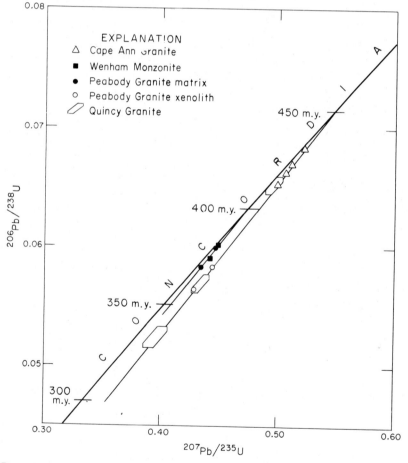

Figure 3   Concordia diagram for zircon from alkalic granite of the eastern Massachusetts province.

recognized by Lyons & Krueger (1976), however, the Cape Ann stock may contain distinctly younger rocks in its western part. If so, the previously determined Rb-Sr whole-rock isochrons seem to be dominated by the older, eastern facies rocks. At

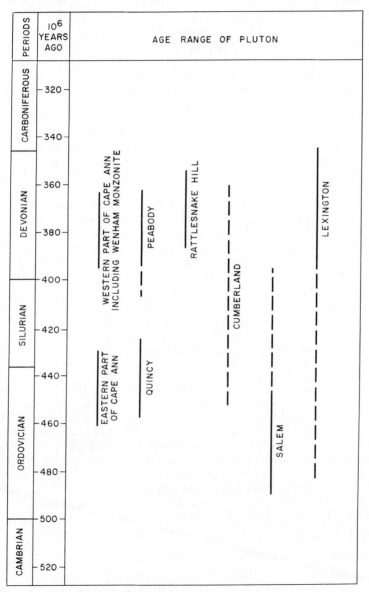

Figure 4 Best estimates for age ranges of the main alkalic plutons of the eastern Massachusetts province (dashed where evidence is equivocal).

present, its appreciably lower U-Th-Pb zircon age makes the Devonian Wenham Monzonite the best candidate for a younger unit.

2. A reexamination of the original sample site of Zartman & Marvin (1971) suggests that a xenolith of slightly more mafic alkalic granite than the typical Peabody Granite was collected for zircon analysis. The new analyses reported here for zircon from the lighter colored matrix give an essentially concordant and younger age of 390 m.y. I do not fully understand this complexity among seemingly cognate units of a simple stock, but the older $^{207}Pb/^{206}Pb$ ages from the xenolith are tentatively interpreted as reflecting some inheritance phenomenon. The good agreement among all other methods at approximately 380 ± 15 m.y., which was previously dismissed by Zartman & Marvin (1971), now becomes the preferred age for the bulk of the Peabody Granite.

3. The 370 ± 15 m.y. age of Lyons & Krueger (1976) is accepted for the Rattlesnake Hill stock on the basis of the tight clustering of their results, although only the K-Ar method was used in the study. The Cumberland, R.I., stock has apparently responded so greatly to later metamorphism (Zartman et al 1970) as to make any meaningful age assignment using existing data impossible.

4. The confused pattern of ages given in Table 2 for the mafic complexes at Salem and Lexington may arise from both disturbed radiometric systems and real age difference within each complex. The nearly concordant K-Ar and Rb-Sr biotite ages of 460 to 490 m.y. for some gabbroic rocks of the Salem complex agrees with field relations, making them among the oldest rocks of the province. Younger K-Ar biotite ages probably reflect heating caused by later intrusion, as is certainly the case with the gabbro-diorite hornfelsed by the Peabody Granite. The observed range of 346 to 570 m.y. in K-Ar amphibole ages for the complex at Lexington is frankly baffling. The dated samples were chosen from field criteria to be among the youngest of the province, but only the granodiorite gives a compatible result. Excess radiogenic argon is known for associated pyroxenes, but X-ray diffraction spectra did not detect this impurity in any of the amphibole separates.

Finally, it is not possible to put the eastern Massachusetts alkalic rocks into any precise temporal or spatial framework relative to their original setting. Figure 4 does suggest a bimodal age distribution, at least for the granitic rocks, but much uncertainty still surrounds these assignments. As modern mapping has shown, this coastal area is in many respects exotic in comparison to the rest of New England and may be best regarded as a fragment of a more easterly terrane. The alkalic rocks are areally restricted, and if correlations with other bodies are to be made, they will probably be with rocks underlying the Gulf of Maine. The original geometry of the plutons has been hopelessly lost during later periods of faulting, and if any linear trend such as that of the White Mountain Plutonic Series once existed, only a small remnant of the structure remains today.

Western Appalachian Mountains

Small dikes and plugs of dominantly mafic to ultramafic alkalic rock occur singly or in clusters along the western Appalachian Mountains in unmetamorphosed Paleozoic rocks of the Valley and Ridge and Appalachian Plateau Provinces (Figure 5).

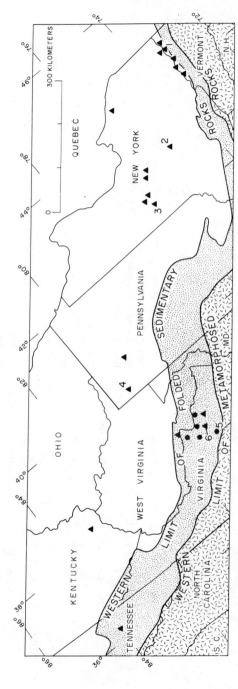

Figure 5   Occurrences of alkalic rocks within the western Appalachian Mountains. Symbols represent individual bodies or localized clusters: circle, felsic rocks, mostly nepheline syenite and phonolite; triangle, mafic rocks, mostly nepheline basalt, mica peridotite, and lamprophyre. Numbers refer to dated localities given in Table 4.

Table 4 Radiometric ages of alkalic rocks from the western Appalachian Mountains, 38th parallel lineament, and Gulf Coastal region (where multiple samples from a rock unit have been analyzed, the number of determinations is indicated in [ ])

Number	Location	Rock type	Mineral	Age, millions of years			Ref.[a]
				K-Ar	Rb-Sr	Fission track	
WESTERN APPALACHIAN MOUNTAINS							
1	Grand Isle, Vt.	Lamprophyre	Biotite	136	—	—	1
2	Manheim, N.Y.	Kimberlite	Biotite, coarse-grained [6]	255–371	118–146	—	1
			Biotite, fine-grained	150	—	—	1
3	Ithaca, N.Y.	Kimberlite	Biotite, coarse-grained [4]	420–493	136	—	1
			Biotite, fine-grained	145	—	—	1
4	Masontown, Pa.	Kimberlite	Biotite, coarse-grained [2]	368–408	—	—	1
			Biotite, unsized [2]	175–188	—	—	2
5	Staunton, Va.	Nepheline syenite	Hornblende	153	—	—	1
			Biotite	145	114	—	1
6	Augusta County, Va.	Teschenite	Hornblende	151	—	—	3
			Biotite	152	—	—	3
38TH PARALLEL LINEAMENT							
1	Highland County, Va.	Porphyritic felsite	Biotite [2]	46.9–47.2	—	—	4
			Whole-rock-biotite isochron	—	37	—	4
2	Elliott County, Ky.	Kimberlite	Biotite [2]	270–279	257	—	1
3	Hicks Dome, Ill.	Kimberlite	Hornblende	281	—	—	1
			Biotite	258	—	—	1

								Ref.
4	Rosiclare, Ill.	Peridotite	Biotite	269	260	—	—	1
5	Crittenden County, Ky.	Peridotite	Biotite	252	279	—	—	1
6	Lake County, Tenn.	Peridotite	Biotite	267	—	—	—	5
7	Avon, Mo.	Kimberlite	Biotite [2]	377–388	396–399	—	—	1
8	Silver City, Kans.	Peridotite	Biotite [2]	90–91	—	—	—	1
9	Rose Dome, Kans.	Peridotite	Biotite	88	—	—	—	1
10	Riley County, Kans.	Kimberlite	Chlorite [6]	142–380	—	—	—	6
			Chloritized biotite	112	—	—	—	6
			Apatite from granite xenolith [2]	—	—	—	115–123	6

GULF COASTAL REGION

								Ref.
1	Magnet Cove, Ark.	Melteigite	Biotite	95	—	—	—	1
		Garnet ijolite	Biotite	97	99	—	—	1
2	Little Rock, Ark.	Nepheline syenite	Biotite [2]	87–91	86	—	—	1
3	Murfreesboro, Ark.	Kimberlite	Phlogopite [2]	97–106	—	—	—	5
4	Pilot Knob, Austin, Tex.	Limburgite	Whole-rock	68	—	—	—	7
5	Uvalde County, Tex.	Nephelinite	Whole-rock	70	—	—	—	7
			Nepheline	73	—	—	—	7
			Pyroxene	69	—	—	—	7

[a] References: (1) Zartman et al 1967; (2) Pimentel et al 1975; (3) Marvin 1968; (4) Fullagar & Bottino 1969; (5) Appendix, this work; (6) Brookins 1969, 1970, Brookins & Naeser 1971; (7) Burke et al 1969.

Kimberlites are ubiquitous and have been identified in New York (Smyth 1902, Matson 1905, Martens 1924), Pennsylvania (Kemp & Ross 1907, Honess & Graeber 1926), Virginia–West Virginia (Watson & Cline 1913, Johnson & Milton 1955), and Tennessee (Hall & Amick 1944). The occurrences in Virginia and West Virginia also contain a number of other alkalic rocks of felsic to mafic composition, which occupy a prominent set of northwest-trending dikes that extend for 70 km across the two states. A swarm of lamprophyre dikes trends northward along Lake Champlain valley in eastern Vermont and northeastern New York (Shimer 1902, Hudson & Cushing 1931), and provides a link with the Monteregian Hills alkalic province in Quebec and the White Mountain Plutonic Series in New Hampshire.

Although somewhat isolated from each other, the alkalic rocks of the western Appalachian Mountains all display sufficient similarities in petrography and radiometric age to be discussed together. In fact, unlike the other provinces defined mainly by geographic proximity or some unifying, obvious structure, the most striking feature that these rocks have in common is uniqueness of age.

The radiometric ages for the western Appalachian Mountains are presented in Table 4. If the anomalously old K-Ar results on coarse-grained, xenocrystic biotites are excluded, all of the dated materials yield Early Cretaceous to Late Jurassic ages of 136–153 m.y. Several younger Rb-Sr biotite ages have apparently been chemically disturbed and are not considered reliable. It is possible that this narrow time range of emplacement may not be valid everywhere, because necessary sample coverage is still incomplete. A glaring omission from Table 4 is the Union County, Tennessee, kimberlite for which I have tried unsuccessfully to find unchloritized mica. Nevertheless, existing data leave no doubt that a substantial portion of the 1500 km–long province experienced magmatism during a period of geologic time previously regarded as quiescent. Earlier speculations about the age of these intrusions had been based upon erroneous correlation with Paleozoic Appalachian deformation, Upper Triassic basalt dikes, or Upper Cretaceous alkalic rocks of the Gulf Coastal region. Again, only after isotopic age techniques became available were these rocks accurately dated.

The coarse-grained biotites from the New York and Pennsylvania kimberlite dikes contain large quantities of excess radiogenic argon, which must already have been in the mica when the magma was intruded. In order not to have lost the inherited argon while still hot, the mica apparently was enveloped by a high external partial pressure of this gas. This observation agrees with the abundant evidence that kimberlites can be highly charged with fluid and injected with explosive velocities. Whether the apparent ages of the xenocrystic biotite have some geologic significance was not clarified in the study by Zartman et al (1967). It is more probable, however, that the amount of excess argon is determined by the gas content of the magma and its partitioning among the mineral phases rather than by the true age of some accidental or cognate inclusion. The fine-grained biotite from the kimberlite matrix did not display this phenomenon, suggesting that it was either completely degassed or only crystallized at the time of intrusion.

Why was this particular interval in the Late Jurassic and Early Cretaceous so favorable for scattered dike emplacement along much of the length of the western

Appalachian Mountains? Perhaps a clue lies in the extensive rifting that just previously had created the Upper Triassic graben basins and related tholeiitic dikes, sills, and flows. As the North Atlantic first began to open, isostatic adjustments near the new continental edge quite reasonably would have occurred. Accompanying movement between the Appalachian orogen and the interior craton could have provided the impetus for alkalic dike invasion. It is noteworthy that these rocks were introduced during a time period essentially devoid of activity in the New Hampshire and Monteregian Hills provinces.

38th Parallel Lineament

A remarkable linear array of alkalic dikes, sills, plugs, and diatremes extends across eastern and central United States, intersecting the western Appalachian Mountains province in Virginia and West Virginia (Figure 6). These predominantly mafic to ultramafic intrusions occur intermittently along a line of connected faults and crypto-volcanic structures affecting both basement and supracrustal rocks over a distance in excess of 2000 km at approximately the 38th parallel of latitude. The relative abundance of alkalic rocks here contrasts strikingly with their paucity elsewhere in the continental interior of the United States. Among the igneous centers cropping out along this zone, from east to west, are (a) the syenite to kimberlite dikes of Virginia and West Virginia, discussed previously as part of the western Appalachian Mountains occurrence, (b) kimberlite dikes and diatremes of Elliott County, eastern Kentucky (Diller 1887), (c) syenite to peridotite dikes, sills, and explosion breccias of Rosiclare, Illinois, and vicinity (Clegg & Bradbury 1956, Koenig 1956), (d) kimber-lite dikes and diatremes of the Avon, Missouri, area (Kidwell 1947), and (e) peridotite dikes of Silver City and Rose Domes, Kansas (Knight & Landes 1932, Franks et al 1971). Several major lead-zinc-fluorite deposits of the midcontinent also are associated with the lineament, especially near its intersection with other major faults and structural trends.

Although the age of the igneous rocks generally cannot be closely determined from field relations, stratigraphic evidence suggests that the 38th parallel lineament, as a structural element, developed over a long period of Phanerozoic time. Except for being younger than their immediate host rock, very little local control over age is available for the intrusive bodies. Rarely, some datable structural feature, such as a later crosscutting fracture or joint system, or an identifiable xenolith from a higher stratigraphic level, places additional constraints on the age. More successful has been the stratigraphic assignment of several alkalic volcanic ash deposits found in the subsurface. Long-distance correlation with the Upper Cretaceous alkalic rocks of the Gulf Coastal region has proven in most instances to be invalid.

Most earlier workers focused their attention on individual occurrences along the lineament, and not until the regional syntheses by Heyl & Brock (1961; see also Heyl et al 1965) and Snyder & Gerdemann (1965) was the continuity of the province fully recognized. An appreciation of the province's long history of igneous activity followed soon afterwards, when Zartman et al (1967), using isotopic ages, docu-mented the existence of several distinct periods of intrusion extending from Middle Devonian to early Late Cretaceous (Table 4). If we include the stratigraphically

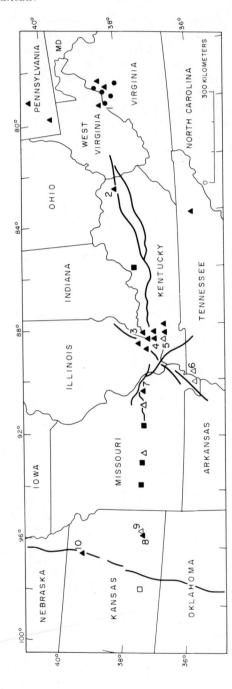

Figure 6 Occurrences of alkalic rocks and related cryptovolcanic features within the 38th parallel lineament. Symbols represent individual bodies or localized clusters (open where found in subsurface only): circle, felsic rocks, mostly nepheline syenite and phonolite; triangle, mafic rocks, mostly nepheline basalt, mica peridotite, and lamprophyre; square, intensely disturbed localized uplift of probably cryptovolcanic origin. Numbers refer to dated localities in Table 4. Heavy lines represent prominent basement faults.

controlled Cambrian ash deposits of Missouri (Snyder & Gerdemann 1965), and the Eocene felsite porphyry dikes at the eastern end of the lineament in Virginia (Fullagar & Bottino 1969), an even longer duration of activity becomes evident.

Perhaps, in hindsight, episodes of igneous activity along the fracture zone that span the entire Phanerozoic Eon are most compatible with other structural evidence. The structural setting of this alkalic province especially exemplifies the contrast with the more rapidly evolving tectonic setting near a plate boundary with its accompanying calc-alkaline suite of igneous rocks. In addition to age differences among the igneous centers, the Virginia–West Virginia occurrence is unique in being the locus of three separate periods of dike intrusion into an anorogenic setting—Upper Triassic tholeiitic basalt, Upper Jurassic to Lower Cretaceous syenite and peridotite, and Eocene felsite porphyry.

Although lying some distance off the 38th parallel lineament, two other dated localities are closely related to important fracture systems that intersect the lineament and share a similar tectonic style. Mica peridotite forms thin sills intruding Paleozoic sedimentary rocks in the subsurface of the Mississippi River embayment of northwestern Tennessee. This locality, together with related occurrences in adjacent southeastern Missouri, lies along the New Madrid fault zone, which may connect with northeast-trending faults in the vicinity of Rosiclare, Illinois, 150 km to the northeast. Because most of the intervening surface rocks are Cretaceous sediments of the embayment, it is not possible to directly relate the peridotite of Rosiclare and vicinity with the peridotite of the northwest corner of Tennessee. However, a newly reported K-Ar biotite age of 267 m.y. for the Tennessee peridotite supports a contemporaneous origin for intrusions within the fault zone.

The second occurrence consists of several kimberlite diatremes situated astride the Nemaha fault in Riley County, northeastern Kansas, about 200 km north of the 38th parallel lineament (Byrne et al 1956). An extensive study of these kimberlites by Brookins (1969, 1970) and Brookins & Naeser (1971) indicates an Early Cretaceous age of 110–120 m.y. on the basis of a K-Ar determination on chloritized biotite and two fission-track determinations on apatite from granite xenoliths. The fact that fission tracks easily anneal in apatite at 100°C or more makes this mineral ideal for dating diatremes in which an inclusion of hot basement rock is rapidly brought to the surface and cooled. In kimberlites, such as those of Riley County, where mica has been strongly chloritized, the fission-track method may well provide the most definitive age information. Chlorites retaining less than 0.5 percent K_2O were shown to be unreliable chronometers because of either original excess radiogenic argon or preferential potassium relative to argon loss during alteration.

Despite the long-term stability of much of the continental interior, which usually experiences only epeirogenic movements, localized stresses of great magnitude do apparently build up from time to time along major fracture systems. Once these systems are initiated, they may remain as weak zones subject to recurring activity. The sporadic emplacement of igneous rocks, particularly of the alkalic mafic to ultramafic clan, is seen to be one manifestation of the process. We are still largely ignorant of the reasons why complex interactions taking place at the boundaries of lithospheric plates can also influence their more stable interiors. As we proceed far inland and into the remote past, our abilities to speculate on the configuration of ancient stress patterns further diminishes. The 38th parallel lineament represents an excellent example of an area where field and laboratory workers are combining resources to document the very long evolution of one such fracture system.

Gulf Coastal Region

Several scattered occurrences of alkalic rocks intruding Cretaceous and adjacent Paleozoic rocks are exposed in outcrop in central Arkansas (Figure 7). The Magnet Cove Complex (Erickson & Blade 1963), the bauxite-forming syenite bodies near Little Rock (Gordon et al 1958), and the diamondiferous kimberlite at Murfreesboro (Miser & Ross 1922) are well-known examples from this suite. A continuation of these alkalic intrusive and related volcanic rocks is found in the subsurface of southern Arkansas, western Mississippi, northern Louisiana, northeastern Texas, and southeastern Oklahoma (Ross et al 1929, Moody 1949). Belonging to the petrographic group is a varied association of plutonic and volcanic rock types— peridotite, olivine basalt, nepheline syenite, phonolite, tinguaite, mochiquite, fourchite, and related pyroclastics. The stratigraphic position of these latter tuffaceous beds in the lower part of the Upper Cretaceous sedimentary rocks offers a clue to an extensive volcanic landscape at that time. More distantly, but

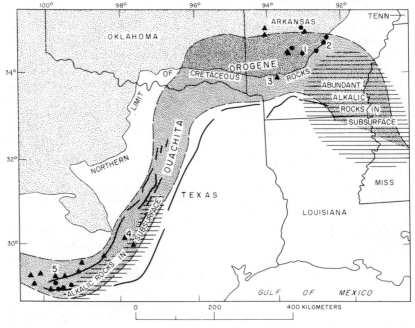

Figure 7   Occurrences of alkalic rocks within the Gulf Coastal region. Symbols represent individual bodies or localized clusters: circle, felsic rocks, mostly nepheline syenite and phonolite; triangle, mafic rocks, mostly nepheline basalt, mica periodotite, and lamprophyre. Numbers refer to dated localities in Table 4.

apparently still related, Upper Cretaceous to Tertiary peridotite, alkalic basalt, and phonolite occur in plugs, dikes, sills, and flows in both outcrop and the subsurface along the Balcones fault zone of south-central Texas (Lonsdale 1927, Spencer 1969).

Together, the rocks of this province rather faithfully follow the northern onlap of sediments of the Gulf Coastal region upon the Paleozoic Ouachita orogen. Evidently, this alignment marks a zone of weakness along the orogen near the old continental edge, which, in places, also is reflected by faults cutting the sediments. Unlike the considerable uncertainty in using field criteria for establishing ages in the preceding four provinces, stratigraphic control here has permitted rather good age assignments to be made. Thus, classical arguments about the positioning of volcanic and plutonic units relative to fossiliferous strata had, prior to geochronologic studies, already restricted the igneous rocks to the Cretaceous and earliest Tertiary. There remained, however, the possibility that bodies like the Magnet Cove Complex intruding Paleozoic rocks in central Arkansas were really not coeval with other members of the province. And rarely are stratigraphic limits so narrow, especially for intrusive rocks, that they cannot benefit from isotopic age measurements.

Previously published and two new analyses for this province are given in Table 4. Indeed, for the first time we see no great surprises in these results compared to ages predicted on the basis of field relations. An apparently real dispersion from 88 m.y.—Turonian—to 106 m.y.—Albian—among the Arkansas occurrences is in good agreement with the stratigraphic range through which volcanic ash is found nearby. The preferred K-Ar biotite age of 106 ± 3 m.y. for the Murfreesboro kimberlite would extend igneous activity back into late Early Cretaceous time, but would not violate bounds set by field relationships. Likewise, the 70 ± 5 m.y.—Maestrichtian to Campanian—age for the south-central Texas localities fall near that deduced from stratigraphic considerations.

The Gulf Coastal region contains some of the youngest alkalic rocks herein discussed and lies at the edge of the craton; nevertheless, it presents no less of an enigma concerning tectonic setting. The fact that igneous activity followed some tens of millions of years after maximum marine transgression in Early Cretaceous time may somehow relate to tensional stresses localized about a hinge line at the continental margin. But such an explanation does nothing to contrast this area with many similar areas where sedimentary loading has caused no equivalent phenomenon. We might also ask if there is any significance to the fact that this province parallels an older orogen, whereas the New Hampshire province transects an older orogen. Perhaps these diversities are a warning not to demand too much uniformity from all the alkalic occurrences. If their petrographic uniqueness is primarily a function of source material, the provision of conduits may be the only important structural factor for emplacement. Accordingly, a variety of mechanisms could satisfy the requirements to create and sustain an appropriate fracture system.

CONCLUSION

A rapid succession of discoveries involving the application of plate tectonic theory to problems in igneous petrology has taken place in recent years. Although most of the attention has focused on activity near plate boundaries, a general excitement has resulted over understanding the causes of magmatic processes wherever they operate. The petrogenesis of alkalic rocks is no exception, despite the tendency of this clan to occur in relatively stable environments at times and places generally regarded as quiescent. Although no completely satisfactory hypothesis has yet emerged to explain the relationship of alkalic rocks to their site of emplacement, some aspects of their occurrence happen so repetitiously as to demand attention. At the same time, obvious differences in many features also exist, which caution us against proposing too unified an origin. However, my purpose here is not to make new petrologic speculations, but rather to summarize the present state of the art regarding the dating of some representative alkalic rocks. Because of the major advances during the past decade in acquiring accurate age control, this seems to be an ideal time to bring together those factual data upon which any theory must ultimately rest.

The age of alkalic rocks tends to be poorly defined stratigraphically, and the establishment of a time framework has often presented a special challenge to geochronologists. Through case histories, this paper attempts to illustrate the complexities and surprises encountered in several eastern and central United States provinces.

The posttectonic nature of these alkalic provinces is definitely confirmed by radiometric age studies, which prove that emplacement occurred long after cessation of the last significant deformation of the host country rock. In four of the five provinces where a Paleozoic or Upper Precambrian orogen constitutes the local bedrock, an intriguingly consistent time delay of 100 to 150 m.y. separates the last stage of orogenic dynamothermal processes from the onset of first alkalic magmatism. The fifth province lies almost entirely in the continental interior where flat-lying Paleozoic sedimentary rocks rest on a considerably older Precambrian crystalline craton. Other alkalic occurrences, as, for example, those of the western cordillera, can be more closely associated with orogeny. Perhaps it is the uniqueness of alkalic rocks in the anorogenic environment rather than their absence elsewhere that should be found noteworthy.

A largely unexpected result that has emerged from recent geochronologic investigations is the dispersion of ages that is found within most provinces. Beforehand, it certainly seemed reasonable to assume that such petrographically closely related rocks should be emplaced during a rather brief interval of time. On the contrary, three of the provinces contain members with ages spanning 100 m.y., and one of these—the 38th parallel lineament—boasts igneous activity extending from Cambrian to Tertiary! A difference in age of some tens of millions of years has been demonstrated for the fourth province, and only once does our intuition seem to have been correct.

No less fascinating than the age dispersion is the manner in which time and space patterns present themselves. Rather than having successive emplacement proceed in an orderly direction along the generally linear trend of a province, we face a geometry more suggestive of a random walk. The intricacy is most evident in the New Hampshire and 38th parallel lineament provinces, for which detailed geologic histories can now be reconstructed. Only on the smallest scale is igneous activity usually, but not always, coeval within local dike swarms or individual plutons. Otherwise, it is difficult to comprehend any systematics underlying the observed distribution of the igneous rocks of either province. Although igneous activity is confined overall to a structural zone of weakness of some regional extent, each actual location of emplacement is apparently governed by a special stress field of more local origin.

Alkalic rocks have shown a propensity for presenting the geochronologist with a variety of geochemical peculiarities that affect the determination of radiometric ages. A few of these peculiarities that cause discrepant results, such as excess radiogenic argon, inherited relict zircons, and subtle postcrystallization disturbances, were briefly mentioned in the case histories. On the bright side, future research efforts may well come to appreciate these phenomena, which must have a natural explanation and hold promise of giving new insight into petrologic processes.

APPENDIX: LOCALITY, DESCRIPTION, AND ANALYTICAL DATA FOR NEW SAMPLES

A. Locality and description

Sample number	Latitude N.	Longitude W.	Location	Rock description
MASSACHUSETTS				
88	42°25′38″	70°54′44″	Outcrop overlooking Nahant Bay at Black Mine, Nahant	Dark grayish-green, fine-grained gabbro (Nahant gabbro)
154	42°28′21″	71°13′05″	Roadcut for future extension of US Route 3, 500 m southeast of intersection with State Route 128, Burlington	Greenish-gray, medium-grained granodiorite
140	42°30′10″	71°06′13″	Roadcut inside southeastern loop of cloverleaf at intersection of State Routes 128 and 28, Reading	Dark bluish-gray, fine-grained diorite
141	42°25′56″	71°10′23″	Outcrop 60 m southeast of water tower atop Turkey Hill, Arlington	Dark bluish-gray, medium-grained diorite
145	42°24′44″	71°11′28″	Outcrop on low knoll 300 m northwest of forested hill at Arlmont Country Club, Belmont	Dark bluish-gray, medium-grained diorite
163	42°41′15″	70°38′14″	Small abandoned quarry on north side of State Route 127 at Folly Cove, Rockport	Light gray, medium-grained granite (Cape Ann granite)
170A	42°34′27″	70°50′24″	Roadcut along State Route 128, 800 m east of underpass at Standley Street, Beverly	Greenish-gray, medium-grained granite (Cape Ann granite)
175	42°35′56″	70°54′40″	Roadcut along State Route 97, 600 m north of intersection with Cedar Street, Wenham	Light pinkish-gray, medium-grained monzonite (Wenham monzonite)

172	42°31'28"	70°58'10"	Abandoned quarry 60 m north of State Route 128, 1400 m southwest of intersection with Forest Street, Peabody	Light greenish-gray, medium-grained granite (Peabody granite)
98	42°14'33"	71°02'46"	Abandoned quarry, 350 m southwest of water tower, West Quincy	Purplish-gray, medium-grained granite (Quincy granite)
210	42°14'51"	71°01'00"	Sahlsten Quarry (abandoned), 120 m north of intersection of Quarry and Smith Streets, North Commons, Quincy	Bluish-gray, medium-grained granite (Quincy granite)
TENNESSEE				
M-1	36°20'52"	89°25'17"	Drill cuttings from 730–745 m depth in A. E. Markham No. 1 well, Lake County	Dark gray, fine-grained mica peridotite intruding Paleozoic limestone
ARKANSAS				
AK-1	34°02'01"	93°40'20"	Exploration trench 180 m southeast of summit of East Hill, Crater of Diamonds mine, Murfreesboro	Greenish-gray kimberlite breccia
AK-8	34°01'57"	93°40'32"	Exploration trench 160 m south of summit of Middle Hill, Crater of Diamonds mine, Murfreesboro	Bluish-gray, fine-grained kimberlite tuff

APPENDIX: LOCALITY, DESCRIPTION, AND ANALYTICAL DATA FOR NEW SAMPLES
(continued)

B.   Potassium-argon analytical data

Sample number	Mineral	K_2O (percent)	$Ar^{40}_{radiogenic}$ (moles gram^{-1})	Percentage of Ar^{40} of radiogenic origin	Age, in millions of years
88	pyroxene	0.138	6.03×10^{-10}	90	1780 ± 55
154	hornblende	1.44	8.08	98	346 ± 10
140	hornblende	0.675	6.22	93	539 ± 16
141	hornblende	0.770	5.30	86	417 ± 13
145	hornblende	1.18	11.58	96	570 ± 17
M-1	biotite	4.59	19.44	94	267 ± 8
AK-1	phlogopite	5.74	8.38	89	97 ± 2
AK-8	phlogopite	9.82	15.78	95	106 ± 3

Literature Cited

Aldrich, L. T., Wetherill, G. W., Davis, G. L., Tilton, G. R. 1958. Radioactive ages of micas from granitic rocks by Rb-Sr and K-A methods. Trans. Am. Geophys. Union 39:1124–34
Armstrong, R. L., Stump, E. 1971. Additional K-Ar dates, White Mountain magma series, New England. Am. J. Sci. 270: 331–33
Ballard, R. D., Uchupi, E. 1975. Triassic rift structure in Gulf of Maine. Am. Assoc. Petrol. Geol. Bull. 59:1041–72
Bell, K. G., Dennen, W. H. 1972. Plutonic series in the Cape Ann area. Geol. Soc. Am. Abstr. with Programs 4(1):2
Billings, M. P. 1945. Mechanics of igneous intrusion in New Hampshire. Am. J. Sci. 234-A:40–68
Billings, M. P. 1956. The geology of New Hampshire: Pt. 2, Bedrock geology. N.H. State Plann. Devel. Comm. 203 pp.
Bottino, M. L., Fullagar, P. D. 1968. The effects of weathering on whole-rock Rb-Sr ages of granitic rocks. Am. J. Sci. 266: 661–70
Bottino, M. L., Fullagar, P. D., Fairbairn, H. W., Pinson, W. H. Jr., Hurley, P. M. 1970. The Blue Hills igneous complex, Massachusetts: Whole-rock Rb-Sr open systems. Geol. Soc. Am. Bull. 81:3739–46
Brookins, D. G. 1969. The significance of K-Ar dates on altered kimberlitic phlogopite from Riley County, Kansas. J. Geol. 77:102–7
Brookins, D. G. 1970. The kimberlites of

Riley County, Kansas. Kans. State Geol. Surv. Bull. 200:1–32
Brookins, D. G., Naeser, C. W. 1971. Age of emplacement of Riley County, Kansas, kimberlites and a possible minimum age for the Dakota Sandstone. Geol. Soc. Am. Bull. 82:1723–26
Burke, W. H., Otto, J. B., Denison, R. E. 1969. Potassium-argon dating of basaltic rocks. J. Geophys. Res. 74:1082–86
Byrne, F. E., Parish, K. L., Crumpton, C. F. 1956. Igneous intrusions in Riley County, Kansas. Am. Assoc. Petrol. Geol. Bull. 40: 377–87
Chapman, C. A. 1976. Structural evolution of the White Mountain magma series. Geol. Soc. Am. Mem. 146:281–300
Chapman, R. W., Williams, C. R. 1935. Evolution of the White Mountain magma series. Am. Mineral. 20:502–30
Clapp, C. H. 1921. Geology of the igneous rocks of Essex County, Massachusetts. US Geol. Surv. Bull. 704:1–132
Clegg, K. E., Bradbury, J. C. 1956. Igneous intrusive rocks in Illinois and their economic significance. Ill. Geol. Surv. Rep. Invest. 197. 19 pp.
Dennen, W. H. 1972. Correlation of igneous rocks by chemical signatures of minerals. Geol. Soc. Am. Abstr. with Programs 4(1): 13
Diller, J. S. 1887. Peridotite of Elliott County, Kentucky. US Geol. Surv. Bull. 38:1–31
Erickson, R. L., Blade, L. V. 1963. Geochemistry and petrology of the alkalic

igneous complex at Magnet Cove, Arkansas. *US Geol. Surv. Prof. Pap.* 425:1–95

Fairbairn, H. W., Moorbath, S., Ramo, A. O., Pinson, W. H. Jr., Hurley, P. M. 1967. Rb-Sr age of granitic rocks of southeastern Massachusetts and the age of the Lower Cambrian at Hoppin Hill. *Earth Planet. Sci. Lett.* 2:321–28

Faul, H., Stern, T. W., Thomas, H. H., Elmore, P. L. D. 1963. Ages of intrusion and metamorphism in the northern Appalachians. *Am. J. Sci.* 261:1–19

Foland, K. A., Faul, H. 1977. Ages of the White Mountain intrusives — New Hampshire, Vermont, and Maine. *Am. J. Sci.* In press

Foland, K. A., Quinn, A. W., Giletti, B. J. 1971. K-Ar and Rb-Sr Jurassic and Cretaceous ages for intrusives of the White Mountain magma series, northern New England. *Am. J. Sci.* 270:321–30

Franks, P. C., Bickford, M. E., Wagner, H. C. 1971. Metamorphism of Precambrian granitic xenoliths in a mica peridotite at Rose Dome, Woodson County, Kansas: Pt. 2, Petrologic and mineralogic studies. *Geol. Soc. Am. Bull.* 82:2869–90

Fullagar, P. D., Bottino, M. L. 1969. Tertiary felsite intrusions in the Valley and Ridge province, Virginia. *Geol. Soc. Am. Bull.* 80:1853–58

Gold, D. P. 1967. Alkaline ultramafic rocks in the Montreal area, Quebec. In *Ultramafic and Related Rocks,* ed. P. J. Wyllie, pp. 288–97. New York: Wiley

Gordon, M. Jr., Tracey, J. I. Jr., Ellis, M. W. 1958. Geology of the Arkansas bauxite region. *US Geol. Surv. Prof. Paper* 229:1–268

Hall, G. M., Amick, H. C. 1944. Igneous rock areas in the Norris region, Tennessee. *J. Geol.* 52:424–30

Heyl, A. V., Brock, M. R. 1961. Structural framework of the Illinois-Kentucky mining district and its relation to mineral deposits. *US Geol. Surv. Prof. Paper* 424-D:3–6

Heyl, A. V., Brock, M. R., Jolly, J. L., Wells, C. E. 1965. Regional structure of southeast Missouri and Illinois-Kentucky mineral districts. *US Geol. Surv. Bull.* 1202-B:1–20

Hoefs, J. 1967. A Rb-Sr investigation in the southern York County area, Maine, *Mass. Inst. Tech. Dept. Geol. Geophys. 15th Ann. Prog. Rept. to USAEC,* pp. 127–29

Honess, A. P., Graeber, C. K. 1926. Petrography of the mica peridotite dike at Dixonville, Pennsylvania. *Am. J. Sci.* 12:484–94

Hudson, G. H., Cushing, H. P. 1931. The dike invasions of the Champlain valley, New York. *New York State Mus. Bull.* 286:81–112

Hurley, P. M., Fairbairn, H. W., Pinson, W. H., Faure, G. 1960. K-Ar and Rb-Sr minimum ages for the Pennsylvanian section in the Narragansett Basin. *Geochim. Cosmochim. Acta* 18:247–58

Jaffe, H. W., Gottfried, D., Waring, C. L., Worthing, H. W. 1959. Lead-alpha age determinations of accessory minerals of igneous rocks (1953–1957). *US Geol. Surv. Bull.* 1097-B:65–148

Johnson, R. W. Jr., Milton, C. 1955. Dike rocks of central-western Virginia. *Geol. Soc. Am. Bull.* 66:1689–90 (Abstr.)

Kemp, J. F., Ross, J. G. 1907. A peridotite dike in the coal measures of southwestern Pennsylvania. *Ann. N Y Acad. Sci.* 17:509–10

Kidwell, A. L. 1947. Post-Devonian igneous activity in southeastern Missouri. *Mo. Geol. Surv. Water Res. Rep. Invest.* 4. 83 pp.

Knight, G. L., Landes, K. K. 1932. Kansas laccoliths. *J. Geol.* 40:1–15

Koenig, J. B. 1956. The petrography of certain igneous dikes of Kentucky. *Ky. Geol. Surv. Ser. 9 Bull.* 21:1–57

Lonsdale, J. T. 1927. Igneous rocks of the Balcones fault region of Texas. *Tex. Univ. Bull.* 2744:1–178

Lyons, J. B., Jaffe, H. W., Gottfried, D., Waring, C. L. 1957. Lead-alpha ages of some New Hampshire granites. *Am. J. Sci.* 255:527–46

Lyons, P. C., Krueger, H. W. 1976. Petrology, chemistry, and age of the Rattlesnake pluton and implications for other alkalic granite plutons of southern New England. *Geol. Soc. Am. Mem.* 146:71–102

Martens, J. H. C. 1924. Igneous rocks of Ithaca, New York, and vicinity. *Geol. Soc. Am. Bull.* 35:305–20

Marvin, R. F. 1968. Transcontinental geophysical survey (35°–39° N). Radiometric age determinations of rocks. *US Geol. Surv. Misc. Geol. Invest. Map* I-537

Matson, G. C. 1905. Peridotite dikes near Ithaca, New York. *J. Geol.* 13:264–75

Miser, H. D., Ross, C. S. 1922. Diamond-bearing peridotite in Pike County, Arkansas. *Econ. Geol.* 17:662–74

Moody, C. L. 1949. Mesozoic igneous rocks of northern Gulf Coastal plain. *Am. Assoc. Petrol. Geol. Bull.* 33:1410–28

Morgan, W. J. 1972. Deep mantle convection plumes and plate motions. *Am. Assoc. Petrol. Geol. Bull.* 56:203–13

Naylor, R. S. 1975. Age provinces in the northern Appalachians. *Ann. Rev. Earth*

Planet. Sci. 3:387–400

Pimentel, N., Bikerman, M., Flint, N. K. 1975. A new K-Ar date on the Masontown dike, southwestern Pennsylvania. *Pa. Geol.* 6(3):5–7

Quinn, A. W., Jaffe, H. W., Smith, W. L., Waring, C. L. 1957. Lead-alpha ages of Rhode Island granitic rocks compared to their geologic ages. *Am. J. Sci.* 255:547–60

Ross, C. S., Miser, H. D., Stephenson, L. W. 1929. Water-laid volcanic rocks of early Upper Cretaceous age in southwestern Arkansas, southeastern Oklahoma, and northeastern Texas. *US Geol. Surv. Prof. Pap.* 154-F:175–202

Shimer, H. W. 1902. Petrographic description of the dikes of Grand Isle, Vermont. *State Geol. Vermont Rep.* 3:174–83

Smyth, C. H. Jr. 1902. Petrography of recently discovered dikes in Syracuse, New York. Petrographic description. *Am. J. Sci.*, Ser. 4. 14:26–30

Snyder, F. G., Gerdemann, P. E. 1965. Explosive igneous activity along an Illinois-Missouri-Kansas axis. *Am. J. Sci.* 263:465–93

Sorensen, H., ed. 1974. *The Alkaline Rocks.* New York: Interscience. 622 pp.

Spencer, A. B. 1969. Alkalic igneous rocks of the Balcones province, Texas. *J. Petrol.* 10:272–306

Tilton, G. R., Davis, G. L., Wetherill, G. W., Aldrich, L. T. 1957. Isotopic ages of zircon from granites and pegmatites. *Trans. Am. Geophys. Union* 38:369–71

Toulmin, P. III. 1961. Geological significance of lead-alpha and isotopic age determinations of "alkalic" rocks of New England. *Geol. Soc. Am. Bull.* 72:775–80

Toulmin, P. III. 1964. Bedrock geology of the Salem quadrangle and vicinity, Massachusetts. *US Geol. Surv. Bull.* 1163-A:1–79

Warren, C. H. 1913. Petrology of the alkali granites and porphyries of Quincy and the Blue Hills, Massachusetts, U.S.A. *Proc. Am. Acad. Arts Sci.* 49:203–31

Warren, C. H., McKinstry, H. E. 1924. The granites and pegmatites of Cape Ann, Massachusetts. *Proc. Am. Acad. Arts Sci.* 59:315–57

Watson, T. L., Cline, J. H. 1913. Petrology of a series of igneous dikes in central western Virginia. *Geol. Soc. Am. Bull.* 24:301–34

Webber, G. R., Hurley, P. M., Fairbairn, H. W. 1956. Relative ages of eastern Massachusetts granites by total lead ratios in zircon. *Am. J. Sci.* 254:574–83

Wyllie, P. J., ed. 1967. *Ultramafic and Related Rocks.* New York: Wiley. 464 pp.

Zartman, R. E., Brock, M. R., Heyl, A. V., Thomas, H. H. 1967. K-Ar ages of some alkalic intrusive rocks from central and eastern United States. *Am. J. Sci.* 265:848–70

Zartman, R. E., Hurley, P. M., Krueger, H. W., Giletti, B. J. 1970. A Permian disturbance of K-Ar radiometric ages in New England: Its occurrence and cause. *Geol. Soc. Am. Bull.* 81:3359–74

Zartman, R. E., Marvin, R. F. 1971. Radiometric age (Late Ordovician) of the Quincy, Cape Ann, and Peabody Granites from eastern Massachusetts. *Geol. Soc.* *Am. Bull.* 82:937–58

Zen, E. 1972. Some revisions in the interpretations of the Taconic allochthon in west-central Vermont. *Geol. Soc. Am. Bull.* 83:2573–88

Ann. Rev. Earth Planet. Sci. 1977. 5: 287–317

TRANSPORT PHENOMENA ×10075
IN CHEMICAL RATE PROCESSES
IN SEDIMENTS

Patrick A. Domenico

Department of Geology, University of Illinois at Urbana-Champaign,
Urbana, Illinois 61801

INTRODUCTION

Transport processes are defined as those processes which take some physical entity from one point to another in an extended array of molecules. The properties carried may or may not be material substances; examples might include a mass of molecules of one particular type or the heat of excess molecular motion. In the case of the former, the transport process is referred to as one of diffusion, dispersion, or convection of a chemical substance. The latter process may be referred to as conduction, thermal dispersion, or convection. Logically included as a transport process is the transport of momentum, which is the exertion of a force on a body, and the conductance of an electrical current. Common to all such irreversible processes are the transport of some entity, a driving force, and a move toward equilibrium.

The principles of transport are synthesized directly from the main ideas of fluid dynamics, thermodynamics, and heat and mass transfer, and are sufficiently basic to cut across specialty areas in science and engineering. These principles are used in reactor design and chromatographic separation, including the dynamics of sorption (Wilhelm 1962, Aris & Amundson 1973), in metallurgical technology (Szekely & Themelis 1971), in the soil sciences (Gardner 1965), and, more recently, in the life sciences (Lightfoot 1974). These same principles apply, more or less, to problems in the earth sciences, including terrestrial heat flow (Lee 1965), diffusion and heat conduction influenced by moving boundary conditions, where the movement of the boundary is generated by the accretion of sediments (Berner 1964, Tzur 1971), uplift or denudation (Benfield 1949, 1950), or where thermal conditions give rise to a change of state of material, such as melting or solidification (Kolodner 1956, Chuang & Szekely 1971). Other examples can be found in the vast literature on diffusion through clays and ocean sediments, the onset of free convection, and compaction currents as transport mechanisms, especially with regard to oil migration and low-temperature mineralization.

The versatility of transport equations and concepts is nowhere better illustrated than in the fourth volume of this review series. Treated there are specific transport models for surface water bodies (Lick 1976), the topside ionosphere (Banks, Schunk & Raitt 1976), and general treatment of multicomponent electrolyte diffusion (Anderson & Graf 1976). In addition, Burst (1976) discusses argillaceous sediment dewatering, which is a special case of a general transport phenomenon where the transported entity is a "fluid." These articles are all of a similar theoretical construct in that they treat the transport of either mass, momentum, or energy, or any two or all three of these entities in various environments. Subtle differences occur in different physical environments of application, or in the mathematical formulation of the balance equations. The latter is particularly true in the irreversible thermo-dynamic approach, which is treated later in this review. This present review, then, shares a similar point of view as those cited above in that interest is focused on some transported entity in a given physical environment. In particular, it will be concerned solely with the influence of sediment properties on mass and energy transfer in a porous medium, and the ensuing intrusion of these physical phenomena on chemical rate processes.

For purposes of discussion and analysis, the topic may be divided into two parts (Figure 1). The domain, or field A, signifies interparticle diffusion, conduction, and dispersions, the latter of which is caused by velocity variations due to the presence of the particles, and occurs in both longitudinal and transverse directions. These processes transport mass and heat, contributing to the overall concentration and temperature distributions within a porous medium. In addition, intraparticle con-duction is recognized as a heat transport mechanism, contributing to the thermal regime of parts of the porous system. Physical phenomena are thus understood as intraparticle and interparticle transport of mass and heat in porous media. The interphase field B refers to transfers between the mobile and stationary phases, which may be processes of exchange, adsorption, absorption, dissolution, precipi-tation, etc. The terms *mobile phase* and *stationary phase* refer to the fluids and the solids, respectively. Mass transfer from one phase to another proceeds at a decreasing rate until the two phases come to equilibrium with each other. Thermodynamic information regarding this equilibrium is basic to an understanding of mass transfer, but the concern here is on the transfer process itself and its encumbrance by physical processes.

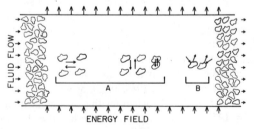

Figure 1   Major transport domains in a sedimentary bed: (A) interparticle diffusion, conduction, and dispersions, intraparticle conduction; (B) interphase transport.

This discussion opens with the empirical relations between the fluxes and the forces causing them in a fluid continuum as proposed by Fourier and Fick, but appropriately modified by convection and thermodynamic cross effects:

$$\mathbf{q} = -\kappa\nabla T + \rho C T\mathbf{v} - A_D\nabla c,$$
$$\mathbf{J} = -D\nabla c + c\mathbf{v} - A_S\nabla T, \tag{1}$$

where $\mathbf{q}$ and $\mathbf{J}$ are the fluxes of heat and mass per unit time per unit area, respectively; ∇c and ∇T are gradients in concentration and temperature; and $\mathbf{v}$, ρ, and C represent the fluid velocity, density, and specific heat. The proportionality constants D and κ are called, respectively, the coefficients of molecular diffusion and thermal conductivity. In addition, various cross effects, such as mass flow in relation to temperature gradient (thermal diffusion) and heat flow in response to a concentration gradient, are accounted for by appropriate terms proportional to these gradients, where A_D and A_S are coefficients representing the Dufour and Soret effects.

A second set of equations for the fluid continuum deal with the conservation of energy and mass:

$$\rho C\, \partial T/\partial t = -\nabla\cdot\mathbf{q} + W,$$
$$\partial c/\partial t = -\nabla\cdot\mathbf{J} + r, \tag{2}$$

where t is time; W is a heat source; and r is a homogeneous reaction term, that is, a reaction that proceeds in the liquid phase. Combining Equations 1 and 2, for constant coefficients, gives the main working equations for energy and mass transfer in a fluid continuum:

$$\partial T/\partial t = (\kappa/\rho C)\nabla^2 T - \nabla\cdot(T\mathbf{v}) + A_D\nabla^2 c + W/\rho C,$$
$$\partial c/\partial t = D\nabla^2 c - \nabla\cdot(c\mathbf{v}) + A_S\nabla^2 T + r, \tag{3}$$

where $\kappa/\rho C$ is the thermal diffusivity, hereafter designated as λ.

An important consideration here is that the mass balance equation given above describes one component only. For a mixture, there will be a system j of such equations of balance, one for each constituent. Further, Equation(s) 3 do not consider the simultaneous transport of mass and energy but treat each entity separately. Equations that combine the two transport phenomena have been presented by Dagan (1972).

MASS TRANSFER IN SEDIMENTS

The ultimate aim of a mathematical description of mass transfer in sediments may be stated as follows: A fluid is introduced into a porous medium. Given the initial concentration of the substances in the fluid, the nature of the interactions between the solid particles and the fluid, along with hydrodynamic, temperature, or other effects that influence motion or reaction, find the spatial distribution of one or more of the substances for any moment of time. In the phenomenological approach to such problems, three methods of attack are possible, all of which are used extensively. As a first consideration, thermodynamic cross effects are usually neglected, and we accept this assumption in order to conform with available literature. Given this condition, the main distinction arises between homogeneous and heterogeneous reactions. When the reaction is homogeneous, the reaction term normally appears as a source term in the governing differential equation, which is then solved directly for the given initial and boundary conditions. Danckwerts (1953) solved such a problem for a first-order consumptive reaction, as has Berner (1964) in treating sulfate distribution in pore waters. Ignoring thermal diffusion, the diffusion equation of Equation(s) 3 is strictly valid for the interior fluid of a pore, and is used directly in these problems, provided D is defined as the diffusion coefficient of the substance in a porous medium (not in water). When the reaction is heterogeneous, the rate of production of a chemical species does not normally appear in the differential equation, but rather as a boundary condition at the interface between the phases. This is because the reaction process takes place at the surface of separation between the liquid and the solid, and involves several steps in series, the slowest of which is rate controlling (Kingery 1959). A main advantage with this latter formulation is that it illustrates that chemical rate processes are often encumbered by physical processes. A main disadvantage is that this information is given on the microscopic, or pore size, scale only. This is demonstrated, for example, in Weyl's (1958) study on the solution kinetics of calcite, where the boundary condition at the solid-liquid interface expresses the saturation concentration of calcite, and the solution concentration is determined within a single capillary. Diffusion-controlled growth or dissolution theories of single crystals (Nielsen 1961, 1964), or concretions (Berner 1968), give similar microscale results that clearly demonstrate the control of transport on precipitation and dissolution.

The two methods described above merely require the direct application of the usual differential equation given in Equation(s) 3. A third method seeks the development of an appropriately averaged equation in which heterogeneous reactions in a porous medium take place at the liquid-solid interfaces distributed throughout the porous structure. If B is any scalar, vector, or second-order tensor associated with the fluid (that is, concentration, velocity, etc), the volume average of B is defined as

$$\bar{B} = (1/V) \int_{V_f} B \, dV, \tag{4}$$

where the overbar indicates an averaged quantity, V is conceived to be a total averaging volume containing both solids and pores, and V_f is the volume of fluid contained in V. The averaging theorem of Whitaker (1966) and Slattery (1967, 1969, 1972) provides the relationship between the gradient of the average and the average of the gradient:

$$\overline{\nabla \cdot \mathbf{B}} = \nabla \cdot \bar{\mathbf{B}} + (1/V) \int_{S_w} \mathbf{B} \cdot \mathbf{n} \, dS, \tag{5}$$

where $\mathbf{n}$ is the inward unit vector normal to the pore surface, and S_w is the interfacial area between the fluid and the solid.

The application of Equation 5 to the transport equations provides a set of suitably averaged conservation equations that may be related to measurable physical properties of the porous medium. The volume-averaged equation of continuity of Equation 2 thus becomes

$$\partial \bar{c}/\partial t + \mathbf{V} \cdot \bar{\mathbf{J}} = \bar{r} + r_s, \tag{6}$$

where r_s is the rate at which some species are produced by chemical reaction at the fluid-solid interface, and has the general form

$$r_s = (1/V) \int (\mathbf{J} \cdot \mathbf{n}) \, dS = (\Sigma/S_w) \int (\mathbf{J} \cdot \mathbf{n}) \, dS, \tag{7}$$

where the integral is over the pore surface enclosed by V, $(\mathbf{J} \cdot \mathbf{n})$ is the flux of some species across the fluid-solid interface, and Σ is the specific surface of the porous medium, defined as the surface area of the pores S_w per unit bulk volume of porous medium V. Substituting the volume average of Fick's Law into Equation 6 (ignoring thermal diffusion) gives (Slattery 1972):

$$\partial \bar{c}/\partial t + \text{div} \, (\bar{c} \bar{\mathbf{v}}) = -\text{div} \, \mathbf{J}_e + \bar{r} + r_s, \tag{8}$$

where

$$\mathbf{J}_e = -D\nabla \bar{c} - \xi. \tag{9}$$

In this formulation, $\mathbf{J}_e$ is an *effective mass flux vector*. ξ includes several terms that account for the fact that the average of a product is not the same as the product of two averages, as assumed in the convective term in Equation 8, and an additional term that accounts for the twisting and turning in the pores through which the fluid moves. Ignoring the former terms, ξ takes on the form of a *tortuosity vector*, and is expressed:

$$\xi = (D/V) \int_{S_w} c\mathbf{n} \, dS. \tag{10}$$

Because an immediate goal is to relate these averaged equations to measurable physical properties of a porous medium, it is now possible to ask for a physical and mathematical interpretation of the transport coefficients and heterogeneous reaction terms in a medium possessing tortuosity.

INTERPARTICLE TRANSPORT

The two main processes involved in the movement of a chemical substance in accordance with Equation 8 are convection and molecular diffusion. The process of diffusion plus convection is more complicated than presented here because the microscopic velocity of the fluid is not the same everywhere. That is, the velocity near the center of a pore is not the same as near the edge, and the velocity in narrow pores is less than that in wider, less constricted pores. Further complicating factors are exchanges of molecules between stream paths through lateral diffusion, and possible flow velocity distortions due to viscosity or density variation. One result of these phenomena is a mixing, or mechanical dispersion process. The term *dispersion* has been suggested to distinguish this type of spreading from that due to diffusion alone (Scheidegger 1961). Where diffusion is due to a random thermal motion of solute molecules, mechanical dispersion is due to variations in velocity of flow through a complex pore structure. Although a dispersive flux is most often treated analogously to a diffusive flux within the framework of a diffusion model (that is, proportional to some concentration gradient), the transport coefficient with convection will not be the same coefficient that applies in the absence of water movement.

Diffusion

For the case of no convection ($v = 0$, Equation 8), the presentation is straightforward. Equation 9 becomes:

$$\mathbf{J}_e = -D\left(\nabla \bar{c} + 1/V \int_{S_w} c\mathbf{n} \, dS\right) = -D_e \nabla \bar{c}, \tag{11}$$

where D_e is an *effective diffusion coefficient* in a porous medium and can account for the hindering of free diffusion by collision and interaction with sediment particles. According to Whitaker (1967), the integral over the interstitial surface area represents the reduction in free molecular diffusive transport owing to the tortuosity of the porous medium. Hence, the effective diffusion coefficient D_e is less than the molecular diffusivity D in water. This fact has been well known for many years, and generally accounted for in most diffusion models applied to porous media. Orders of magnitude for diffusion coefficients in various environments have been given by Lerman (1971); a general guide to values of diffusion coefficients in sediments has been given by Manheim (1970).

It is readily accepted that every porous rock is somewhat unique with respect to porosity, the spatial arrangement of the continuous channels, pore size, and degree of pore interconnection. An empirical evaluation of an effective diffusion coefficient is called for, therefore necessitating experimental measurements. General attempts along this line have been to relate diffusion coefficients in porous media to those in aqueous solutions. In an aqueous solution, a ratio of the diffusion coefficient to the effective diffusion coefficient gives a quantity called the *diffusion ratio*, which expresses the total effect of pore structure on the rate of diffusion through the void spaces of the solid. For example, the simple relationship

$$D_e/D = 1/(2)^{1/2} = 0.707 \tag{12}$$

is reported as reasonably supported by the data of several investigators (Perkins & Johnson 1963). A more physically based approach would entail consideration of some functional relations for ξ in Equation 11. For a nonoriented porous solid in the absence of convection, Slattery (1972) suggests:

$$\xi = \xi(L, D, \Psi, \nabla \bar{c}), \tag{13}$$

where L is a characteristic length and Ψ is porosity. This leads, eventually, to

$$D_e = DD^*,   \tag{14}$$

where

$$D^* = D^*(\Psi, L, \nabla \bar{c} \, \bar{\rho}^{-1}).   \tag{15}$$

These equations merely state that the effective diffusion coefficient in a porous medium is the product of the diffusion coefficient D in aqueous solutions and some parameter D^* which is a function of the above-cited scalar and vector quantities. In this case, D^* serves the purpose of a retardation factor, several of which have been proposed under different premises. Most retardation factors incorporate porosity and/or some dimensionless length such as tortuosity, which is a ratio of the length of flow channel for a fluid particle with respect to the length of the porous medium. Helfferich's review (1966) reveals that most effective diffusion coefficients used for ion exchange columns fall between the extremes

$$D_e = D(\Psi/2)   \text{and}   D[\Psi/(2-\Psi)]^2.   \tag{16}$$

In these equations, the retardation factor depends only on porosity. Similar relations have been proposed by soil scientists (Kirkham & Powers 1972) where, in general,

$$D_e = u\Psi D,   \tag{17}$$

where u is some number less than one. Other investigators, mainly chemical engineers (Evans, Watson & Mason 1961, Greenkorn & Kessler 1972) have incorporated tortuosity τ, and porosity

$$D_e = (\Psi/\tau)D,   \tag{18}$$

where Ψ/τ is taken as some characteristic constant that provides for retardation. In some geochemical work, use is made of an expression similar to Equation 18 except that tortuosity is taken to the second power (Berner 1971). Li & Gregory (1974), for example, suggest a retardation factor u/τ^2, where u is a constant less than one, but may approach one in highly porous deep sea sediments. Bear (1972), on the other hand, defines all transport coefficients as the product of porosity, tortuosity, and the conductive property in the bulk (that is, for this case, the diffusion coefficient in the absence of a porous matrix). In view of these rather diverse approximations for D^*, effective diffusion coefficients used in model studies are somewhat ambiguous, and can only serve as crude estimates. Further, at this stage of our ignorance of actual pore structure, there appears to be some justification for utilizing constant effect diffusion coefficients in model studies in the interests of mathematical tractability, rather than those that are complicated functions of concentration, temperature, or viscosity.

Dispersion

When convective transport takes place simultaneously with diffusion, most investigators have focused attention on the convective term in Equation 8. The procedure followed by Whitaker (1967), Bear & Bachmat (1967), and Dybbs & Schweitzer (1973) in an analogous study of energy transfer is to consider variations from the average value of velocity and concentration in the convective term. The net result is a dispersive flux that is different from the diffusive flux discussed above, and is referred to as mechanical dispersion. In this sense, mechanical dispersion is similar to turbulent diffusivity in a viscous fluid, with the exception that turbulent diffusivity is created by velocity fluctuations from the average time velocity, whereas mechanical dispersion is created by velocity fluctuations from the mean space velocity (Saffman 1959, 1960).

It follows that the phenomenon of mixing, accompanying movement of a chemical species in porous media, is commonly handled by the diffusion-convection equation where the diffusion coefficient is replaced by a dispersion coefficient. As in the case of effective diffusion coefficients, it is appropriate to consider some functional relations for ξ in Equation 10 in the interest of providing a physical basis for the dispersion coefficient. For a nonoriented porous solid in which the solids undergo no movement, Slattery (1972) suggests a functional relationship which depends on the parameters already cited for D^*, as well as the velocity of flow, the gradient of an average concentration, and the Peclet number for mass transport, expressed as:

$$N_{Pe} = vL/D_e. \tag{19}$$

These statements merely indicate that the dispersion process is composed of molecular diffusion plus some contribution from the velocity of flow, a fact that has been well known for several years (Taylor 1953, Aris 1956, Saffman 1960). The combined effect of molecular diffusion and mechanical dispersion is referred to as hydrodynamic dispersion (Bear 1972).

The dependences cited above have been substantiated by numerous experimental and theoretical studies. Some workers emphasize the correlations between mechanical dispersion and dimensionally identical quantities. Bear and Todd (Bear 1972), for example, suggest that the coefficient of mechanical dispersion is equal to the product of some characteristic length of the medium and the mean velocity. The characteristic length is referred to as the *dispersivity* of the medium. Raimondi, Gardner & Petrick (1959), working with packs of beads, express mechanical dispersion as proportional to velocity of flow, particle diameter, and an additional factor that is a measure of pack inhomogeneity. The greater the degree of inhomogeneity, the greater the dispersion. In some theoretical work, Taylor (1953) showed that the convective contributions to dispersion can be approximated by $a^2v^2/48D$, where a is the radius of capillary. For a porous medium, the coefficient of hydrodynamic dispersion has been approximated to be (Collins 1961):

$$D_H = D_e + Ga^2v^2/D_e, \tag{20}$$

where D_H is the coefficient of hydrodynamic dispersion, and G is a geometrical factor. Note that when the velocity goes to zero, the dispersion coefficient is replaced by an effective diffusion coefficient.

Further experimental verification of the parameters contributing to dispersion has been accomplished by studies that show the relation between velocity, molecular diffusion, and hydrodynamic dispersion. The main aspects of this work are readily demonstrated by diagrams that plot the ratio of hydrodynamic dispersion to molecular diffusion versus the Peclet number (Blackwell, Rayne & Terry 1959). Additional correlations have been prepared between the Peclet number and the Reynolds number (Perkins & Johnson 1963). At low Peclet numbers corresponding to low velocities (or low Reynold's numbers), molecular diffusion predominates (Figure 2). At high Peclet numbers, the main mixing is due to mechanical dispersion, suggesting that molecular diffusion can be neglected. These relations have been verified theoretically (Rumer 1972).

Other variables important to the dispersion process include particle size distribution and particle shape. In general, a decrease in grain size increases the inhomogeneity factor introduced by Raimondi and his coworkers (1959), increasing dispersion. Dispersion increases with wide particle size distribution, presumably because small particles can easily reside in the pore spaces between the large particles (Perkins & Johnson 1963). Further, dispersion is generally greater for packs of nonspherical particles than for spherical particles. In reservoir rocks, the large-scale inhomogeneities, ranging from a few inches to a few feet, appear to contribute more to dispersion than the pore to pore variations (Byevich, Leonov & Safrai 1969). The macroscopic factors which cause these inhomogeneities result from stratification and preferred flow channels in the media, and from viscosity and density differences in the media's fluid. This is supported by the work of Skibitzke & Robinson (1963), who relate dispersion to the refraction of flow at heterogeneity boundaries, and by field dispersion measurements at the Hanford reactor site, which are several times larger than laboratory measurements, a factor attributed to the wide distribution of permeability (Theis 1963).

As indicated above, the microscopic factors of dispersion predominate in natural rocks only in the case of a very homogeneous medium. With regard to dispersion in natural rocks, most authorities would agree on the following points:

1. Dispersion can be divided into two components which are parallel and perpendicular, respectively, to the velocity vector. The component of dispersion parallel

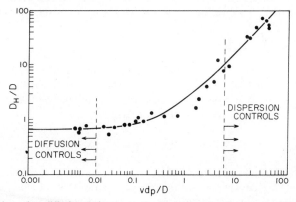

Figure 2 Dispersion-diffusion ratio versus the Peclet number for unconsolidated random packs of uniform size sand or beads where dp is the average diameter of the particles (Perkins & Johnson 1963). Reprinted with permission of *J. Soc. Pet. Eng.*

to the velocity is greater than the lateral or transverse component, as is evidenced by the characteristic elongate shape of tracer material injected at large-scale experimental sites (Barraclough, Teasdale & Jensen 1965). The ratio of longitudinal to transverse dispersion generally falls between 5 to 7 although values as high as 24 have been reported (Bear 1972).

2. The amount of dispersion is directly related to large-scale inhomogeneity of the medium. The inhomogeneities result from stratification and are often reflected by a wide distribution of permeability, both horizontally and vertically. Because of these large-scale inhomogeneities, there is currently no effective way to measure the coefficient of dispersion in the field.

3. The rate of dispersion is directly related to the fluid velocity. Although molecular diffusion is important under conditions of stagnation or for very slow flow, the contribution of molecular diffusion to dispersion in many porous flow situations can be neglected.

As indicated above, hydrodynamic dispersion is a well-investigated phenomenon. Indeed, according to one account (Banks & Jerasate 1962), more than 100 papers appeared on one aspect or another of dispersion during the period 1953–1962. Complete reviews on the subject applicable to porous media have been prepared by Perkins & Johnson (1963) and, more recently, Bear (1972). Details of computation of the dispersion coefficient from experimental data are given by Kirkham & Powers (1972). Dispersion in other environments, such as estuaries and open channels, has been examined by Holley, Harleman & Fischer (1970), and Sooky (1969).

INTERPHASE TRANSPORT

Insight into the physical meaning of the reaction term r_s of Equation 7 and the mathematical models that may be used to represent it may be obtained by considering a material balance for an elementary layer. In one-dimensional form we have:

$$\partial c/\partial t + (1/\Psi)\, \partial n/\partial t = D_H\, \partial^2 c/\partial x^2 - v\, \partial c/\partial x, \tag{21}$$

where c is the concentration of solute in the liquid phase and n is the concentration of solute in the solid phase. As material in solution is generated at the expense of the solid portion of the medium (or removed from solution with a commensurate increase in the volume of solid material) we have:

$$r_s = (1/\Psi)\, \partial n/\partial t = (\Sigma/S_w) \int (\mathbf{J} \cdot \mathbf{n})\, dS. \tag{22}$$

Equation 22 involves no assumptions about the mechanism of reaction. If there is a finite rate of transfer of solute from the solid material to solution, or vice versa, it can be assumed that the rate depends upon the pertinent parameters of the problem. These include the concentration of all the components both in the mobile phase $(c_1, c_2 \ldots c_j)$ and in the solid $(n_1, n_2 \ldots n_j)$, the velocity, temperature, and density of the moving fluid, and other parameters that determine whether the reaction is rate controlled (K_{chem}) or diffusion controlled (K_{diff}). For a single component, then, we have:

$$r_s = F(c, n, v, T, \rho, K_{chem}, K_{diff}). \tag{23}$$

If the process is reversible, the resultant rate is equal to the difference between the forward and reverse rates:

$$r_s = (\partial n/\partial t)_{forw} - (\partial n/\partial t)_{rev}. \tag{24}$$

When the physical factors v, T, ρ, K_{chem}, K_{diff} remain constant throughout the process, we have:

$$r_s = f(c, n). \tag{25}$$

If the rate of reaction is very fast in comparison to the rate of convective and dispersive transport, equilibrium may be established immediately between the dissolved ions and solid minerals at every point, giving

$$n = g(c). \tag{26}$$

The function $g(c)$ is called the adsorption isotherm because it is fixed when the temperature is fixed. As applied to Equation 21, the assumption of equilibrium means that $n(x) = g[c(x)]$, indicating that the derivative of the function $g(c)$ will be contained in the differential equation in place of $\partial n/\partial t$. In short, the $\partial n/\partial t$ terms are converted into terms containing only time derivatives of the solute concentrations, enormously simplifying the mathematical problem.

From Equation 23 through 26 it is clear that either of two approaches may be followed to express the dynamics of heterogeneous interphase transport: equilibrium and nonequilibrium. In the case of the former, reaction rates are assumed to be fast with respect to transport of constituents through the system; in the latter case, a further distinction is required between transport and rate controlled reactions.

Equilibrium Interphase Transport

If pointwise equilibrium is established at every point, it is necessary to formulate the appropriate adsorption isotherm $n = g(c)$. A common assumption is that $n = Kc$, or $n = K_1 c + K_2$, where K is a constant. Equation 21 then becomes:

$$(1 + K/\Psi)\, \partial c/\partial t = D_H\, \partial^2 c/\partial x^2 - v\, \partial c/\partial x, \tag{27}$$

which readily demonstrates that equilibrium assumptions transfer the $\partial n/\partial t$ terms into a tractable $\partial c/\partial t$ term. Aris & Amundson (1973) cite several other forms for the adsorption isotherm, all of which have this property.

Equilibrium assumptions have been used primarily by geochemists in diagenesis models of marine sediments (Berner 1974), by chemical engineers interested in the effects of diffusion and longitudinal dispersion in ion exchange and chromatographic columns (Lapidus & Amundson 1952, Kasten, Lapidus & Amundson 1952), and by soil scientists interested in the effects of dispersion and ion exchange in porous media (Lai & Jurinak 1971, Rubin & James 1973). The intrusion of physical phenomena on equilibrium-controlled ion exchange is readily demonstrated by the theoretical breakthrough curves given in Figure 3. A breakthrough curve is a plot of c/c_o versus the number of pore volumes of effluent collected, where c is the concentration of the solute found in the effluent, and c_o is the initial concentration of the solute in the displacing fluid. The ratio c/c_o will be zero at first and then approach one as the concentration of the effluent approaches that of the displacing fluid; that is, as the displacing fluid *breaks through*. The curves A, B, C, and D are for decreasing dispersivity with fixed velocity, or for increasing velocity with fixed dispersivity. Hence, when dispersion is small (or velocity is large) the breakthrough is nearly instantaneous, so that the concentration ratio c/c_o quickly approaches unity and remains so as long as the displacing solution is passed through the tube (Curve D). The effect of increased dispersion with fixed velocity is readily seen in curves C, B, and A, which show a gradual rise in the ratio c/c_o with increased pore volumes (time). Curves E, F, and G illustrate the effect with a lower capacity adsorbent.

It is mentioned that the results of Figure 3 are theoretical, and do not include the effects of velocity on dispersion. Further, the equilibrium assumption means that although dispersive transport acts to spread and therefore dilute the concentration front, any ensuing reduction in chemical potential at a point in no way affects the equilibrium rate of exchange. For situations where the rate of exchange is infinitely fast with respect to transport, the Peclet number is the only important dimensionless group affecting the outcome. If dispersion or convective transport is large, the equilibrium assumption becomes questionable, and some rate process must be used for the reaction mechanism. It follows that the equilibrium assumption appears most useful in situations where the velocity of flow is small, thereby accompanied by diffusion rather than dispersion as the concurrent transport mechanism.

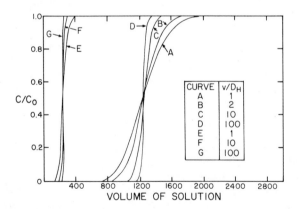

Figure 3 Plot showing the effect of longitudinal dispersion in a column in which equilibrium is established locally. Initial adsorbate concentration is zero (Lapidus & Amundson 1952). Reprinted with permission of *J. Phys. Chem.* Copyright by the American Chemical Society.

Nonequilibrium Interphase Transport

The general case of a reaction between a solid and a liquid may comprise as many as five steps in a series: (a) transport of solute molecules to the interface; (b) adsorption at the interface; (c) reaction at the surface; (d) desorption of products at the interface; (e) transport of products from the interface. In this scheme, steps (b), (c), and (d) constitute an interaction between the solid and the liquid, and are termed chemical processes (Equation 23). If these steps are considerably more rapid than those controlling transport [(a) and (e)], the transport of material and products is considered to be rate controlling. In general, then, it is possible to classify non-equilibrium processes on the basis of the rate-determining step (Bircumshaw & Riddiford 1952):

1. Transport control, where the transport of material to and from the interface is much slower than the rate of chemical reaction at the interface.
2. Chemical control, where the rate of reaction at the interface is slower than the transport.
3. Intermediate control, where both reaction and transport are of the same order.

The intermediate case is obviously the general one, with either transport or chemical control being of the nature of limiting cases. The immediate problem is to choose a mathematical form for r_s in Equation 7 that describes the mechanism of the solid-liquid interaction.

DIFFUSION AS THE RATE-CONTROLLING STEP   According to Nernst (Kingery 1959) it is likely that chemical processes at the interface of solids dissolving in a liquid are very much faster than the transport of materials to or away from the interface, which indicates that the observed rate is transport controlled. If the reaction at the interface is sufficiently rapid so as to maintain an effective concentration corresponding to saturation, the observed rate is simply the rate at which materials diffuse across a thin layer of solution adhering to the solid surface. For solids dissolving in a liquid, the concentration gradient across the layer is assumed to be linear, and is expressed $(c_{eq} - c)/\sigma$, where c_{eq} is the saturation concentration at the surface, c is the concentration in the bulk solution, and σ is the thickness of the stationary layer attached to the interface. The flux $\mathbf{J}$ in moles vol^{-1} time^{-1} is then expressed:

$$\mathbf{J} = (DA/\sigma)(c_{eq} - c), \tag{28}$$

where A is the surface area of the material per unit volume of water. The ratio D/σ is referred to as the mass transfer coefficient k_d, and has the units of velocity. In this form, the rate of solution is the product of a driving force, which is the distance from saturation $(c_{eq} - c)$, a mass transfer coefficient, and the surface area of the material per unit volume. The stationary layer varies inversely with velocity of flow, presumably getting thinner with increased velocity.

The flux equation cited above has been used to describe diffusion-controlled crystal growth (Frank 1950, Nielson 1961), concretion formation (Berner 1968), and the weathering of feldspars (Wollast 1967). It has been used in transport equations to describe the occurrence of silica in pore waters due to the dissolution of opaline silica (Berner 1974). Other systems that are currently accepted as diffusion-controlled include the oxidation rate for certain metals (Kingery 1959) and the dissolution of calcite (Weyl 1958).

When the reaction takes place at the liquid-solid interfaces distributed throughout a porous structure, diffusion-controlled reactions are readily expressed from the general form of r_s:

$$r_s = (\Sigma/S_w) \int (\mathbf{J} \cdot \mathbf{n})\, dS = \overline{(\mathbf{J} \cdot \mathbf{n})}\Sigma = (D\Sigma/\sigma)(c_{eq} - \bar{c}). \tag{29}$$

Equation 29 has been employed in transport equations to examine the physical and kinetic processes that tend to promote an approach to equilibrium in carbonate systems (Palciauskas & Domenico 1976). The total reaction term incorporates Equation 29 plus some further contributions to calcium in solution from reactions other than at the fluid-calcite interface. Distance to saturation in a porous carbonate system is calculated to be:

$$x_s = \frac{-2}{(v/D_H)[1 - (1 + 4D\Sigma D_H/\sigma v^2)^{1/2}]} \log \frac{c_{ss} - c_o}{c_{ss} - c_{eq}}, \tag{30}$$

where c_{ss} is the concentration of calcium when its rate of production is just balanced by its rate of consumption, i.e. the concentration gradient goes to zero, and c_o is the concentration at the formation entrance. As the bracketed quantity in the denominator becomes more negative, the saturation distance decreases. Hence, when the spatial dependence of the ionic constituents is examined from the point of view of these processes, the results indicate that distance to attainment of saturation, with respect to an individual mineral, increases with increasing rates of dispersion and velocity of flow, and decreases with increasing rates of $D\Sigma/\sigma$. The process of diffusion in the above formulation is thus significant to interparticle transport in the sense that it augments the spreading due to mechanical dispersion, and to interphase transport in that it exerts direct control on the liquid-solid exchanges that take place on the pore size scale. This suggests a mechanism of self-regulation for diffusion-controlled reactions in porous systems. As velocity and dispersion get large with respect to kinetics, distance to saturation is increased, thereby tending to increase porosity development in a larger part of the formation. A high rate of flow or dispersion, however, speeds up the kinetics through its influence on the diffusive layer, increasing dissolution and tending to decrease the distance to saturation. The system is thereby recognized as self regulating in its attempts to approach equilibrium. An overall view of self regulatory or cybernetic chemical systems is given by Frank-Kamenetskii (1969).

ADSORPTION-DESORPTION AT THE INTERFACE   Surface reactions can be processes of adsorption whereby a component is extracted from the liquid stream and becomes part of the solid matrix, or a reverse process, desorption, whereby a soluble component of the solid becomes part of the solute. Absorption and desorption are thus processes that occur at an interface. Several forms for the process are available, both for equilibrium and nonequilibrium dynamics. According to Klotz (1946), one of the earliest formulations assumes that the irreversible local rate of removal is governed by

$$(1/\Psi)\, \partial n/\partial t = K_1 c(N_o - n), \tag{31}$$

where K_1 is a constant and N_o is the saturation concentration of the solid. The analogy between this form and that of diffusion control, at least with respect to the *driving forces* (Equation 28), is obvious. Adsorption with Langmuir-type kinetics can be expressed:

$$(1/\Psi)\, \partial n/\partial t = K_1 c(N_o - n) - K_2 n, \tag{32}$$

where K_1 and K_2 are constants. The reference to Langmuir-type kinetics enters because, in the limit that $\partial n/\partial t$ equals zero, we have

$$n = K_1 N_o c/(K_2 + K_1 c), \tag{33}$$

which is the Langmuir isotherm. This case has been considered by Thomas (1944) for ion exchange, where he stipulated that rate processes are slow compared to transport processes. Once it became known that ion exchange is controlled by diffusion rather than chemical reaction, the coefficients in adsorption isotherms have been considered to be overall mass transfer coefficients relating to film thickness (Helfferich 1966). Amundson (1948, 1950) has presented solutions to the transport equations utilizing both adsorption terms given above but ignoring dispersion.

To be physically realistic as well as mathematically tractable, a reversible adsorption process might best be presented in the form:

$$r_s = \Sigma(k_1 c - k_2 n), \tag{34}$$

where k_1 and k_2 possess the units of a velocity, and are forward (sorption) and backward (desorption) rate constants, respectively. In general, Equation 34 states that sorption and desorption occur simultaneously and at different rates. It is noted that Equation 34 is readily converted to the case of diffusion control at the particle surface. If n is taken as $k_1 c_{eq}$, where c_{eq} is the saturation concentration, Equation 34 becomes

$$r_s = \Sigma k_d(c - c_{eq}), \tag{35}$$

where k_d is a mass transfer coefficient.

The utilization of the reversible adsorption process described above is widespread in transport problems. The transport equations have been solved first for the condition of convection only (Amundson 1950) and later for both convection and dispersion (Lapidus & Amundson 1952). Deju (1971), supported by earlier experimental work (Deju & Bhappu 1966, 1970), utilized this form of adsorption to describe the chemical weathering of silicate minerals. Some results of these studies are shown in Figure 4, which is a plot of the concentration ratio c/c_o versus a dimensionless group consisting of velocity, time, and distance. For infinitely fast reactions, ignoring dispersion, the change in concentration due to reaction is

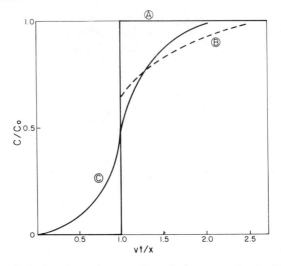

Figure 4   Plot of concentration ratio versus dimensionless group showing (*A*) an infinitely fast reaction in the absence of dispersion, (*B*) a finite rate reaction in the absence of dispersion, and (*C*) dispersion in the absence of sorption (Deju & Bhappu 1970). Reprinted with permission from *Trans. Am. Inst. Min. Metall. Pet. Eng.*

preserved without distortion as the interface moves through the porous bed (Curve *A*). The concentration ratio jumps from zero to unity after one pore volume is displaced, and remains so as long as the displacing fluid is passed through the bed. This, of course, is the assumption of equilibrium dynamics under plug flow conditions, and the result is similar to curve *D* of Figure 3, with the exception that dispersion in Figure 3 tends to slightly smear the sharp breakthrough. For finite rate reactions without dispersion, curves of type *B* are obtained. These results can be compared with curves of type *C*, where dispersion but not sorption is considered. Hence, the effect of dispersion with sorption would be to delay further the attainment of unity for the concentration ratio. As a general rule, the reaction rate constant in chemically controlled reactions varies exponentially with temperature, but is independent of the transport mechanisms of fluid flow and dispersion.

SURFACE REACTIONS AS SPECIAL CASES OF HOMOGENEOUS REACTIONS   In reaction controlled precipitation and dissolution, an analogy may be made to the kinetics of homogeneous solution reactions. The reaction rate constant for a first-order homogeneous reaction has the units $(time)^{-1}$; that is, volume reacted/(unit volume)(unit time). Reaction rate constants for heterogeneous reactions have the units $(length)(time)^{-1}$; that is (volume reacted)/(unit surface area)(time). Clearly, the rate equation for heterogeneous reactions must incorporate the reactive surface area per unit volume of porous material. Thus, one may convert a homogeneous solution reaction to one representing surface reaction by incorporation of the *concentration* of the solid through the surface area. For interfaces distributed throughout a porous structure, this may be accomplished by incorporating the specific surface. A homogeneous solution rate may thus be expressed for surface reactions as

$$dc/dt = k\Sigma c^m, \tag{36}$$

where k is a rate constant with the units of a velocity, and m is the order of the reaction.

In the most general case, the reaction rate of Equation 36 is a function of the concentrations of all the components taking part in the reaction. It has been shown, however, that when the dispersion coefficients of all the components taking part in the reaction are identical, the relations between the concentrations are uniquely defined by the stoichiometry of the reaction and their initial concentration values (Burghardt & Zaleski 1968). In this regard, whenever the Peclet numbers of all the components taking part in the reaction are equal, it is permissible to express the reaction rate as a function of the concentration of one component only. Reaction forms equivalent to Equation 36 have been used extensively in the examination of diagenetic phenomena in accumulating sediments, and in reactor design.

Numerous one–dimensional diagenetic models have been proposed to assess the effects of diffusion, advection, sediment accumulation, reaction rates, and compaction on the chemical composition of pore water in marine sediments undergoing iso-thermal diagenesis. First-order kinetics of the form given by Equation 36 have been assumed for bacterial sulfate reduction (Berner 1964) and for the production of ammonia from decomposition of organic nitrogen compounds (Berner 1974). First- and zero-order kinetics in combination with diffusion control have been employed to examine anaerobic phosphate diagenesis (Berner 1974) and diagenesis in general (Tzur 1971). Two things are worth pointing out concerning these studies. First, the transport equations are generally applied to accumulating sedimentary piles so that the convective term is viewed as a rate of burial of sediment grains below the sediment-water interface. Effects of compaction are usually ignored if porosity-depth profiles are not available. Second, a main purpose of these studies parallels the purposes of the experimental kineticist who seeks chemical kinetic knowledge but must first disentangle the intruding physical effects. This is accomplished via a curve-fitting procedure utilizing the solution to the transport equations and actual concentration versus depth profiles in marine sediments.

If geochemists are interested in disentangling the physical effects in order to understand kinetic phenomena, chemical engineers are more concerned with ascer-taining how these same physical effects enter in their calculations of reactor output. A reactant is presumed to be supplied to the system inlet and can undergo consumption in accordance with a first-order reaction of the form given by Equation 36 (Danckwerts 1953; Wehner & Wilhelm 1956; Levenspiel & Bischoff 1959, 1961). The transport equations are then solved for the case where dispersion is important and for the case of plug flow ($D_H = 0$). The greater the dispersive effects, for a given velocity, the larger the volume required for the reactor to achieve a given fractional conversion. This is demonstrated in Figure 5, where the volume required for a given fractional conversion (c/c_o) is expressed as a multiple of the volume required under

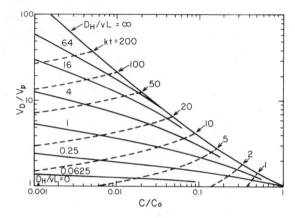

Figure 5 Comparison of performance for plug flow and dispersed flow models with a first-order consumptive reaction (Levenspiel & Bischoff, 1959, 1961). Reprinted with permission from *Ind. Eng. Chem.* Copyright by the American Chemical Society.

ideal plug flow conditions, where V_D represents the volume required when dispersion is important, and V_p represents the volume required in the absence of dispersion. The solid lines represent various measures of dispersion, represented by the inverse of the Peclet number, whereas the dotted lines are characteristic of chemical reaction rate constants, represented by the product of the rate constant and the average residence time for plug flow reactors, determined by the ratio of reactor length to average velocity. For large values of the inverse of the Peclet number and high conversion ($c/c_o < 0.1$, low fraction of reactant remaining), a significantly larger volume is required than under conditions of ideal plug flow. For a given reaction rate constant, a decrease in dispersivity results in an increase in reactant consumption over a progressively smaller volume of porous media. Hence, as with the physical-chemical models discussed thus far, interest is centered on the effects of transport on chemical rate processes.

The idea of converting solution reactions to reactions at the surface can be extended to reversible surface reactions for crystal growth and dissolution in a porous media (Nancollas & Purdie 1964). If Equation 36 is accepted as a precipitation rate, dissolution may be expressed as $k'\Sigma$. In a saturated solution, the rates are equal, giving $k' = kc_{eq}^m$. In an undersaturated solution, the net rate of dissolution is expressed:

$$dc/dt = k\Sigma(c_{eq}^m - c^m).\tag{37}$$

As pointed out by Nancollas & Purdie (1964), this equation does not fit experimental results, and is replaced by:

$$dc/dt = k\Sigma(c_{eq} - c)^m.\tag{38}$$

Diffusion, rather than surface control, is indicated when $m = 1$.

REACTIONS OF THE INTERMEDIATE TYPE   Intermediate control, where both reaction at the surface and transport are of the same order, has been examined and reviewed by several investigators (Bircumshaw & Riddiford 1952, Wilhelm 1962, Frank-Kamenetskii 1969). The rate of the chemical process may be obtained by setting the equation for diffusion control equal to that of chemical control which, for a first-order reaction, results, eventually, in (Bircumshaw & Riddiford 1952):

$$r_s = [kk_d/(k+k_d)]\Sigma c, \tag{39}$$

where the bracketed quantity is designated an overall reaction rate constant. If $k_d \gg k$, the rate is chemically controlled, whereas the condition $k_d \ll k$ indicates diffusion control.

We recall that k varies exponentially with temperature whereas, by contrast, the mass transfer coefficient k_d has a small temperature dependence and a substantial dependence on fluid velocity. For heterogeneous chemical processes, it can be argued that it is not possible for the overall rate of reaction to increase without limit if heat is added to the porous system. If surface reactions are normally small compared to diffusion rates at a given temperature, the overall rate is chemically controlled. An increase in temperature will cause an exponential rise in the surface reaction in accordance with Arrhenius' equation, but will have a small effect on the mass transfer coefficient. Surface kinetics will thus speed up, eventually overhaul, and become fast with respect to transport, thereby forcing the reaction process from the kinetic to the diffusion–controlled regime (Frank-Kamenetskii 1969). Homogeneous liquid–liquid reactions, on the other hand, may increase without limit in high-temperature environments. These same ideas apply to exothermic reactions taking place in the absence of outside energy. At least in theory, chemically controlled exothermic reactions can pass from the kinetic to the diffusion regime, whereas endothermic reactions, which are self regulating in a thermal sense, cannot.

Reactions as Attenuation Mechanisms in Subsurface Contaminant Transfer

Underground disposal of radioactive wastes implies the placement of a complex chemical solution within a complex natural environment. The most desirable natural environment would isolate the waste product until the radioactivity decays to a safe level. The mathematical approach to this problem has elements that parallel the concerns of chemical engineers in reactor design. The porous medium, like the reactor, is presumed to be supplied with a liquid product that is subject to various physical and chemical processes. Dispersion, for example, can thoroughly spread a contaminant front through a porous media, increasing the size of the affected area while diluting the contaminant in the process. Dispersion is thus recognized as an important concentration attenuation, or thinning, mechanism. Ion exchange and radioactive decay are two of the most important chemical processes. Column studies (Kaufman, Orcutt & Klein 1956) indicate that ion exchange reactions may retard the advance of radiostrontium concentration fronts by as much as 1/40 of that of the liquid front. Ion exchange is thus another important attenuation mechanism, as is radioactive decay, and its role in radioactive waste disposal has been reviewed extensively by Robinson (1962). The potential significance of other reactions and various mineral species in subsurface environments has been reviewed by Roedder (1959).

The attenuation aspects of dispersion, ion exchange with a variable cation exchange capacity, ion charge, and spontaneous radioactive decay in subsurface contaminant transfer has been examined in model studies by Schwartz (1975). The pertinent conclusions here are somewhat analogous to those arrived at in the study of reactor behavior and summarized in Figure 5. Namely, the extent or spread of the liquid contaminant is fixed by the physical transport process. The chemical processes act to reduce the size of the contaminated area to some fraction of that which is attributable to physical processes alone. In addition, the heat generation aspect of radioactive waste disposal presents critical problems (Skibitzke 1961, Roedder 1959). Excellent reviews of these topics are presented in the readings of the *Second International Symposium on Underground Waste Management and Artificial Recharge* (Braunstein 1973).

SYSTEMS WITH TEMPERATURE GRADIENTS

As the analogy between mass and energy transfer permits direct utilization of the averaging theorem introduced earlier, an appropriate starting place for this section is an averaged equation for energy transport in a porous medium. Unlike mass transfer, however, it is now necessary to distinguish between energy transfer in the fluid phase, and in the solid. Although equations have been developed for both phases, we will not be concerned here with small temperature differences between them. In a final form provided by Green & Perry (1961), and in an auxiliary form given by Slattery (1972), the equation for energy transport in the two-phase mixture is:

$$[\Psi\rho_f C_f + (1-\Psi)\rho_s C_s]\, \partial T_m/\partial t + \Psi\rho_f C_f \,\mathrm{div}\,(T_m \bar{\mathbf{v}}_f) = -\,\mathrm{div}\,\mathbf{q}_e, \tag{40}$$

where the subscripts f and s refer to the fluid and the solid, respectively, T_m is the average temperature of the mass of both phases, and

$$\mathbf{q}_e = -\kappa_e \nabla T_m. \tag{41}$$

In this form, $\mathbf{q}_e$ is an *effective energy flux vector,* analogous to the mass flux vector of Equation 9, and κ_e is an *effective thermal conductivity* of the two-phase mixture. Transport by radiation has been ignored in this development.

Inter- and Intraparticle Transport

Several relations have been proposed for the effective thermal conductivity stated in Equation 41. Lewis & Rose (1970) suggest

$$\kappa_e = \kappa_s(\kappa_f/\kappa_s)^\Psi. \tag{42}$$

A more commonly accepted relationship is (Dagan 1972):

$$\kappa_e = \Psi\kappa_f + (1-\Psi)\kappa_s. \tag{43}$$

The conductivity of a multiphase material is thus apparently completely determined by the volume fractions and conductivities of the individual phases. As the statistics of the phase distribution will often influence the result, it is sometimes better to propose upper and lower bounds for this quantity. For macroscopically homogeneous and isotropic two-phase materials, such limits may be determined by application of the method of Hashin & Shtrikman (1962). For the upper and lower limits, respectively, we have:

$$\kappa_{e_u} = \kappa_s + \Psi/[1/(\kappa_f - \kappa_s) + (1 - \Psi)/3\kappa_s] \tag{44}$$

and

$$\kappa_{e_l} = \kappa_f + (1 - \Psi)/[1/(\kappa_s - \kappa_f) + \Psi/3\kappa_f]. \tag{45}$$

If porosity equals zero, the upper and lower limits go to the conductivity of the solids. If porosity equals one, both limits go to the conductivity of the fluid. These results would appear to hold also for the electrical conductivity as well as other two-phase conductive properties of porous media.

From the foregoing, we see that the mathematical approach to describing the temperature distribution in a porous media has given rise to the concept of an effective thermal conductivity. This concept attains acceptability provided that the effective conductivity is expressed as a function of all the variables that affect it. Identification of these pertinent variables is accomplished, at least in part, through the development of a *thermal tortuosity vector* analogous to that given in Equation 9 for mass transfer, along with some functional relationships for it. In this development, Slattery (1972) makes an important distinction between porous bodies filled with either stagnant or flowing fluid. For a stagnant fluid in a porous medium, the effective energy flux vector becomes

$$\mathbf{q}_e = [\Psi\kappa_f + (1 - \Psi)\kappa_s - \kappa_s K^*]\nabla T_m, \tag{46}$$

where

$$K^* = K^*(\kappa_f/\kappa_s, \Psi). \tag{47}$$

The functional relationship for K^* must represent the reduction in free transport expressed through the effect of tortuosity on fluid conduction, and particle to particle contact between the solid grains. Note that the empirical thermal conductivity of Equation 42 is of a form suggested by this functional relationship.

When convection is considered, the effective energy flux contains other functional relationships, K^*, which depend on parameters already cited, the velocity of flow, the gradient of the average temperature, and the Peclet number for heat transport, expressed as:

$$N_{Pe} = vL/\lambda_f, \tag{48}$$

where λ_f is the thermal diffusivity of the fluid phase.

Virtually dozens of experimental results attest to the validity of the relationships given above. The effect of porosity on conductivity, as suggested in the limiting Equations 44 and 45, is demonstrated in plots of effective conductivity versus porosity for limestones and sandstones (Kunii & Smith 1960). The results show that, instead of decreasing linearly, conductivity decreases with increasing void volume in a concave downward fashion. When tortuosity factors are included, the evidence is equally conclusive. Slattery (1972) cites as many as seven papers in support of the stagnant case, and four for the convective case. Experimental data summarized by Green & Perry (1961) demonstrates that the value of the effective thermal conductivity with a flowing fluid is greater than the value at zero velocity and, in general, increases with increasing velocity. A summary of theoretical results by Yagi & Kunii (1957) indicates that the effective thermal conductivity can be separated into two terms, one of which is independent of fluid flow, and the other dependent on the lateral mixing of fluid in a porous medium. Later papers by Yagi & Wakao (1959) indicate that the convective term is a function of the Peclet, Reynolds, and Prandtl numbers, the total relationship simplified to:

$$\kappa_e = \kappa_o + v\rho C L/N_{Pe}^*, \tag{49}$$

where κ_o is the conductivity in the absence of convection, and N_{Pe}^* is a modified Peclet number wherein the thermal diffusivity is defined for the two-phase mixture, not just the fluid. The similarity in form to mass transport coefficients is obvious (see Equation 20).

In assessing the contribution of each transport mechanism to the effective thermal conductivity, the following mechanisms require consideration (Singer & Wilhelm 1950): (a) molecular conduction through the fluid phase, κ_f, (b) solid particle to particle molecular conduction κ_s, (c) particle to particle radiation κ_r, (d) a series mechanism from fluid to solid to fluid, etc κ_{sf}, (e) convection through the fluid phase κ_c.

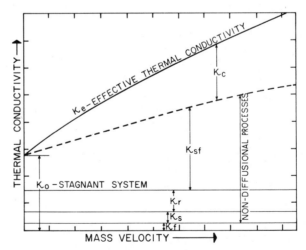

Figure 6   Relationship between the effective thermal conductivity and the mechanisms that contribute to it. (Schuler, Stallings & Smith 1952.) Reprinted with permission from *Chem. Eng. Prog. Symp. Ser.*

Contributions (a) through (c) have been discussed at length by Schuler, Stallings & Smith (1952). None of these three mechanisms is affected by fluid flow. They may be plotted as horizontal lines in a plot of thermal conductivity versus velocity of flow (Figure 6). Mechanisms (d) and (e) depend on fluid velocity and become predominant at high velocities. As velocity increases, the increase in mechanism (e) appears to be greater than in mechanism (d). On the other hand, the "dispersivity of heat" in a porous media is larger than the dispersivity of mass because the dispersion of heat is augmented by solid conduction and radiation whereas no analogous processes are effective in mass transport (Wicke 1972). This is generally confirmed by experiments at low Reynolds numbers where the Peclet numbers for mass transfer are larger than those for heat transfer (Bernard & Wilhelm 1950; Schuler, Stallings & Smith 1952).

Temperature and Chemical Rate Processes

In the previous sections, the effect of temperature on rate processes has been more or less ignored, although there are several geologic problems for which an understanding of heat transfer characteristics is important. The solution alteration of carbonate rocks, for example, takes place because of point-to-point solubility variations which can be attributed to mineralogical nonuniformity, as might be expected in calcite, aragonite, and magnesium calcite mixtures in emerged carbonate muds, and to temperature variations in deep basins. With regard to the latter, it has been suggested that permeability augmentation or destruction might be predictably understood through a better understanding of basinal flow paths and their effect on temperature distribution (Sipple & Glover 1964). Inherent in this problem is the understanding that the geothermal gradient and the velocity of flow are independent variables only when the velocity vector is oriented perpendicular to the geothermal gradient. Maximum convective alterations of the geothermal gradient are achieved when the hydraulic and conductive thermal gradients are collinear. Even moderate vertical flow can thus significantly alter the geothermal gradient, as observed in field studies (Van Orstrand 1934, Schneider 1964) and deduced theoretically (Sipple & Glover 1964, Donaldson 1962, Domenico & Palciauskas 1973). Critical parameters contributing to the convective alteration of the temperature distribution in regional groundwater flows include the geometry of the flow medium, expressed in terms of a basin depth-to-length ratio, and the dominance of convective transport over conduction, expressed in terms of the Peclet number for heat transport (Domenico & Palciauskas 1973).

Ideas on convective heat transfer are also of importance to the general theories of petroleum formation and migration as discussed by Beskrovnyi et al (1973) and Durmish'yan (1973). Chemical reactions involved in the transformation of organic source material to petroleum are recognized as being strongly temperature dependent (Phillipi 1969). The organic substances are changed into paraffinic hydrocarbons at temperatures between 80 and 170 degrees centigrade. At temperatures in excess of 170 degrees, the process goes too far, only methane being produced (Dozy 1970). Related ideas pertaining to the formation of sulfide and other ore deposits from rising hot brines in sedimentary basins have also received some attention (Dunham

1970). Excess pressures are generally invoked as the brine movement mechanism, and certain critical temperatures must be preserved during mineralization. Interest is thus focused on the occurrence of a specific temperature range at various points in time and space in sedimentary basins undergoing diagenesis. Most studies to date, however, sidestep this difficult problem by merely extrapolating the present-day geothermal gradient. For example, linear geothermal gradients have been extrapolated in the Illinois Basin to determine whether temperatures have been high enough to agree with those necessary for the precipitation of ores from brines, and for the generation of oil (Dozy 1970). For Gulf Coast petroleum reservoirs, the change in the degree of cracking with progressively changing temperature has been determined by assuming first-order temperature-dependent kinetics and a steady-state geothermal gradient (Johns & Shimoyama 1972). Of importance here is that these studies focus on existing or former excess pressure environments where the thermal evolution of the basin, expressed in terms of its pressure–temperature history, is the important factor in economic mineral and petroleum genesis. The sediments in the Gulf of Mexico geosyncline, for example, have been geopressured since the Late Cretaceous, with temperatures at any given depth currently assuming their lowest values since that time (Sharp & Domenico 1976). It follows that if temperature-dependent reactions do not go to completion at elevated temperatures in effect over significant portions of geologic time, there is little likelihood that they ever will do so at lower temperatures.

Dimensionless Groups and Chemical Rate Processes

The physical significance of dimensionless groups in chemical rate phenomena are of practical interest. For mass transfer, the Nusselt number (also referred to as the Sherwood number) represents the ratio of the total rate of mass transfer per unit area to the rate of mass transfer by diffusion, and has been used in expressions treating mineral precipitation and dissolution in a sediment (Berner 1974). The dependence of the thickness of the stationary layer in diffusion controlled reactions on both the Reynolds and the Prandtl numbers has been demonstrated by Bircumshaw & Riddiford (1952). The Prandtl number, better referred to as the Schmidt number when dealing with mass transfer, is the ratio of the fluid property governing momentum transport (due to a velocity gradient) to the fluid property governing mass transport (due to a concentration gradient). The Reynolds number may be regarded as the ratio of fluid momentum to viscous drag. The dependence of the thickness of the surface layer on fluid velocity and dispersion is thus apparent.

The Peclet number for mass transport (Equation 19) represents the ratio of mass transfer by bulk flow to that of diffusion. Its role in the situation where the rate of reaction is infinitely fast with respect to transport has already been discussed in the section on equilibrium interphase transport. For nonequilibrium conditions, two additional groups arise, kL/v and kL^2/D_H, which are referred to as the Damköhler numbers I and II. The first is a measure of the rate of reaction in comparison to the rate of convective transport, while the second compares the rate of chemical reaction to the rate of dispersive transport. The relative values of these numbers, when compared to the Peclet number, are of some help in determining whether reaction or transport will dominate in a porous system. The dominating phenomena—transport or reaction—will depend on whether rates of dispersion are fast or slow compared to rates of reaction at low values of the Peclet number, and the comparison between velocity of flow and rates of reaction at high Peclet numbers (Palciauskas & Domenico 1976).

Dimensionless groups are of particular interest in situations where heat and mass transfer are simultaneously involved, as in the thermal conditions associated with reaction. For exothermic first-order reactions, the reaction term r as given, for example, in Equation(s) 3 becomes:

$$r = kc = A\,e^{-E/RT}\,c, \tag{50}$$

where the rate constant is replaced by the Arrhenius expression and where A is a constant of proportionality, E is the energy of activation, and R is a gas constant. The energy generation term in the energy Equation 3 is the rate at which energy is liberated by chemical reaction, or

$$W = \Delta H r = \Delta H A\,e^{-E/RT}\,c, \tag{51}$$

where ΔH is the enthalpy of reaction. Equation(s) 3 are thus coupled through the reaction and heat generation term. Dimensional analysis of the coupled equations reveals the following dimensionless groups:

$$
\begin{array}{llll}
LA\,e^{-E/RT}/v & \text{(a)} & L\Delta H A\,e^{-E/RT}/T\rho Cv & \text{(c)} \\[6pt]
L^2 A\,e^{-E/RT}/D_H & \text{(b)} & L^2 \Delta H A\,e^{-E/RT}/T\kappa_e & \text{(d)}.
\end{array}
\tag{52}
$$

Equation(s) 52(a) and (b) are the Damköhler numbers I and II, now modified for temperature-dependent reactions. Equation(s) 52(c) and (d) are referred to as Damköhler numbers III and IV, respectively. Damköhler III is a ratio of the rate at which heat is generated to the rate of convective transport, while Damköhler IV is the ratio of heat generation to conductive transport. The relative values of these numbers, when compared to the Peclet number for heat transport, are of some aid in determining whether heat will accumulate at a reaction site in a porous media, or whether transport might rapidly disperse the heat. By analogy to what has been said about mass transport, the dominating phenomenon—heat accumulation or transport—will depend upon whether the rate of thermal dispersion is fast or slow compared to the rate of heat generation at low values of the Peclet number. At high values of the Peclet number, it will also depend on a comparison of velocity of flow and heat generation. Note that the ratio between numbers I and II yields the Peclet number for mass transport, while the ratio between numbers III and IV yields the Peclet number for heat transport.

Similar insights into other coupled phenomena are easily provided by dimensional analysis. The role of heat of friction in driving chemical reactions may be qualitatively assessed, for example, through dimensional analysis of the coupled energy and momentum balance. On the other hand, there are obvious practical applications pertaining to critical temperature rise associated with subsurface radioactive waste disposal. In general, in these problems as well as all others to which this type of analysis is applied, the variables involved in a physical process are known, but the relations between them are not. Dimensional analysis provides the relationships required. For experimental work, it is easier to focus on relationships between groups by varying one variable of the group, rather than subjecting all variables to some controlled variation. Classic examples of this procedure include the establishment of correlations between the Peclet, Prandtl, and Reynolds numbers in experimental work on mass and heat dispersion (Perkins & Johnson 1963, Yagi & Wakao 1959), and in ascertaining that the dispersion of heat is greater than the dispersion of mass in porous media (Bernard & Wilhelm 1950; Schuler, Stallings & Smith 1952).

THE IRREVERSIBLE THERMODYNAMIC APPROACH

According to the thermodynamics of irreversible processes, there exists a contribution to each flux from each driving force in a system, and therefore not all transport processes may simply be described as a proportionality between a flux and one force. Mass flow may occur, for example, in response not only to a concentration gradient, but to a temperature gradient. The resulting thermal diffusion is referred to as *cross phenomena* and is described mathematically by way of an additional driving force of a temperature gradient in Fick's law (Equation 1). Similarly, in multicomponent systems, there can exist a diffusion of one component on the concentration gradient of another. Although the Onsager Reciprocal Relations gives information as to the interrelations of the coupled effects in each instance above, the mathematical approach in each case differs somewhat.

The term *irreversible thermodynamics* is usually employed when cross effects or coupling occur between two or more entities. In this regard, available literature suggests three rather distinct categories of study that have been described under the banner of irreversible thermodynamics. The first two categories entail the development of an equation of transfer similar to Equation 3 from an appropriate conservation statement. In one case, the flux in the continuity statement is not represented by the product of a transport property and a driving force, but by an *entropy flux vector*. This permits the development of an equation describing the rate of generation of entropy in nonisothermal systems. The description of irreversible processes in terms of entropy production, however interesting from the viewpoint of thermodynamics, is of little value in the study of the actual physical processes. The relationship between the forces and fluxes must be known, and therefore we discover the importance of the phenomenological linear laws attributed to Fick and Fourier, and the extensions provided by Onsager. Several references are available on the development of irreversible thermodynamics from the viewpoint of entropy production (Kirkwood & Crawford 1952; Bird, Curtiss & Hirschfelder 1955; Katchalsky & Curran 1967).

In the second category, cross transport phenomena are included in transport equations in addition to the phenomena known already as Fourier's and Fick's laws. Most generally these are Soret or Dufour effects. The special designation *irreversible thermodynamics* is probably not warranted here, as mathematical procedures and principles are identical to the conventional treatment of irreversible phenomena in transport problems. Much of the theoretical work in this area has been conducted by soil scientists interested in water movement in response to temperature gradients in soils (Cassel, Nielsen & Biggar 1969) and saturated porous media (Philip & DeVries 1957), and the simultaneous flow of water and salt in unsaturated porous media (Abd-El-Aziz & Taylor 1965) and clays (Greenberg & Mitchell 1971). Experimental work along these lines is more in the spirit of what most scientists would refer to as irreversible thermodynamics, and includes the experimental verification of Onsager's Reciprocal Relations (Letey & Kemper 1969, Olsen 1969), the determination of clay membrane efficiency (Dirksen 1969, Kemper & Rollins 1966), and the influence of pore size and electrokinetic phenomena on ion transport. With regard to the latter, electrokinetic phenomena have been demonstrated to either enhance or impede convective ion transport, depending on the properties of the medium. Transport enhancement has been explained in terms of anion repulsion, whereby the interaction of dissolved salts with negatively charged minerals, particularly clay minerals, gives rise to a maximum salt concentration in the center of individual pores. As this position corresponds with the maximum microscopic velocity, the dissolved salts move more rapidly than the average velocity of the water in which they are dissolved (Krupp, Biggar & Nielsen 1972). The opposite effect of dissolved constituents moving at a lower velocity relative to that of the water takes place when the porous media behaves like a permeable selective membrane. A permeable selective membrane is a charged medium with pores too large to mechanically interfere with ion migration, but sufficiently small so that the membrane is considered as a homogeneous mixture of fixed and mobile charges. Pore size is thus too large to promote ultrafiltration; however, because of the fixed negative charge of the membrane, anions are selectively excluded at low ionic strengths (Merten 1966). Anion repulsion in permeable selective membranes is thus recognized as another important concentration attenuation mechanism in subsurface contaminant transport (Schwartz 1975).

The third category deals with multicomponent systems, characterized by the diffusion of one component on the concentration gradient of another. The procedures and theoretical framework in this research are also in line with what most researchers consider to be irreversible thermodynamics. This topic has recently been reviewed in an earlier *Annual Review* (Anderson & Graf 1976). A related problem with a similar mathematical construct occurs in the study of anisotropic material, where there exists a contribution to flow in one direction from components of hydraulic force in another direction.

CONCLUDING STATEMENT

The theoretical aspects of transport examined in this treatment are scattered widely among the literature of several fields. In this present contribution, I have focused on literature that favors the mechanistic-theoretical approach. A major problem, always, is to find ways to measure accurately the various coefficients and to obtain reliable chemical rate data in the laboratory. The simplified rate equations that have been examined here are not proposed as an adequate theory of chemical rate processes, but merely as a qualitatively correct demonstration of the intrusion of physical phenomena on such processes. Similar statements can be made concerning the field coefficient of dispersion and the need for reliable correlations between field dispersivity and physical properties of the porous medium. Although progress has been made toward the goal of understanding rate processes in geologic materials, a need for continuing research in these areas is indicated.

With regard to actual applications of the transport theory to geologic problems, it is clear that the theory is best employed when what is occurring in sediments is more adequately described in terms of transport and reaction, rather than just equilibrium processes. Unfortunately, the complexity of many geologic systems precludes such a description. Dimensional analysis may be useful in some cases, but the full emergence of transport theory as a scientific tool in the analysis of geologic problems depends to a large extent on a deeper understanding of the phenomena occurring in sediments and the translation of these phenomena into differential equation form.

Literature Cited

Abd-El-Aziz, M. H., Taylor, S. A. 1965. Simultaneous flow of water and salt through unsaturated porous media, I, Rate equations. *Soil Sci. Soc. Am. Proc.* 29:141–43

Amundson, N. R. 1948. A note on the mathematics of adsorption in beds. *J. Phys. Colloid Chem.* 52:1153–57

Amundson, N. R. 1950. Mathematics of adsorption in beds, II. *J. Phys. Colloid Chem.* 54:812–20

Anderson, D. E., Graf, D. L. 1976. Multicomponent electrolytic diffusion. *Ann. Rev. Earth Planet. Sci.* 4:95–121

Aris, R. 1956. On the dispersion of a solute in a fluid flowing through a tube. *Proc. R. Soc. London* A235:67–77

Aris, R., Amundson, N. R. 1973. *Mathematical Methods in Chemical Engineering,* II. Englewood Cliffs, NJ: Prentice-Hall. 369 pp.

Banks, P. M., Schunk, R. W., Raitt, W. J. 1976. The topside ionosphere: a region of dynamic transition. *Ann. Rev. Earth Planet. Sci.* 4:381–440

Banks, R. B., Jerasate, S. 1962. Dispersion in unsteady porous media flow. *Am. Soc. Civ. Eng. J. Hydraul. Div.* 88:1–21

Barraclough, J. T., Teasdale, W. E., Jensen, R. G. 1965. *US Geol. Survey Open File Rep.,* issued Feb. 1967, Water Res. Div. 107 pp.

Bear, J. 1972. *Dynamics of Fluids in Porous Media.* New York: American Elsevier. 764 pp.

Bear, J., Bachmat, Y. 1967. *Int. Assoc. Sci. Hydrol. Symp., Artificial recharge and management of aquifers, Haifa, Isr.* Publ. 72:7–16

Benfield, A. E. 1949. The effect of uplift and denudation on underground temperature. *J. Appl. Phys.* 20:66–70

Benfield, A. E. 1950. The temperature in an accreting medium with heat generation. *Q. Appl. Math.* 7:436–39

Bernard, R. A., Wilhelm, R. H. 1950. Turbulent diffusion in fixed beds of packed solids. *Chem. Eng. Prog.* 46:233–44

Berner, R. A. 1964. An idealized model of sulfate distribution in recent sediments. *Geochim. Cosmochim. Acta.* 28:1497–1503

Berner, R. A. 1968. Rate of concretion growth. *Geochim. Cosmochim. Acta.* 32:477–83

Berner, R. A. 1971. *Principles of Chemical Sedimentology.* New York: McGraw-Hill. 240 pp.

Berner, R. A. 1974. In *The Sea,* ed. E. D. Goldberg, 5:427–50. New York: Wiley. 895 pp.

Beskrovnyi, N. S., Glavatskhk, S. G., Yermakova, V. I., Lebedev, B. A., Talyev, S. D. 1973. Presence of oil in hydrothermal systems associated with volcanism. *Int. Geol. Rev.* 15:384–93

Bircumshaw, L. L., Riddiford, A. C. 1952. Transport control in heterogeneous reactions. *Chem. Soc. London Q. Rev.* 6:157–85

Bird, R. N., Curtiss, C. F., Hirschfelder, J. O. 1955. Fluid mechanics and the transport phenomena. *Chem. Eng. Prog. Symp. Ser.* 51:69–85

Blackwell, R. J., Rayne, J. R., Terry, W. M. 1959. Factors influencing the efficiency of miscible displacement. *Trans. Am. Inst. Min. Metall. Pet. Eng.* 216:1–8

Braunstein, J., ed. 1973. *Underground Waste Management and Artificial Recharge*, Vols. 1, 2. Am. Assoc. Pet. Geol. 931 pp.

Burghardt, A., Zaleski, T. 1968. Longitudinal dispersion at small and large Peclet numbers in chemical reactors. *Chem. Eng. Sci.* 23:575–91

Burst, J. F. 1976. Argillaceous sediment dewatering. *Ann. Rev. Earth Planet. Sci.* 4:293–318

Byevich, Y. A., Leonov, I., Safrai, V. M. 1969. Variations in filtration velocity due to random large scale fluctuations of porosity. *J. Fluid Mech.* 37:371–81

Cassel, D. K., Nielsen, D. R., Biggar, J. W. 1969. Soil water movement in response to imposed temperature gradients. *Soil Sci. Soc. Am. Proc.* 33:443–500

Chuang, Y. K., Szekely, J. 1971. The use of Green's functions for solving melting or solidification problems. *Int. J. Heat, Mass Transfer* 14:1285–94

Collins, R. E. 1961. *Flow of Fluids Through Porous Materials.* New York: Reinhold. 270 pp.

Dagan, G. 1972. In *Fundamentals of Transport Phenomena in Porous Media, Dev. Soil Sci.* 2:55–64. Amsterdam: Elsevier. 392 pp.

Danckwerts, P. V. 1953. Continuous flow systems. *Chem. Eng. Sci.* 2:1–13

Deju, R. A. 1971. A model of chemical weathering of silicate minerals. *Geol. Soc. Am. Bull.* 82:1055–62

Deju, R. A., Bhappu, R. 1966. *N. M. Bur. Mines, Min. Resour. Circ. 89,* Socorro, N.M. 13 pp.

Deju, R. A., Bhappu, R. 1970. Relationship of chemical reactions between water and a silicate bed to the flow equation. *Trans. Am. Inst. Min. Metall. Pet. Eng.* 247:115–19

Dirksen, C. 1969. Thermo-osmosis through compacted saturated clay membranes. *Soil Sci. Am. Proc.* 33:821–26

Domenico, P. A., Palciauskas, V. V. 1973. Theoretical analysis of forced convective heat transfer in regional groundwater flow. *Geol. Soc. Am. Bull.* 84:3803–14

Donaldson, I. G. 1962. Temperature gradients in the upper layers of the earth's crust due to convective water flows. *J. Geophys. Res.* 67:3449–59

Dozy, J. J. 1970. A geological model for the genesis of the lead-zinc ores of the Mississippi Valley, USA. *Inst. Min. Metall. Trans. Sec. B. Appl. Earth Sci.* 79:163–72

Dunham, K. C. 1970. Mineralization by deep formation waters, A review. *Inst. Mining Met. Trans. Sec. B. Appl. Earth Sci.* 79:127–36

Durmish'yan, A. G. 1973. Role of anomalously high formation pressures in development of traps for accumulations of oil and gas in the southern Caspian Basin. *Int. Geol. Rev.* 15:508–16

Dybbs, A., Schweitzer, S. 1973. Conservation equations for nonisothermal flow in porous media. *J. Hydrol.* 20:171–80

Evans, R. B., Watson, G. M., Mason, E. A. 1961. Gaseous diffusion in porous media at uniform pressure. *J. Chem. Phys.* 35:2076–83

Frank, F. C. 1950. Radially symmetric phase growth controlled by diffusion. *Proc. R. Soc. London* A201:586–99

Frank-Kamenetskii, D. A. 1955. *Diffusion and Heat Transfer in Chemical Kinetics.* Transl. J. P. Appleton, 1969. New York: Plenum Press. 574 pp. (from Russian)

Gardner, W. R. 1965. In *Soil Nitrogen, Am. Soc. Agron. Monogr. No. 10,* ed. W. V. Bartholomew, F. E. Clark, pp. 550–72

Green, D. W., Perry, R. H. 1961. Heat transfer with a fluid flowing through porous media. *Chem. Eng. Prog. Symp. Ser.* 57:61–68

Greenberg, J., Mitchell, J. K. 1971. *Calif. Dept. Water Res. Bull. No. 63–64. Append. D:397–569.* Sacramento. 569 pp.

Greenkorn, R. A., Kessler, D. P. 1972. *Transfer Operations.* New York: McGraw-Hill. 548 pp.

Hashin, Z., Shtrikman, S. 1962. A variational approach to the theory of effective magnetic permeability of multiphase materials. *J. Appl. Phys.* 33:3125–31

Helfferich, F. 1966. In *Ion Exchange, A Series of Advances,* ed. J. A. Marinsky, pp. 65–100. New York: Dekker. 424 pp.

Holley, E. R., Harleman, D. R., Fischer, H. B. 1970. Dispersion in homogeneous estuary flow. *Am. Soc. Civ. Eng. J. Hydraul. Div.* 96:1691–1709

Johns, W. D., Shimoyama, A. 1972. Clay minerals and petroleum forming reactions during burial and diagenesis. *Am. Assoc. Pet. Geol. Bull.* 56:2160–67

Kasten, P. R., Lapidus, L., Amundson, N. R. 1952. Mathematics of adsorption in beds, V, Effect of intraparticle diffusion in flow systems in fixed beds. *J. Phys. Chem.* 56:683–88

Katchalsky, A., Curran, P. F. 1967. *Nonequilibrium Systems in Biophysics.* Cambridge: Harvard Univ. Press. 248 pp.

Kaufman, W. J., Orcutt, R. G., Klein, G. 1956. *US AEC Prog. Rep. 1,* AECU-3115. 87 pp.

Kemper, W. D., Rollins, J. B. 1966. Osmotic efficiency coefficients across compacted clays. *Soil Sci. Soc. Am. Proc.* 30:529–34

Kingery, W. D. 1959. In *Kinetics of High Temperature Processes,* ed. W. D. Kingery, 1:1–7. Cambridge: MIT Press. 326 pp.

Kirkham, D., Powers, W. L. 1972. *Advanced Soil Physics.* New York: Wiley. 534 pp.

Kirkwood, J. G., Crawford, B. L. 1952. The macroscopic equations of transport. *J. Phys. Chem.* 56:1048–51

Klotz, I. M. 1946. The adsorption wave. *Chem. Rev.* 39:241–68

Kolodner, I. 1956. Free boundary problem for the heat equation with application to problems of change of phase. *Commun. Pure Appl. Math.* 9:1–31

Krupp, H. K., Biggar, S. W., Nielsen, D. R. 1972. Relative flow rates of salt in water in soil. *Soil Sci. Soc. Am. Proc.* 36:412–17

Kunii, D., Smith, J. M. 1960. Heat transfer characteristics of porous rocks. *Am. Inst. Chem. Eng. J.* 6:71–77

Lai, S. H., Jurinak, J. J. 1971. Numerical approximation of cation exchange in miscible displacement through soil columns. *Soil Sci. Soc. Am. Proc.* 35:894–99

Lapidus, L., Amundson, N. R. 1952. Mathematics of adsorption in beds, VI, The effect of longitudinal diffusion in ion exchange and chromatographic columns. *J. Phys. Chem.* 56:984–88

Lee, W. H., ed. 1965. *Terrestrial Heat Flow, Am. Geophys. Union Monogr. No. 8.* Am. Geophys. Union Publ. 276 pp.

Lerman, A. 1971. *Adv. Chem. Ser.* 106:30–76

Letey, J., Kemper, W. D. 1969. Movement of water and salt through a clay-water system: Experimental verification of Onsager Reciprocal Relations. *Soil Sci. Soc. Am. Proc.* 33:25–29

Levenspiel, O., Bischoff, K. B. 1959. Backmixing in the design of chemical reactors. *Ind. Eng. Chem.* 51:1431–34

Levenspiel, O., Bischoff, K. B. 1961. Reaction rate constant may modify the effects of backmixing. *Ind. Eng. Chem.* 53:313–14

Lewis, C. R., Rose, S. C. 1970. A theory relating high temperatures and overpressures. *J. Pet. Tech.* 22:11–16

Li, Y. H., Gregory, S. 1974. Diffusion of ions in sea water and in deep sea sediments. *Geochim. Cosmochim. Acta.* 38:703–14

Lick, W. 1976. Numerical modeling of lake currents. *Ann. Rev. Earth Planet. Sci.* 4:49–74

Lightfoot, E. N. 1974. *Transport Phenomena in Living Systems.* New York: Wiley. 495 pp.

Manheim, F. T. 1970. The diffusion of ions in unconsolidated sediments. *Earth Planet. Sci. Lett.* 9:307–9

Merten, U. 1966. In *Desalination by Reverse Osmosis,* ed. U. Merten, 1:15–24. Cambridge: MIT Press. 289 pp.

Nancollas, G. H., Purdie, N. 1964. The kinetics of crystal growth. *Chem. Soc. London Q. Rev.* 18:1–20

Nielsen, A. E. 1961. Diffusion controlled growth of a moving sphere. The kinetics of crystal growth in potassium perchlorate. *J. Phys. Chem.* 65:46–49

Nielsen, A. E. 1964. *The Kinetics of Precipitation.* New York: Macmillan. 151 pp.

Olsen, H. W. 1969. Simultaneous fluxes of liquid and charge in saturated kaolinite. *Soil Sci. Soc. Am. Proc.* 33:338–44

Palciauskas, V. V., Domenico, P. A. 1976. Solution chemistry, mass transfer, and the approach to chemical equilibrium in porous carbonate rocks and sediments. *Geol. Soc. Am. Bull.* 87:207–14

Perkins, T. K., Johnson, O. C. 1963. A review of diffusion and dispersion in porous media. *J. Soc. Pet. Eng.* 3:70–84

Philip, J. R., DeVries, D. A. 1957. Moisture movement in porous materials under temperature gradients. *Trans. Am. Geophys. Union.* 38:222–32

Philippi, G. T. 1969. In *Advances in Organic Geochemistry, Int. Ser. Monogr. Earth Sci.* 31:21–46. New York: Pergamon

Raimondi, P., Gardner, G. H., Petrick, C. B. 1959. *Effect of pore structure and molecular diffusion on the mixing of miscible liquids flowing in porous media.* Presented at Am. Inst. Chem. Eng.–Soc. Pet. Eng. Joint Symp., San Francisco. Preprint 43

Robinson, B. P. 1962. *US Geol. Surv. Water Supply Pap. 1616,* Washington, D.C. 132 pp.

Roedder, E. 1959. *Geol. Survey Bull. 1088,* Washington, D.C. 65 pp.

Rubin, J., James, R. V. 1973. Dispersion affected transport of reacting solutes in saturated porous media. Galerkin method applied to equilibrium controlled exchange in unidirectional steady water flow. *Water Resour. Res.* 9:1332–56

Rumer, R. R. 1972. In *Fundamentals of Transport Phenomena in Porous Media, Dev. Soil Sci.* 2:268–75. Amsterdam: Elsevier, 392 pp.

Saffman, P. G. 1959. A theory of dispersion in a porous medium. *J. Fluid Mech.* 6:321–49

Saffman, P. G. 1960. Dispersion due to molecular diffusion and macroscopic mixing in flow through a network of capillaries. *J. Fluid Mech.* 7:194–208

Scheidegger, A. 1961. General theory of dispersion in porous media. *J. Geophys. Res.* 66:3273–78

Schneider, R. 1964. Relation of temperature distribution to groundwater movement in carbonate rocks of central Israel. *Geol. Soc. Am. Bull.* 75:209–16

Schuler, R. W., Stallings, V. P., Smith, J. M. 1952. Heat and mass transfer in fixed bed reactors. *Chem. Eng. Prog. Symp. Ser.* 48:19–30

Schwartz, F. W. 1975. On radioactive waste management: an analysis of the parameters controlling subsurface contaminant transfer. *J. Hydrol.* 27:51–71

Sharp, J. M., Domenico, P. A. 1976. Energy transport in thick sequences of compacting sediment. *Geol. Soc. Am. Bull.* 87:390–400

Singer, E., Wilhelm, R. H. 1950. Heat transfer in packed beds. Analytical solution and design methods. *Chem. Eng. Prog.* 46:343–57

Sipple, R. F., Glover, E. D. 1964. The solution alteration of carbonate rocks, the effects of temperature and pressure. *Geochim. Cosmochim. Acta.* 28:1401–17

Skibitzke, H. 1961. *US Geol. Surv. Prof. Pap. 386-A,* Washington, D.C. 8 pp.

Skibitzke, H., Robinson, G. M. 1963. *US Geol. Surv. Prof. Pap. 386-B,* Washington, D.C. 3 pp.

Slattery, J. C. 1967. Flow of viscoelastic fluids through porous media. *Am. Inst. Chem. Eng. J.* 13:1066–71

Slattery, J. C. 1969. Single phase flow through porous media. *Am. Inst. Chem. Eng. J.* 15:866–72

Slattery, J. C. 1972. *Momentum, Energy, and Mass Transfer in Continua.* New York: McGraw-Hill. 679 pp.

Sooky, A. A. 1969. Longitudinal dispersion in open channels. *Am. Soc. Civ. Eng. J.*

Hydraul. Div. 95:1327–46

Szekely, J., Themelis, N. J. 1971. *Rate Phenomena in Process Metallurgy.* New York: Wiley. 784 pp.

Taylor, G. I. 1953. Dispersion of soluble matter in solvent flowing through a tube. *Proc. R. Soc. London* A219:186–203

Theis, C. V. 1963. *IAEA Symp. Tokyo, Radioisotopes in Hydrology,* pp. 193–206. Vienna: IAEI. 459 pp.

Thomas, H. C. 1944. Heterogeneous ion exchange in a flowing system. *J. Am. Chem. Soc.* 66:1664–66

Tzur, Y. 1971. Interstitial diffusion and advection of solute in accumulating sediments. *J. Geophys. Res.* 76:4208–11

Van Orstrand, C. E. 1934. In *Problems of Petroleum Geology,* pp. 989–1021. Tulsa: Am. Assoc. Pet. Geol. 1073 pp.

Wehner, J. F., Wilhelm, R. H. 1956. Boundary conditions of flow reactor. *Chem. Eng. Sci.* 6:89–93

Weyl, P. K. 1958. The solution kinetics of calcite. *J. Geol.* 66:163–76

Whitaker, S. 1966. The equations of motion in porous media. *Chem. Eng. Sci.* 21:291–300

Whitaker, S. 1967. Diffusion and dispersion in porous media. *Am. Inst. Chem. Eng. J.* 13:420–27

Wicke, E. 1972. *Adv. Chem. Ser.* 109:183–208

Wilhelm, R. H. 1962. Progress towards the a priori design of chemical reactors. *Pure Appl. Chem.* 5:403–21

Wollast, R. 1967. Kinetics of the alteration of K-feldspar in buffered solutions at low temperature. *Geochim. Cosmochim. Acta.* 31:635–48

Yagi, S., Kunii, D. 1957. Studies on effective thermal conductivities in packed beds. *Am. Inst. Chem. Eng. J.* 3:373–81

Yagi, S., Wakao, N. 1959. Heat and mass transfer from wall to fluid in packed beds. *Am. Inst. Chem. Eng. J.* 5:79–85

Ann. Rev. Earth Planet. Sci. 1977. 5:319–55

THE HISTORY OF THE EARTH'S SURFACE TEMPERATURE DURING THE PAST 100 MILLION YEARS[1]

×10076

Samuel M. Savin

Department of Earth Sciences, Case Western Reserve University, Cleveland, Ohio 44106

INTRODUCTION

For well over a century geologists have attempted to infer from studies of ancient sediments and geomorphic features something of the nature of the earth's past climates. As early as the first half of the nineteenth century it had been recognized by many that present-day temperatures in Europe were lower than those of Late Cretaceous time (Lyell 1854), and that, relatively recently in geologic history, large portions of the continents of the Northern Hemisphere had been covered by vast sheets of ice (Agassiz 1840). The last 10 or 15 years have seen an enormous increase in our knowledge of the history of climate during late Mesozoic and Cenozoic times. The conclusions of early geologists that during that time the earth underwent significant net cooling have not been altered. The highlights of the most recent results reviewed here, however, indicate that temperatures have not decreased monotonically during Cenozoic time nor, apparently, have all parts of the globe undergone significant net cooling.

The increased tempo of research on Cenozoic paleoclimatology in recent years can be attributed to a number of factors. (a) An increased number of samples suitable for paleoclimatic research has become available through expanded programs of ocean sediment sampling. Extensive piston coring programs in all of the world's oceans have recovered large numbers of sediment cores, primarily of Holocene and Pleistocene age. Since 1968 the Deep Sea Drilling Project (DSDP) has recovered many Tertiary and older sedimentary sequences especially well suited for paleoclimatic study. (b) There has been substantial refinement in biostratigraphic zonations and correlations of Cenozoic and Cretaceous sediments, based on information derived from the newly available ocean sediment samples. Improved biostratigraphy has permitted more precise correlation of climatic changes in different

[1] Contribution No. 120, Department of Earth Sciences, Case Western Reserve University.

parts of the world as well as better resolution of climatic events of short duration. (c) An increase in the number of laboratories equipped to make oxygen isotope paleotemperature measurements has led to a corresponding increase in the amount of isotopic paleotemperature data. Paleontologic techniques for evaluating ancient climates have also been developed and refined.

The word *climate* denotes several environmental variables, including temperature, rainfall, evaporation, wind patterns, etc. All of these variables have annual mean values as well as seasonal ranges of variation. Most paleoclimatic data now available pertain to the oceans and are, in fact, paleotemperature data. Ideally, in the interpretation of such data it is possible to specify the seasonal weighting (e.g. mean annual temperature, summer maximum, or winter minimum temperature) but this cannot always be done as successfully as is desirable. Most indicators of climatic conditions on land reflect variations in more than one climatic variable. For example, glaciations require not only low temperatures but an appropriate seasonal range of temperatures as well as sufficient snowfall. Evidence for past glaciations is abundant, but the exact range of climatic conditions implied by glaciation is not clearly defined at this time. Sedimentological evidence for other climatic extremes, such as extreme aridity or high rainfall, is sometimes available, but it is seldom possible to resolve quantitative estimates of the basic climatic variables from the geologic data.

The work reviewed in this paper is concerned almost entirely with the history of temperatures at the surface of the earth, and especially at the sea surface. The development of glacial icecaps is discussed, primarily in terms of its relationship to the earth's temperature history. The history of other climatic variables is largely ignored. This emphasis parallels that of most recent paleoclimatic research.

The emphasis of this paper is on Tertiary temperatures, including the history leading up to the development of the Pleistocene-Holocene glacial-interglacial fluctuations. The enormous literature on the climates of Pleistocene and Holocene times is reviewed only briefly and selectively. The temperature history of Late Cretaceous time, the oldest time for which abundant isotopic paleotemperature data exist, is also briefly discussed. Stress is placed on the deep-sea sedimentary record, the basis of so many recent paleoclimatic advances. In addition to outlining Cenozoic climatic history, the following questions are examined: 1. What was the nature of temperature change at the time of the Tertiary-Cretaceous boundary, during which so many life forms became extinct? 2. What was the time of onset of significant continental glaciation in Antarctica and in the Northern Hemisphere? 3. Do the rapidly fluctuating glacial and interglacial cycles of the Pleistocene-Holocene climatic regime have any counterpart in the nonglacial regime of earlier Cenozoic time? 4. What are the relations between climatic changes and changes in oceanic circulation induced by crustal plate motions?

METHODS OF STUDYING PALEOCLIMATES

Schwarzbach (1963) wrote in his book, *Climates of the Past*: "There are many climatic indicators and imaginative geologists are always finding new ones. Unfortunately, however, most of them are extraordinarily indefinite." While the situation has improved since that was written, it is probably safe to say that there still exist no broadly applicable quantitative indicators of paleoclimates whose interpretations are widely accepted without controversy. There are, however, a number of techniques that seem to give reasonable and consistent results and that have therefore become widely used. Most of these involve some aspect of the study of sedimentary rocks or the fossils they contain.

One of the difficulties encountered in comparing the paleoclimatic results of different workers is in correlating the ages of rocks studied in different parts of the world. This is especially true for rocks from nonmarine deposits and for marine sequences studied prior to the late 1960s. In this review I attempt to replot the paleoclimatic results from different studies on the Cenozoic time scale of Berggren & van Couvering (1974) and the Cretaceous time scale of Moberly, Gardner & Larson (1975), using their suggested correlations among the stages defined in different parts of the world.

Authors of some papers cited may take exception to the way some of their isotopic data have been plotted here. The common practice of connecting isotopic data points for adjacent samples from a core by straight line segments has not been followed. As pointed out by Ledbetter & Ellwood (1976), when the sampling interval for a core is longer than or comparable to the interval over which changes occur in the parameter studied, the shape of the curve obtained by connecting adjacent analyses is an artifact of the sampling interval. Therefore in most instances smoothed curves have been visually fitted to plots of isotopic data from single cores.

Sedimentary Rocks as Climatic Indicators

It is usually only the extremes of climate that are evident from strictly sedimentological features. For example, ancient glaciations may be recognized by the presence of tillites, fluvioglacial deposits, glaciomarine deposits, ice-rafted debris in marine sediments, etc (Schwarzbach 1964, Crowell 1964, Margolis & Kennett 1971). Arid climates may be evident from evaporites, desert features, etc (McKee 1964). While many attempts have been made to draw climatic inferences from the existence of red beds and laterites, these seem to be able to form under a variety of climatic conditions. Within normal ranges of climatic variation of present-day tropical and temperate latitudes there seem few uncontroversial and clear-cut climatic indicators of strictly sedimentologic nature.

Fossils as Climatic Indicators

Most successful paleoclimatic studies have been based on some aspect of the chemistry or biology of plant or animal fossil remains.

PALEONTOLOGICAL APPROACHES   Paleoclimatic inferences can be drawn from data on fossil flora or fauna using a variety of approaches. The most common approach involves selecting living taxa that inhabit restricted climatic regimes, identifying these (or closely related taxa) in fossil assemblages, and drawing analogies between the climatic conditions in which the modern taxa live and those that prevailed at the time of deposition of the fossils. Implicit in most paleoclimatic studies of this type is the assumption that the optimum climatic regime for a taxon or closely related group of taxa has not changed throughout the time period studied. This assumption is somewhat tenuous at least when used to infer climatic changes over time scales that are long compared to times required for significant biological evolution. Therefore, most attempts to evaluate paleoclimates from data on fossil distribution rely on evidence from many taxa, in hope that errors introduced by evolution will cancel each other. The application of paleobotanical data to the determination of climatic change has been reviewed by Dorf (1970) and Kräusel (1961). The application of data on invertebrate distribution to paleoclimatic studies has been discussed by many authors including Craig (1961), Berger & Roth (1975), and in volumes edited by Hallam (1973) and Hughes (1973).

In the past several years data on fossil distribution have been analyzed for their paleoclimatic significance using increasingly sophisticated techniques. Advancing beyond calibration of parameters such as the relative abundance of one or more taxa in a fossil assemblage with a climatic variable of interest, Imbrie & Kipp (1971 and elsewhere) have described a factor analysis–transfer function technique useful for calculating a number of climatically related variables from a relatively complete taxonomic description of a sediment sample. Hecht (1973) reported a technique that used differences between the distance coefficients calculated from taxonomic descriptions of samples as the basis for paleotemperature determinations. Both of these techniques yield quantitative data on ancient climates after calibration of data on fossil assemblages from modern sediment samples with present-day climatic conditions. The calibrations are valid only to the extent that evolution does not change the response of the flora or fauna to climate. So far, these techniques have been applied primarily to Holocene and Pleistocene sediments. Temperature and salinity estimates obtained using transfer function techniques were compared with oxygen isotopic compositions of planktonic foraminifera from the same samples in a Pleistocene Caribbean core by Imbrie, van Donk & Kipp (1973). They concluded that the results of the isotopic and faunal methods were in agreement with one another if the isotopic data were interpreted as reflecting both temperature and local and worldwide variations in the isotopic composition of seawater. Relationships between measured surface temperatures and those determined using paleontological and geochemical techniques have been discussed by Berger & Gardner (1975).

To avoid complications related to evolution when drawing paleoclimatic inferences from data on fossil distribution, a number of techniques that are independent of the climatic preferences of individual taxa have been developed for interpretation of climatic conditions. Sinnott & Bailey (1915) pointed out relationships between climate and the shape and structure of angiosperm leaves that have continued to serve as the basis of many paleoclimatic studies of terrestrial deposits. The taxonomic diversity of living organisms usually decreases from the equator to either pole (Stehli 1970), presumably in response to poleward temperature decrease. Correlations have been recognized between diversity of fossil planktic organisms and paleotemperatures estimated by other methods (Roth 1974, Jenkins 1968a,

Haq 1973). Factors affecting the diversity of microfossils and nannofossils in ocean sediments have been reviewed by Berger & Roth (1975).

GEOCHEMICAL APPROACHES   The technique most widely and successfully applied to the study of Cenozoic climates in the past several years has been the oxygen isotope paleotemperature method of Urey et al (1951) and Epstein et al (1951). This technique is based on the fact that the magnitude of the calcium carbonate–water oxygen isotope fractionation (i.e. the difference between $^{18}O/^{16}O$ of water and $^{18}O/^{16}O$ of calcium carbonate deposited from that water under conditions of isotopic equilibrium) varies as a function of depositional temperature. Therefore, by measuring $^{18}O/^{16}O$ in fossil carbonate shells, estimating $^{18}O/^{16}O$ of the water in which they grew, and knowing the manner in which the isotopic fractionation between calcium carbonate and water varies with temperature, it is possible to estimate the temperature at which the carbonate was deposited. Temperature estimates made in this way are referred to in this review as *isotopic temperatures*. Measurements of $^{18}O/^{16}O$ of calcium carbonate can be routinely made with a precision of 0.1 per mil[2] which corresponds to a precision in the temperature estimate of approximately 0.5°C. The precautions necessary for valid applications of the technique and interpretation of results have been discussed by several authors, including Urey et al (1951), Bowen (1966), and Savin, Douglas & Stehli (1975).

Many organisms secrete calcium carbonate either in isotopic equilibrium with seawater or very close to equilibrium. Molluscs and foraminifera are the groups that have been most frequently used in Cenozoic isotopic paleotemperature studies. The attainment of isotopic equilibrium has been checked carefully for planktic foraminifera, and there is some indication that these may exhibit small departures from equilibrium, especially at tropical temperatures (van Donk 1970, Shackleton, Wiseman & Buckley 1973, R. G. Douglas and S. M. Savin, unpublished data). While further investigation of this is desirable, it is most likely that these small departures from isotopic equilibrium in planktic foraminifera will be of greater importance in studies of foraminiferal biology and ecology than in paleotemperature studies.

Benthic foraminifera in deep-sea sediments frequently exhibit departures from isotopic equilibrium (Duplessy, LaLou & Vinot 1970, Shackleton 1974, F. Woodruff, unpublished data). Shackleton (1974) reported that *Uvigerina* deposits its tests at or near isotopic equilibrium in the temperature range 0.8° to 7°C. The magnitudes of disequilibrium oxygen isotopic fractionations between different benthic taxa remain relatively constant throughout single cores, permitting empirical intercalibration of the isotopic temperatures obtained from different taxa.

---

[2] $^{18}O/^{16}O$ ratios are commonly reported in δ notation as deviations of the ratio of the sample from that of an arbitrary standard, i.e.

$$\delta^{18}O = \left[\frac{(^{18}O/^{16}O)_{sample} - (^{18}O/^{16}O)_{std}}{(^{18}O/^{16}O)_{std}} \right] \times 1000.$$

Units are per mil or parts per thousand. The most commonly used standard in paleotemperature studies is the PDB standard (Urey et al 1951, Craig 1965).

Coccoliths, a second major calcium carbonate component of ocean sediments, have recently received increasing attention from isotope paleoclimatologists because of their presence in many ocean sediments in which foraminifera are scarce or absent (Margolis et al 1975, Anderson & Cole 1975, Douglas & Savin 1975, Dudley 1976, Lindroth et al 1977). Because the coccolithophorids, which produce coccoliths, require light for photosynthesis, their isotopic compositions should reflect environmental conditions closer to the ocean surface than many taxa of planktic foraminifera. Results to date indicate that coccoliths do not always deposit calcium carbonate in isotopic equilibrium with water. The magnitude of carbonate-water isotopic fractionations of coccoliths varies with temperature, but in different fashion for different taxa. The isotopic compositions of coccoliths frequently vary in sub-parallel fashion to those of planktic foraminifera from the same samples. The isotopic compositions of coccoliths, therefore, clearly do contain a climatic signal. However, because of apparent disequilibrium precipitation as well as preservational problems discussed below, the interpretation of the isotopic compositions of coccoliths is not always clear at this time. Until the isotopic behavior of this group is better understood, coccoliths must be considered useful but less desirable than foraminifera for paleotemperature research.

The isotopic composition of biogenic calcium carbonate must remain unaltered by diagenesis if it is to have paleoclimatic significance. Evidence of any recrystallization or chemical alteration may be taken as sufficient grounds for the elimination of a sample from any isotopic paleotemperature study (Urey et al 1951). Criteria for recognizing the state of preservation of belemnite guards have been reviewed by Stevens & Clayton (1971) and Spaeth, Hoefs & Vetter (1971). Buchardt (1977) has discussed preservation of the isotopic record in a variety of molluscs. In addition to recrystallization, other diagenetic processes can cause important modifications in the isotopic temperature record of materials from deep-sea sediments. Species-related requirements restrict the range of water depths occupied by individuals of a single species of planktic foraminifera living at a single location. Within that range of depths, however, individuals may form with a range of isotopic compositions corresponding to the in situ temperature variation. Shallow-dwelling individuals of a species are probably more subject to solution, a process ubiquitous in deep marine basins. Preferential loss of the "warmer" fraction of a foraminiferal assemblage through selective solution biases isotopic temperatures of the remaining foraminifera to lower values (Savin & Douglas 1973). This effect may be as much as 2° or 3°C in Recent tropical sediments. In the tropics the effect is greater for shallower-dwelling species of foraminifera and is more pronounced for samples in sediments from deeper waters. The importance of this solution effect is probably less for Tertiary and Cretaceous sediments than for Recent ones, since during those times the thermocline in tropical regimes was less pronounced than at present (Savin, Douglas & Stehli 1975). Another diagenetic effect is the encrustation of coccoliths with secondarily precipitated calcite. This encrustation, which can be readily noted by scanning electron microscopy, can bias isotopic temperatures of coccoliths to colder values (Douglas & Savin 1975).

There are also ecological factors that must be considered in the interpretation of isotopic paleotemperature data. Even assuming that temperatures of carbonate deposition can be accurately estimated from isotopic data, these temperatures do not have quantitative climatic significance until interpreted further. In regions with seasonal temperate variations, carbonate precipitation may occur predominantly at one time of the year, as pointed out in molluscs by Epstein & Lowenstam (1953). Different taxa occupy water of different average depths and hence different temperatures. Additionally individuals of many taxa migrate into waters of different depths or temperatures either seasonally or as a function of growth stage. Temperatures of carbonate deposition do not, therefore, in general indicate temperatures at the ocean surface. However, surface temperatures can frequently be estimated from isotopic temperatures when the growth habits of the organisms are understood.

The depth stratification of planktic foraminifera and its expression in $^{18}O/^{16}O$ of their calcite was discussed initially by Emiliani (1954a) and subsequently by Lidz, Kehm & Miller (1968), Shackleton (1968), Oba (1969), Emiliani (1971a), Hecht & Savin (1972), Duplessy (1972), Savin & Douglas (1973), Savin & Stehli (1974), and others. Depth habitats of extinct Tertiary and Upper Cretaceous planktic foraminifera have been reconstructed using isotopic evidence (Douglas & Savin 1977). Extinct taxa that are morphologically similar to present-day planktic species exhibited depth stratification similar to that of their modern counterparts. This information, when combined with data on the differences between surface temperatures and isotopic temperatures of foraminifera from low-latitude Recent ocean sediments (Savin & Douglas 1973), has been used to infer Tertiary surface temperatures from the isotopic temperatures of Tertiary foraminifera (Savin, Douglas & Stehli 1975). Such inferences, however, are based on speculative assumptions concerning the past thermal structure of the upper portion of the water column. The relationship between isotopic temperatures of foraminifera and surface temperatures requires further exploration.

To relate temperatures of carbonate secretion to surface temperatures and especially to the surface temperature fluctuations between glacial and interglacial times, it is necessary to understand how the depth habitats of foraminifera change as climatic conditions change. Every species of living foraminifera appears preferentially to inhabit water of a particular density (Emiliani 1954a). If the density preferences of the species have remained constant from glacial to interglacial period, then dilution of the oceans by icecap runoff and the resultant lowering of the density of seawater would result in the downward migration of foraminifera. Conversely, icecap growth would result, other things being equal, in upward migration of foraminifera. The net effect for foraminifera that migrate in this fashion would be to minimize glacial-interglacial differences in isotopic temperatures (Shackleton 1968). Savin & Douglas (1973) suggested that osmotic interaction with the surrounding water is an important factor in determining the depth habitats of foraminifera. If this is so, or if other factors such as the availability of nutrients are important in determining their depth habitats, the vertical migration of foraminifera between glacial and interglacial time may be more complex than suggested by simple density models (Savin & Stehli 1974).

Finally, in determining temperatures of shell secretion from isotopic data, $^{18}O/^{16}O$

of the water in which the shell grew must be estimated. Compared with fresh waters, the oxygen isotopic compositions of ocean waters of normal salinity fall within a fairly narrow range (about 2.5 per mil). Within this range, $^{18}O/^{16}O$ is affected primarily by processes of evaporation and precipitation, and in certain regions, influx of meltwater (Epstein & Mayeda 1953). These factors and the manner in which they affect the isotopic composition of the modern oceans have been discussed in detail by Craig & Gordon (1965). In the interpretation of isotopic paleotemperature data, variation of the isotopic composition of seawater from two processes must be considered: 1. Large-scale polar icecap formation results in the preferential removal of ^{16}O from the oceans into the icecaps, and the corresponding enrichment of the oceans in ^{18}O. The periodic advance and retreat of polar icecaps during the Pleistocene is associated with corresponding periodic fluctuations in $^{18}O/$ ^{16}O of the oceans. The magnitude of these fluctuations, discussed later, remains one of the most controversial topics in the interpretation of Pleistocene isotopic paleotemperature data. 2. Local variations in evaporation and precipitation produce local variations in $^{18}O/^{16}O$ of surface seawater. To a first approximation, evaporation and precipitation are controlled by large-scale atmospheric circulation, which in turn is affected predominantly by the mass of the atmosphere, the speed of rotation of the earth, etc. Therefore, for lack of a better alternative, there has been a tendency among isotope paleoclimatologists to ignore variations of evaporation/precipitation and its effect on the isotopic composition of seawater when interpreting paleo-temperature data. As our picture of past oceanographic and climatic conditions becomes more complete it should become possible to adjust data to account for local variations in oceanic $^{18}O/^{16}O$.

Because application of the oxygen isotope paleotemperature technique permits the undiscriminating user to calculate growth temperatures to any number of significant figures, some have tended to regard the absolute isotopic temperature values obtained as being accurate water temperatures. Others, because of the un-certainty in the factors discussed above, have been skeptical of the utility of isotopic paleotemperature measurements. It is my feeling that the truth lies in the middle ground. The isotopic compositions of many organically precipitated carbonates contain climatic information, obscured only in part by the factors discussed above. The more accurately the effects other than temperature can be determined, the more realistically the isotopic data can be interpreted. It is unlikely that ancient surface temperatures can be estimated from isotopic data with an accuracy of better than 3° or 4°C except when estimates can be tied to modern surface temperature measurements. Surface temperature gradients over an area at a single time and temperature fluctuations with time at a single place may be estimated much more accurately. Because of the very uniform temperature structure and isotopic composition of deep ocean water as well as the slow rate of evolution of benthic foraminifera it may ultimately become possible to estimate bottom water temperatures from the isotopic compositions of benthic foraminifera with an accuracy of 1° or 2°C.

In addition to the large amount of work done on carbonates, oxygen isotope techniques have been applied to both phosphate (Longinelli & Nuti 1968a, b,

1973a, b) and silica (Labeyrie 1974, Knauth & Epstein 1975, Kolodny & Epstein 1976) in attempts to obtain paleoclimatic information. Both of these materials may yield useful paleoclimatic data, and they may find increasing use in paleotemperature studies, especially in situations where suitable carbonates do not exist. However, at present little application has been made of either of these geothermometers to the study of Cenozoic climate.

A number of investigators have attempted to relate the Mg content of carbonates to growth temperature. The environmental significance of Mg in calcium carbonate was reviewed by Dodd (1967). While there is some correlation between Mg content and growth temperature for many groups of organisms, factors other than temperature also affect Mg content. As a result most workers have concluded that this parameter seems to have limited utility as a paleothermometer. Recent work by Savin & Douglas (1973) on planktic foraminifera and Durazzi (1975) on ostracods has led to similar conclusions. However, Kilbourne & Sen Gupta (1973) reported a correlation between Mg content and depositional temperature in planktic foraminifera from a Pleistocene Caribbean core and suggested the use of Mg content in this group as a paleotemperature indicator. Several articles in the Soviet literature have reported results of the development of a Ca/Mg geothermometer for carbonates (Berlin & Khabakov 1966, 1967, 1974, Berlin et al 1966, and others). This technique has been applied primarily to Mesozoic sediments and chiefly to belemnites.

THE PALEOCLIMATIC RECORD

Deep-sea sediments contain the most complete stratigraphic sections of Cenozoic age available for paleoclimatic study. Because they are deposited far from the continents, effects of local geography and topography on factors such as the positions of currents, evaporation, and precipitation on the paleoclimatic record are minimized. Information concerning changes in the earth's climate is therefore more readily obtained from the study of deep-sea sediments than of other types of sedimentary deposits. The paleoclimatic record of shallow-water, nearshore marine sediments may be related to worldwide climatic change with somewhat more uncertainty. The terrestrial climatic record is most difficult of all to interpret and to place in the context of worldwide climatic events because of the extreme local variability of continental climates and the scarcity of terrestrial sedimentary sequences representing long periods of deposition as well as the difficulty of making accurate biostratigraphic correlations between these rocks and marine sequences.

The approach followed in this section is to use the available oxygen isotope paleotemperature data to develop as complete as possible a marine temperature record. Other climatic information is then reviewed and integrated with the isotopic data.

Cretaceous Paleotemperatures

A number of early isotopic paleotemperature studies, including the classic investigations of Urey et al (1951) and Lowenstam & Epstein (1954), were focused on Upper

Cretaceous sediments. These early studies were based almost entirely on analyses of megafossils from shallow-water sediments. It was only after the initiation of the Deep Sea Drilling Project that Tertiary and Cretaceous microfossils from deep-sea sediments became sufficiently available to be important in paleotemperature studies.

The range in $^{18}O/^{16}O$ that may be encountered in specimens from uplifted marine sediments from a restricted region is indicated in the data for belemnites from Northern Europe (Sweden, Denmark, England, Holland, Belgium, and France) published by Lowenstam & Epstein (1954) and shown in Figure 1a. Isotopic analyses of megafossils from uplifted sediments in other parts of the world show similar variations. Scatter in the isotopic data for samples such as these can readily result from alteration in the presence of ^{18}O-depleted fresh waters, with a resultant increase in the apparent isotopic temperature. Because of this, most investigators, in interpreting isotopic paleotemperatures of fossils from uplifted marine sediments, have placed greatest meaning on the coldest isotopic temperatures obtained, and that is the interpretation used in the following discussion. However, it is clear that where scatter in the data reflects differences in the life histories of individual organisms and/or seasonal deposition of carbonate, average or even maximum values of isotopic temperature might have more climatic significance than minimum values. The best method for interpreting paleotemperature data from such samples is therefore not always clear.

In Figure 1b is shown a series of paleotemperature curves for Cretaceous time derived from isotopic compositions of belemnites published by several authors for different parts of the world. Each curve has been drawn through the most ^{18}O-rich analyses of a sample suite, an interpretation with which the authors of the original papers may not always agree. In Figure 1c are plotted isotopic compositions of

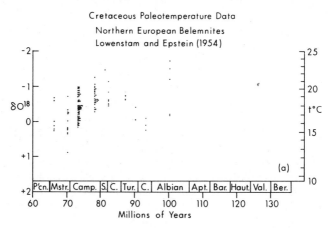

Figure 1a Isotopic analyses of Northern European belemnites by Lowenstam & Epstein (1954) plotted on the Cretaceous time scale of Moberly, Gardner & Larson (1975). Each point represents one analysis. Relationship between $\delta^{18}O$ scale and temperature scale was calculated assuming $\delta^{18}O$ of water was -1.00 per mil relative to standard mean ocean water.

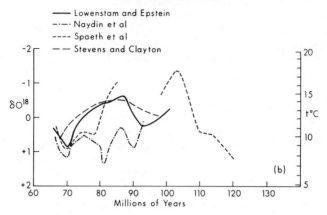

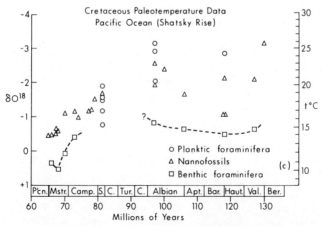

Figure 1b Cretaceous paleotemperature curves drawn by this reviewer from published isotopic analyses of belemnites. Each curve was drawn through the most ^{18}O-rich values for that data set (see text) after replotting on the time scale of Moberly, Gardner & Larson (1975). For example, compare the solid line of Figure 1b with the data plotted in Figure 1a. These curves are not necessarily in agreement with those of the original authors. [Note: one exceptionally ^{18}O-rich analysis of early Albian age was ignored in drawing the curve through the data of Spaeth, Hoefs & Vetter (1971).] Published $\delta^{18}O$ values were used in plotting all data except those of Naydin et al (1966), who published paleotemperatures rather than $\delta^{18}O$ values. Relationship between $\delta^{18}O$ scale and temperature scale was calculated assuming $\delta^{18}O$ of water was -1.00 per mil relative to SMOW.

Figure 1c Paleotemperature data of Cretaceous nannofossils, planktic foraminifera, and benthic foraminifera from the Shatsky Rise (Douglas & Savin 1975). The upward trend in temperature in Cenomanian time is suggested by a single analysis of benthic foraminifera from the Hess Rise (not shown). Relationship between $\delta^{18}O$ scale and temperature scale was calculated assuming $\delta^{18}O$ of water was -1.00 per mil relative to SMOW.

nannofossils and planktic and benthic foraminifera from the central Pacific Ocean (Douglas & Savin 1975). The curves derived from the belemnite data differ in detail from one another. This probably reflects in part the different stratigraphic ages of the samples studied. There are also differences between the curves through the belemnite data and the trends through the Pacific Ocean microfossil and nanno-fossil data. However, some generalizations can be drawn. Temperatures appear to have been especially warm during Albian time in Northern Europe and in the

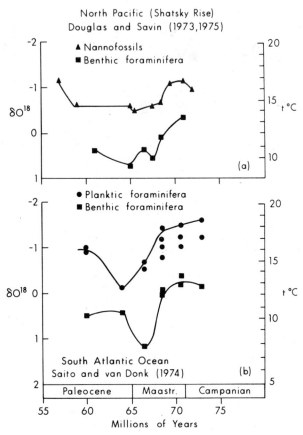

Figure 2 Isotopic analyses of detailed sections across the Cretaceous-Tertiary boundary. Data from Douglas & Savin (1973, 1975) and Saito & van Donk (1974). The Pacific data are, with the exception of the 61 m.y. point, from the Shatsky Rise. The 61 m.y. point is from the Magellan Rise. The small time lag between the temperature minimum of the South Atlantic planktic foraminifera and that of the benthic foraminifera may be an artifact of sampling. The 64 m.y. South Atlantic sample is from 13° latitude further south than all the other South Atlantic sites. Relationship between $\delta^{18}O$ scales and temperature scales was calculated assuming $\delta^{18}O$ of water was -1.00 per mil relative to SMOW.

tropical Pacific Ocean, but there is only a hint of that warm peak in the deep ocean water, as evident from the few available analyses of benthic foraminifera. Deep-sea bottom temperatures must have reflected surface conditions in some high latitude region. Sometime during Albian or Cenomanian time, temperatures dropped. A temperature rise and a warm period followed in Turonian and/or Coniacian time. There was then a general lowering of temperatures, although perhaps not uniformly and monotonically, until a temperature minimum was reached in early Maastrichtian time. Temperatures in late Maastrichtian and early Tertiary time are discussed in the following section.

The differences between the isotopic paleotemperature curves in different parts of the world and between isotopic paleotemperature curves and climatic events inferred from fossil distributions have been discussed at great length (Lowenstam 1964, Voigt 1965, Naydin, Teis & Zadorozhny 1966, Bowen 1966, Stevens 1971, Scheibnerova 1973, and others). While the details of late Cretaceous climatic history in different parts of the world are still somewhat uncertain, it does seem clear that at least most of the world experienced a significant net cooling (perhaps as much as 8° or 10°C in both Northern Europe and the tropical Pacific Ocean) between Albian time and the end of the Cretaceous period.

The Cretaceous-Tertiary Boundary

The Cretaceous-Tertiary boundary was a time of major extinctions of both terrestrial and marine taxa. Among the many postulated causes for the dramatic biological revolution at that time is sudden temperature change. Isotopic analyses of planktic and benthic fossils from two detailed sections across the boundary are shown in Figure 2. The temperature histories of the two areas, the South Atlantic (Saito & van Donk 1974) and the North Pacific (Douglas & Savin 1975) are quite similar. At both sites, the isotopic temperatures recorded by both planktic and benthic fossils dropped by a few degrees in early to middle Maastrichtian time, and then recovered. The drop was slightly greater for the South Atlantic samples than for the North Pacific. The magnitude of the temperature drop identified near the boundary is small compared to the drop from Santonian through Maastrichtian time (Figure 1) or that near the Eocene-Oligocene boundary (Figure 3). The drop occurred prior to the end of Maastrichtian time and hence earlier than the great extinctions of the Cretaceous-Tertiary boundary. It is of course possible that a more marked, very short-lived cooling occurred near the Cretaceous-Tertiary boundary and has not yet been recognized either because of insufficiently close sample spacing or because portions were missing from the stratigraphic sections studied. Barring these possibilities, however, it seems doubtful that the biological revolution at the time of the Cretaceous-Tertiary boundary was predominantly the result of purely temperature effects.

Tertiary Paleotemperatures

ISOTOPIC EVIDENCE   Most Tertiary isotopic paleotemperature data have been obtained using samples from the Pacific basin. In Figure 3 are shown isotopic data for New Zealand (Devereux 1967), the North Pacific (Douglas & Savin 1971, 1973,

1975, and unpublished data; Savin, Douglas & Stehli 1975), and the South
Pacific (Shackleton & Kennett 1975a). Dorman (1966) has published a Tertiary
isotopic paleotemperature curve for Australia that is quite similar to Devereux's
New Zealand curve. The New Zealand analyses are of fossils from nearshore,
shallow-water sediments subsequently uplifted above sea level. Interpretation of
those data is therefore more subject to uncertainties related to the isotopic composition of the water and fossil preservation, as discussed previously, than is the interpretation of data from the DSDP cores. In addition, at the time the New Zealand
studies were done it was not possible to achieve the refinement in biostratigraphic
correlation now possible with the DSDP cores.

Emiliani (1954b) pointed out that bottom water temperatures anywhere in the
deep ocean, as interpreted from the isotopic compositions of benthic foraminifera,
should be indicative of surface water temperatures at high latitudes (where dense,
cold water sinks to form bottom water). The isotopic compositions of Tertiary
benthic foraminifera seen in Figure 3b from the North and South Pacific (and in a
few instances, from the Atlantic) all lie close to a single smooth curve. The benthic
data in the figure support Emiliani's model, and indicate the temperature history
of high-latitude oceans during Tertiary time. This curve is similar to a curve that
could be drawn through the benthic data of Shackleton & Kennett (1975a) although

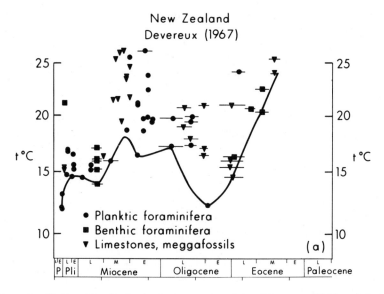

Figure 3a Tertiary New Zealand paleotemperature data of Devereux (1967) replotted on
the time scale of Berggren & van Couvering (1974). The temperatures are those calculated
by Devereux and incorporate his estimated corrections to compensate for latitudinal
temperature variations within the data set and to relate all the analyses to temperatures
at Wellington. The curve, drawn by this reviewer, is essentially through the coldest isotopic
temperatures.

there are differences in detail. These differences may be due to problems of bio-stratigraphic correlation, to artifacts of sampling interval, to possible shallow-water depths for the South Pacific sites, or to oceanographic complications in the Australia–New Zealand–Antarctica sector of the South Pacific related to crustal plate motions and the development of the Antarctic Circumpolar Current. There has been a net lowering of isotopic temperatures at intermediate and high latitudes

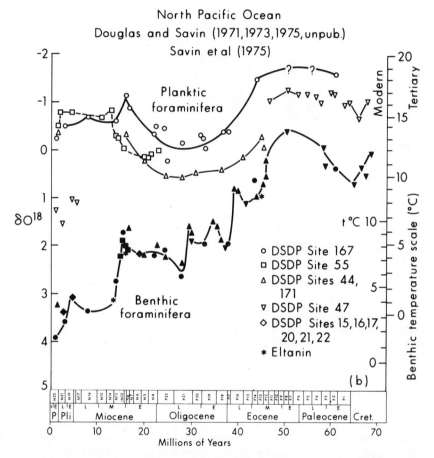

Figure 3b Isotopic paleotemperature data for Tertiary planktic foraminifera (open symbols) and benthic foraminifera (closed symbols) primarily from the North Pacific, from Douglas & Savin (1971, 1973, 1975, and unpublished) and Savin, Douglas & Stehli (1975). A few points are from the South Atlantic and South Pacific. The curve through the benthic data was drawn to best fit the points from all the sites. The "Modern" and "Tertiary" benthic temperature scales were calculated assuming water $\delta^{18}O$ values of -0.08 per mil and -1.00 per mil respectively. To estimate isotopic temperatures of planktic foraminifera add approximately 2.5°C to appropriate benthic temperature scale.

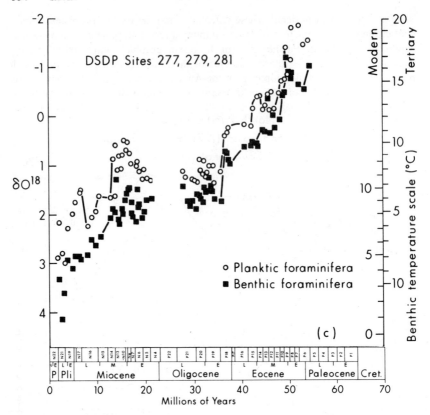

Figure 3c Isotopic paleotemperature analyses of planktic and benthic foraminifera from the Campbell Plateau and the Macquarie Ridge (Subantarctic Pacific) by Shackleton & Kennett (1975a). Analyses have been replotted on the time scale of Berggren & van Couvering (1974). The Modern and Tertiary benthic temperature scales were calculated assuming water $\delta^{18}O$ values of -0.08 and -1.00 per mil respectively. Because of the extreme oceanographic changes in the area during Tertiary time a separate temperature scale for planktic foraminifera was not estimated.

since the beginning of Tertiary time (indeed, since Coniacian-Santonian time). However, the isotopic data indicate that the cooling has not preceeded smoothly, and that there have been significant periods in which temperatures increased. This is in accord with the results of some, but not all, paleoclimatic studies based on paleontologic evidence discussed below.

The absolute values of isotopic temperatures assigned the benthic foraminifera depend upon the assumptions made concerning the effect of polar icecap formation on the isotopic composition of ocean water. Uncertainties arise regarding the isotopic composition of the icecaps and the manner in which both the volume and the isotopic composition have changed with time. Craig (1965) suggested that if all the present-day icecaps were melted and added to the oceans the average $\delta^{18}O$ of the oceans would decrease by about 0.52 per mil, but he did not give details about how he estimated the mean $\delta^{18}O$ of the polar icecaps. Shackleton (1967) calculated that the decrease in $\delta^{18}O$ upon melting the polar icecaps would be 0.92 per mil. An error of 0.4 per mil in the isotopic composition of deep ocean water would correspond to a systematic error of about 2°C in the calculation of absolute isotopic temperatures. Prior to the time of significant icecap formation, estimates of temperature gradients and of temperature changes as a function of time are, however, unaffected by this uncertainty in the mean isotopic composition of the ocean. Difficulty is encountered in trying to estimate the magnitude of temperature changes at times when icecaps were undergoing major growth (or contraction). The resolution of the isotopic signal into a temperature effect and an ice-volume effect cannot yet be adequately done for all times in the Tertiary. This is important in interpreting isotopic compositions of samples of middle Miocene through Pleistocene age. Two temperature scales have been provided on the diagrams of Figure 3. One is labeled "Tertiary" and has been calculated assuming seawater was uniformly depleted in ^{18}O by 0.92 per mil relative· to modern values. The other is labeled "Modern" and has been calculated assuming modern $\delta^{18}O$ values for seawater. Temperatures from middle Miocene through Pleistocene times fall between those indicated by these two scales (or a bit warmer than the Modern temperatures for samples deposited at some times during Pleistocene and probably late Pliocene times when the extent of the icecaps was greater than now).

Two marked and rapid declines in isotopic temperature are evident from the data in Figure 3. These occur in the vicinity of the Eocene-Oligocene boundary and in early middle Miocene time. The early middle Miocene event is more apparent from the North Pacific benthic analyses and the South Pacific planktic analyses than from the South Pacific benthic data. The rapidity with which a drop in isotopic temperature can occur is illustrated in Figure 4. Kennett & Shackleton (1976) have estimated that a bottom water temperature drop of approximately 4°C occurred in earliest Oligocene time in a time span of 75,000 to 100,000 years, and have argued that this reflects the formation of the system of cold bottom waters characteristic of the modern oceans. The magnitude of the temperature drop in bottom waters in early middle Miocene time is probably similar, although, as noted, there is some uncertainty in the absolute value of the drop because of uncertainty in the ice volume effect at that time. The cooling in middle Miocene time very likely correlates with the onset of significant stable icecap formation in Antarctica.

The location of the low-latitude Pacific sites, when tracked back to compensate for crustal plate motion, appear to have occupied tropical positions throughout late Cretaceous and Tertiary time (Lancelot & Larson 1975). Therefore the isotopic temperatures derived from the planktic fossils from those sites should be indicative of low-latitude surface temperatures throughout late Cretaceous and Tertiary time. Isotopic temperature trends of planktic organisms of pre-middle Miocene age both in the tropical North Pacific and the Subantarctic Pacific are roughly parallel to those of benthic organisms (Figure 3).

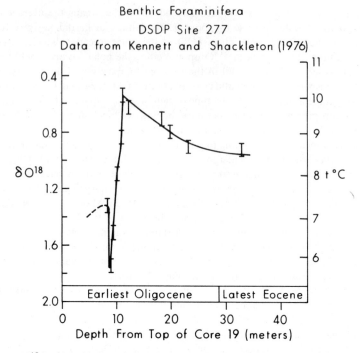

Benthic Foraminifera
DSDP Site 277
Data from Kennett and Shackleton (1976)

Figure 4 $\delta^{18}O$ of benthic foraminifera from DSDP Site 277, after Kennett & Shackleton (1976). They estimated the temperature drop of approximately 4°C to have occurred in 75,000 to 100,000 years, on the basis of sedimentation rate.

In early middle Miocene time, surface temperatures in high and low latitudes (the former, as evident from isotopic temperatures of benthic foraminifera) began to change in markedly different ways. At that time, while high-latitude surface temperatures dropped rapidly, low-latitude surface temperatures stayed relatively constant or perhaps increased. [The uncertainty here is related to the uncertainties in the manner in which surface temperatures can be estimated from the growth temperatures of planktic foraminifera (Savin, Douglas & Stehli 1975).] The climatic event in early middle Miocene time involved a marked steepening of the equator-to-pole surface temperature gradient in the oceans, and presumably on land. The thermocline in tropical regions also became more pronounced at that time and the intensity of abyssal oceanic circulation probably increased as well.

A detailed isotopic paleotemperature curve for planktic foraminifera for late Miocene and Pliocene time [DSDP Site 284 on the Challenger Plateau west of New Zealand (Shackleton & Kennett 1975b)] is shown in Figure 5. Also shown are planktic and benthic isotopic curves from the Pliocene to early Pleistocene time that are derived from sediments from New Zealand (Devereux, Hendy & Vella 1970). The data from both localities indicate relatively stable conditions during early Pliocene time and a cooling in late Pliocene time. Shackleton & Kennett (1975b) have suggested that the late Pliocene cooling corresponded to the initiation of glaciation in the Northern Hemisphere.

PALEONTOLOGICAL EVIDENCE   Large numbers of estimates of climatic conditions during Tertiary time based on some aspect of the fossil record have been published. Most of these are imprecise, semiquantitative estimates of climate in a single area at a single time or over a short span of time, and integration of the conclusions, derived using a variety of different methods, is extremely difficult.

In a classic study, Smith (1919) examined large numbers of fossil faunas of Tertiary age and presented a fairly detailed interpretation of the climatic history of the west coast of North America. He concluded that there had been a fairly steady cooling trend from Eocene to late Pliocene time, followed by a warming culminating in a Quaternary temperature maximum and a subsequent cooling. Durham (1950) interpreted Tertiary climates in western North America using data

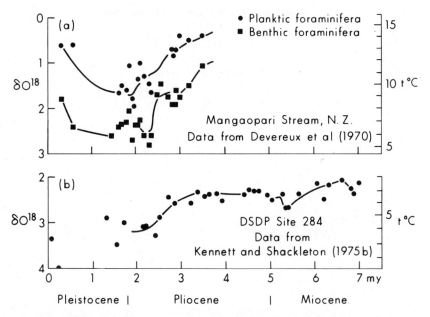

Figure 5a $\delta^{18}O$ of planktic and benthic foraminifera from Mangaopari Stream, New Zealand. Data from Devereux, Hendy & Vella (1970). Sample ages estimated by this reviewer from stratigraphic information provided by Devereux, Hendy & Vella.
Figure 5b $\delta^{18}O$ of planktic foraminifera Neogloboquadrina pachyderma from DSDP Site 284 (west of New Zealand). Isotopic data from Shackleton & Kennett (1975b). Sample ages estimated in part by this reviewer from information in Kennett et al (1975). Curves through the data of Figures 5a and 5b are those of this reviewer, not the original author. Relationship between $\delta^{18}O$ scale and temperature scale was calculated assuming water $\delta^{18}O$ values of −0.08 per mil.

338   SAVIN

on marine fauna, especially corals. His results have been widely quoted and are in many ways similar to Smith's. They indicate a fairly steady temperature decrease from Eocene through middle Pliocene time, a minor warming in late Pliocene time, and a subsequent cooling prior to the end of the Tertiary period. There is no indication in either Smith's or Durham's results of the warming in late Oligocene and early Miocene time found in the isotopic data. The number of late Oligocene and early Miocene samples they studied, however, was quite small. Addicott (1969, 1970) concluded, on the basis of molluscan distribution in Pacific Coast Tertiary sediments, that such a warming did occur in that region, but suggested that the inferred temperature changes could have reflected changes in local oceanographic conditions, especially changes in upwelling and the development of shallow embayments.

Perhaps no area of the world has been the object of as many paleoclimatic investigations as New Zealand with its well-preserved section of Tertiary marine sediments. In Figure 6 are shown Tertiary paleoclimatic curves derived from a

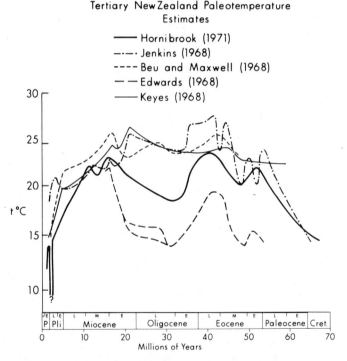

Figure 6 Tertiary paleotemperature curves for New Zealand based on paleontologic evidence. Hornibrook (1971), variety of evidence; Jenkins (1968b), coiling ratios of *G. pachyderma* (planktic foraminifera); Beu & Maxwell (1968), molluscs; Edwards (1968), nannoplankton; Keyes (1968), corals. Data have been replotted using correlations given by Berggren & van Couvering (1974).

variety of data on marine fauna (Hornibrook 1971), nannoplankton distributions (Edwards 1968), coral distributions (Keyes 1968), coiling directions of planktic foraminifera (Jenkins 1968b), and diversity of molluscs (Beu & Maxwell 1968). The differences between the curves are indicative of the difficulties encountered in drawing paleoclimatic conclusions from data on marine fauna. However, in spite of these difficulties, it appears from most of the curves shown that cooling occurred some time during late Eocene or early Oligocene time and again during Miocene time, in agreement with the isotopic data. Other features of some of the curves, especially those of Hornibrook (1971) and Edwards (1968), are also in agreement with the isotopic curves. Further work, however, is needed to refine both paleoclimatic estimates and biostratigraphic correlations before the significance of the fine details of the curves can be evaluated.

Paleoclimatic curves for Tertiary time based on the nature of terrestrial flora have been derived for several parts of the world. Shown in Figure 7 are curves for western Europe (Dorf 1964), western North America (Dorf 1964, Wolfe & Hopkins 1967), and Japan (Tanai & Huzioka 1967). The paleoclimatic curves of western North America have been the subject of some controversy (Wolfe & Hopkins

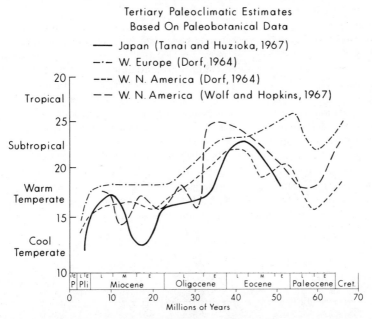

Figure 7 Climatic estimates for western Europe, western North America, and Japan, based on paleobotanical data. Curves have been replotted on the time scale of Berggren & van Couvering (1974). To keep the vertical scale comparable to that in other figures, the following temperature relationships between climatic zones have been assumed: sub-antarctic–cool temperate, 10°C; cool temperate–warm temperate, 15°C; warm temperate–subtropical, 20°C; subtropical–tropical, 25°C.

1967, Axelrod & Bailey 1969, Wolfe 1971). Axelrod & Bailey have argued that the paleobotanical data are not indicative of any periods of warming during Tertiary time, while Wolfe & Hopkins have concluded that there were indeed such periods. The marine data, of course, indicate that warming periods did occur in the oceans.

Quaternary Paleotemperatures

Since the pioneering studies of Emiliani (1955), isotopic paleotemperature measurements have been made on a great number of ocean cores of Quaternary age. These studies have been in large part responsible for a complete revision of the history of the Ice Ages, including a reassessment of both the number of major glaciations and their chronology. From the isotopic paleotemperature curves derived from cores primarily from the Caribbean and Equatorial Atlantic, Emiliani (1966) published and subsequently revised a "Generalized Temperature Curve" for Pleistocene time. A revised version of this (Emiliani & Shackleton 1974) is shown in Figure 8. The chronology is based in part on magnetic stratigraphy (Shackleton & Opdyke 1973). The data indicate that there have been eight glacial-interglacial cycles during the past 700,000 years. Results of van Donk (1970) indicate that fluctuations between glacial and interglacial regimes have been occurring considerably longer than that.

Compared to Tertiary samples, a very large number of oxygen isotopic analyses of samples very closely spaced in time have been done on fossils of Quaternary age. This has permitted a far more detailed and much more highly resolved picture to be drawn of isotopic temperature variations of Quaternary samples. Because of this and the much greater similarity of Quaternary, than of Tertiary, taxa to modern ones, attempts have been made to interpret the Quaternary isotopic data in greater detail than has generally been warranted for the interpretation of isotopic compositions of Tertiary samples. Detailed interpretation is necessary when the isotopic data are used as the basis for climatic models. Among the factors involved in interpreting isotopic data in terms of surface temperatures, none has been given as much attention as has the change in the average isotopic composition of the oceans between glacial and interglacial time. The question of the magnitude of fluctuation of seawater $^{18}O/^{16}O$ has been considered from a number of different points of view: paleontological (Emiliani 1971b), glaciological (Olausson 1965,

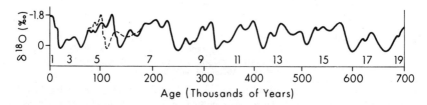

Figure 8 Generalized paleotemperature curve for the Caribbean for the past 700,000 years, from Emiliani & Shackleton (1974). The two curves in the time range 75,000 to 175,000 years indicate two different interpretations. Copyright 1974 by the American Association for the Advancement of Science.

Dansgaard & Tauber 1969), and differences between the variations in isotopic compositions of planktic and benthic foraminifera (Emiliani 1955, Shackleton 1967, Shackleton & Opdyke 1973, Savin & Stehli 1974). While the question is not completely resolved to the satisfaction of all workers involved, most recent workers have estimated that the variations ranged between 0.8 and 1.3 per mil, with more estimates lying closer to 1.2 per mil. Superimposed upon the glacial-interglacial shift in the isotopic composition of average ocean water is the variation resulting from local changes in evaporation and precipitation. The relationship between isotopic temperatures and surface temperatures is further complicated by the ability of foraminifera to migrate vertically in the water column and to alter their preferred depth habitats as climate changes.

While most workers agree that the variations in the isotopic compositions of Quaternary foraminifera reflect climatic events, controversy concerning the extent to which this is a "temperature" signal as opposed to a "continental ice volume" signal has caused many to be hesitant to interpret these isotopic records in terms of quantitative temperature variations. It has long been recognized that changes in the isotopic composition of benthic foraminifera from Quaternary deep-sea sediments must reflect to a close approximation changes in the isotopic composition of seawater. This is because the temperature of bottom water is held relatively constant by the presence of ice in the polar regions in which the bottom water forms. A strong case has been made by Shackleton & Opdyke (1973) that in the western tropical Pacific the isotopic record for shallow-dwelling planktic foraminifera reflects, almost in its entirety, the growth and retreat of continental ice and the resultant change in the mean isotopic composition of the oceans. The results of Shackleton & Opdyke (1973) have frequently been misinterpreted by others as indicating that the isotopic records of Pleistocene planktic foraminifera from all the world's oceans reflect changes in ice volume rather than changes in temperature. Shackleton & Opdyke did not state this, and their data do not imply it. However, isotopic variations in planktic foraminifera certainly do reflect in part (and often in large part) variations in ice volume. Even in regions where the isotopic record of planktic foraminifera reflects in large part temperature change, the temperature and ice volume signals are probably fairly close to synchronous with one another. To the extent that this is true, glacial maxima and minima can be identified as maxima and minima in the $^{18}O/^{16}O$ curves and can be used as stratigraphic markers in correlating ocean sediment cores.

Using a technique based on factor analysis of microfossil populations from ocean sediments, and transfer functions to relate the biologic data to oceanographic variables (Imbrie & Kipp 1971), McIntyre et al (1976) produced a map showing estimated positions of sea surface isotherms during August 18,000 years BP (Figure 9), the time of the maximum extent of continental ice during the last Ice Age. Stratigraphic control was provided by oxygen isotope ratios, as outlined above. The difference between August sea surface temperatures 18,000 years ago and present-day values is shown in Figure 10. It is clear that the oceans were not uniformly cooler during the last Ice Age than at present. In some restricted regions they may actually have been warmer.

Figure 9 Summer sea surface temperatures 18,000 years ago estimated by McIntyre et al (1976), at the height of the last glacial stage. Shaded regions are areas covered by snow or ice. Copyright 1976 by the American Association for the Advancement of Science.

It is possible to use climatic information on a map such as that in Figure 9 to provide boundary conditions applicable for mathematical modeling of ancient climates. Gates (1976) has published the results of such a study. Using the sea surface temperatures of Figure 9 and other climatic data, Gates calculated mean sea level barometric pressures for July 18,000 years BP as well as surface air temperatures on land. Where independent estimates of air temperatures are available, for example from paleobotanical studies, they are generally in agreement with the calculated values. The ability to calculate atmospheric circulation models is an important step in understanding how climate changes.

Relationships between Quaternary and Tertiary Climatic Fluctuations

In searching for causes of the glacial-interglacial fluctuations that characterize the Quaternary period, it is of interest to know whether the fluctuations had any counterpart in pre–middle Miocene times when there was essentially no ice cover. Savin, Douglas & Stehli (1975) analyzed closely spaced samples of early Miocene age (foraminiferal zones N6 and N7) from DSDP Site 15 (South Atlantic). The results (Figure 11) indicate that during early Miocene time, there were quasi-periodic fluctuations with amplitudes between 0.2 and 0.5 per mil (corresponding to temperature fluctuations of approximately 1° to 2.5°C). The average period was approximately 80,000 to 90,000 years. This period is similar to that of the Pleistocene glacial-interglacial cycles (Emiliani & Shackleton 1974) and suggests a similarity in cause. The much smaller amplitude of the early Miocene cycles suggests either that the factors that produced them did not operate as strongly in Miocene time or

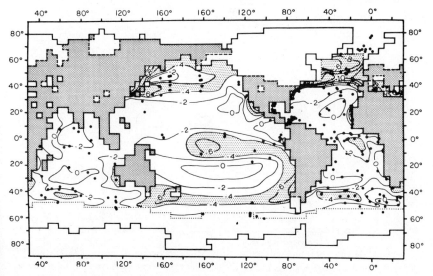

Figure 10 Estimated differences between summer sea surface temperatures 18,000 years ago and modern ones. From McIntyre et al (1976). Copyright 1976 by the American Association for the Advancement of Science.

that the earth's climatic regime was more unstable in Pleistocene time and hence more subject to major change as the result of a small variation in one or more variables.

The History of High-Latitude Ice Cover

Antarctica and Greenland are now almost completely covered by ice sheets, and the surface of the Arctic Ocean is largely ice-covered throughout the year. These massive accumulations of ice affect the world's climate and are in turn affected by it. It is clear that such extensive ice cover did not exist in early Tertiary time. From paleontologic, sedimentologic, and isotopic evidence of deep-sea sediments, as well as geologic evidence found in rocks of polar regions, it is possible to outline the history of the high-latitude Tertiary ice sheets.

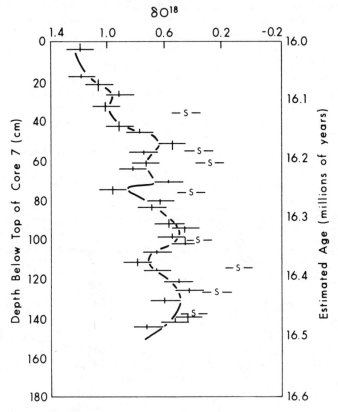

Figure 11 High-resolution study of an early Miocene section of core 7, DSDP Site 15. Crosses indicate analyses of *Globoquadrina dehiscens*. S indicates analyses of *Spheroidinellopsis seminulina*. Estimated ages were calculated by assuming an age of 15.0 m.y. for the top of core 7 and a sedimentation rate of 0.3 cm 10^{-3} yr. From Savin, Douglas & Stehli (1975).

SOUTHERN HEMISPHERE GLACIATION   The isotopic compositions of benthic foraminifera from deep-sea sediments almost certainly reflect surface conditions in high
latitudes. It has frequently been assumed in discussions of Tertiary paleotemperatures
that throughout Tertiary time it has been high-latitude surface temperatures in the
Southern Hemisphere that are reflected in benthic isotopic data. While the validity
of this assumption remains undemonstrated, it is assumed valid in the discussion
that follows.

The isotopic evidence shown in Figures 3a and 3b indicates that bottom water
temperatures prior to latest Eocene time were warmer than 7°C. Assuming that
the temperature of bottom water was similar to that of surface water around the
coast of Antarctica, extensive Antarctic glaciation at sea level and ice shelf development before late Eocene–early Oligocene time are inconceivable. The isotopic
temperatures of 2° to 5°C obtained on Oligocene benthic foraminifera are not
inconsistent with the existence of extensive Antarctic ice cover at high altitudes
during at least part of Oligocene time. However, those temperatures are warmer
than would be prevalent if sea ice were forming or if massive ice shelves existed.
The precipitous drop in isotopic temperatures (increase in $\delta^{18}O$) of benthic
foraminifera in middle Miocene time certainly in part reflects a major period of
ice cap growth as well as a drop in Antarctic coastal surface temperatures.
Shackleton & Kennett (1975a) concluded that by late Miocene time bottom
temperatures were near freezing and that the icecap covering East Antarctica had
achieved its present form by that time. Savin, Douglas & Stehli (1975) suggested
that with the data available it was not possible to satisfactorily resolve the two
causes of change of isotopic composition of the benthic foraminifera (ice formation
vs temperature change) during the middle Miocene climatic event. However, it is
certain from the isotopic record that by late Miocene time a large Antarctic ice
sheet had formed.

The geologic evidence for the warm nature of early Tertiary climates in Antarctica
has been reviewed by Denton, Armstrong & Stuiver (1971). Eocene, Oligocene, and
Miocene rocks containing plant fossils crop out on the Antarctic Peninsula. The
flora suggests climates too warm for massive glaciation in that region, at least into
early Miocene time. Kemp & Barrett (1975) have summarized palynological evidence
that vegetation existed in the Ross Sea area at least until late Oligocene time,
and have suggested that vegetation in that region disappeared during Miocene
time. On the other hand, some evidence points to early Tertiary glaciation. Le
Masurier (1970) obtained Eocene age dates on hyaloclastites from West Antarctica,
which he interpreted as having formed under at least 300 to 400 m of glacial
ice. The onset of glaciation in high latitudes has been inferred by a number of
workers from the presence of presumably ice-rafted sand-sized material with
diagnostic surface features in ocean sediments far from shore. Geitzenauer, Margolis
& Edwards (1968) and Margolis & Kennett (1971) identified poorly sorted
continental detritus and sand-sized quartz with glacial surface features in early and
middle Eocene, Oligocene, early Miocene, Pliocene, and Pleistocene deep-sea
sediments from the Subantarctic Pacific Ocean. The apparently conflicting evidence
for Eocene glaciations probably indicates that during Eocene time glaciers accumu-

lated in limited regions of Antarctica, perhaps at high altitudes, and flowed only locally to the sea. Sedimentological studies done on DSDP cores drilled on Legs 28 and 29 shed further light on the glacial history of East Antarctica. Margolis (1975) found evidence of ice-rafted sediments in middle Miocene and younger cores recovered on DSDP Leg 29, south of New Zealand. Barrett (1975) concluded that in the south-central Ross Sea, sediment-laden ice had been melting from approximately 25 m.y. BP. On the basis of changing sediment patterns in that area he estimated that in early Pliocene time a substantial cooling caused the Ross Ice Shelf to extend beyond its present northern limit. Subsequently, the shelf disappeared (Gauss magnetic epoch) and reformed (Matuyama epoch). Deep-water sediments of early Oligocene and older age on the continental margin off East Antarctica (Piper & Brisco 1975) contain no sand grains. A middle and upper Oligocene section contains occasional rare sand grains. The concentration of sand is low but increases upward in the section. Only in upper Miocene and younger sediments does sand become abundant.

Antarctic glaciation probably commenced locally and was of little global climatic significance during Eocene time. Ice cover became more important during Oligocene time. However, ice shelves and massive ice sheet formation on a scale such as exists today probably did not develop until middle Miocene time.

NORTHERN HEMISPHERE GLACIATION   There is little published evidence from Arctic Ocean sediments concerning the nature of Tertiary climates and the existence of ice in that region. Only two pre-Pliocene sediment cores recovered from the Arctic Ocean have been reported (Ling, McPherson & Clark 1973). Micropaleontologic evidence in these cores indicates that the region was warmer and ice-free until Eocene time and perhaps later. The temperate nature of Paleogene Arctic climates is apparent from the study of fossils from rocks in Arctic regions (summary in Frakes & Kemp 1972, Dawson et al 1976). Clark (1971) concluded from faunal analysis of foraminifera in Pliocene and younger sediment cores that the Arctic Ocean had been continually ice-covered, at least since middle Pliocene time.

Ice-rafted debris has been identified in a number of North Pacific sediment cores as old as 2.4 m.y. by Kent, Opdyke & Ewing (1971). An increase in the amount of ice-rafted debris at high latitudes in the North Pacific and Bering Sea at approximately the same time (middle Pliocene) is suggested from the study of cores from DSDP Leg 19 (Scholl & Creager 1973). Krinsley (1973) identified glacial surface features on middle Miocene quartz grains at DSDP Site 178 on the Alaskan Abyssal Plain. Because limited amounts of glaciation can result in the ice-rafting of debris into ocean sediments it is not possible to specify, from the data discussed above, when the freezing of the Arctic Ocean or the onset of major high-northern-latitude glaciation occurred. It would appear, however, that by middle Pliocene time the Arctic Ocean was frozen and there was significant ice-rafting of sand. As noted earlier, Shackleton & Kennett (1975b) inferred that the drop in isotopic temperatures at DSDP Site 284 (Figure 5) could be correlated with the onset of significant glaciation in the Northern Hemisphere.

RELATIONSHIP BETWEEN PLATE TECTONICS AND CLIMATIC CHANGE

Many theories have been offered over the years as to the causes of climatic changes. Certainly the changes that have occurred in climate are the result of many factors, and no single simple theory can explain the late Mesozoic and Cenozoic paleo-temperature history discussed in the previous sections.

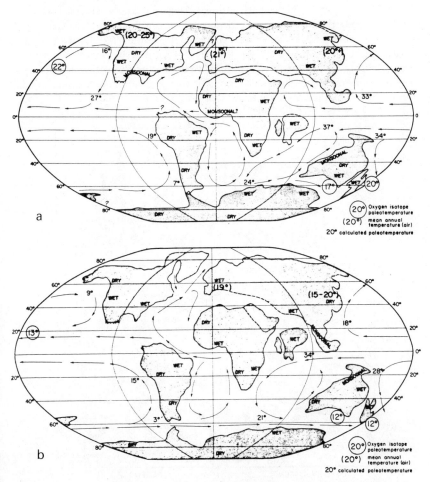

Figure 12 Reconstructions of Paleogene continental positions, surface currents, and winds, by Frakes & Kemp (1972). (*a*) Eocene, (*b*) Oligocene. Reprinted by permission of *Nature*.

Recent work on the causes of climatic change has concentrated on two main problems: the causes of the long-term average temperature variations of the past, and the causes of the relatively rapid quasi-periodic fluctuations from glacial to interglacial conditions that are evident, at least since the end of early Pliocene time, and that may be related to fluctuations of small amplitude identified in early Miocene sediments.

Much recent work on the causes of climatic change has focused on the effect of tectonic motions on climate. It is not a new idea that changing continental positions, areas and locations of epicontinental seas, and locations of large mountain ranges can produce major climatic changes. However, the development of plate tectonic models during the past decade has provided a framework through which to examine the records of temperature change in different parts of the world.

At present, as estimated by Sverdrup (1955) for the Northern Hemisphere, approximately 20% of meridional heat transport occurs through ocean currents. The positions of the continents play a major role in determining oceanic circulation patterns. In Figure 12 are shown reconstructions of Eocene and Oligocene

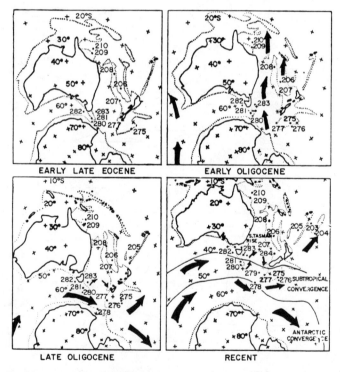

Figure 13 Reconstructions of Paleogene continental positions in the Australia–New Zealand–Antarctica region, by Kennett et al (1974). Arrows show directions of bottom currents. Copyright 1974 by the American Association for the Advancement of Science.

continental positions and ocean currents published by Frakes & Kemp (1972). The Eocene reconstruction is markedly different from the modern ocean circulation pattern, especially in the Southern Hemisphere. The proximity of Australia to Antarctica and the inferred connection between Antarctica and South America, where the Drake Passage now exists, prevented circumpolar flow around Antarctica during Eocene time. The Antarctic Circumpolar Current is one of the major features of the modern ocean. The Eocene South Pacific Ocean is dominated by a massive gyre in which surface water circulates between high and low latitudes.

The Oligocene reconstruction of Frakes & Kemp bears many resemblances to present-day circulation patterns. The movement of Australia away from Antarctica and the opening of the Drake Passage have permitted the formation of an Antarctic Circumpolar Current.

Kennett et al (1974) concluded on the basis of sedimentation patterns and unconformities in the Australian sector of the southern ocean that the Antarctic Circumpolar Current developed appoximately 30 m.y. BP (late Oligocene). Their reconstructions of bottom water circulation are shown in Figure 13.

If a significant surface current flowed eastward around the northern edge of Australia and subsequently southward along Australia's east coast during Paleogene time as shown in Figure 12 (or as depicted in Figure 13 for bottom water circulation) the surface temperature gradient from high to low latitudes would probably have been lower than at present. It is possible that the major change in bottom water circulation during Oligocene time described by Kennett et al (1974) is related to the sharp temperature drop in the vicinity of the Eocene-Oligocene boundary documented in isotopic and geobiologic studies and discussed previously. Savin, Douglas & Stehli (1975) pointed out that the history of crustal plate motion in the region of the Drake Passage was poorly understood. The oldest oceanic crust identified in the passage by Pitman, Larson & Herron (1974) was Miocene in age. Savin, Douglas & Stehli speculated that the Drake Passage did not open sufficiently to permit the establishment of a completely circumpolar current carrying a substantial volume of water until the beginning of middle Miocene time. The thermal isolation of high and low latitudes resulting from the establishment of the circumpolar current in early middle Miocene time could have produced the cooling of high-latitude surface waters and the warming of low-latitude waters apparent from the isotope data. This, in turn could have resulted in the onset of major Antarctic icecap formation. The time of opening of the Drake Passage, however, is still in doubt. Many geophysical estimates place the opening much earlier than middle Miocene time (Dalziel & Elliot 1971, Barker & Griffiths 1972, and others). Foster (1974) argued that the Drake Passage opened approximately 40 m.y. BP on the basis of fossil echinoid distributions.

FUTURE WORK

Progress in the quantitative evaluation of Cretaceous climatic variations has been limited largely by the scarcity of suitable samples. While future deep-sea drilling can be expected to yield well-preserved samples of Cretaceous age from time to time, our understanding of Cretaceous climates will probably increase only slowly in the next few years. There exists, however, a much larger number of well-preserved Tertiary (and especially Neogene) samples and there is a growing interest in paleoclimatic research and other aspects of Tertiary paleo-oceanography. In addition to evaluating climatic variations with time at single locations, sample availability is now sufficient to permit the reconstruction of global surface temperature maps for different time intervals in the past. This enhanced picture of climatic history will undoubtedly become much more fully integrated with the histories of other paleo-oceanographic parameters and models of plate tectonics. In the next few years we can expect to achieve not only a better knowledge of Tertiary climatic variations but also a better understanding of the causes of those variations and the cause of large-scale formation of icecaps.

ACKNOWLEDGMENTS

Many of the ideas and opinions in this paper developed during conversations with my colleague Robert Douglas. Stanley Margolis, Peter Roth, and Chiye Wenkam reviewed the manuscript and their suggestions were most helpful. My special thanks go to Stanley Margolis, who was host for my Hawaiian sabbatical during which this paper was written and who saw to it that I had the facilities I needed to make my stay at the University of Hawaii both productive and enjoyable. Financial support was provided by the National Science Foundation under grants DES 75-20431 and OCE 76-01457.

Literature Cited

Addicott, W. A. 1969. Tertiary climatic change in the marginal northeastern Pacific Ocean. *Science* 165:583–86

Addicott, W. A. 1970. Latitudinal gradients in Tertiary molluscan faunas of the Pacific Coast. *Palaeogeogr. Palaeoclimatol. Palaeoecol.* 8:287–312

Agassiz, L. 1840. *Etudes sur Les Glaciers.* Neuchatel, Switzerland: Jent & Gassmann. 346 pp.

Anderson, T. F., Cole, S. A. 1975. The stable isotope geochemistry of marine Coccoliths: a preliminary comparison with planktonic foraminifera. *J. Foraminiferal Res.* 5:188–92

Axelrod, D. I., Bailey, H. P. 1969. Paleotemperature analysis of Tertiary floras. *Palaeogeogr. Palaeoclimatol. Palaeoecol.* 6:163–95

Barker, P. F., Griffiths, D. H. 1972. The evolution of the Scotia Sea. *Philos. Trans. R. Soc. London Ser. A* 271:151–83

Barrett, P. J. 1975. Textural characteristics of Cenozoic preglacial and glacial sediments at Site 270, Ross Sea, Antarctica. In *Initial Reports of the Deep Sea Drilling Project,* D. E. Hayes, L. A. Frakes et al, 28:757–67. Washington DC:GPO

Berger, W. H., Gardner, J. V. 1975. On the determination of Pleistocene temperatures from planktonic foraminifera. *J. Foraminiferal Res.* 5:102–13

Berger, W. H., Roth, P. H. 1975. Oceanic micropaleontology: progress and prospect. *Rev. Geophys. Space Phys.* 13:561–85

Berggren, W. A., van Couvering, J. A. 1974. The Late Neogene: Biostratigraphy, geochronology and paleoclimatology of the last 15 million years in marine and continental sequences. *Palaeogeogr. Palaeoclimatol. Palaeoecol.* 16:1–216

Berlin, T. S., Khabakov, A. V. 1966. Analytical chemical determination of the ratio of calcium and magnesium in belemnoid rostra as a method of estimating environmental temperatures of existence in seas of the Cretaceous period in the U.S.S.R. (in Russian). *Geokhimiya,* pp. 1359–64. Trans. in *Geochem. Int.* 3:1087–88 (Abstr.)

Berlin, T. S., Khabakov, A. V. 1967. Comparison of paleotemperatures of late Cretaceous seas determined through the calcium-magnesium ratio indices and oxygen isotope mass spectrometer data for belemnitellid rostra. *Dokl. Akad. Nauk. SSSR* 175:450–51 (In Russian). Transl. available from Am. Geol. Inst.

Berlin, T. S., Khabakov, A. V. 1974. Ca-Mg paleotemperature determination for carbonate fossils and country rocks. *Geochem. Int.* 11:427–33

Berlin, T. S., Naydin, D. P., Saks, V. N., Teis, R. V., Khabakov, A. V. 1966. Jurassic and Cretaceous climate in northern U.S.S.R. from paleotemperature determinations. *Int. Geol. Rev.* 9:1080–92

Beu, A. G., Maxwell, P. A. 1968. Molluscan evidence for Tertiary sea temperatures in New Zealand: a reconsideration. *Tuatara* 16:68–74

Bowen, R. 1966. *Paleotemperature Analysis.* Elsevier: Amsterdam. 265 pp.

Buchardt, B. 1977. Oxygen isotope ratios from shell material from the Danish middle Paleocene (Selandian) deposits and their interpretation as paleotemperatures. *Palaeogeogr. Palaeoclimatol. Palaeoecol.* In press

Clark, D. L. 1971. Arctic Ocean ice cover and its late Cenozoic history. *Geol. Soc. Am. Bull.* 82:3313–24

Craig, G. Y. 1961. Palaeozoological evidence of climate. (2) Invertebrates. In *Descriptive Paleoclimatology,* ed. A. E. M. Nairn, pp. 207–26. New York: Interscience

Craig, H. 1965. The measurement of oxygen isotope paleotemperatures. *Proc. Spoleto Conf. Stable Isotopes in Oceanographic Studies and Paleotemperatures* 3:1–24

Craig, H., Gordon, L. 1965. Deuterium and oxygen 18 variations in the ocean and marine atmosphere. *Proc. Spoleto Conf. Stable Isotopes in Oceanographic Studies and Paleotemperatures,* pp. 9–130

Crowell, J. C. 1964. Climatic significance of sedimentary deposits containing dispersed megaclasts. In *Problems in Paleoclimatology,* ed. A. E. M. Nairn, pp. 86–99. London-New York-Sydney: Interscience

Dalziel, I. W. D., Elliot, D. H. 1971. Evolution of the Scotia Arc. *Nature* 233: 245–52

Dansgaard, W., Tauber, H. 1969. Glacier oxygen-18 content and Pleistocene ocean temperatures. *Science* 166:499–502

Dawson, M. R., West, R. M., Langston, W. Jr., Hutchinson, J. H. 1976. Paleogene terrestrial vertebrates: northernmost occurrence, Ellesmere Island, Canada. *Science* 192:781–82

Denton, G. H., Armstrong, R. L., Stuiver,

M. 1971. The late Cenozoic glacial history of Antarctica. In *Late Cenozoic Glacial Ages,* ed. K. K. Turekian, pp. 267–306. New Haven: Yale Univ. Press

Devereux, I. 1967. Oxygen isotope paleotemperature measurements on New Zealand Tertiary fossils. *New Z. J. Sci.* 10: 988–1011

Devereux, I., Hendy, C. H., Vella, P. 1970. Pliocene and early Pleistocene sea temperature fluctuations, Mangaopari Stream, New Zealand. *Earth Planet. Sci. Lett.* 8:163–68

Dodd, J. R. 1967. Magnesium and strontium in calcareous skeletons: a review. *J. Paleontol.* 41:1313–29

Dorf, E. 1964. The use of fossil plants in paleoclimatic interpretation. In *Problems in Paleoclimatology,* ed. A. E. M. Nairn, pp. 13–31. London-New York-Sydney: Interscience

Dorf, E. 1970. Paleobotanical evidence of Mesozoic and Cenozoic climatic changes. *Proc. North American Paleontological Convention, Chicago, 1969,* pp. 323–47

Dorman, F. H. 1966. Australian Tertiary paleotemperatures. *J. Geol.* 74:49–61

Douglas, R. G., Savin, S. M. 1971. Isotopic analyses of planktonic foraminifera from the Cenozoic of the Northwestern Pacific, Leg 6. In *Initial Reports of the Deep Sea Drilling Project,* A. G. Fisher et al, 6: 1123–27. Washington DC: GPO

Douglas, R. G., Savin, S. M. 1973. Oxygen and carbon isotope analyses of Cretaceous and Tertiary foraminifera from the central North Pacific. In *Initial Reports of the Deep Sea Drilling Project,* E. L. Winterer, J. I. Ewing et al, 17:591–605. Washington DC: GPO

Douglas, R. G., Savin, S. M. 1975. Oxygen and carbon isotope analyses of Tertiary and Cretaceous microfossils from Shatsky Rise and other sites in the North Pacific Ocean. In *Initial Reports of the Deep Sea Drilling Project,* R. L. Larson, R. Moberly et al, 32:509–20. Washington DC: GPO

Douglas, R. G., Savin, S. M. 1977. Depth stratification in Tertiary and Cretaceous foraminifera based on oxygen isotope ratios. *Marine Micropaleontol.* In press

Dudley, W. C. 1976. *Paleoceanographic application of oxygen isotopic analyses of calcareous nannoplankton grown in culture.* PhD thesis. Univ. Hawaii, Honolulu. 168 pp.

Duplessy, J. C. 1972. *La geochimie des isotopes stables du carbone dans la mer.* Thesis. Univ. Paris VI, Paris. 196 pp.

Duplessy, J. C., LaLou, C., Vinot, A. C. 1970. Differential isotopic fractionation in

352 SAVIN

benthic foraminifera and paleotemperatures reassessed. *Science* 168:250–51

Durazzi, J. T. 1975. *The shell chemistry of ostracods and its paleoecological significance.* Thesis. Case Western Reserve Univ., Cleveland, Ohio. 192 pp.

Durham, J. W. 1950. Cenozoic climates of the Pacific coast. *Geol. Soc. Am. Bull.* 61: 1243–63

Edwards, A. R. 1968. The calcareous nannoplankton evidence for New Zealand Tertiary marine climate. *Tuatara* 16:26–31

Emiliani, C. 1954a. Depth habitats of some species of pelagic foraminifera as indicated by oxygen isotope ratios. *Am. J. Sci.* 252: 149–58

Emiliani, C. 1954b. Temperatures of Pacific bottom waters and polar superficial waters during the Tertiary. *Science* 119: 853–55

Emiliani, C. 1955. Pleistocene temperatures. *J. Geol.* 63:538–78

Emiliani, C. 1966. Paleotemperature analysis of Caribbean cores P6304–8 and P6304–9 and a generalized temperature curve for the past 425,000 years. *J. Geol.* 74:109–26

Emiliani, C. 1971a. Depth habitats and growth stages of pelagic foraminifera. *Science* 173:1122–24

Emiliani, C. 1971b. The amplitude of Pleistocene climatic cycles at low latitudes and the isotopic composition of glacial ice. In *Late Cenozoic Glacial Ages,* ed. K. K. Turekian, pp. 183–97. New Haven: Yale Univ. Press

Emiliani, C., Shackleton, N. J. 1974. The Brunhes Epoch: Isotopic paleotemperatures and geochronology. *Science* 183: 511–14

Epstein, S., Buchsbaum, R., Lowenstam, H. A., Urey, H. C. 1951. Carbonate-water isotopic temperature scale. *Geol. Soc. Am. Bull.* 62:417–26

Epstein, S., Lowenstam, H. A. 1953. Temperature-shell growth relations of Recent and Interglacial Pleistocene shoalwater biota from Bermuda. *J. Geol.* 51: 424–38

Epstein, S., Mayeda, T. 1953. Variation of O^{18} content of waters from natural sources. *Geochim. Cosmochim. Acta* 4: 213–24

Foster, R. J. 1974. Eocene echinoids and the Drake Passage. *Nature* 249:751.

Frakes, L. A., Kemp, E. M. 1972. Influence of continental positions on Early Tertiary climates. *Nature* 240:97–100

Gates, W. L. 1976. Modeling the Ice-Age climate. *Science* 191:1138–43

Geitzenauer, K. R., Margolis, S. V., Edwards, D. S. 1968. Evidence consistent with Eocene glaciation in a South Pacific deep sea sedimentary core. *Earth Planet. Sci. Lett.* 4:173–77

Hallam, A., ed. 1973. *Atlas of Paleobiology.* Amsterdam: Elsevier. 531 pp.

Haq, B. U. 1973. Transgressions, climatic change and the diversity of calcareous nannoplankton. *Marine Geol.* 15:M25–30

Hecht, A. D. 1973. Faunal and oxygen isotopic paleotemperatures and the amplitude of glacial/interglacial temperature changes in the equatorial Atlantic, Caribbean Sea and Gulf of Mexico. *Quat. Res.* 4:671–90

Hecht, A. D., Savin, S. M. 1972. Phenotypic variation in oxygen isotope ratios in Recent planktonic foraminifera. *J. Foraminiferal Res.* 2:55–67

Hornibrook, N. de B. 1971. New Zealand Tertiary climate. *NZ Geol. Surv. Rep.* 47. 19 pp.

Hughes, N. F., ed. 1973. *Organisms and Continents through Time, Spec. Pap. Paleontol.,* Vol. 12. 334 pp.

Imbrie, J., Kipp, N. G. 1971. A new micropaleontological method for quantitative paleoclimatology: application to a late Pleistocene Caribbean core. In *Late Cenozoic Glacial Ages,* ed. K. K. Turekian, pp. 71–181. New Haven: Yale Univ. Press

Imbrie, J., van Donk, J., Kipp, N. G. 1973. Paleoclimatic investigation of a Late Pleistocene Caribbean deep-sea core: comparison of isotopic and faunal methods. *Quat. Res.* 3:10–38

Jenkins, D. G. 1968a. Variations in the numbers of species and subspecies of planktonic Foraminiferida as an indicator of New Zealand Cenozoic paleotemperatures. *Palaeogeogr. Palaeoclimatol. Palaeoecol.* 5:309–13

Jenkins, D. G. 1968b. Planktonic Foraminiferida as indicators of New Zealand Tertiary paleotemperatures. *Tuatara* 16: 32–37

Kemp, E. M., Barrett, P. 1975. Antarctic glaciation and early Tertiary vegetation. *Nature* 258:507–8

Kennett, J. P., Houtz, R. E., Andrews, P. B., Edwards, A. R., Gostin, V. A., Hajos, M., Hampton, M. A., Jenkins, D. G., Margolis, S. V., Ovenshine, A. T. 1974. Development of the Circum-Antarctic Current. *Science* 186:144–47

Kennett, J. P., Houtz, R. E., Andrews, P. B., Edwards, A. R., Gostin, V. A., Hajos, M., Hampton, M., Jenkins, D. G., Margolis, S. V., Ovenshine, A. T., Perch-Nielsen, K. 1975. Site 284. In *Initial Reports of the*

Deep Sea Drilling Project, J. P. Kennett, R. E. Houtz et al, 29 : 403–45. Washington DC: GPO

Kennett, J. P., Shackleton, N. J. 1976. Oxygen isotopic evidence for the development of the psychrosphere 38 Myr ago. Nature 260 : 513–15

Kent, D., Opdyke, N. D., Ewing, M. 1971. Climatic change in the North Pacific using ice-rafted detritus as a climatic indicator. Geol. Soc. Am. Bull. 82 : 2741–54

Keyes, T. W. 1968. Cenozoic marine temperatures indicated by scleractinian coral fauna of New Zealand. Tuatara 16 : 21–25

Kilbourne, R. T., Sen Gupta, B. K. 1973. Elemental composition of planktonic foraminiferal tests in relation to temperature-depth habitats and selective solution. Geol. Soc. Am. Abstr. with Programs 5 : 408–9

Knauth, L. P., Epstein, S. 1975. Hydrogen and oxygen isotope ratios in silica from the Joides Deep Sea Drilling Project. Earth Planet. Sci. Lett. 25 : 1–10

Kolodny, Y., Epstein, S. 1976. Stable isotope geochemistry of deep sea cherts. Geochim. Cosmochim. Acta. 40 : 1195–1209

Kräusel, R. 1961. Palaeobotanical evidence of climate. In Descriptive Paleoclimatology, ed. A. E. M. Nairn, pp. 227–54. New York : Interscience

Krinsley, D. H. 1973. Surface features of quartz sand grains from Leg 18 of the Deep Sea Drilling Project. In Initial Reports of the Deep Sea Drilling Project, R. Kulm, R. von Huene et al, 18 : 925–33. Washington DC : GPO

Labeyrie, L. 1974. New approach to surface seawater paleotemperatures using O^{18}/O^{16} ratios in silica of diatom frustules. Nature 248 : 40–42

Lancelot, Y., Larson, R. 1975. Sedimentary and tectonic evolution of the Northwestern Pacific. In Initial Reports of the Deep Sea Drilling Project, R. L. Larson, R. Moberly et al, 32 : 925–39. Washington DC : GPO

Ledbetter, M. T., Ellwood, B. B. 1976. Selection of sample intervals in deep-sea sedimentary cores. Geology 4 : 303–4

LeMasurier, W. E. 1970. Volcanic evidence for Early Tertiary glaciations in Marie Byrd Land. Antarct. J. US 5 : 154–55

Lidz, B., Kehm, A., Miller, H. 1968. Depth habitats of pelagic foraminifera during the Pleistocene. Nature 217 : 245–47

Lindroth, K. J., Miller, L. G., Durazzi, J. T., McIntyre, A., von Donk, J. 1977. Coccoliths as isotopic temperature indicators : a preliminary investigation. In press

Ling, H. Y., McPherson, L. M., Clark, D. L. 1973. Late Cretaceous (Maestrichtian?) silicoflagellates from the Alpha Cordillera of the Arctic Ocean. Science 180 : 1360–61

Longinelli, A., Nuti, S. 1968a. Oxygen-isotope ratios in phosphate from fossil marine organisms. Science 160 : 879–82

Longinelli, A., Nuti, S. 1968b. Oxygen isotopic composition of phosphorites from marine formations. Earth Planet. Sci. Lett. 5 : 13–16

Longinelli, A., Nuti, S. 1973a. Revised phosphate-water isotopic temperature scale. Earth Planet. Sci. Lett. 19 : 373–76

Longinelli, A., Nuti, S. 1973b. Oxygen isotope measurements of phosphate from fish teeth and bones. Earth Planet. Sci. Lett. 20 : 337–40

Lowenstam, H. A. 1964. Palaeotemperatures of the Permian and Cretaceous Periods. In Problems in Palaeoclimatology, ed. A. E. M. Nairn, pp. 227–48. New York : Interscience

Lowenstam, H. A., Epstein, S. 1954. Paleotemperatures of the post-Aptian Cretaceous as determined by the oxygen isotope method. J. Geol. 62 : 207–48

Lyell, C. 1854. Principles of Geology. New York : Appleton. 834 pp.

Margolis, S. V. 1975. Paleoglacial history of Antarctica inferred from analysis of Leg 29 sediments by scanning-electron microscope. In Initial Reports of the Deep Sea Drilling Project, J. P. Kennett, R. E. Houtz et al, pp. 1039–48. Washington DC : GPO

Margolis, S. V., Kennett, J. P. 1971. Cenozoic paleoglacial history of Antarctica recorded in Subantarctic deep-sea cores. Am. J. Sci. 271 : 1–36

Margolis, S. V., Kroopnick, P. M., Goodney, D. E., Dudley, W. C., Mahoney, M. E. 1975. Oxygen and carbon isotopes from calcareous nannofossils as paleooceanographic indicators. Science 189 : 555–57

McIntyre, A. et al (35 other CLIMAP Project Members) 1976. The surface of the Ice-Age Earth. Science 191 : 1131–37

McKee, E. D. 1964. Problems on the recognition of arid and of hot climates of the past. In Problems in Paleoclimatology, ed. A. E. M. Nairn, pp. 367–77. New York : Interscience

Moberly, R., Gardner, J. V., Larson, R. L. 1975. Introduction. In Initial Reports of the Deep Sea Drilling Project, R. L. Larson, R. Moberly et al, pp. 32 : 5–14. Washington DC : GPO

Naydin, D. P., Teis, R. V., Zadorozhny, I. K. 1966. Isotopic palaeotemperatures of the Upper Cretaceous in the Russian Platform and other parts of the U.S.S.R.

Geochem. Int. 3:1038–51

Oba, T. 1969. Biostratigraphy and isotopic paleotemperature of some deep-sea cores from the Indian Ocean. *Sci. Rep. Tohoku Univ. Sendai 2nd Ser. (Geol.)*, Vol. 41, No. 2:129–95

Olausson, E. 1965. Evidence of climatic changes in North Atlantic deep-sea cores, with remarks on isotopic paleotemperature analysis. In *Prog. Oceanogr.* 3:221–52

Piper, D. J. W., Brisco, C. D. 1975. Deep-water continental-margin sedimentation, DSDP Leg 28, Antarctica. In *Initial Reports of the Deep Sea Drilling Project,* E. D. Hayes, L. A. Frakes et al, 28:727–55. Washington DC: GPO

Pitman, W. C. III, Larson, R. L., Herron, E. M. 1974. The age of the ocean basins (two charts and summary statement). *Geol. Soc. Am.*

Roth, P. H. 1974. Calcareous nannofossils from the northwestern Indian Ocean, Leg 24, Deep Sea Drilling Project. In *Initial Reports of the Deep Sea Drilling Project,* R. L. Fisher, E. T. Bunce et al, 24:968–84. Washington DC: GPO

Saito, T., van Donk, J. 1974. Oxygen and carbon isotope measurements of late Cretaceous and early Tertiary foraminifera. *Micropalaeontology* 20:152–77

Savin, S. M., Douglas, R. G. 1973. Stable isotope and magnesium geochemistry of Recent planktonic foraminifera from the South Pacific. *Geol. Soc. Am. Bull.* 84:2327–42

Savin, S. M., Douglas, R. G., Stehli, F. G. 1975. Tertiary marine paleotemperatures. *Geol. Soc. Am. Bull.* 86:1499–1510

Savin, S. M., Stehli, F. G. 1974. Interpretation of oxygen isotope paleotemperature measurements: effect of the O^{18}/O^{16} ratio of sea water, depth stratification of foraminifera, and selective solution. *Colloq. Int. CNRS No. 219, Les Methodes Quantitative d'Etudes des Variations au cours du Pléistocène,* pp. 183–91

Scheibnerova, V. 1973. The value of paleotemperature measurements and paleobioprovincial analysis of fossil animals for paleogeographical reconstructions. *NSW Geol. Surv. Rec.* 15: Pt. 1, pp. 5–18

Scholl, D. W., Creager, J. S. 1973. Geologic synthesis of Leg 19 (DSDP) results: far north Pacific, and Aleutian Ridge, and Bering Sea. In *Initial Reports of the Deep Sea Drilling Projects,* J. S. Creager, D. W. Scholl et al, 19:897–913. Washington DC: GPO

Schwarzbach, M. 1963. *Climates of the Past.* London: Van Nostrand. 328 pp.

Schwarzbach, M. 1964. Criteria for the recognition of ancient glaciations. In *Problems in Palaeoclimatology,* ed. A. E. M. Nairn, pp. 81–85. New York: Interscience

Shackleton, N. J. 1967. Oxygen isotope analyses and Pleistocene temperatures reassessed. *Nature* 215:15–17

Shackleton, N. J. 1968. Depth of pelagic foraminifera and isotopic changes in Pleistocene oceans. *Nature* 218:79–80

Shackleton, N. J. 1974. Attainment of isotopic equilibrium between ocean water and benthonic foraminifera genus *Uvigerina:* isotopic changes in the ocean during the last glacial. *Colloq. Int. CNRS No. 219, Les Méthodes Quantitative d'Etudes des Variations du Climat au cours du Pléistocène,* pp. 203–9

Shackleton, N. J., Kennett, J. P. 1975a. Paleotemperature history of the Cenozoic and the initiation of Antarctic glaciation: oxygen and carbon isotope analyses in DSDP sites 277, 279 and 281. In *Initial Reports of the Deep Sea Drilling Project,* J. P. Kennett, R. E. Houtz et al, 29:743–55. Washington DC: GPO

Shackleton, N. J., Kennett, J. P. 1975b. Late Cenozoic oxygen and carbon isotopic changes at DSDP site 284: implications for glacial history of the Northern Hemisphere. In *Initial Reports of the Deep Sea Drilling Project,* J. P. Kennett, R. E. Houtz et al, 29:801–7. Washington DC: GPO

Shackleton, N. J., Opdyke, N. 1973. Oxygen isotope and palaeomagnetic stratigraphy of equatorial Pacific core V28-238: oxygen isotope temperatures and ice volumes on a 10^5 year and 10^6 year scale. *Quat. Res.* 3:39–55

Shackleton, N. J., Wiseman, J. D. H., Buckley, H. A. 1973. Nonequilibrium isotopic fractionation between seawater and planktonic foraminiferal tests. *Nature* 242:177–79

Sinnott, E. W., Bailey, I. W. 1915. Investigations on the phylogeny of the angiosperms. 5: Foliar evidence as to the ancestry and early climatic environment of the angiosperms. *Am. J. Bot.* 2:1–22

Smith, J. P. 1919. Climatic relations of the Tertiary and Quaternary faunas of the California Region. *Calif. Acad. Sci., Proc. Ser. 4,* 9:123–73

Spaeth, C., Hoefs, J., Vetter, U. 1971. Some aspects of isotopic composition of belemnites and related paleotemperatures. *Geol. Soc. Am. Bull.* 82:3139–50

Stehli, F. G. 1970. A test of the Earth's magnetic field during Permian time. *J. Geophys. Res.* 75:3325–42

Stevens, G. R. 1971. The relationship of isotopic temperatures and faunal realms to Jurassic-Cretaceous paleogeography, particularly of the S.W. Pacific. *J. R. Soc. NZ* 1:145–58

Stevens, G. R., Clayton, R. N. 1971. Oxygen isotope studies on Jurassic and Cretaceous belemnites from New Zealand and their biogeographic significance. *NZ J. Geol. Geophys.* 14:829–97

Sverdrup, H. V. 1955. Discussions on the relationship between meteorology and oceanography. *J. Mar. Res.* 14:501–3

Tanai, T., Huzioka, K. 1967. Climatic implications of Tertiary floras in Japan: Tertiary correlations and climatic changes in the Pacific. *Symp. No. 25, 11th Pac. Sci. Congr. Proc.,* pp. 77–87. Sendai, Japan: Sasaki

Urey, H. C., Lowenstam, H. A., Epstein, S., McKinney, C. R. 1951. Measurement of paleotemperatures and temperatures of the upper Cretaceous of England, Denmark, and the southeastern United States. *Geol. Soc. Am. Bull.* 62:399–416

van Donk, J. 1970. *The oxygen isotope record in deep sea sediments.* PhD thesis. Columbia Univ., New York. 228 pp.

Voigt, E. 1965. Zur temperatur-kurve der oberen Kreide in Europa. *Geol. Rundsch.* 54:270–317

Wolfe, J. A. 1971. Tertiary climatic fluctuations and methods of analysis of Tertiary floras. *Palaeogeogr. Palaeoclimatol. Palaeoecol.* 9:25–57

Wolfe, J. A., Hopkins, D. M. 1967. Climatic changes recorded by Tertiary land floras in Northwestern North America: Tertiary correlations and climatic changes in the Pacific. *Symp. No. 25, 11th Pac. Sci. Congr. Proc.,* pp. 67–76. Sendai, Japan: Sasaki

Ann. Rev. Earth Planet. Sci. 1977. 5:357–69
Copyright © 1977 by Annual Reviews Inc. All rights reserved

LASER DISTANCE-MEASURING TECHNIQUES

×10077

Judah Levine

Time and Frequency Division, National Bureau of Standards, Boulder, Colorado 80302

INTRODUCTION

The need to measure accurately distance on the surface of the earth arises in many different areas of science and engineering. The applications of geodesy range from simple surveys over distances of tens of meters to studies of the motions of lithospheric plates using measurements of transcontinental baselines.

It is clear that no one technique can hope to satisfy these widely disparate needs. Nevertheless some general principles emerge from a survey of the field. For distances longer than a few thousand meters the human surveyor and the steel tape are being replaced by automated machines that measure distance using electromagnetic radiation. The laser has proven very useful for these measurements because of its high brightness (i.e. its ability to produce intense beams whose divergence is limited largely by diffraction) and its excellent frequency stability. In this review I examine the various distance-measuring techniques currently in use or under development.

TECHNIQUES OF LENGTH METROLOGY

At the outset we must distinguish two broad areas of length metrology. First we have absolute length measurements in which the given result must be expressed in absolute meters. These measurements must be made using a length standard traceable to the definition of the meter. The "ruler" used in these measurements must be accurate and stable, and the measurement technique must be free from systematic errors. As an example of such measurements we may mention surveying. Second, we may think of measurements in which the change in the length of the baseline with time is more important than the actual length itself. Such measurements place a premium on stability rather than on absolute accuracy both in the length standard and in the measurement technique. We may not need to know the magnitudes of the various systematic corrections, being satisfied instead to know that they are constant in time. As examples of these measurements we may mention measurements of strain tides and studies of motions near or across plate boundaries.

Although this difference is clear in principle, the distinction between absolute

357

and relative measurements becomes somewhat blurred in practice. The construction of a length standard of the requisite long-term stability for differential measurements near plate boundaries, for example, is nearly as difficult as construction of an absolute length standard. Furthermore, an investigation into the systematic errors of a measurement that is sufficiently detailed to insure that they are constant in time is often as difficult as calculating their magnitudes and removing them. Nevertheless the distinction is useful, as we see below. Accordingly we first discuss differential measuring instruments, principally strainmeters, and then absolute instruments.

DIFFERENTIAL MEASUREMENTS

Estimates of the Minimum Signal-to-Noise Ratio Required

The most common differential length measurement is a measurement of the fractional change in the length of a given baseline. We refer to this quantity as local strain.

The relationship between stress applied to the surface of the earth and the resultant strain is generally described in terms of a spherical coordinate system whose origin is at the center of the earth. The applied stress and the resulting strain are not usually colinear, and the local stress field is described in terms of a 3×3 matrix. If we limit our discussion to the surface of the earth, and if we assume that the matrix is symmetric (i.e. that the stress-strain relationship does not depend on the angle between applied stress and the local meridian), then the number of independent components is reduced from nine to three. Thus a complete specification

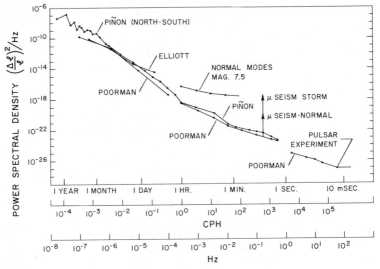

Figure 1   An estimate of the broadband strain noise measured at Piñon Flat, California, and at the Poorman mine near Boulder, Colorado.

of local strain requires three noncolinear measurements at every point. This is usually done by deploying three separate instruments at every site.

The magnitudes of the strains encountered in practice vary over a wide range. Earth-tide strain may be as large as 10^{-7} peak-to-peak while strains produced by teleseismic earthquakes rarely exceed 10^{-9}. We consider a strainmeter to be adequate if it is limited by earth noise over its entire operating frequency range.

The strain noise at quiet sites has been measured by Berger & Levine (1974), who used data from very different instruments located about 2000 km apart in very different geology. Nevertheless, their results are in good agreement over the frequency range extending from approximately 1 cycle month^{-1} to several cycles sec^{-1}. It is reasonable to postulate that their results represent a reasonable estimate of the earth noise (see Figure 1).

The measurement of strain at frequencies below 1 cycle hr^{-1} presents considerable experimental difficulty. In addition to the increasing level of broadband noise, measurement techniques using physical "rulers" (fused silica rods, for example) are limited by their relatively large coefficients of thermal expansion and by their unknown long-term stability outside of controlled laboratory environments. Furthermore, the difficulty of constructing a fused silica strainmeter increases at least as fast as the length, making long instruments (i.e. those greater than tens of meters) impractical.

Laser Strainmeter Technology

A laser strainmeter is simply an optical interferometer attached to the earth. The two most commonly used types are the Michelson interferometer and the Fabry-Perot interferometer.

In the Michelson interferometer (see Figure 2), the incoming radiation is split by beamsplitter B into two equal halves. The beams traverse the two paths L_1 and L_2, are reflected by mirrors m_1 and m_2, and are then recombined at the beamsplitter. The intensity of the recombined light is measured by detector D and is given by

$$I = \frac{I_0}{2}\left\{1 + \cos 2\pi\left(\frac{L_1 - L_2}{\lambda/n}\right)\right\}$$

where I_0 is the incident intensity, λ is the vacuum wavelength of the incident radiation, and n is the index of refraction of the air between the mirrors.

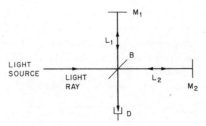

Figure 2 A Michelson interferometer; m_1 and m_2 are mirrors, B is a beamsplitter, and D is the detector.

In practice, L_2 is kept constant, and it is the variations in L_1 that are recorded. The quantity $(L_1 - L_2)/(\lambda/n)$ is simply the difference in path measured in units of the wavelength in the system. It may be rewritten in the form

$$\frac{L_1 - L_2}{\lambda/n} = N + \varepsilon,$$

where N is an integer and ε is a small number between zero and one. The intensity is then given by

$$I = \frac{I_0}{2}(\cos 2\pi\varepsilon + 1).$$

The intensity I thus reaches a maximum for $\varepsilon = 0$, i.e. when the difference in path length is an integral number of wavelengths.

The Fabry-Perot interferometer is shown in Figure 3. It consists of two partially transmitting mirrors (usually spherical) separated by a distance L. Light is incident from the left and bounces back and forth between the two mirrors. The intensity transmitted through the far mirror to the detector, D, is given by

$$I = I_0 \left\{ \frac{1}{1 + [4r/(1-r)^2] \sin^2 [2\pi L/(\lambda/n)]} \right\},$$

where r is the reflectivity of the mirrors and the other symbols are as defined above. The shape of the transmission maximum (the "fringe") is somewhat more complicated than the Michelson geometry, but the general idea is the same. The quantity $L/(\lambda/n)$ is the length of the path in units of the wavelength in the system. As above, we can choose an integer N such that

$$L/(\lambda/n) = N + \varepsilon.$$

The transmission thus is proportional to

$$\frac{1}{1 + [4r/(1-r)^2] \sin^2 2\pi\varepsilon},$$

and the intensity reaches a maximum when $\varepsilon = 0$ or $\varepsilon = 1/2$, i.e. when the path length is any integral number multiple of a half-wavelength.

The two interferometers differ mainly in the width of the transmission fringes. The Michelson interferometer fringe has a cosine shape whose width depends only on the path difference, $L_1 - L_2$. The width of the fringe in a Fabry-Perot interferometer depends on its length *and* on the reflectivity of the mirrors. The fringe in a typical Fabry-Perot interferometer has a width which is 1/15 or less of that obtained in a Michelson instrument of comparable size. We can use this fact to increase

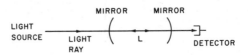

Figure 3   A Fabry-Perot interferometer.

the sensitivity of Fabry-Perot–type instruments by using a fringe-locking technique to be discussed later.

The two interferometers are quite similar in other respects. As described, neither system provides any direct way of measuring the integer number of wavelengths in the long path. Thus neither system can measure lengths absolutely and both are confined to relative measurements. (There are ways of measuring the absolute length of the interferometer, but none of them have been used in a geophysical field instrument to my knowledge.) In addition, both instruments are sensitive in first order to three problems: long term changes in the physical properties of the mirror coatings, changes in the index of refraction of the air between the mirrors, and changes in the wavelength of the incident light.

The condition for maximum transmission is really a statement about the phase shift suffered by the light in traversing the interferometer. In addition to the phase shift suffered by the light due to the transit time through the system, there is an additional, usually unknown, phase shift on reflection from the multilayer dielectric coated surface used as the mirror. Changes in this phase shift with time produce corresponding changes in the apparent optical length of the interferometer arm. This phase-shift change is proportionally less serious in longer instruments since it represents a smaller fraction of the phase change through the system, but it could be quite serious indeed for instruments only a few meters long.

The index of refraction of the air between the mirrors can also present a problem. The interferometers are usually evaluated to a few tenths of a N/m^2 (a few millitorr). Nevertheless, a significant amount of air remains.

The index of refraction of air at standard temperature and pressure differs from unity by about 300×10^{-6} in the red part of the spectrum. Thus, a change in pressure of 0.13 N/m^2 (1 millitorr) would change the index of refraction by about 4×10^{-10}, producing a spurious strain signal of this amplitude. This is a rather large change in pressure to remain undetected, often amounting to 5% of the system base pressure, but it can be easily generated by a small leak. The residual air in the path can also result in the instrument's being sensitive to ambient atmospheric pressure. This effect is very small in well-designed systems.

Finally, we must deal with the stability of the laser wavelength itself. The frequency stability of a laser is governed by the stability of the optical cavity and by the gain profile of the amplifying medium. In a well-designed system, the instantaneous frequency is governed by the optical cavity whose fringe width (i.e. the width in frequency units over which the cavity will support oscillations) is about 5 MHz. Expressed in terms of the laser frequency, 5 MHz is a fractional stability of about 1×10^{-8} and is therefore marginally acceptable for strainmeters. However, this instantaneous narrow line will shift its frequency in response to any change in ambient temperature or in other laser-operating parameters. The drift in oscillation frequency will be limited by the gain profile of the amplifying medium, which is of the order of 1 GHz. Thus the long-term fractional frequency stability of the laser will probably be no better than 5×10^{-6} and may approach 1×10^{-5} unless special precautions are taken to stabilize the cavity.

Although it is theoretically possible to stabilize a laser by directly stabilizing the length of its cavity, such systems have never been used in geophysical applications.

The most commonly used stabilizers are passive cavity references and saturated absorption stabilizers. A passive cavity reference is simply an ancillary Fabry-Perot cavity. The laser wavelength is stabilized by tuning the laser (usually by changing the length of the laser cavity piezoelectrically) until the laser wavelength corresponds to one of the fringes of the reference cavity. A servo system then maintains the laser at the top of the transmission maximum by locking to the point of zero first derivative of the transmission function.

At first sight instruments using a laser stabilized to a passive reference cavity would seem to be no better than instruments using a fused silica bar for the strain-meter itself since the reference cavity, which is a physical "ruler," is subject to all of the thermal expansion and long-term creep problems of fused silica strainmeters. However, the passive cavity can be made short and rigid, and the stabilization of its ambient temperature and pressure is easier than trying to do the same stabilization for the entire strainmeter. Indeed, a well-designed reference cavity can have adequate stability for earth-tide measurements. The behavior of a passive cavity over periods of months is, however, not well understood, and it is probably not adequate for secular strain measurements.

A saturated absorption stabilizer exploits the overlap between a molecular absorption line and the laser frequency. The absorber is placed inside the laser cavity (Lee & Skolnick 1968). When the laser is tuned to the center of the absorption profile, the two oppositely directed running waves that make up the standing wave in the cavity interact with the same velocity group of absorbing molecules—namely those with velocity purely transverse to the laser beam. When the laser is not tuned to the center of the absorption line, the two running waves interact with two different groups of molecules whose velocity components along the laser beam provide the requisite Doppler shifts to bring them into resonance. Thus, there are twice as many absorbing molecules available off of the line center as there are at line center.

By proper choice of operating conditions, the optical power saturates the absorption at the center of the line, resulting in a power *increase* there. This feature is quite narrow. Its width is limited by cavity parameters and is usually of the order of 100 kHz for field-type systems. Its position is determined by the absorption line in the stabilizer molecule and is essentially independent of temperature, pressure, or cavity parameters. (There are certain systematic offsets that must be dealt with, but they are of the order of 10^{-11} or less in well-designed systems.)

The best saturated absorption-stabilized laser is obtained by stabilizing a helium-neon laser oscillating at 3.39 μm to a transition in methane; the parameters quoted above apply to this system (Barger & Hall 1969, Levine & Hall 1972). Other saturated absorption-stabilized lasers exist. The only other one used in the geophysical context exploits an overlap between I_2 and a helium-neon laser oscillating at 632.8 nm (Ezekiel & Weiss 1968). It is somewhat more difficult to use in a field situation, but has an accuracy capability almost as good as methane-stabilized devices. The full accuracy capability is often not realized in field devices, however, because the absorption profile must be measured in the presence of a baseline whose slope changes with time.

With either stabilization system, it is necessary to read out the length of the interferometer in terms of the wavelength. Again we may distinguish two types of readout systems.

The simplest system is a fringe counter. Each time the intensity passes through a preset value corresponding to some chosen value of ε, a counter is either increased or decreased by increments depending on the direction of motion. The direction of motion can be determined in various ways (Rowley 1966). The value in the counter at any time is converted to strain by noting that each count represents a strain increment of $\lambda/2Ln$. Although the fringe-counting system is simple to implement, its fundamental limitation is that it has a sensitivity limited by the least count. For example, a 100-m-long instrument illuminated by a 632.8-nm light source would have a least count of about 3×10^{-9}. Therefore, fringe-counting systems are only viable in very long interferometers where the least-count noise does not dominate the system response.

It is possible to reduce the least-count noise of a system by counting fractional fringes. If a system can be constructed to divide a fringe into m parts, then the smallest detectable strain is reduced by the same factor m. It is difficult to make m larger than about 10 without introducing problems with long-term stability. Levine and Hall (1974 unpublished) have proposed techniques to divide fringes into 500 parts using digital techniques, but these techniques have never been used in field instruments.

A second, more complex system employs fringe locking. A free running laser is tuned so that its frequency corresponds to the peak of the fringe of the system. A servo system then maintains the laser locked to this fringe by seeking the zero of the first derivative of the transmission function. As the path length changes, the servo tunes the laser to keep the integer N a constant. The frequency of the laser is thus given by

$$f = \frac{Nc}{2Ln},$$

where c is the velocity of the light and N is now a constant (note—it is an unknown constant). The frequency of this laser is then beat against the frequency of a second, stabilized laser. The difference frequency is extracted. If f_0 is the frequency of the standard, the beat frequency is given by

$$f_b = f_0 - \frac{Nc}{2Ln},$$

and the fractional change in the beat frequency is related to strain by

$$\frac{\Delta f_b}{f_0} = \frac{\Delta L}{L}.$$

Since f_0 is known to a high degree of accuracy, the system reads strain directly without any adjustable constants or calibration factors.

Although the fringe-locking system is more complicated than a fringe counter in that it requires two lasers and roughly twice as much electronics, it has several distinct advantages. First of all, it possesses no least count as such, because its minimum detectable strain is limited instead by noise in the electronics. Second, it can take full advantage of the increased fringe sharpness obtainable with Fabry-Perot interferometers and can easily resolve changes in the length of the path of order $10^{-5} \times \lambda$ with a reasonable quality Fabry-Perot. Third, its output is in the form of a frequency. The measurement of a frequency is very easy and can be implemented using digital counting circuits that can be made almost totally insensitive to drifts in gain or changes in offset voltage. This greatly increases their reliability.

Typical Laser Strainmeters

The principles I have outlined have been used to construct several laser strainmeters. Two of the best designed instruments are the 800-m Michelson instruments operated by the University of California near La Jolla (Berger & Lovberg 1970) and the 30-m Fabry-Perot instrument operated by the National Bureau of Standards near Boulder, Colorado (Levine & Hall 1972).

Applications of Laser Strainmeters

Although laser strainmeters have extremely wide band widths extending from tens or hundreds of Hz down to d.c., they are not usually used at frequencies above about 0.01 Hz because this frequency range can be covered adequately using conventional instruments. [An exception is the attempt by Levine & Stebbins (1972) to detect gravitational radiation from various pulsars using a 30-m instrument.]

They are most useful for investigations of the earth normal modes in the frequency range from 1 to 30 cycles hr^{-1}, in earth-tide studies, and in measurements of secular strain.

These investigations require instruments that are long enough to average over the local inhomogeneities inevitably present at any site. At the same time, the instruments must be as insensitive as possible to the large thermal and pressure effects present at these frequencies.

Detailed analyses of the earth tides have been carried out by Levine & Harrison (1976) and by Beaumont & Berger (1975). Analyses of long tidal series have shown that the strain tides are sensitive to local topography, local crustal inhomogeneities, and to various other site effects, and that incorporation of these effects into the theory substantially improves the agreement between theory and experiment. The residuals in the published data are of the order of 1% of the amplitude of the tides and are almost certainly the result of inadequacies in the estimation of the contamination of the record by local effects.

Long-term strain accumulation has also been monitored (Berger & Wyatt 1973). This is especially significant near the San Andreas fault system where long-term strain data may prove useful in understanding earthquakes. Continuous records of long-term strain can also be analyzed to measure steps in the baseline associated with earthquakes.

Limitations of Strainmeters

The fundamental limitation of laser strainmeters is that they are not absolute length-measuring instruments. They achieve their excellent sensitivity not by eliminating the various systematic corrections (refractive index of the residual air in the path, for example), but by eliminating or at least recording the changes of these corrections with time. Thus, a measurement of long-term changes in strain requires that the instrument be operated at a given site for long periods of time. A breakdown of any part of the system makes it impossible, at least in principle, to recover the baseline. [In practice, data gaps of a few days or less can be patched using the earth tides and the measured parameters at the site (Levine & Harrison 1976).] The instrument must be extremely reliable. In practice, the entire system must have a mean time between failures of several thousand hours with a mean time to repair of a day or two. This is not a trivial requirement, and it, more than any other single factor, has limited the use of laser instruments.

The differential nature of laser strainmeter measurements has a second, equally serious implication in the measurement of secular strain. Although these strain rates are very slow and could be adequately monitored by only occasional measurements at any one site, the differential nature of strainmeters requires that the strainmeter be operated continuously at a single site. This is an extremely inefficient use of a very complicated and expensive apparatus. A portable instrument of equal complexity capable of making absolute measurements could be used much more efficiently in secular strain determinations, since it could be used to monitor the secular strain at many sites simultaneously.

The ultimate limitation to accurate strain measurement is due to our ignorance of local crustal inhomogeneities. The only practical solution to this problem is to average these effects by making instruments with longer and longer baselines. Although this increases the cost of the pipe and hence the cost of the instrument, the real difficulty with increasing the length of the instrument is the difficulty of finding suitable sites, and the cost of site preparation. It is clear that, except in exceptional cases (e.g. old railway tunnels or mines), finding sites for instruments much longer than 800 m would be a very difficult task indeed.

ABSOLUTE MEASUREMENTS

The previous discussion suggested that secular-strain measurements could be done more efficiently using absolute measuring techniques capable of measuring long baselines directly through the atmosphere. There are, in fact, several ways by which this may be accomplished. The techniques fall into two broad categories—totally ground-based and extraterrestrial.

Ground-Based Measurements

INTRODUCTION   The idea of measuring distances using the time-of-flight of electromagnetic signals is not new. Indeed, for applications requiring ranging to an uncooperative target, and where resolution of a few meters is acceptable, the familiar radar time-of-flight techniques can be used. These techniques are not sufficiently accurate for geophysically motivated measurements.

It is difficult to estimate the minimum acceptable resolution needed for geophysical measurements. Experience with the application of existing techniques to secular strain determination suggests that the minimum acceptable range is 10–20 km with an absolute fractional accuracy of less than 1×10^{-6}. Furthermore, increases of factors of 2 to 3 in range and 50 in accuracy would be most welcome and could substantially improve our knowledge of secular strain rates in seismically active areas. As I show below, various systematic errors will probably conspire to limit ground-based measurements to ranges less than about 100 km with fractional accuracies that will probably never be better than 1×10^{-8}. With these limits in mind, we can proceed to evaluate the various techniques.

TIME-OF-FLIGHT MEASUREMENT   The time-of-flight of a light beam is about 3 μsec/km. Thus measurement of distances by optical time-of-flight measurements with an accuracy of 1 part per million or better requires absolute timing accuracies in the picosecond range. This is not impossible in principle, but is well beyond the current state of the art. An absolute timing accuracy of 25 psec has been reported in a prototype instrument (Querzola 1974), but the ranges measured were only 5 to 10 km long so that the fractional accuracy was just slightly better than 10^{-6}.

MODULATION TECHNIQUES   Most of the work in electromagnetic distance measurement has used various modulation schemes. The basic idea is quite simple: The light is modulated at some high frequency, and the distance is determined by measuring the phase shift of the modulation due to the transit time along the path.

Evaluation of systematic errors   The most serious limitation to the measurement of distances through the atmosphere is the correction required to account for the deviation of the index of refraction from unity. Bender & Owens (1965) suggested that the index might be determined most easily by measuring the difference in the apparent distance as measured by light of two different wavelengths. Thus, the index is determined by measuring the dispersion it produces.

We can quantitatively evaluate this method using the formulae for the dependence of the index on wavelength and on composition (Edlen 1953).

The group index of refraction of dry air at standard temperature and pressure is

$$N_g = 1 + 10^{-6} \times \left\{ 83.4213 + 24060.3 \left(\frac{130 + \sigma^2}{(130 - \sigma^2)^2} \right) + 59.97 \left(\frac{38.9 + \sigma^2}{(38.9 - \sigma^2)^2} \right) \right\},$$

where σ is the wave number in inverse micrometers. For any other temperature T (in K) or pressure P (in mm Hg) we have:

$$n(P, T) - 1 = (n_g - 1) \left\{ \frac{P/760}{T/288} \right\}.$$

The index of refraction differs from unity by about 282×10^{-6} at 632.8 nm and 299×10^{-6} at 441.6 nm, so that the dispersion results in a fractional change in the measured length of about

$$17 \times 10^{-6} \left\{ \frac{P/760}{T/288} \right\}.$$

Thus, we can see that the index must be determined to 1 part in 10^4 if the length is to be determined to 2 parts in 10^8. The fractional requirement on the index translates into an identical fractional requirement on P/T. This in turn places a heavy burden in the dispersion measurement since the magnitude of the dispersion is only about 8% of the index itself.

This relatively simple approach must be modified somewhat in actual usage since the atmosphere contains some water vapor. The presence of a partial pressure of f torr of water vapor results in an additional term in the index whose magnitude is (Barrell & Sears 1940)

$$-f\{5.722 - 0.1371\sigma^2\} \times 10^{-8}.$$

Thus, the water vapor contributes about 8×10^{-6} to the index of refraction in the visible for reasonable values of f, but the contribution is almost nondispersive across the visible, making its determination by optical dispersion impossible. However, the water vapor contribution to the index is sufficiently small that it may be estimated by directly measuring the dew point at the ends of the path. Alternatively, several investigators have proposed adding a third microwave-frequency measurement of the distance. The index of refraction at microwave frequencies depends strongly on the water vapor in the atmosphere so that a measurement of the microwave-optical dispersion can be used to measure the water-vapor density.

There are other systematic errors associated with multiple-wavelength systems. These include errors produced by the curvature of the path and by the fact that the different wavelengths do not traverse the same path and hence may sample different atmospheric conditions. These and other errors have been analyzed in detail. Thayer (1967) concluded that ranges of the order of 50 km can be measured with an accuracy of several parts in 10^8 using triple-wavelength devices and with perhaps a factor of ten lower accuracy using only two optical wavelengths and estimating the water density from ancillary measurements at the endpoints.

Results   Although multiple-wavelength systems have been in the literature for several years, most of the measurements have been done with less sophisticated instruments. The U.S. Geological Survey (Savage & Prescott 1973) uses single-wavelength devices that must be corrected for the index of refraction using ancillary measurements. These measurements are often made from an airplane flying as closely as practical to the line of sight. The Geological Survey has reported a reproducibility of about 2 parts in 10^7 for ranges up to 35 km (Savage & Prescott 1973). At that level of precision, determination of the strain accumulation at sites along the San Andreas fault system will require annual observation of many line lengths over a period of at least 5 years.

Two- and three-wavelength optical and optical/microwave distance-measuring instruments have been constructed by various groups (Earnshaw & Hernandez 1972, Huggett & Slater 1975). Standard deviations as low as 1 part in 10^7 over 10 km baselines have been reported.

Summary   The majority of distance measurements using electromagnetic distance-measuring equipment do not exploit the various multiple wavelength schemes proposed in the literature. Several groups have constructed two- and three-wavelength devices, but so far they have not been widely adopted. This situation will almost certainly change as the multiple wavelength devices are made simpler and more reliable.

Extraterrestrial Techniques

Finally I discuss below extraterrestrial techniques. These include satellite ranging, lunar ranging, and very long baseline interferometry (VLBI). Although VLBI is similar in concept to satellite ranging and lunar ranging, it uses microwave radio antennas and so is technically outside of the scope of this paper. Nevertheless, many of our discussions concerning the use of the other techniques for distance measurement on the surface of the earth can be applied to VLBI with little change.

GENERAL PRINCIPLES   The general principles of these techniques may be simply stated. Distances on the surface of the earth are determined using triangulation by measuring the distance from the extraterrestrial object to several stations at different points on the surface of the earth. These observations may be made simultaneously, or, if the orbit is sufficiently well known, at different times. The data obtained in this way may be analyzed to yield the distance between the two stations.

EVALUATION OF SYSTEMATIC ERRORS   I do not discuss the systematic errors in these techniques in detail. Rather we concern ourselves with the uncertainties that arise in the extraction of the distance between the two stations from the raw ranging data.

Extraterrestrial techniques do not directly measure the line-of-sight distance between the stations. The distance between the states is extracted by (loosely speaking) taking the difference between two ranges which are the same to first order. The determination of the station separation is therefore very sensitive to small errors in the range measurement made at either station.

Published analyses of the various techniques have shown that the distance between two points on the surface of the earth can be determined to a few centimeters (Bender & Silverberg 1975, Coates et al 1975).

If we take 1 cm as an optimistic estimate of the accuracy capability of these techniques, then measurement of strains at the level of one part in 10^7 can only be achieved over baselines exceeding 100 km in length.

SUMMARY   Measurements of distances on the surface of the earth using these techniques have just begun in the last few years. The development of mobile ground stations should make it possible to measure accurately motions in tectonically or seismically active regions in the near future.

CONCLUSIONS

The development of saturated absorption-stabilized lasers has resulted in a substantial improvement in the state of the strainmeter art to the point that instruments are now limited by our ignorance of the properties of the site. For baselines less than 1000 m long, these instruments can reliably measure strain to an accuracy limited by earth noise and site inhomogeneities.

In the next few years we should see a substantial improvement in the techniques for measuring longer baselines. Multiple wavelength geodesy will be used to provide strain measurements over baselines up to 50 or 100 km with accuracies exceeding 1 part in 10^7. The various extraterrestrial techniques will be used to provide comparable accuracies over baselines longer than a few hundred kilometers.

It is not yet clear if a significant range or accuracy gap exists between the maximum range capability of ground techniques and the minimum range (given the accuracy requirement) obtainable with extraterrestrial ones. We will almost certainly see substantial progress in these areas in the next few years.

Literature Cited

Barger, R. L., Hall, J. L. 1969. Pressure shift and broadening of methane line at 3.39 microns studied by laser saturated molecular absorption. *Phys. Rev. Lett.* 22:4–8

Barrell, H., Sears, J. E., Jr. 1940. The refraction and dispersion of air for the visible spectrum. *Phil. Trans. R. Soc. London. Ser. A* 238:1–9

Beaumont, C., Berger, J. 1975. An analysis of tidal strain observations from the United States of America: I. The laterally homogeneous tide. *Bull. Seismol. Soc. Am.* 65:1613–29

Bender, P., Owens, J. C. 1965. Correction of optical distance measurements for the fluctuating atmospheric index of refraction. *J. Geophys. Res.* 70:2461–62

Bender, P. L., Silverberg, E. C. 1975. Present tectonic plate motions from lunar ranging. *Tectonophysics* 29:1–7

Berger, J., Lovberg, R. H. 1970. Earth strain measurement with a laser interferometer. *Science* 170:296–303

Berger, J., Levine, J. 1974. The spectrum of earth noise from 10^{-8} to 10^{+2} Hz. *J. Geophys. Res.* 79:1210–14

Berger, J., Wyatt, F. 1973. Some observation of earth strain tides in California. *Phil. Trans. R. Soc. London. Ser. A* 274:267–77

Coates, R. J., Clark, T. A., Counselman, C. C., III, Shapiro, I. I., Hinteregger, H. F., Rogers, A. Z., Whitney, A. R. 1975. Very long baseline interferometry for centimeter accuracy geodetic measurements. *Tectonophysics* 29:9–18

Earnshaw, K. B., Hernandez, E. N. 1972. Two-laser optical distance measuring instrument that corrects for the atmospheric index of refraction. *Appl. Opt.* 11:749–54

Edlen, B. 1953. The dispersion of standard air. *J. Opt. Soc. Am.* 43:339–44

Ezekiel, S., Weiss, R. 1968. Laser-induced fluorescence in a molecular beam of iodine. *Phys. Rev. Lett.* 20:91

Huggett, G. R., Slater, L. E. 1975. Precision electromagnetic distance measuring instrument for determining secular strain and fault movement. *Tectonophysics* 29:19–27

Lee, P. H., Skolnick, M. L. 1968. Saturated neon absorption inside a 6328 Å laser. *Appl. Phys. Lett.* 10:303–5

Levine, J., Hall, J. L. 1972. Design and operation of a methane absorption stabilized laser strainmeter. *J. Geophys. Res.* 77:2595–2609

Levine, J., Harrison, J. C. 1976. Earth tide strain measurements in the Poorman mine near Boulder, Colorado. *J. Geophys. Res.* 81:2543–55

Levine, J., Stebbins, R. 1972. Upper limit on the gravitational flux reaching the earth from the Crab Pulsar. *Phys. Rev. D* 5:1465–68

Querzola, B. 1974. *High accuracy distance measurement by a two wavelength pulsed laser.* Paper presented at Int. Symp. Terr. Electromagn. Distance Meas. Atmos. Eff. Angular Meas., Royal Inst. Technol., Stockholm, Sweden

Rowley, W. R. C. 1966. Some aspects of fringe counting in laser interferometers. *IEEE Trans. Instrum. Meas.* IM-15:146–49

Savage, J. C., Prescott, W. H. 1973. Precision of geodolite distance measurements for determining fault movements. *J. Geophys. Res.* 78:6001–8

Thayer, G. D. 1967. Atmospheric effects on multiple frequency range measurement. *ESSA Tech. Rep. IER 56-ITSA 53.* Washington, D.C.: GPO. 57 pp.

Ann. Rev. Earth Planet. Sci. 1977. 5:371-96

AULACOGENS AND CONTINENTAL BREAKUP

✕10078

Kevin Burke

Department of Geological Sciences, State University of New York at Albany, Albany, New York 12222

INTRODUCTION

Shatski and the Recognition of Aulacogens

Aulacogens (Greek *aulax,* a furrow) are long troughs extending into continental cratons from fold belts. They contain accumulations of sediment that are commonly more than three times as thick as neighboring contemporary cratonic sequences. Recognition that sequences in fold belts are generally thicker than those on cratons had been an essential element in the development of geosynclinal theory in the 19th century, and stratigraphic studies accompanying development of the petroleum industry in the 1920s and 30s confirmed this idea and led to further recognition of systematic variations within sedimentary cratonic cover. Thick sedimentary sequences were recognized as characteristic of a group of basins (foreland basins or exogeosynclines) lying parallel to and on the cratonic side of fold belts, and another group of basins, of which the Michigan and Paris basins are typical, was distinguished as being apparently randomly distributed within cratonic areas. By the middle of this century classifications of sedimentary basins, such as those of Kay (1951) and Umbgrove (1947), recognized these kinds of intracratonic basins as well as the geosynclines of folded areas but did not yet distinguish aulacogens.

The Soviet geologist Nicholas Shatski (sometimes transliterated Schatski), in the course of petroleum exploration during the Russo-German war of 1941–45, recognized long sediment-filled troughs in the cratonic Russian platform and later called the troughs aulacogens (Shatski 1946a,b, 1947, 1955; translated into German in Schatski 1961). The type examples were the Pachelma and Dnieper-Donetz aulacogens west of the Urals and north of the Black Sea. Aulacogens were clearly distinct from other areas with thick sediments on cratons, and their properties generally included (*a*) extension from an orogen far into a craton with gradual diminution in size, (*b*) a thick unfolded or gently folded sedimentary sequence, (*c*) location of the junction between the aulacogen and the orogen at a re-entrant, or deflection, in the fold belt, (*d*) long duration of the aulacogen as an active structure normally corresponding to the active period of the related orogen, (*e*) an early history of the

371

aulacogen as a narrow fault-bounded graben and a later history as a broader superimposed basin itself sometimes involved in later faulting, (f) the occurrence of horst-like features within the aulacogen that influenced sedimentation, especially in the early history, (g) the occurrence of evaporites, (h) occurrences of igneous rocks often in a bimodal rhyolitic and basaltic association, the igneous rocks being commonly, but not exclusively, formed at the beginning of the aulacogen's history, (i) a tendency to reactivation marked by renewed faulting and thick sedimentation long after the associated orogen completed its cyclical development. Although Shatski (1946b) suggested that structures outside the Soviet Union were aulacogens, citing in North America the Belt terrain of Montana, the Ottawa-Bonnechere graben, and the structure now widely known as the southern Oklahoma aulacogen, and although Soviet and eastern European geologists continued to describe and analyze them, aulacogens received very little attention in other countries for the next 25 years (see references to Soviet literature in Hoffman, Dewey & Burke 1974, Siedlecka 1975).

Aulacogens Related to Plate Structure

In North America the major breakthrough in the recognition of aulacogens came with Hoffman's work in the East Arm of the Great Slave Lake. Hoffman (1973) studied the stratigraphy, sedimentology, and structure of an area of Proterozoic rocks that had been mapped by Stockwell (1936), and contacts with John Rodgers, the Yale stratigrapher, who was familiar with the aulacogen studies of Shatski and other Soviet geologists, helped Hoffman in recognizing that the structure in the East Arm of Great Slave Lake had many of the typical features of an aulacogen. This recognition in about 1970 came at an opportune time. The plate structure of the lithosphere (Wilson 1965) was becoming widely appreciated and geologists were beginning to follow Wilson's lead (1968) in interpreting the record of older earth history in terms of complex interwoven cycles of opening and closing of oceans (Dewey & Bird 1970). Hoffman (1973) showed that relations between the Athapuscow aulacogen (as he named the Great Slave Lake structure), the cratonic Epworth basin to the north, and the Coronation orogen to the west could be related to the Wilson cycle (Dewey & Burke 1974) of opening and closing of oceans. An episode of continental rupture was represented in the development of a well-preserved continental margin or miogeoclinical sequence in the lower part of the Epworth basin and Coronation orogen successions. During this episode sediments were carried from the Canadian shield westward toward the (then) continental margin. Sediments in the narrow, rift-structured lower part of the aulacogen were similarly derived from the east. Later the direction of sediment derivation reversed and clastic sediments were eroded off the mountains of the Coronation orogen and deposited in a foreland trough on the cratonic area of the Epworth basin and in the broad basinal upper part of the aulacogen. The westernmost outcrops of these sediments became involved in the folding and thrusting, calc-alkaline igneous activity, and formation of a foreland trough or exogeosyncline, that are typical of geological development at a convergent plate margin (Dewey & Bird 1970), although there is, as yet, no certainty as to whether the particular plate boundary involved was Andean,

Himalayan, or an arc-collision boundary. Hoffman had shown that the aulacogens recognized by Soviet scientists, many of whom interpreted world structure without considering major horizontal displacements, could be interpreted in plate-tectonic terms as structures leading away from ruptured continental margins that became involved later in convergent plate margin phenomena.

At the time that Hoffman was unravelling the history of the Athapuscow aulacogen, Burke and Whiteman, working in Africa, were comparing the rifts of the Neogene African rift system with those developed in Africa at the time of the breakup of Gondwana. By considering a fairly large number of rifts they were able to show (Burke & Whiteman 1973) that twice in the last 200 m.y. there had been a common sequence of development in Africa that went from doming, through the development of three-armed rift systems on the crests of domes, to continental rupture by the development of two crestal rifts as a plate margin or margins, with preservation of the third or "failed arm" as a rift striking into the continent (Figure 1).

Burke & Dewey (1973) related this sequence of events to the interpretation of aulacogens. For the first time clearly grasping in plate tectonic terms an idea that Alpine geologists of an earlier generation (Suess 1909, Argand 1924) had groped for, Wilson (1968) had made the observation that orogenic belts generally mark places where oceans have opened and closed. Once this is appreciated, aulacogens, striking at high angles into orogenic belts, can be recognized as "failed" rifts in which the later episodes of the Wilson cycle of opening and closing of oceans are recorded (Figure 1). A similar idea occurred to Siedlecka (1975, especially Figure 7 and p. 341) as a means of interpreting relations between the Timan aulacogen and the oceans whose sites are now marked by the northern Urals and the Caledonides of northern Norway, but publication of her work was delayed.

Rifts, Graben, Failed Rifts, and Aulacogens

Intersecting rift patterns and their developed forms, with a failed rift system striking into a continent from an ocean embayment or an aulacogen striking away from a fold belt at a re-entrant, are distinctive and readily recognized. Burke & Dewey (1973) outlined the characteristics of a number of these and many more have been described in the last three years. A complex and as yet unstabilized nomenclature is emerging, but it seems desirable to limit the use of the word aulacogen to structures meeting Shatsky's definition, that is, to structures with thick sedimentary sequences striking into fold belts. The term *failed rift* (also *failed rift arm* or *failed arm*) is available for rifts or rift complexes striking from oceans into continents, and these terms also seem appropriate for old rift systems that strike into fold belts but have no associated thick sediment sequence (for example, the Great Dike system of Rhodesia). The German word *graben* (both singular and plural) has been applied to a number of rift structures ("Viking graben," "Rhine graben") and, like rift, seems useful as a general descriptive term. It has sometimes been used adjectivally (for example: "graben facies" for the sediments and igneous rocks typical of early rifting).

374

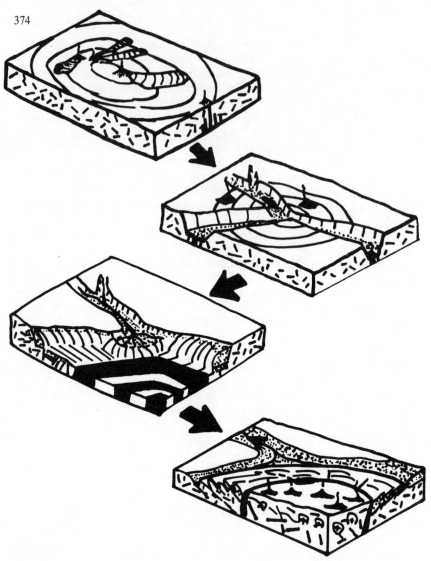

Figure 1 The sequence of swell, rift, continental rupture, and collision in aulacogen formation. A swell within a continent (top left) is crested with volcanoes from which lavas flow and evolves into a three-rift system (upper right) with volcanoes, sediment fill (stippled), and axial dikes. Two of the rifts develop to a continental margin (lower left) and ocean (striped) forms on their sites. A miogeoclinal wedge (stippled) is deposited along the new continental margin and a delta progrades down the failed rift to lie on ocean floor. Later (bottom right) another continent crosses the closing ocean and two continents collide. Rocks of the miogeoclinal wedge are thrust onto the continents and eroded by a river flowing down the aulacogen in the reversed direction. A suture zone (vertical black lines) marks the former site of the ocean and igneous rocks develop in the thickened continent on one side.

Aim and Scope of Review

Aulacogens are structures that record a great deal of earth history, and because they commonly escape the extreme tectonism experienced in fold belts at ocean closure, this record is often better preserved than in the fold belts themselves. However, emphasis in this review is less on the details of the historical record than on the information that aulacogens yield in their distribution in space and time and in their contained sediments, igneous rocks, and structures about the processes that have operated in their origin and development. For this reason, discussion of the initial rifting process, an evolutionary stage through which all aulacogens have passed, is included, as well as consideration of the later phases of aulacogen development.

RIFT DEVELOPMENT AND CONTINENTAL RUPTURE

The earliest stage in the development of an aulacogen is one of intracontinental rifting, and it is at this stage that the peculiarities of crustal and lithospheric structure are established which govern the later history. East Africa has been suffering this kind of rifting for the last 25 m.y. (since the beginning of the Neogene) and is therefore of special interest.

Neogene Rifts in East Africa

Although many features of the East African Rift System are relevant to the study of aulacogens, one of the most striking is the elaborate areal pattern of the active rifts (Figure 2). If the African continent splits into two in the future by the formation of new ocean along a linked set of rifts, many rifts will be left as failed arms in one or other of the two new continents.

The best developed three-armed rift system is around the Afar, where the Gulf of Aden and Red Sea are oceanic. Ocean floor, or something very like it, lies on the bottom of much of the Afar, and the Ethiopian rift is at present a failed arm. If, at some time in the future, ocean floor develops along this rift, it will lose its failed status and evidence of the earlier rifting event may be preserved along the sides of the new ocean.

The northern and southern Gregory rifts in Kenya form a three-rift system with the Kavirondo rift, which is centered on the town of Nakuru and is at an early evolutionary stage. Studies of the elevation of erosion surfaces on the rift shoulders around the Nakuru rift junction have shown that the topographic dome, centered on Nakuru at the crest of which the rifts meet, developed before the rifts themselves (Baker & Wohlenberg 1971). This is of interest because it has been suggested that intracontinental rifts commonly develop by horizontal propagation of vertical lithospheric cracks with uplift following later as a response to the emplacement of hot rock in the crack. While this process certainly happens (Dewey & Burke 1974), present evidence indicates that intracontinental rift systems have developed more commonly by linking uplifted areas on which individual crestal rift systems have formed [see discussion in Burke (1976a)].

Failed rift systems and aulacogens within continents often contain several kilometers of sediments that accumulated in the early stages of rifting. Familiar examples are the Triassic rifts of eastern North America, which formed during the rifting episode that culminated in the opening of the central Atlantic between the east coast of North America and the bulge of Africa. Lee (1976), for example, has reported an estimate of more than 10 km of Triassic sediments in the Culpeper graben. Comparable sedimentary accumulations are not known in rifts of the East African system, but it seems likely that the African rifts have evolved to a structural state in which such an accumulation could occur, and it is appropriate to consider what evidence there is of the structure underlying the rifts.

UNDERLYING STRUCTURE OF EAST AFRICAN RIFTS   It is becoming widely appreciated that the structural peculiarities of the African Plate [its anomalous elevation, basin and swell structure, large amount of intraplate (hot spot) vulcanism, and active rift system] are associated with Africa's position at rest for about the last 25 m.y. with respect to the underlying mantle and the earth's spin-axis (see, for example, Burke & Wilson 1972, Briden & Gass 1974, Burke & Dewey 1974, Richter & Parsons 1975,

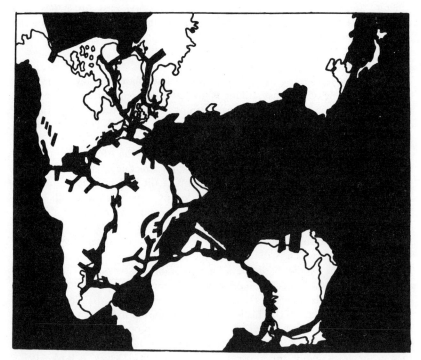

Figure 2   Mesozoic and Cenozoic rift systems and oceans are shown in black on a Triassic earth (based on Smith, Briden & Drewry 1972) to illustrate the abundance of failed rift systems associated with continental breakup. In Africa the rift systems have been active twice, in Mesozoic and Cenozoic times.

McKenzie & Weiss 1975). Because there is no lateral motion between the litho-sphere and the underlying asthenosphere, the present topography of the African Plate seems likely to bear an unusually simple relation to underlying mantle convection. Elevated areas, many of them capped by volcanoes (e.g. the Cape Verde islands and the Ahaggar), are likely to overlie rising convection currents or plumes and depressions (e.g. the Chad basin) to mark the areas between the rising currents. The elevations mark underlying areas of mass deficiency related to partial melting of the mantle, which is produced by the hot rising currents. As long as the rising convection current continues to reach the base of the lithosphere in the same place, elevation is dynamically maintained as a response to underlying mass deficiency. Ages of volcanic rocks that go back to 25 m.y. ago on African uplifts indicate that these areas have been elevated, although perhaps intermittently, throughout the Neogene.

The rift valleys of East Africa occur within a broad uplifted area (the East African swell) and are associated with discrete uplifts, as at Nakuru, within the swell. Geophysical observations on the rift system have suffered from difficulty in distinguishing effects related to the rift valleys from those related to the broad uplifts. The active rift systems of East Africa overlie large volumes of hot, partly melted asthenosphere that have their topographic manifestation in high ground, and most teleseismic effects in the Rift system, for example, the poor transmission of Sn (Gumper & Pomeroy 1970), are no doubt related to this material. Gravity observations are particularly ambiguous and susceptible to assumptions about regional fields as well as the effects of sediment accumulations in the rifts (Figure 3).

Axial dikes   Perhaps the most structurally significant geophysical observations made in the East African Rift System were refraction seismic studies in the Kenya Rift reported by Griffiths et al (1971). Although their work was not unambiguous, Griffiths and his colleagues showed that their results could have been caused by an axial dike of basaltic material about 10 kilometers wide lying roughly in the middle of the Gregory rift, reaching nearly to the surface and extending along the rift valley for a distance of over 100 kilometers. Khan & Mansfield (1971) showed that gravity observations in the Kenya rift were compatible with the presence of such an axial dike, and Searle & Gouin (1972) showed that the gravity field in the 150 km wide Ethiopian rift at the latitude of Addis Ababa could be interpreted as related to three similar axial basaltic dikes together about 75 km wide.

Three other lines of evidence support the idea that axial dikes are an important element in rift valley structure. These are evidence from the igneous rocks of both ancient and modern rifts and the outcrop and gravity pattern of ancient rifts.

A great deal has been learned about the igneous rocks of ancient and modern rift systems. The most structurally significant conclusions from these studies are that rift igneous rocks have been produced by partial melting of the mantle, probably at depths in the range 60–120 km, and that although fractional crystal-lization and remelting have been important in producing the wide range of alkaline and, to a lesser extent, tholeiitic igneous rocks reported from the rifts, both isotopic and other geochemical evidence show that generally reaction with continental crustal rocks has been unimportant. The implication is that rift igneous rocks are

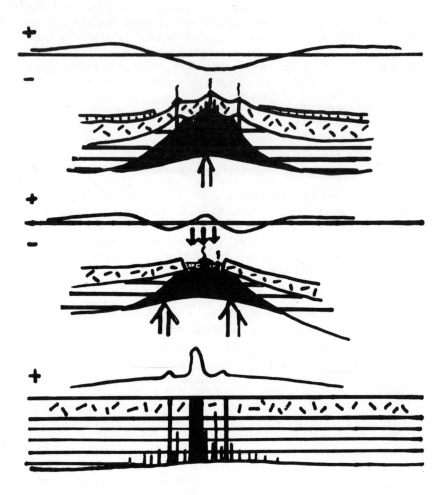

Figure 3   Structure and Bouguer anomalies at swell, active, and failed rift structures. Bouguer anomaly over a swell (top) is negative and attributable to mass deficient, partly melted asthenosphere (black) rising (open arrow) through the lithospheric mantle (lined) and doming continental crust (dashes). Volcanoes crest the dome and overlying sediments (dots) form a bordering scarp (based on the Ahaggar). A central positive Bouguer anomaly is related to axial dikes in an active rift (middle figure). A broad negative anomaly is attributed to the persistent asthenosphere swell (black, open arrows). The greater mass of the axial dike system due to mafic and ultramafic rocks makes the rift sink (solid arrows) (based on Kenya Rift). In an inactive rift (bottom) with no sediments the axial dikes are exposed cutting the continental crust but are more abundant in the lithospheric mantle. A sharp positive Bouguer anomaly marks the axial dike (based on the Great Dike of Rhodesia).

discrete bodies with intrusive contacts and without significant reaction zones with their country rocks throughout the thickness of the continental crust. This result is entirely compatible with the axial dike model.

Gravity surveys over inactive rift systems are especially illuminating because the mass deficient part of the asthenosphere that dominates the gravity field over an active rift is not present under an inactive rift, where the gravity field mainly results from the changes effected in the crust and lithosphere at the time of rift formation. Jones (1956), reporting a gravity survey of the Cretaceous Canellones graben in Uruguay, showed an axial positive anomaly. Oil wells drilled over this anomaly and on its flank confirmed that an axial dike model can satisfy the gravity observations. Positive Bouguer anomalies attributable to axial dikes are known over other fossil rift structures (see Burke & Whiteman 1973, Figure 2) and in the *Great Dike* of Rhodesia it appears (Grantham 1957) that the responsible igneous rocks outcrop. The *Great Dike* is about 10 km wide, has vertical sides (Weiss 1940), and is composed of layered mafic and ultramafic rocks of approximately tholeiitic composition.

The floor of a rift valley lies in isobaric equilibrium at a lower elevation than the shoulders because of the greater density of the basaltic and ultramafic rocks of its underlying axial dike system. The Neogene structure of Africa suggests that during active rifting, as sediments are eroded from elevated rift shoulders and deposited on the rift floor loading it and causing subsidence, the shoulders are elevated anew because of the persistent underlying mass deficiency in the asthenosphere. It is very likely that this process has allowed the accumulation of sedimentary sections up to 10 km thick in such graben as the Culpeper basin. When the underlying mass-deficient object is no longer there, general subsidence leaves the rift with its thick sediment section and a thinned underlying continental crust, the latter being an expression of the dike injection within the continent (Figure 3). The widely mapped variations in the thickness of continental crust (from 35–45 km over most continents) indicate that the lithosphere is strong enough to maintain the lateral stresses set up by the irregular mass distribution across the fossil rift.

Other Active Continental Rift Systems

Although the East African rifts are the best developed of active intracontinental rifts, there are both active and Neogene rifts in most other continents. Many of these are small-scale or poorly known features, but the Rhine graben in Europe and the Baikal graben in Asia are major features that have been thoroughly studied. The Baikal and Rhine graben have both been attributed to tension set up within the Eurasian continent as a result of Alpine-Himalayan plate collisions farther south (see, for example, Molnar & Tapponnier 1975; Sengör 1976). It is sometimes possible to determine whether or not old rifts or aulacogens originated in this way by dating the time of their inception. If rift formation coincides with neighboring collisional phenomena, there is a good possibility that the mechanism considered to have produced the Baikal rift may have operated. For example, J. Tuzo Wilson has suggested (personal communication) that the Ottawa-Bonnechere graben was formed in association with Appalachian collisional events of mid-Ordovician or later age, but Rickard's (1973) study of Lower Paleozoic sediment thicknesses in New York State shows that the Ottawa structure was already there in Cambrian times some m.y. earlier and is therefore more likely to be a feature produced by African-type continental rupture than by a continental rupture produced as a result of collisional orogeny.

Rifting events during collisional orogeny are well known in the Alpine System (Dewey, in press), the best documented being those associated with the Miocene rotation of Corsica and Sardinia away from the Mediterranean coasts of France and Spain (Dewey et al 1973). Because the products of this kind of rifting phenomenon are normally soon caught up in mountain belts, they are outside the scope of this review. Tensional processes perhaps associated with lateral plate motion have produced the rift-like features of the Basin and Range province.

Conditions at Rifting

Evidence from active and fossil rifts indicates that although continental rifts can develop in a variety of environments, the largest scale rift systems are associated with continental rupture. The rupturing process leaves inactive rifts within continents underlaid by abnormal lithosphere, which controls the further structural develop-ment of the rifts. Although the nature of the abnormality in the lithosphere underlying rifts is not generally established, evidence of the presence of an axial dike system has been recognized under both ancient and active rifts and axial dikes reaching high into the crust may be a general feature of the lithosphere underlying fully developed continental rift valleys. The lithosphere below the axial dikes in ancient rifts is likely to contain pyrolite depleted by the removal of both the axial dike basalt and the alkaline and tholeiitic volcanic rocks of the rift system. This depleted pyrolite is material that would have formed part of the asthenosphere while the rift system was active. Some of the basaltic material in ancient axial dikes and in other structures under rifts is likely to have undergone a transition to eclogite which, because of its great density, will contribute to subsidence of the rift. How much basalt becomes eclogite in this kind of environment depends on thermal history. Experimental work shows that eclogite is a stable form of rocks of basaltic com-position over a much wider pressure and temperature range than its scarcity at outcrop suggests (Ringwood & Green 1966). The scarcity is commonly interpreted to indicate that the basalt to eclogite transition is kinetically controlled and normally takes place at too slow a rate to be significant over geological time. A key to some of the variation in behavior of failed rift arms may therefore lie in the basalt-to-eclogite transition. Where the axial dike system is subjected to pro-longed elevated temperatures, dike material becomes eclogite and additional subsidence ensues. Where the axial dike material persists as basalt, renewed subsidence, after the initial loading of the active rift, may not occur. Since only part of the dike system material may suffer the phase changes involved in the eclogite transition, varied amounts of subsidence will take place even where the transition to eclogite occurs. If a part of the dike rock persists in basaltic mineralogy, then renewed subsidence may accompany additional later formation of eclogite.

FAILED RIFT SYSTEMS EXTENDING FROM OCEAN MARGINS

The most studied rift systems in the world are those formed at continental rupture and now located within the continents bordering Atlantic-type oceans. Rift complexes are best developed close to the continental margins, but in some cases, for example the North Sea and the Benue trough, they extend far into the continent. The reason these rifts are so well known is that they have been explored for petroleum because they contain great thicknesses of Mesozoic and Cenozoic marine and nonmarine sediments and because major rivers have prograded down them to produce oil-bearing deltas.

Many of the major oil provinces of the world occur within these rift systems and, although their structural and stratigraphic history is well known, underlying deep crustal and lithospheric structure is as yet little known.

Distribution of Rifts Striking into Continents

Since the short-lived continental assembly of Pangea began to break up about 200 m.y. ago, major oceans have formed in the North and South Atlantic, the Indian Ocean, the Arctic, and the Southern Ocean, and many smaller oceans have developed elsewhere. Figure 2 shows a Triassic reconstruction of the world on which both the sites of oceans and major Mesozoic and Cenozoic rift systems associated with continental rupture are shown in black. There are hundreds of rifts in the systems sketched on Figure 2 and it is likely that many more remain to be mapped.

The best known rift systems are those bordering the Atlantic Ocean (Burke 1976a). These fall into three groups: an old population of Triassic age around the site of the central Atlantic between the bulge of Africa and North America, a population associated with the opening of the South Atlantic at the beginning of the Cretaceous, and a very varied and complex group on the flanks of the North Atlantic largely related to the complex Permian and later rifting events of Northwest Europe.

In the Arctic development of a spreading boundary adjacent to the Canadian Arctic islands in the Jurassic was accompanied by transform motion parallel to the north coast of Alaska and development of a failed rift system on the site of the Mackenzie delta (Herron, Dewey & Pitman 1974). This development implies that unrecognized failed rift systems may exist at the continental margin north of Siberia.

Around the Indian Ocean Mesozoic failed rift systems are well developed on the east coast of Africa, the west coast of Madagascar (UNESCO-ASGA 1968), the east coast of India, and the coasts of Sri Lanka. The west and northwestern coasts of Australia have well-developed failed rift systems (Veevers 1974, Veevers & Cotterill 1976). No rift systems have been described from Antarctica or the east coast of Madagascar, and the Khambat rift system of the west coast of India appears of early Cenozoic age. On the south coast of Australia there are well-developed failed rift systems associated with the opening of the ocean between

Australia and Antarctica, but no failed rifts have yet been reported from the Antarctic side. Between Australia and New Zealand the Tasman sea has one well-developed failed rift in the Gippsland basin.

Mesozoic rifts opening into the Tethys are discernible in places, although continuing convergent activity obscures their development. The Sirte and Hon graben of Libya, having so far escaped convergent phenomena, are the most prominent Tethyan rifts.

Late Triassic and Early Jurassic rifts around the Caribbean and Gulf of Mexico help to date the formation of the ocean floor in those areas. The Maracaibo structure has suffered later convergent and strike-slip movement and is partly obscured, but the three rifts of the Boa Vista system (Berrangé & Dearnley 1975) and the Mississippi and Seewannee embayments are well-preserved rifts striking at high angles into the South and North American continents. Failed rift systems have not generally been recognized in association with the numerous marginal basins of the western Pacific, maybe because they originate in a dominantly convergent environment. However, the Seoul-Wonsan graben (Figure 4) striking north-northeast across the Korean peninsula is an apparent exception. This graben is of Miocene to Recent age and contains Quaternary volcanoes. If it formed, as the illustration indicates is possible, in association with the development of the Japan Sea, it suggests a Miocene age for that small ocean basin.

The numerous failed rift systems extending into the Russian Platform from the

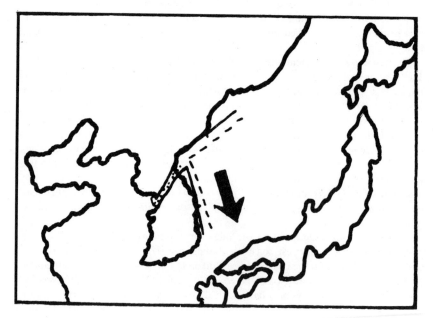

Figure 4 The Neogene and Recent Seoul-Wonsan graben (stippled) may be a failed rift associated with the opening of the Japan Sea.

North Caspian depression (*Tectonic Map of Europe* 1962) should perhaps be included here. These failed rifts, including the type aulacogen of the Dneiper valley, all strike into the North Caspian depression (Figure 5), an area that is underlaid by about 14 km of sediment including, within the lower half of the succession, a salt sequence probably of Upper Paleozoic age. This salt sequence

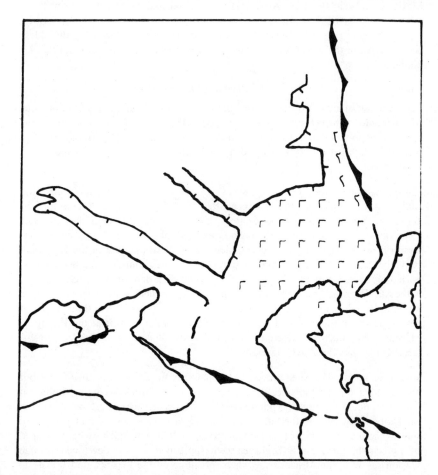

Figure 5 Area north of the Caspian (right) and Black (left) seas. The North Caspian depression is underlain by evaporites (inverted L's) and these salts are tectonized in the southern Urals (distorted inverted L's). Aulacogens including the Dneiper-Donetz aulacogen strike into the Urals and the depression. Analogy with the Gulf of Mexico suggests that the North Caspian depression may be underlain by pre-Permian ocean floor that has escaped obduction in both the Ural continental collision (thrust symbols on right of figure) and the Caucasus active collision (thrust symbols at bottom of map) (based on *Tectonic Map of Europe*, 1962).

has been tectonized on the western side of the Ural mountains by Permian convergent processes (as indicated on Figure 5). What kind of rocks underlie the North Caspian depression is not known. Seismic velocities, reported by Soviet scientists, lie in the ambiguous range between continental and oceanic, and in view of the increasing recognition that major salt deposits, such as those of the Gulf of Mexico (Ladd et al 1976), the Atlantic (Burke 1975a), and the Mediterranean (Ryan & Hsü 1973), have been laid down on oceanic crust, it seems probable that oceanic crust underlies the North Caspian depression. If this is so, it is probably the oldest oceanic crust in the world neither subducted nor caught up in a mountain belt by obduction.

Characteristics of Failed Rifts Extending into Continents

There are too many failed rift systems that extend from the margins of oceans into continents for it to be practicable to catalogue them here. A more useful course seems to be to distinguish categories and outline the features of examples of each. Accordingly, failed rift systems are discussed here roughly in order of increasing complexity.

ETHIOPIAN-TYPE RIFTS   The simplest of rifts facing oceans are those like the Ethiopian. In most cases, time has permitted rifts of this type to develop further, and older rifts that have developed no further than the Ethiopian stage are not commonly preserved. Perhaps the Oslo graben (Ramberg 1972) is an example of a fossil rift in this early stage of development. A significant feature of the Ethiopian rift is that sediments derived from erosion of the African continent are being transported along it, mainly by the Awash river, and being deposited on oceanic crust in the Afar. These sediments are close analogues of the clastics that bury ocean floor topography below the salt of the Gulf of Mexico (Ladd et al 1976) (Figure 4) and below the salt in the marginal graben of the South Atlantic. It is at the Ethiopian rift stage that much of the copper mineralization of failed rifts and aulacogens discussed by Sawkins (1976) develops.

SUEZ-TYPE RIFTS   The Suez graben has apparently no ocean floor at its base but has subsided about 5 km since the early Miocene (Heybroek 1965). T. Thompson (personal communication) has pointed out that the occurrence of Cretaceous Nubian sandstone at the base of the Suez section makes it unlikely that doming preceded subsidence, inasmuch as the thin Nubian section (< 400 m) would have been stripped off a precursor dome. The Suez graben seems more likely to have developed through tension set up in the African plate when spreading on the Red Sea and transform motion on the Dead Sea rift began and perhaps date these events as early Miocene. The Lower Cretaceous graben of Cape Province in South Africa apparently developed in a similar way when the Malvinas plateau, part of South America, moved along a transform fault past the south coast of Africa (Rigassi & Dixon 1972, Burke 1976a).

NEWARK-TYPE RIFTS   The Triassic rifts of the eastern seaboard of North America, although they contain thick nonmarine sequences deposited before continental rupture, had practically no further development once the Central Atlantic began to open. Almost the latest event in their development was a basaltic emplacement episode best known in the Palisades sill, and this may have coincided with the onset of Atlantic sea floor spreading (Cousminer & Manspeizer 1976). This does not seem a very common style of rift behavior and it will be particularly interesting to learn, as petroleum exploration develops on the neighboring continental shelf, whether buried Triassic rifts in that area behaved in the same way.

SERGIPE-TYPE RIFTS   Rifts of this class contain several kilometers of nonmarine sequences overlain by an evaporite blanket that commonly drapes older structures. The type examples are in the Lower Cretaceous Sergipe-Alagoa basin of coastal Brazil (Asmus & Ponte 1973), and other examples from the South Atlantic, all developed at continental rupture, include those of the Cuanza, Cabinda, and Gabon basins on the African side. All these produce hydrocarbon from below the evaporite horizon. The rift systems below the salt often show horst and graben structures that invite comparison with those of the active Afar. The unconformable relation of the salt is sometimes ascribed to a specific tectonic event but may generally be explained as a result of spills of saline waters over irregular topography into subaerial subsealevel graben (Burke 1975a).

In the North Atlantic, Sergipe-type graben with probable salt spills into subsealevel structures occur in the Permian of the North Sea (Taylor & Colter 1975), the uppermost Triassic of Portugal (Zbyszewski 1959), and the Aquitaine basin (BRGM et al 1974). Much of the Keuper salt of northwest Europe can be regarded as deposited in Sergipe-type graben. On the North American side of the Atlantic the best described Sergipe-type graben is the Triassic Orpheus graben of offshore Nova Scotia (Jansa & Wade 1975).

The history of the type Sergipe graben throughout the later Cretaceous and Cenozoic (Asmus & Ponte 1973) is one of continued miogeoclinal deposition and subsidence at the continental margin. Renewed tectonic activity of the graben has been apparently restricted to halokinesis.

RIO SALADA-TYPE RIFTS   The Sergipe-type graben characterized by salt deposits is a climatically controlled environment because salt is only precipitated in graben where evaporation rates are exceptionally high. Graben formed at continental rupture without a hypersaline phase pass straight from nonmarine to marine environments. The southern part of the South Atlantic was in more temperate latitudes at the time of Cretaceous continental rupture than the area of Sergipe-type graben north of the Walvis ridge and Rio Grande Rise. Graben in the southern area show a simple nonmarine to marine transition. The Rio Salada–graben of Argentina (Urrien & Zambrano 1973) is proposed as the type.

Graben of this type can be the site of major petroleum resources if the first marine shales provide both source rocks and traps and the underlying nonmarine sands provide reservoirs.

BARREIRINHAS-TYPE RIFTS   Dewey (1975) showed that rifts associated with intracontinental transforms may be expected to exhibit compressional structures because of the imperfection of strike-slip motion along transforms, which is inevitable on a many-plate earth. The Barreirinhas basin of northern Brazil (Asmus & Ponte 1973) shows folds developed in mid-Cretaceous times as the shoulder of Brazil slid past the Guinea Coast of Africa and is the type example of this kind of rift.

LIMPOPO-TYPE RIFTS   All the rift types distinguished so far developed as the result of a single rifting event, but it has long been a recognized feature of rift systems that old rifts are the sites of renewed rifting (McConnell 1974). The simple case in which an Ethiopian-type rift is reactivated later is widely represented in eastern Africa, where rifts formed during the breakup of Gondwana have been rejuvenated in the Neogene. A simple example is the Limpopo rift (Cox 1970), which had an igneous-dominated history in Triassic and Jurassic time and has undergone dominantly topographic rejuvenation in the Neogene. In both Mesozoic and Neogene times delta progradation along the Limpopo rift ensued. The Godavari and Cuttack graben of India, active in the Permo-Triassic and Cretaceous, are examples of the Limpopo-type rift.

LUSITANIAN-TYPE RIFTS   A common situation is for an initial intracontinental or Ethiopian type rift to be reactivated so that the intracontinental rift later becomes a site of rifting that leads to ocean development. The Lusitanian basin of Portugal (Wilson 1975, Zbyszewski 1959, Montadert et al 1974) is the proposed type example. Late Triassic rifting along the border between the Grand Banks and Iberia led to the formation of salt-filled Triassic intracontinental graben in the Lusitanian basin. Throughout the Jurassic shelf sediments were deposited over the graben that rifted again in lower Cretaceous time to form part of the North Atlantic ocean. The Early Cretaceous ocean-forming event was preceded by a doming and rifting episode at the end of the Jurassic with faulting, basaltic igneous activity, and accentuated halokinesis.

NORTH SEA-TYPE RIFTS   The rift system of the North Sea is highly complex and individual rifts show varied styles of development, but the distinctive feature of the system is that the rifts, after establishment in Permian times, have experienced repeated renewed activity without ever developing ocean floor (Whiteman et al 1975a, b; Ziegler 1975b; Woodland 1975). The Viking graben, for example, was initiated in the Permian and was active in Triassic, Middle and Late Jurassic, and Lower Cretaceous times. The Middle Jurassic (Bathonian) episode was the most important phase of renewed rifting in the North Sea. During this phase a classic dome system centered on the area underlying the Piper oil field with basaltic vulcanism, and three radial rifts were developed (Figure 6).

During the varied Mesozoic rifting episodes more than three kilometers of sediment accumulated in North Sea graben, but continental rupture between Greenland and Norway and the British Isles at the end of the Paleocene was accompanied by a radical change in the style of North Sea structural development. The individual rifts ceased to subside independently and a broad trough coaxial with the present North Sea began to develop, filling with sediment largely eroded from the high ground around the igneous centers of the west of Scotland. This change from a narrow subsiding rift system to a later broad basin is like that recognized by Soviet geologists in aulacogens. In the Athapuscow and Southern Oklahoma aulacogens the change to a broader basinal structure correlates with the onset of collisional orogeny (Hoffman, Dewey & Burke 1974). Evidence from the North Sea shows that the change may occur in a failed rift system extending toward a continental margin and is not restricted to the onset of convergent processes.

Other areas in northwest Europe provide the best described examples of North Sea-type rifts. The Mesozoic developments of the More and Voring basins of offshore Norway (Ronnevik et al 1975) and the Aquitaine basin in France (BRGM et al 1974) are comparable. Episodes of repeated rifting without ocean formation

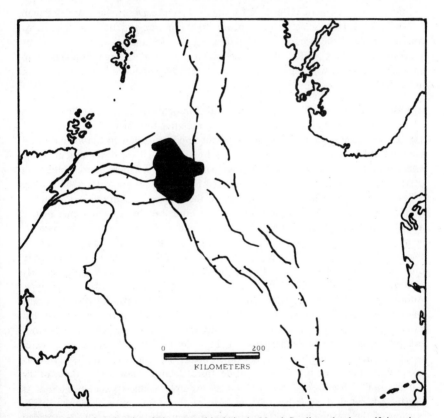

Figure 6   Jurassic volcanics of Piper area (black) in the North Sea lie at the three-rift junction of Central, Viking (top of map), and Moray (left of map) graben. Uplifted structure within the Moray graben is Halibut horst. North Sea graben developed in Permian times and were repeatedly reactivated in the Mesozoic. In the most violent reactivation in mid-Jurassic times, whose effects are depicted here, basaltic volcanism, uplift, erosion, and deposition of thick marine and nonmarine graben facies sediments all occurred (based on papers in Woodland 1975).

similar to those that have struck northwestern Europe since the Permain have not been described from other parts of the world, and they may be very unusual phenomena.

BENUE-TYPE RIFTS   There is clearly a gradation from rift to ocean as axial dike emplacement increases. Very narrow oceans that have opened and soon after shut are not different from wide rifts involved in convergence. This generalization is significant in the interpretation of aulacogens as it implies that there is a gradation, in the more tectonized of aulacogens, into fold belts.

The type example of a rift system that has large-scale compressional folds parallel to its axis is the Benue rift system of Cretaceous age striking into Africa from the shoulder of South America (Burke & Dewey 1974; Burke, Dessauvagie & Whiteman 1971). This is a large-scale structure nearly 1000 km long and 100 km wide. At its southeastern end the occurrence of andesitic volcanics and associated intrusions suggests that where widest the Benue ocean contained enough lithosphere to subduct to a depth where partial melting could induce andesite formation, that is, about 100 km.

MISSISSIPPI-TYPE RIFTS   The early history of the Mesozoic rift occupying the southern Mississippi embayment is of Sergipe type with continental clastics below Lou Ann salt (Lowest Jurassic?), and this development appears related to the formation of the Gulf of Mexico ocean. After continental-margin, shelf-sea development in the later Mesozoic, with carbonate reefs fringing much of the northern shore of the Gulf of Mexico, Cenozoic history was dominated by deltaic progradation (McGookey 1975). In the Paleogene, deltas formed by the erosion of Laramide mountains developed in the northwestern corner of the Gulf, but by the Neogene, deltaic activity was concentrated in the Mississippi and the river prograded down its embayment until its delta lay on oceanic crust. Progradation of major rivers in this way down failed rift systems is very widespread and many examples were discussed by Burke & Dewey (1973). Different kinds of rifts with variable early histories are likely to develop prograding deltaic systems of Mississippi type. Additional examples include the Gabon and Cuanza basins (Burke 1976a) and the Mackenzie delta. Special cases of deltas that prograde along rifts are those that reach oceanic crust that is young, hot, and still rapidly subsiding. The most obvious example of this kind of delta is that of the Colorado river. In some older cases where sediment supply has been cut off to deltas deposited on rapidly subsiding ocean floor, the entire delta may lie at depths of 1 km or more. The "delta of the Barentz sea" (Briseid & Mascle 1975) is an example of this style of behavior. As Greenland slid past Svalbard by transform motion (Talwani & Eldholm 1972), erosion of the "Alpine" mountains of Spitzbergen (Birkenmaier 1972) provided large volumes of sediment that were carried by river across the Hammerfest Basin (Ronnevik et al 1975) to lie on the young ocean floor of the Norwegian Sea.

Once Greenland was free of the coast of Spitzbergen, the Alpine mountains, no longer dynamically maintained, were eroded away and sediment supply to the Barentz delta was cut off. As the delta lay largely on ocean floor, less than 10 m.y. old, it has sunk in the last 35 m.y. to its present depth.

AULACOGENS AND FAILED RIFT SYSTEMS STRIKING INTO FOLD BELTS

The distinction of 11 different kinds of failed rift systems striking into continents from ocean margins gives an idea of the diversity of development that can be expected in aulacogens since the oceans in which the rift systems end can be expected to close at some time in the future, turning these failed rift systems into aulacogens. Because aulacogens are subject to further tectonic and erosional events, it is unlikely that the rocks in them record the same diversity as rift systems striking into oceans. For this reason, no attempt at subdivision of aulacogens is made here. Instead, after a brief discussion of what happens to a failed rift system at a collision, the properties of some examples are reviewed, roughly in order of increasing age.

The Later History of Aulacogens

Events that happen in aulacogens as a result of continental or island arc collision have been well described from the Athapuscow (Hoffman 1973) and Southern Oklahoma (Ham & Wilson 1967) structures. In both these areas thick, coarse clastic sediments, marine below, nonmarine above, eroded from the rising collisional mountains, have been deposited in a trough about three times as wide as the rift structure active in earlier history. Local sediment derivation is also important within both troughs and indicates the existence of upfaulted blocks. Wickham et al (1975) recognized large-scale left-lateral strike-slip faulting active in the Southern Oklahoma aulacogen during the deformation associated with the Ouachita collisional orogeny at the end of the Paleozoic. This kind of within-plate deformation at collision is becoming widely recognized (e.g. Molnar & Tapponnier 1975). The abnormal lithosphere in the aulacogen provides a locus for displacement and deformation. Similar strike-slip faulting and related deposition may have occurred in the Athapuscow structure, and the great McDonald fault forming the southeastern boundary of the aulacogen could well be an analogue of the strike-slip faults of Asia that Molnar & Tapponnier (1975) associate with the Indo-Asian collision.

Aulacogens and Failed Rift Systems of Phanerozoic Fold Belts

TETHYAN SYSTEMS   Because continental collision in the Alpine-Himalayan mountain system is still incomplete, fully developed aulacogens are not common. Rifts developed in association with the opening of Tethys have been widely recognized (see Dewey et al 1973, especially p. 3156) and in the Upper Triassic of western Tethys Sergipe-type rifts were particularly common. Only the Upper Rhine graben, whose site was occupied by the Hessian-Thur depression in the Triassic (Ziegler 1975a, Figure 8; Ziegler 1975b, Figure 11), appears to be completing the aulacogen cycle as the Rhine carries detritus northward from the Alps.

IAPETAN SYSTEMS   The ocean that opened and closed in forming the Northern Appalachians and Caledonides has been called Iapetus (after the putative father of Atlantis). Rift systems striking into North America at the same time as the opening of this ocean include the site of the Southern Oklahoma aulacogen and the Ottawa-

Bonnechere graben. Rankin (1975) has identified late Precambrian volcanics at the Sutton mountains (Quebec), South Mountain (Pennsylvania), and Mt. Rogers (Virginia), marking places where rifts that evolved to become Iapetus met the Ottawa-Bonnechere graben and two more southerly failed rifts. Kay (1975) suggested that the Ottawa-Bonnechere graben might be a later feature but the early Paleozoic ages obtained by Doig (1970) on alkaline intrusions along the westward extension of its strike and the Cambrian and Lower Ordovician thick sections near 'its mouth (Rickard 1973) are good evidence for its origin at the time of the opening of Iapetus.

The Ottawa-Bonnechere graben illustrates how incomplete the record in an old failed rift system may be. Although thick clastics derived from the Taconic arc collision are known in the Appalachians of Quebec and of New York State, to the north and south, there is no preserved thick section representing the later stage of aulacogen development in the graben itself. The reason for this appears to lie in the fact that the Ottawa-Bonnechere graben has been the site of Cretaceous hot-spot (nonplate margin) vulcanism, presumably localized by its abnormal lithosphere. Uplift associated with this vulcanism has led to erosion of the higher sequences from the graben, which seem likely to have contained evidence of derivation from the Appalachians.

Farther east the history of failed rift systems on this side of Iapetus is, as yet, poorly known. Doig (1970) reported early Paleozoic dates on alkaline intrusions at Saguenay, in Labrador, and in West Greenland that may mark the sites of failed rifts, and a pattern of continental margin headlands and embayments northeast of the site of Iapetus that is discernible on the *Tectonic Map of North America* (King 1969) from Marathon and the Mexican border can be extended as far as Newfoundland. Across the Atlantic Kidd (1975) has interpreted the Moine-Dalradian sediment wedge of Scotland as being related to two-stage rifting of Iapetus, but much remains to be done in this area.

In places on other sides of the Iapetus ocean the beginning of the Paleozoic appears to have been mainly marked by convergent phenomena (Burke 1975b). Rifting is indicated by the greatly increased thickness of the Eocambrian Sparagmite formation north of the Oslo graben (Strand & Kulling 1972, p. 18) and the contemporary alkaline intrusions of the Fen area. Aki, Husebye & Christofferson (1975) mapped anomalous lithosphere beneath the Norsar array in this area, but it is not at present possible to distinguish whether this material formed at the opening of Iapetus or in the later Permian rifting on the same site that formed the Oslo graben. Whittington & Hughes (1972) detected major differences between early Ordovician faunas of Scandinavia and southern Britain, and this may imply the existence of a barrier to faunal migration, possibly a rift roughly on the site of the North Sea at this time. The Charnwood Forest late Precambrian of southern Britain in the same area strikes at a high angle to Iapetus and may, like the Benue trough, be the locus of the opening and closing of a narrow ocean.

Aulacogens and Failed Rift Systems of Proterozoic Age

NORTH AMERICA   Proterozoic failed rift systems have been widely recognized in North America. Burke & Dewey (1973) concluded that a prominent set of rifts now exposed in tectonized condition within the Cordillera, but striking at right angles to its general north-south trend, marked an episode of continental rupture about 1200 m.y. ago and thus dated the origin of the Pacific ocean. But Stewart (1976) interpreted these structures as intracontinental considering that a continental rupture episode was better dated by occurrences of miogeoclinal sediments in the Cordilleran, Innuitian, and Appalachian mountains with ages ranging from 850 to 540 m.y. This difference in interpretation arises, at least in part, from difficulties in dating rocks, especially sediments, in the Precambrian. Dating imprecision is a general constraint on Precambrian aulacogen studies. For example, the Amaragosa aulacogen of Death Valley (Wright et al 1974) contains a sequence of several kilometers of sediment, but it is not yet possible to refine the timing of its distinct depositional episodes to within several hundred million years.

An exceptionally well-dated Precambrian failed sift system is the Keweenawan, which, on zircon dating, appears about 1140 m.y. old (L. T. Silver, cited in Symposium 1972). The Keweenawan rift is mainly filled with volcanics at outcrop in upper Lake Superior (Symposium 1972). It continues southwestward as the midcontinent structure producing the prominent gravity and magnetic anomalies typical of ancient rift systems with axial dikes (Burke & Whiteman 1973). Chase & Gilmer (1973), noting that offsets of the midcontinent gravity high can be described as small circles about a pole in Mexico, have ingeniously suggested that the midcontinent gravity high marks a spreading ridge and that the offsets are transform faults at right angles, but the structure of younger intracontinental rifts shows that concentric transforms develop only when true ocean floor is formed. The Keweenawan appears not to have reached this stage and it seems unlikely that the midcontinent structure, closer to the proposed descriptive pole, could have done so.

In a triple-rift junction at the lower end of Lake Superior, the Keweenawan rift joins the Kapuskasing failed rift of Ontario and a buried structure striking under the Michigan basin that Burke & Dewey (1973) first interpreted as a rift from its gravity and magnetic signature. Deep drilling on this structure has revealed the presence below the oldest Cambrian rocks of 1500 m of what may be Keweenawan graben fill with diabase in a Newark-type rift (Sloss 1976). The development of the broad Paleozoic Michigan basin over this Proterozoic failed rift suggests comparison with the relations between the Cenozoic North Sea broad basin and underlying Mesozoic rifts and also with the Neogene Chad basin and its underlying Cretaceous rifts (Burke 1976b).

The development of the Michigan basin may have had a more direct relationship to Late Proterozoic rifting. Ervin & McGinnis (1975) have shown that a Late Proterozoic rift on the future site of the Mississippi embayment received early Paleozoic sediment and there is evidence that thick accumulations of Cambrian and Early Ordovician sediments extend northward into the Michigan basin (Shell Oil Co. 1975).

EASTERN ASIA AND AUSTRALIA   Because of their position on the opposite side of the Pacific from North America, these continents are places to look for Proterozoic rifts that might be complementary to those of western North America. Rifting episodes of about the right age exist in both continents and P. Washington has suggested (personal communication, 1976) that the Proterozoic continental margin of Baikalia with its aulacogens fits very well against that of contemporary Western North America.

In Australia, where Rutland (1973) first identified the Adelaide "geosyncline" as an aulacogen, there are complex rift systems with thick sedimentary sections of Proterozoic age striking into the Tasman fold belt that occupies the whole eastern part of the country. These features thus satisfy the definition of aulacogen used in this review and published descriptions of them (in Brown, Campbell & Crook 1968) present intriguing features. For example, rift-type bimodal volcanics lie at the bottom of 15 km of shallow water sediments in the Adelaide aulacogen. In the latest Precambrian and earliest Paleozoic the Adelaide structure appears to have opened to the south into the Kanmantoo fold belt, which shows a typical re-entrant at the aulacogen mouth, but isopach maps on older lithounits within it indicate that the aulacogen may have opened earlier into the ocean at its northern end. Although the Proterozoic of Australia contains structures, especially the Amadeus basin and Adelaide geosyncline, that are clearly aulacogens, there has been, as yet, no published discussion of their development in relation to Proterozoic Wilson cycles of opening closing of oceans.

Aulacogens and Failed Rift Systems of Archaean Age?

Although the rigidity of the oldest continents is demonstrated by the more than 3 b.y. old Ameralik dike swarm of western Greenland, there are no very old recognized failed rift systems. Burke, Dewey & Kidd (1976, p. 116) considered the southern Huronian, about 2.3 b.y. old, as the earliest candidate for deposition in an aulacogen and the Great Dike of Rhodesia, intruded 2.65 b.y. ago (Allsop 1965), is a likely candidate for a very old rift structure. With its satellite dikes it forms a feature capable of producing the geophysically recognized properties of a rift-system axial dike complex (Figure 3), and it ends both to the north and the south in fold belts.

Coward et al (1976, Figure 5) have suggested that the Tuni ultramafic belt of eastern Botswana (?2.7 b.y. old) may represent a "failed arm," but it seems rather too strongly tectonized to meet the definition. In general, although Archaean greenstone belts show features typical of the Wilson cycle of opening and closing of oceans (Burke, Dewey & Kidd 1976), all rupturing events appear to have been successful in making ocean and no failed systems older than the Great Dike are preserved. The general success of rupture events seems a likely consequence of the then mechanical relations of lithosphere and asthenosphere caused by the need to dissipate more heat from the early earth by making plate boundary and oceanic lithosphere at a faster rate than now (McKenzie & Weiss 1975).

If the absence of old rifts is not just an accident of preservation, the onset of failed rift and aulacogen formation about 2.6 to 2.3 b.y. ago is a secular change brought about as the earth became older. It is one of very few such structural changes recognized to date.

CONCLUSION

Continental rifting can be the start of a highly complex sequence of events that culminates in the development of aulacogens. The lithospheric peculiarity of the rift environment is established early in its history, but further development can be very diverse and it is not yet possible, for example, to explain why some rifts form oceans while others do not.

It has only been since the recognition of the plate structure of the world and the realization that this kind of structure has existed for at least the latter half of earth history that the behavior of old and new rifts has been seen as broadly similar and relatable to cycles of opening and closing of oceans. Further appreciation of this unity will help in the understanding of rift structure. Although, for instance, the stratigraphy and shallow structure of the southern Oklahoma aulacogen are well known from over 75,000 oil wells, the petrology of the igneous rocks at the base of the rift and its deep structure are virtually unknown. Studies of these phenomena will help to show how close the analogy is between the Oklahoma structure and the rifts of East Africa. With intensified searches for fluid hydrocarbons attention will focus on the thermal histories of rift systems, and this should help in understanding their structural development.

It is clear that many young rift systems are associated with the breakup of Pangea (Figure 2). Study of ancient rift ages and distribution should help to show whether such continental unions are the statistically likely result of random motion at plate speeds of continental objects covering one third of the earth or have some other cause. Sawkins (1976) has identified an episode of widespread continental rifting about 1.1 b.y. ago and suggests there may have been another at about 2 b.y.

Literature Cited

Aki, K., Husebye, K., Christofferson, A. 1975. Three dimensional seismic images on the lithosphere. *Eos.* 56:456

Allsopp, H. L. 1965. Rb-Sr and K-Ar age measurements in the Great Dyke of Rhodesia. *J. Geophys. Res.* 70:977–84

Argand, E. 1924. La tectonique de l'Asie. *C.R. Int. Geol. Congr., 13th* 1:171–372

Asmus, H. E., Ponte, F. C. 1973. The Brazilian marginal basins. In *The Ocean Basins and Margins, Vol. 1, South Atlantic,* ed. A. E. M. Nairn, F. G. Stehli, pp. 87–132. New York: Plenum. 567 pp.

Baker, B. H., Wohlenberg, J. 1971. Structure and evolution of the Kenya Rift Valley. *Nature* 229:538–42

Berrangé, J. P., Dearnley, R. 1975. The Apoteri volcanic formation—tholeiitic flows in the North Savannas graben of Guyana and Brazil. *Geol. Rundsch.* 64:883–99

Birkenmaier, K. 1972. Alpine fold belt of Spitsbergen. *Int. Geol. Congr., 24th* 3:282–92

BRGM, ELF-Re, ESSO-REP, SNPA. 1974. *Geologie du Bassin d'Aquitaine* (Atlas). Bureau de Recherches Géologique et Minière

Briden, J., Gass, I. 1974. Plate movement and continental magmatism. *Nature* 248:650–53

Briseid, E., Mascle, J. 1975. Structure de la marge continentale Norveggienne. *Mar. Geophys. Res.* 2:231–41

Brown, D. A., Campbell, K. S. W., Crook, K. A. W. 1968. *The Geological Evolution of Australia and New Zealand.* London: Pergamon. 409 pp.

Burke, K. 1975a. Atlantic evaporites formed by evaporation of water spilled from Pacific, Tethyan and Southern oceans. *Geology* 3:613–16

Burke, K. 1975b. Hot spots and aulacogens of the European margin: Leicester/Shropshire; Oslo/Fen: Alno. *Abstr. Programs Geol. Soc. Am.* 7:34–35

Burke, K. 1976a. Development of graben associated with the initial ruptures of the AtlanticOcean. *Tectonophysics* 36:93–112

Burke, K. 1976b. The Chad Basin: an active intra-continental basin. *Tectonophysics.* 36:197–206

Burke, K., Dessauvagie, T. F. J., Whiteman, A. J. 1971. Opening of the Gulf of Guinea and geological history of the Benue depression and Niger delta. *Nature Phys. Sci.* 233:51–55

Burke, K., Dewey, J. F. 1973. Plume generated triple junctions: key indicators in applying plate tectonics to old rocks. *J. Geol.* 81:406–33

Burke, K., Dewey, J. F. 1974. Two plates in Africa during the Cretaceous? *Nature* 249:313–16

Burke, K., Dewey, J. F., Kidd, W. S. F. 1976. Dominance of horizontal movements, arc and microcontinental collisions in the later permobile regime. In *The Early History of the Earth*, ed. B. F. Windley, pp. 113–129. London: Wiley. 619 pp.

Burke, K., Wilson, J. T. 1972. Is the African Plate stationary? *Nature* 239:387–90

Burke, K., Whiteman, A. J. 1973. Uplift rifting and the breakup of Africa. In *Implications of Continental Drift to the Earth Sciences*, ed. D. H. Tarling, S. K. Runcorn, pp. 735–55. London: Academic. 1184 pp.

Chase, C. G., Gilmer, T. H. 1973. Precambrian plate tectonics: the midcontinent gravity high. *Earth Planet. Sci. Lett.* 21:70–78

Cousminer, H. L., Manspeizer, W. 1976. Triassic pollen date High Atlas and incipient rifting of Pangea as Middle Carnian. *Science* 191:943–44

Coward, M. P., Lintern, B. C., Wright, L. I. 1976. The pre-cleavage deformation of the sediments and gneisses of the Northern part of the Limpopo Belt. In *The Early History of the Earth*, ed. B. F. Windley, pp. 323–30. London: Wiley. 619 pp.

Cox, K. G. 1970. Tectonics and vulcanism of the Karroo period and their bearing on the postulated fragmentation of Gondwanaland. In *African Magmatism and Tectonics*, ed. T. N. Clifford, I. G. Gass, pp. 211–36. Edinburgh: Oliver & Boyd. 461 pp.

Dewey, J. F. 1975. Finite plate implications: some implications for the evolution of rock masses at plate margins. *Am. J. Sci.* 275:260–84

Dewey, J. F. 1977. Suture zone complexities: a review. *Tectonophysics.* In press

Dewey, J. F., Bird, J. M. 1970. Mountain belts and the new global tectonics. *J. Geophys. Res.* 75:2625–47

Dewey, J. F., Burke, K. 1974. Hot spots and continental breakup: some implications for collisional orogeny. *Geology* 2:57–60

Dewey, J. F., Pitman, W. C., Ryan, W. B. F., Bonnin, J. 1973. Plate tectonics and the evolution of the Alpine system. *Geol. Soc. Am. Bull.* 84:3137–80

Doig, R. 1970. An alkaline rock province linking Europe and North America. *Can. J. Earth Sci.* 7:22–28

Ervin, P. C., McGinnis, L. D. 1975. Reelfoot Rift: reactivated precursor to the Mississippi Embayment. *Geol. Soc. Am. Bull.* 86:1287–95

Grantham, D. R. 1957. Personal communication

Griffiths, D. H., King, R. F., Khan, M. A., Blundell, D. J. 1971. Seismic refraction line in the Gregory Rift. *Nature Phys. Sci.* 229:69–75

Gumper, F., Pomeroy, P. W. 1970. Seismic wave velocities and earth structure on the African continent. *Bull. Seismol. Soc. Am.* 60:651–68

Ham, W. E., Wilson, J. L. 1967. Paleozoic epeirogeny and orogeny in the Central United States. *Am. J. Sci.* 265:332–407

Herron, E. M., Dewey, J. F., Pitman, W. C. 1974. Plate tectonics model for the evolution of the Arctic. *Geology* 2:377–80

Heybroek, F. 1965. The Red Sea Miocene evaporite basin. In *Salt Basins Around Africa*, ed. D. C. Ion, pp. 17–40. London: Inst. Petrol. 105 pp.

Hoffman, P. F. 1973. Evolution of an early Proterozoic continental margin: the Coronation Geosyncline, and associated aulacogens of the NW Canadian Shield. *R. Soc. London Phil. Trans. A* 273:547–81

Hoffman, P., Dewey, J. F., Burke, K. 1974. Aulacogens and their genetic relation to geosynclines with a Proterozoic example from Great Slave Lake, Canada. In *Modern and Ancient Geosynclinal Sedimentation*, ed. R. H. Dott, R. H. Shaver, pp. 38–55. SEPM Spec. Publ. 19. 380 pp.

Jansa, L. F., Wade, J. A. 1975. Geology of the continental margin off Nova Scotia and Newfoundland. *Geol. Surv. Can. Pap. 74-30* 2:51–105

Jones, G. 1956. Some deep Mesozoic basins recently discovered in southern Uruguay. *Int. Geol. Congr., 20th* 11:53–72

Kay, M. 1951. North American geosynclines. *Geol. Soc. Am. Mem. 48.* 143 pp.

Kay, M. 1975. Ottawa-Bonnechere graben: tectonic significance of an aulacogen.

Geol. Soc. Am. Abstr. with Programs 7:82

Khan, M. A., Mansfield, J. 1971. Gravity measurements in the Gregory Rift. Nature Phys. Sci. 229:72–75

Kidd, W. S. F. 1975. Tectonic environment of Torridonian-Moine-Dalradian deposition. Geol. Soc. Am. Abstr. with Programs 7:84–85

King, P. B. 1969. Tectonic Map of North America. Washington: U.S. Geol. Surv.

Ladd, J., Buffler, R. T., Watkins, J. S., Worzel, J. L., Carranza, A. 1976. Deep seismic reflection results from the Gulf of Mexico. Geology 4:365–68

Lee, K. Y. 1976. Triassic geology in the Culpeper Basin. Geol. Soc. Am. Abstr. with Programs 8:215–16

McConnell, R. B. 1974. Evolution of Taphrogenic lineaments in continental platforms. Geol. Rundsch. 63:389–430

McGookey, D. P. 1975. Gulf Coast Cenozoic sediments and structure: an excellent example of extra-continental sedimentation. Trans. Gulf Coast Assoc. Geol. Soc. 25:104–20

McKenzie, D., Weiss, N. 1975. Speculations on the thermal and tectonic history of the earth. Geophys. J. R. Astron. Soc. 42:131–74

Molnar, P., Tapponnier, P. 1975. Cenozoic tectonics of Asia: effects of a continental collision. Science 189:419–26

Montadert, L., Winnock, E., Delteil, J-R., Grau, G. 1974. Continental margins of Galicia-Portugal and Bay of Biscay. In The Geology of Continental Margins, ed. C. A. Burk, C. L. Drake, pp. 323–42. New York: Springer. 1010 pp.

Ramberg, I. B. 1972. Crustal structure across the Permian Oslo graben from gravity measurements. Nature Phys. Sci. 240:149–53

Rankin, D. W. 1975. Opening of the Iapetus ocean: Appalachian salients and recesses as Precambrian triple junctions. Geol. Soc. Am. Abstr. with Programs 7:1238

Richter, F. M., Parsons, B. 1975. On the interaction of two scales of convection in the mantle. J. Geophys. Res. 80:2529–41

Rickard, L. V. 1973. Stratigraphy and structure of the subsurface of New York. Albany, N.Y.: State Mus. Sci. Serv. Map Chart Ser.: 18

Rigassi, D. A., Dixon, G. E. 1972. Cretaceous of Cape Province Republic of South Africa. In African Geology, Ibadan 1970, ed. T. Dessauvagie, A. Whiteman, pp. 513–27. Ibadan: Univ. Press. 870 pp.

Ringwood, A. E., Green, D. H. 1966. Petrological nature of the stable continental crust. Geophys. Monogr. Am. Geophys.

Union 10:611–19

Ronnevik, H., Bergsager, E. I., Moe, A., Øvrebø, O., Navrestad, T., Stangenes, J. 1975. The geology of the Norwegian continental shelf. In Petroleum and the Continental Shelf of N.W. Europe, ed. A. W. Woodland, 1:117–30. London: Halsted-Wiley. 501 pp.

Rutland, R. W. R. 1973. In Implications of Continental Drift to the Earth Sciences, ed. D. H. Tarling, S. K. Runcorn, 2:1011–33. London: Academic. 1184 pp.

Ryan, W. B. F., Hsü, K. J. 1973. Initial Reports of the Deep Sea Drilling Project. 13. Washington: GPO. 514 pp.

Sawkins, F. J. 1976. Widespread continental rifting: some considerations of timing and mechanism. Geology 4:427–30

Searle, R. C., Govin, P. 1972. A gravity survey of the central part of the Ethiopian Rift Valley. Tectonophysics 15:15–29

Sengör, A. M. C. 1976. Collision of irregular continental margins: implications for foreland deformation of Alpine type orogens. Geology 4:779–82

Shatski, N. S. 1946a. Basic features of the structures and development of the East European Platform. SSR Akad. Nauk Izv. Geol. Ser. 1:5–62

Shatski, N. S. 1946b. The great Donets basin and the Wichita system: comparative tectonics of ancient platforms. SSR Akad. Nauk Izv. Geol. Ser. 6:57–90

Shatski, N. S. 1947. Structural correlations of platforms and geosynclinal folded regions. SSR Akad. Nauk Izv. Geol. Ser. 5:37–56

Shatski, N. S. 1955. On the origin of the Pachelma trough. Mosk. O-va. Lyubit. Prir. Byull. Geol. Sec. 5:5–26

Schatski, N. 1961. Vergleichende Tektonik alter Tafeln. Berlin: Akademie. 220 pp.

Shell Oil Co. 1975. Stratigraphic Atlas, North & Central America. Houston: Shell Oil Co.

Siedlecka, A. 1975. Late Precambrian stratigraphy and structure of the northeastern margin of the Fennoscandian Shield. Nor. Geol. Unders. 316:313–48

Sloss, L. L. 1976. Deep drill hole in the cratonic interior. Eos. 57:326

Smith, A. G., Briden, J. C., Drewry, G. E. 1972. Phanerozoic world maps. In Organisms and Continents Through Time, ed. N. F. Hughes, pp. 1–42. London: Pal. Assoc. 334 pp.

Stewart, J. H. 1976. Late Precambrian evolution of North America. Geology 4:11–15

Stockwell, C. H. 1936. Eastern portion of Great Slave Lake. Geol. Surv. Can., Maps

377A, 378A
Strand, T., Kulling, O. 1972. *Scandinavian Caledonides,* p. 17. London : Wiley. 302 pp.
Suess, E. 1909. *Das Antlitz der Erde,* Vol. 3, Pt. 2. Leipzig: Freytag. 789 pp.
Symposium. 1972. Late Precambrian geology of the Lake Superior area. *Geol. Soc. Am. Abstr. with Programs* 4:XXV
Talwani, M., Eldholm, O. 1972. The continental margin off Norway: a geophysical study. *Geol. Soc. Am. Bull.* 83:3575–3606
Taylor, J. C. M., Colter, V. S. 1975. Zechstein of the English sector of the southern North Sea basin. In *Petroleum and the Continental Shelf of N.W. Europe,* ed. A. W. Woodland, pp. 249–64. London : Halsted-Wiley. 501 pp.
Tectonic Map of Europe. 1962. 1:2.5m. Moscow: State Publ. House
Umbgrove, J. H. F. 1947. *The Pulse of the Earth.* The Hague: Martinius Nijhoff. 358 pp.
UNESCO-ASGA. 1968. *International Tectonic Map of Africa.* 1:5m. Paris
Urrien, C. M., Zambrano, J. 1973. The geology of the basins of the Argentine continental margin. In *The Ocean Basins and Their Margins,* ed. A. E. M. Nairn, F. G. Stehli, pp. 135–66. New York: Plenum. 567 pp.
Veevers, J. J. 1974. Western continental margin of Australia. In *The Geology of Continental Margins,* ed. C. A. Burk, C. L. Drake, pp. 605–16. New York: Springer. 1009 pp.
Veevers, J. J., Cotterill, D. 1976. Western margin of Australia: a Mesozoic analog of the East African rift system. *Geology* 4:713–17
Weiss, O. 1940. Gravimetric and earth magnetic measurements on the Great Dyke of Southern Rhodesia. *Trans. Geol. Soc. S. Afr.* 43:143–53
Whiteman, A. J., Rees, G., Naylor, D., Pegrum, R. M. 1975a. North Sea troughs and plate tectonics. *Nor. Geol. Unders.* 316:137–61
Whiteman, A. J., Naylor, D., Pegrum, R. M., Rees, G. 1975b. North sea troughs and plate tectonics. *Tectonophysics* 26:39–54
Whittington, H. B., Hughes, C. P. 1972. Ordovician trilobite distribution and geography. In *Organisms and Continents Through Time,* ed. N. F. Hughes, pp. 235–40. London : Pal. Assoc. 334 pp.
Wickham, J., Pruatt, M., Reiter, L., Thompson, T. 1975. The Southern Oklahoma aulacogen. *Geol. Soc. Am. Abstr. with Programs* 7:1332
Wilson, J. T. 1965. A new class of faults and their bearing on continental drift. *Nature* 207:343–48
Wilson, J. T. 1968. Static or mobile earth and current scientific revolution. *Proc. Am. Philos. Soc.* 112:309–20
Wilson, R. C. L. 1975. Atlantic opening and Mesozoic continental margin basins of Iberia. *Earth Planet. Sci. Lett.* 25:33–43
Woodland, A. W. ed. 1975. *Petroleum and the Continental Shelf of Northwest Europe.* London: Halsted-Wiley. 501 pp.
Wright, L. A., Troxel, B. W., Williams, E. G., Roberts, M. T., Diehl, P. E. 1974. Precambrian sedimentary units of the Death Valley Region. In *Guidebook, Death Valley Region 27-35.* Shoshone, Calif.: Death Valley Publ. Co. 106 pp.
Zbyszewski, G. 1959. Étude structurale de l'aire typhonique de Caldas da Rainha. *Serv. Geol. Port. Mem. 3.* pp. 1–182
Ziegler, P. A. 1975a. North Sea basin history in the tectonic framework of N-W Europe. In *Petroleum and the Continental Shelf of N.W. Europe,* ed. A. W. Woodland, 1: 131–50. London : Halsted-Wiley. 501 pp.
Ziegler, W. H. 1975b. Outline of the geological history of the North Sea. In *Petroleum and the Continental Shelf of N-W Europe,* ed. A. W. Woodland, 1:165–90. London: Halsted-Wiley. 501 pp.

Ann. Rev. Earth Planet. Sci. 1977. 5:397–447

METAMORPHISM OF ALPINE PERIDOTITE AND SERPENTINITE

×10079

Bernard W. Evans

Department of Geological Sciences, University of Washington,
Seattle, Washington 98195

INTRODUCTION

The emplacement of peridotite into continental crust results in changes in its mineralogical, structural, textural, and bulk chemical properties. Accompanying deformation may partially or completely remove from it the evidence of its original stratigraphy and any envelope of contact alteration it might once have possessed. Such rocks have long been known as alpine-type peridotites (Benson 1926) and their field relationships have been well characterized (Wyllie 1967).

With advancing metamorphism, many clues as to the original nature of alpine-type peridotites are therefore effaced, but the loss need not be total if the processes involved are understood. To gain this understanding, we need to examine the inter-relationships between temperature, pressure, stress, mineral transformations and reactions, textures, and diffusion, etc, in a relatively simple chemical system in the earth's crust. It is not the purpose of this review to study the origin of peridotite; melting and fractionation in the upper mantle; upwelling of peridotite at mid-ocean ridges, at fracture zones, or behind island arcs; the anatomy or significance of ophiolites; or the crystallization of mafic and ultrabasic magma in the crust. The ample literature on these subjects has been well reviewed elsewhere (Wyllie 1967, 1969, 1970, O'Hara 1968, Miyashiro 1975). Further, the emplacement, tectonic environment, and different basic types of alpine peridotite have been extensively discussed in a number of recent papers (e.g. Coleman 1971a, Dewey & Bird 1971, Jackson & Thayer 1972, Moores & MacGregor 1972, Moores 1973).

In his textbook on metamorphism, Miyashiro was obliged to comment (1973, p. 30): "In contrast to other classes of metamorphic rocks, however, no detailed petrographic data are available on the progressive changes to take place [sic] in this (ultrabasic rock) class." It is hoped that, by gathering and reviewing here some of the recent pertinent literature, this data vacuum may at least partially be filled. This lack of systematic information is in curious contrast to the volume of literature on peridotites exotic to their surroundings—xenoliths in kimberlite, alkali basalt, and nephelinite, alpine "cold-slab" peridotites, serpentinites in melange zones, and samples dredged from the ocean floor. However, there do exist studies of ultrabasic rocks deformed and metamorphosed in harmony with their host rocks, a class that we shall here call *isofacial peridotite*. Alpine peridotites not fulfilling this criterion may be termed *allofacial*. In the context of metamorphism, this subdivision of alpine peridotite is more useful than ones based on tectonic associations (e.g. den Tex 1969, Moores 1973) or other criteria.

The first three sections of this review are concerned principally with the petrographic facts. These are followed by an evaluation of current methods by which the pressures and temperatures of equilibration of metamorphic ultrabasic rocks may be estimated. We then examine ultramafic rocks in three contrasting metamorphic environments, and conclude with a look at some related rocks formed by mass transfer processes accompanying ultrabasic rock metamorphism.

METAMORPHIC ASSEMBLAGES OF ULTRABASIC ROCKS AND THEIR CORRELATION WITH THE METAMORPHIC FACIES

The oxide components MgO, SiO_2, and H_2O commonly constitute between 80 and 90% by weight of peridotites and their metamorphic equivalents. Although FeO is a significant additional component, the partitioning of Fe^{2+} and Mg among the minerals at equilibrium is such that it is seldom necessary to ascribe an additional mineral to this component. On the other hand, it is only in the metamorphic products of some of the dunites and harzburgites that CaO and Al_2O_3 do not give rise to additional phases. If we make the convenient, but by no means universally valid, assumption that H_2O activity is held at maximum or externally fixed values, the principal mineral assemblages exhibited by ultramafic rocks in the various metamorphic facies can be depicted in the tetrahedron $CaO–MgO–SiO_2–Al_2O_3$. To illustrate this in two dimensions, we might take recourse to a projection from a universally present phase such as olivine. In all except the highest-temperature metamorphic facies, however, chlorite effectively represents the Al_2O_3 component, and so in practice it has been found convenient to use the triangle $CaO–MgO–SiO_2$ and note the identity of the alumina phase separately (Figure 1). Assemblages containing a greater variety of aluminous phases (e.g. cordierite, gedrite, sapphirine, yoderite, kornerupine) can be synthesized in the $MgO–Al_2O_3–SiO_2–H_2O$ system (Schreyer 1970), but they are seldom encountered in metamorphosed peridotites.

The compilation of compatibilities (Figure 1) is based on the field occurrence of metamorphic ultrabasic parageneses and the indications supplied by isofacial country rocks as to their correlation with the commonly accepted metamorphic facies. Isochemically metamorphosed alpine ultrabasics will ordinarily fall within the compositional triangle Fo–Di–En. However, the much greater frequency of occurrence of dunite (ol), harzburgite (ol + opx), and lherzolite (ol + opx + cpx), relative to wehrlite (ol + cpx), websterite (cpx + opx), and clino- and orthopyroxenite, for example, should be borne in mind in regard to the frequency of occurrence of the metamorphic parageneses. The compilation of Figure 1 is based on literature references too numerous to cite here, although many are cited elsewhere in this chapter. Facies that do not give rise to unique ultrabasic assemblages are grouped with adjacent facies showing the same assemblage. In some cases, more than one univariant reaction is depicted as separating metamorphic facies. The diagram attempts to synthesize a large amount of data, and not all assemblage boundaries coincide nicely with the reactions that are conventionally used to distinguish the facies. Nevertheless, this procedure is preferred to that of defining "ultrabasic mineral facies" (O'Hara 1967a), since the valuable concept of the metamorphic facies encompassing all bulk compositions is retained. Furthermore, the source of data is

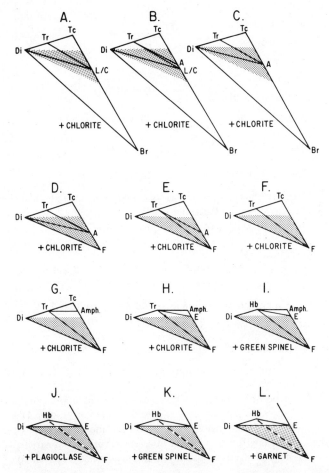

Figure 1 Ultrabasic mineral assemblages in a part of the triangle $CaO–MgO–SiO_2$. Correlation with metamorphic facies is given in the text (p. 400). The principal Al phase has been added beneath each diagram; in K, it refers only to compositions within the Di-E-F triangle. Di, diopside; Tr, tremolite; Tc, talc; L, lizardite; C, chrysotile; Br, brucite; A, antigorite; F, forsterite; Amph, Ca-poor amphibole; E, enstatite; Hb, hornblende. Stippled, the normative triangle Di-E-F.

the geological occurrence of topologically distinct sets of assemblages, rather than the inferred, theoretical, or experimental sets.

The correlation between the metamorphic facies and the *diagnostic* ultrabasic rock is as follows (letters refer to Figure 1):

AB	zeolite pumpellyite }	lizardite/chrysotile serpentinite
C	blueschist "low" greenschist }	brucite-antigorite serpentinite
D	"low" eclogite "high" greenschist }	diopside-antigorite peridotite
EFGHI	amphibolite hornblende hornfels }	tremolite peridotite
J	pyroxene hornfels	plagioclase lherzolite
K	pyroxene granulite { opx + plag cpx + gt	spinel lherzolite { Seiland subfacies Ariegite subfacies
L	"high" eclogite	garnet lherzolite

Because the eclogite facies encompasses such a large range of physical conditions, it is necessary to equate garnet lherzolite with a "high" eclogite facies, and diopside-antigorite peridotite (or olivine-serpentinite) with a "low" eclogite facies and greenschist facies. It is clear from the field that the eclogite and garnet lherzolite facies are not totally equivalent.

Plagioclase is not found in peridotite in the hornblende hornfels or amphibolite facies (cf O'Hara 1967a, p. 18), because the reaction

$$2 \text{ tremolite} + 5 \text{ spinel} + \text{diopside} = 8 \text{ forsterite} + 5 \text{ anorthite} + 2H_2O \qquad (1)$$

takes place very close to the reaction

$$\text{forsterite} + \text{tremolite} = 2 \text{ diopside} + 5 \text{ enstatite} + H_2O, \qquad (2)$$

which divides the amphibolite and hornblende hornfels facies from the granulite and pyroxene hornfels facies.

The correlation between tremolite peridotite and the amphibolite and hornblende hornfels facies is good. Although the forsterite-diopside join (Figure 1) is as stable as forsterite itself, the three-mineral paragenesis diopside + forsterite + Ca-free Mg-silicate phase (enclosing the commonest bulk composition) is restricted to either high-grade or low-grade facies, above reaction (2) and below reaction (3):

$$\text{antigorite} + \text{diopside} = \text{forsterite} + \text{tremolite} + H_2O. \qquad (3)$$

Orthopyroxene, it will be noted, is an important phase in the upper parts of the amphibolite facies, corresponding roughly with the sillimanite zone, as well as in the higher-temperature facies that extend to the liquidus field. Two configurations (F and G, Figure 1) intervene between enstatite and serpentine (cf Rost 1968). Cordierite is not included in the ultrabasic parageneses in any facies for reasons discussed below.

The entry of Al_2O_3, alkalis, and TiO_2 into Ca-amphibole at high grades tends to produce four-mineral assemblages instead of the maximum three implied by the triangular chemography of Figure 1 under divariant conditions. In addition, some blurring of the distinction between plagioclase lherzolite, spinel lherzolite, and garnet lherzolite arises in rocks containing appreciable Cr_2O_3 relative to Al_2O_3. However, in the high amphibolite facies, the spinel is a relatively aluminous variety even in chromiferous rocks, because the pyroxene under these temperature conditions is unable to accommodate much Al_2O_3.

TEXTURES OF METAMORPHIC ULTRABASIC ROCKS

If we can interpret them correctly, the textures of metamorphic rocks are a valuable source of information on geological history, and an indicator of the probable degree of attainment of chemical equilibrium. It is necessary to consider the textures of the isofacial ultrabasic rocks occurring in metamorphic complexes separately from those out of equilibrium with their present surroundings, the allofacial class. Textures in the former are mostly the result of synkinematic or postkinematic annealing or recrystallization, whereas in the latter, the textures tend to retain a complex history of deformation. These differences might be attributed to the prograde path of metamorphism in the first category, with attendant release of volatiles into the grain boundary regions, and to the retrograde path in the second.

Deformation has played the most important role in developing the texture of most allofacial alpine peridotites. Least deformed varieties are coarse-grained (e.g. 0.5 to 5 mm) and possess a xenoblastic, granular texture with interlocking grain boundaries. Kink bands are common in olivine and enstatite, but peripheral partial recrystallization of olivine to a fine-grained mosaic may usually be observed around the kinked olivines. More highly deformed varieties tend to contain strips of large kink-banded olivines (porphyroclasts) in a matrix of recrystallized olivines (e.g. Ragan 1969). Fine-grained blastomylonites, in which may be set augen of enstatite, represent extreme deformation (e.g. Nicolas et al 1971). In general, the enstatite is less affected by deformation in this way than olivine. In many cases it is clear that the deformation took place in the mantle prior to emplacement in the crust (Jackson, Green & Moores 1975). A similar sequence of textures is found in peridotite xenoliths in basaltic tuffs and kimberlites (Boullier & Nicolas 1975, Mercier & Nicolas 1975). It is speculated that the earliest texture, the coarse "protogranular" texture, in which there tends to be a spatial association of diopside and spinel with large enstatite grains, is related to a recrystallization process accompanying partial melting in the mantle. Recrystallization following the cataclastic or porphyroclastic stage results in an equant mosaic with a mildly anisotropic fabric or a tabular mosaic texture with a strongly anisotropic fabric.

The textures that result from grain-boundary equilibrium in isofacial ultrabasics, metaserpentinites in very many cases, are analogous to those described for metamorphic rocks of other bulk compositions. Monomineralic or multiphase aggregates of olivine and/or pyroxene tend to form relatively coarse (e.g. 0.5 mm) polygonal mosaics with 120° interfacial angles and straight or gently curved grain boundaries. The abundance at lower metamorphic grades of phases with anisotropic surface energies, e.g. chlorite, talc, antigorite, and anthophyllite, introduces more rational crystal faces into the texture and a conspicuous shape orientation of the minerals in rocks that have responded to directed stress. Doubts about the metamorphic crystallization of olivine and pyroxene in such metaperidotites may be quickly laid aside when, for example, they enclose planar, often folded, sets of inclusions (e.g. Lappin 1967, p. 186; Trommsdorff & Evans 1974, Plate IC, E). More generally, the same may be inferred when a homogeneous texture resulting from solid-state equilibrium crystallization clearly involves, in addition to olivine, subsolidus minerals such as chlorite, talc, tremolite or antigorite. Equilibrium solid-state textures alone, or tectonite fabrics, or evidence of strain, are, however, not sufficient to rule out the essentially magmatic origin of some ultramafic rocks, some cumulates, for example.

Progressive metamorphism of an ultrabasic rock, such as serpentinite, induces a sequence of synkinematic or postkinematic recrystallization textures that are primarily a function of its mineralogy.

The texture of the common very low grade serpentinite, containing lizardite and usually minor chrysotile, is largely governed by the pseudomorphic preservation of the pre-existing texture, and an equilibrium texture can be regarded only as having developed on a local scale. The characteristic mesh texture of such serpentinites has been amply described elsewhere (e.g. Francis 1956, Peters 1963, Coats 1968, Coleman & Keith 1971, Wicks & Whittaker 1976).

The mesh texture of chrysotile-lizardite serpentinites becomes destroyed as a result of antigorite growth, irrespective of whether the latter is of random orientation as in contact aureoles, or strongly oriented as in regional metamorphism. Like most other sheet silicates, antigorite forms thin flakes whose boundaries with less anisotropic phases like olivine and diopside are comparable to those between mica and quartz, namely, strong development of antigorite (001) faces, and 90° junctions where olivine/olivine grain contacts meet antigorite (001) faces. Strongly microfolded antigorite tends to polygonize, and several textural generations of antigorite may be recognized in deformed antigorite serpentinites (e.g. Bucher & Pfeifer 1973). An optically unstrained metamorphic olivine mosaic may be found in the hinges of microfolded antigorite. Olivine may form megacrysts, in which case they poikiloblastically enclose idioblastic flakes of antigorite (Figure 2A), and become augenshaped if their crystallization was prekinematic or synkinematic. Diopside differs in habit from that in high-grade ultrabasics in its tendency to be elongate parallel to c—unless, as is often the case, it is a pseudomorph of a pre-existing high-temperature clinopyroxene (usually highly charged with opaque oxide)—in which case it may exhibit prismatic extensions (e.g. Trommsdorff & Evans 1974, Plate IA). When growth has occurred in veins, diopside may be quite fibrous.

Although rocks composed principally of olivine + talc often exhibit the expected lepidoblastic-granular texture, this paragenesis, frequently along with magnesite, seems to be a prerequisite for the development of a distinctive texture in which olivine forms large thin plates parallel to (100) (elongation always fast, Evans & Trommsdorff 1974a). This texture bears a superficial resemblance to the spinifex texture of quenched ultrabasic lava flows. In regional metamorphic rocks, the texture characterizes layers or schlieren which grade into rock of more normal texture. It also occurs in crosscutting replacement veins, but is perhaps most common in contact metamorphic environments (e.g. Matthes 1971, Figure 4). The factors that promote this unusual texture are not known. In rocks with the same mineral paragenesis, Vance & Dungan (1977) report idioblastic olivine with hexagonal sector-twinning, analogous to monticellite.

With the progressive decrease in importance of the sheet silicates, the higher-grade ultrabasic rocks approach more ideal granoblastic aggregates. Tremolite occurs as stout prisms and shows mostly nonrational faces in equilibrium contact with olivine, although {110} faces or boundaries close to these planes are often favored (Figure 2B). Anthophyllite and cummingtonite, however, show considerably greater shape anisotropy than tremolite and typically occur as thin needles or sprays of needles penetrating neighboring olivines, even under equilibrium textural conditions (Figure 2C). Similarly, chlorite may be partially enclosed by olivine, with which it is in textural equilibrium (Figure 2D). Cummingtonite characteristically replaces or nucleates on and overgrows the calcic amphibole.

Metamorphic enstatite in the amphibolite facies is typically elongate parallel to c, and often much coarser grained than coexisting olivine (e.g. Evans & Trommsdorff 1969, Figure 1). Faces $\|\{110\}$ and (100) develop well against chlorite and talc, and, to a lesser extent, against olivine, a fact consistent with estimates of surface energies (Spry 1969, Table 7). An interstitial role for olivine relative to orthopyroxene is, of course, the reverse of that in cumulates. Enstatite in contact hornfelses, veins, and metasomatic rocks is often extremely coarse-grained and radiating (Matthes 1971, Figure 7; Evans & Trommsdorff 1974b, Plate 3A; Frost 1975, Plate 2B).

Isofacial ultrabasic rocks in the granulite facies naturally tend to show the closest approach to textural equilibrium (Vernon 1970). Spinels are usually described as idiomorphic when opaque, granular when translucent red or brown, and interstitial when green, analogous to the texture-composition relations of spinel in xenoliths (Basu & MacGregor 1975).

PERIDOTITE PETROFABRICS

For a good many years, the comparative petrofabric analysis of ultrabasic and enclosing country rocks has been used in investigations of the metamorphic and tectonic history of alpine peridotites (e.g. Ave' Lallement 1967, Möckel 1969). More recently, theoretical and controlled experimental work and new observational techniques (e.g. transmission electron microscopy) applied to peridotite tectonites have tended to focus attention more onto a kinematic understanding and application of the subject.

The preferred orientation of olivine in natural peridotite tectonites, as measured with the Universal Stage, involves in most cases an $\alpha = [010]$ point maximum, sometimes within a partial girdle, normal to the schistosity plane and maxima of

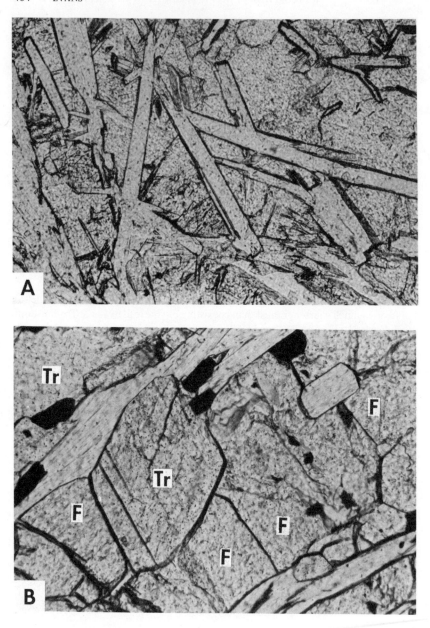

Figure 2 Equilibrium textures in metamorphic ultrabasic rocks. All photomicrographs are in plane polarized light, X170. A: Antigorite-olivine (subpoikiloblastic) paragenesis, Val Malenco, Prov. Sondrio, Italy. B: Tremolite (Tr)-chlorite-olivine (F)-chrome magnetite

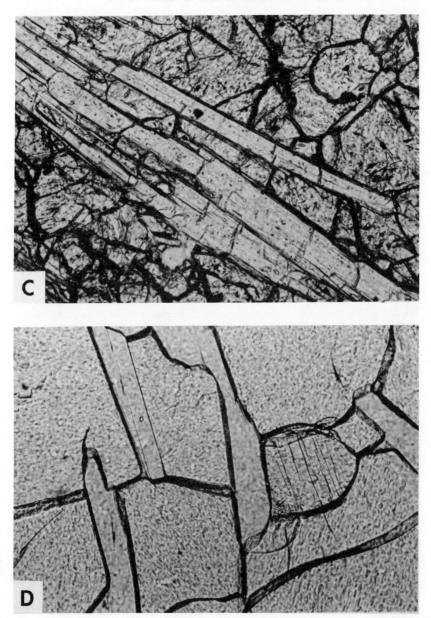

paragenesis, Val Moleno, Ticino, Switzerland. C: Anthophyllite-olivine paragenesis, Val Malenco, Prov. Sondrio, Italy. D: Chlorite-enstatite-olivine paragenesis, Val Cama, Ticino, Switzerland.

$\gamma = [100]$ and $\beta = [001]$ in the schistosity plane. This has been called the principal α-olivine fabric (den Tex 1969). It is essentially the same as that resulting from the settling in magma chambers of inequigranular olivine crystals, which are typically $[001]$-elongate or flattened parallel to (010). In the majority of peridotite tectonites, γ-maxima are aligned parallel to structural b elements, namely mineral lineations and fold axes (Battey 1960, Raleigh 1965a, den Tex 1969, Lappin 1971), but sometimes β-axes are recorded in this orientation. The same principal α-olivine orientation is found in ultramafic xenoliths in alkali basalts (Mercier & Nicolas 1975) and kimberlite pipes (Boullier & Nicolas 1975). This observation is consistent with the customary assumption that the olivine orientation in most allofacial alpine peridotites is derived from the mantle. The importance of this similarity, however, is diminished by the occurrence of the same principal α-olivine fabric in peridotites obviously recrystallized by a metamorphic event in the continental crust (Möckel 1969), including the olivine + antigorite paragenesis in Figure 2A (J. M. Rice, unpublished).

Orientation of the α-maximum normal[1] to the axial planes of isoclinal folds (Ave' Lallement 1967, Loney, Himmelberg & Coleman 1971) indicates that during compression α tends to become parallel to the axis of principal finite compressive strain ε_1. This conclusion has been amply supported by experimental work (Ave' Lallement & Carter 1970, Carter & Ave' Lallement 1970, Nicolas, Boudier & Boullier 1973), which has shown that α tends to become parallel to ε_1 or σ_1 during compressive deformation, apparently by two independent mechanisms: 1. intragranular glide on $T = (010)$, $t = [100]$ and 2. syntectonic recrystallization. Agreement has not been reached, however, on the relative importance of these two processes in developing the observed olivine fabric in natural peridotite tectonites. The tendency for $[100] = \gamma$ to align parallel to structural b (ε_3) in tectonites has been duplicated by Ave' Lallement (1975) in extrusion experiments on compacted olivine powders.

The presence of kink bands in olivine in alpine peridotites, and to a lesser degree deformation lamellae, is, however, evidence that some intragranular glide occurred (Raleigh 1968). Kink bands originate by inhomogeneous slip on planes in directions initially normal to the band boundaries. Measurement of the optical indicatrix on each side of the kink-band boundary, usually approximating (100), gives the axis of bending from which the slip system can be inferred. Measured axes of external rotation mostly concentrate around $[001]$ or scatter between $[001]$ and $[010]$ (Raleigh 1968, Raleigh & Kirby 1970, Mercier & Nicolas 1975, Loney & Himmelberg 1976), indicative of $[100]$ slip on glide planes (010) and $\{0kl\}$ respectively. Carter & Ave' Lallement (1970) found that the sequence of slip systems — $\{110\}[001]$, $\{0kl\}[100]$ (pencil glide), and $(010)[100]$ — was favored by increasing temperature, increasing pressure, and decreasing strain rate. They concluded, however, that, at natural strain rates of the order of 10^{-14} sec^{-1}, far below the experimental range,

---

[1] Slight obliquity of the fabric relative to the structural elements is sometimes observed, and has been interpreted by Nicolas et al (1971) as indicative of a component of simple shear flow.

syntectonic recrystallization could be the operative process responsible for the principal α-olivine fabric. There is the possibility that, as with other minerals in tectonites such as quartz and mica, the formation of kink bands is a reflection of late strains within an already oriented fabric. However, kink-band boundaries are often observed to terminate in a triple-point grain junction, indicating at least that the kinking preceded an episode of recrystallization (Ragan 1969; Nicolas et al 1971). In any case, use of the slip mechanism in natural peridotites for determining the approximate temperatures of deformation events is made highly uncertain by the strong strain-rate dependency of the mechanism transition temperature.

Less common orientations of olivine include one in which $\gamma = [100]$ maxima are in the schistosity plane and $\alpha = [010]$ and $\beta = [001]$ make girdles normal to it. This fabric is probably in part produced by glide on the system $\{0kl\}[100]$. Another important orientation is a principal γ-olivine fabric (den Tex 1969), in which $\gamma = [100]$ forms a point maximum normal to the foliation, $\beta = [001]$ axes lie in a girdle parallel to the foliation, and $\alpha = [010]$ is diffuse (Möckel 1969). This fabric is theoretically predicted to result from recrystallization in a nonhydrostatic stress field (Hartman & den Tex 1964), but experiments have as yet failed to reproduce it.

Pyroxenes in peridotites are generally weaker in fabric anisotropy than olivine. In enstatite, $\gamma = [001]$ is invariably parallel to the foliation plane and the lineation in naturally deformed peridotites. The glide mechanism has long been known to be $T = (100)$, $t = [001]$ (Turner et al 1960, Raleigh 1965b). The predominant orientation of $\beta = [100]$ normal to the foliation (Nicolas et al 1973) would be consistent with this glide mechanism. Enstatite fabrics with $\alpha = [010] \perp$ foliation, apparently similar to those produced by experimental synkinematic recrystallization (Carter 1971, but cf Carter 1975, p. 346), are much less common. Nicolas et al (1973) believe that enstatite fabrics in peridotites are more compatible with an origin by translation gliding than by syntectonic recrystallization. Significantly, experiments on lherzolites (Carter 1971) have shown that higher temperatures and lower strain rates are required for syntectonic recrystallization of pyroxene than that of olivine. Consistent with the foregoing, enstatite fabrics with $\beta = [100]$ parallel to $\alpha = [010]$ in olivine predominate over those with $\alpha = [010]$ parallel to $\alpha = [010]$ in olivine.

Thin lamellae of clinoenstatite ($\gamma \wedge c \approx 30°$ as compared to $40°$ for diopside) parallel to (100), induced by a shear transformation from enstatite, commonly accompany kinking in enstatites (Trommsdorff & Wenk 1968), particularly in peridotites of the amphibolite facies (Grover 1972, Evans & Trommsdorff 1974b, Frost 1975). Deformation of enstatite by shear transformation to clinoenstatite as opposed to slip on the system (100)[001] is favoured by lower temperatures and higher strain rates (Raleigh et al 1971). Extrapolation to the low strain rates in nature ($\approx 10^{-14}$ sec^{-1}) predicts the change in mechanism at $500°–600°C$.

B. R. Frost (unpublished) has shown that clinoenstatite was best developed in hornblende hornfels facies enstatite in crystals oriented with (100) and [001] parallel to the greatest resolved shear stress; the latter was determined from the fabrics of twinned calcite and quartz lamellae in tectonites in the same late fault zone. Coe & Kirby (1975) have indicated that, if temperature and mean pressure during deformation are independently known, estimates of the magnitude as well as the orientation of the stress fields producing clinoenstatite in nature are possible. The method is based on estimates of the effect of resolved shear stress on (100)[001] on the temperature of the transition of clino- to orthoenstatite.

SUBSOLIDUS STABILITIES OF ULTRABASIC ASSEMBLAGES

System $CaO-MgO-Al_2O_3-SiO_2-H_2O$

The multiplicity of possible assemblages and reactions in this system becomes more manageable if we confine our attention largely to reactions affecting compatibility polyhedra that include the composition Mg_2SiO_4. The phase diagram (Figure 3) is the conventional projection onto the plane $P_{solids} = P_{H_2O}$, where the curves give upper temperature limits for the reactions. For the dehydration reactions, of course, these limits are lowered by a reduction in the activity of H_2O, caused for example by high salt concentrations or the presence of fluid components such as CO_2, H_2, CH_4, etc.

Experimental work in this system has had to overcome exceedingly slow reaction rates. The result is that, nearly 30 years after one of the classic papers of experimental petrology (Bowen & Tuttle 1949), the phase diagram for the simple system $MgO-SiO_2-H_2O$ is still undergoing substantial revision. Consequently, Figure 3 differs in

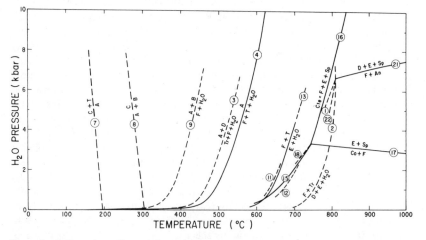

Figure 3 T-P_{H_2O} phase diagram for reactions in the system $CaO-MgO-Al_2O_3-SiO_2-H_2O$ involving the composition Mg_2SiO_4. Full lines, experimentally reversed; dashed lines, calculated, inferred, or provisional curves. For key to reaction numbers, see text. Sources of data: (7), (8), (9), (4), Evans et al 1976; (3), Trommsdorff & Evans (1977); (13), Chernosky (1976); (11), (12), Hemley, Montoya & Shaw (1976); (15), Chernosky, (1974); (16), Zen (1972), after Fawcett & Yoder (1966); (17), Fawcett & Yoder (1966), Seifert (1974); (18), Seifert (1974); (2), Evans & Trommsdorff (1970); (21), K. E. Windom & A. L. Boettcher, personal communication; (1), (22), Frost (1976a). Mineral abbreviations: C, chrysotile; A, antigorite; T, talc; B, brucite; F, forsterite; D, diopside; Tr, tremolite; E, enstatite; Cte, chlorite; Sp, spinel; Co, cordierite; An, anorthite.

many respects from previous summaries (O'Hara 1967a, Turner 1968, Evans & Trommsdorff 1970, Winkler 1974, Essene, Wall & Shettel 1973 [Figure 4.12 in Vernon 1976]). Some of the newer results are discussed below.

A recent successful reversal of the breakdown reaction of antigorite (Johannes 1975),

$$\text{antigorite} = 18 \text{ forsterite} + 4 \text{ talc} + 27 \text{ H}_2\text{O}, \tag{4}$$

confirmed what metamorphic petrologists have long suspected, namely that the reaction

$$5 \text{ chrysotile} = 6 \text{ forsterite} + \text{talc} + 9 \text{ H}_2\text{O} \tag{5}$$

is a metastable one. Reaction (4) is at equilibrium some 70°C higher than (5), and remains positive in slope to at least 15 kbar. Figure 3 contains a set of self-consistent reaction curves for serpentinites, calculated on the basis of experimental reversals of reactions (4) and (5) (Scarfe & Wyllie 1967, Chernosky 1973) and of the metastable reaction

$$\text{chrysotile} + \text{brucite} = 2 \text{ forsterite} + 3 \text{ H}_2\text{O} \tag{6}$$

(Johannes 1968), augmented by field and other information (Evans et al 1976). The model is based on the compositional colinearity of brucite, chrysotile, antigorite, and talc, and the average structural formula of antigorite: $Mg_{48}Si_{34}O_{85}(OH)_{62}$ (Kunze 1961). As a consequence, chrysotile is believed to give place to antigorite with rise in temperature across two solid-solid reactions with steep negative slopes on the P-T diagram

$$15 \text{ chrysotile} + \text{talc} = \text{antigorite} \tag{7}$$

and

$$17 \text{ chrysotile} = \text{antigorite} + 3 \text{ brucite.} \tag{8}$$

This model is compatible with the distribution of serpentine minerals, brucite, talc, and olivine in the field, with all the reliable experimental data, and with temperatures based on $^{18}O/^{16}O$ fractionations between serpentine minerals and magnetite in natural samples (Wenner & Taylor 1971). Unambiguous evidence for the direct growth of chrysotile from antigorite, either from experiment or from the field, would be a valuable confirmation of a stability field for chrysotile. The stable low-temperature limit for forsterite in the presence of H_2O, it should be noted, is given by the reaction

$$\text{antigorite} + 20 \text{ brucite} = 34 \text{ forsterite} + 51 \text{ H}_2\text{O}, \tag{9}$$

even though in nature olivine normally decomposes through an irreversible version of the metastable reaction (6), involving partial oxidation of iron to magnetite (Page 1967a, Moody 1976).

Field and experimental data (Essene, Wall & Shettel 1973) suggest that talc + H_2O is always more stable than serpentine + quartz, and, accordingly, a curve for this equilibrium is not shown.

The existence of a stability field for lizardite, like chrysotile, remains to be established, although it is clear that the mineral is stabilized by Al_2O_3 (Chernosky 1975) and probably also by Fe_2O_3, and it may well occur stably, given the appropriate bulk chemistry. However, the chemically equivalent pair antigorite + chlorite occurs widely in thoroughly recrystallized serpentinites, so that lizardite probably occurs stably, if at all, only at very low temperatures.

Reaction (3) is the sum of the antigorite breakdown reaction (4) and the H_2O-missing reaction:

$$2 \text{ diopside} + \text{talc} = \text{tremolite}. \tag{10}$$

Reaction (3) has been located in Figure 3 using Skippen's (1974) estimate of the free energy for reaction (10) and Johannes' (1975) data for (4). The temperature width of the forsterite + tremolite + antigorite assemblage is probably an underestimate. ΔV for reaction (10) is positive and ΔG is not large, and so the pair diopside + talc could be encountered at high pressures (Vernon 1976, Figure 4.12), perhaps in the blueschist facies.

The relatively narrow stability field ($\simeq 30°C$ at 2 kbars) for anthophyllite + forsterite (Greenwood 1963) in the system $MgO–SiO_2–H_2O$, limited by the reactions

$$9 \text{ talc} + 4 \text{ forsterite} = 5 \text{ anthophyllite} + 4 \text{ } H_2O \tag{11}$$

and

$$\text{anthophyllite} + \text{forsterite} = 9 \text{ enstatite} + H_2O, \tag{12}$$

has been confirmed in a recent study at 1 and 2 kbar (Hemley, Montoya & Shaw 1977), using a technique involving measurement of aqueous silica. These new data (Figure 3) are consistent with Chernosky's (1976) reversal of the metastable reaction

$$\text{talc} + \text{forsterite} = 5 \text{ enstatite} + H_2O \tag{13}$$

in the range 0.5 to 4 kbar; the newer curves are flatter than those of Greenwood (1963) and in better agreement with the thermochemical parameters (Zen 1971). Possible intersection of the H_2O-missing reaction

$$4 \text{ enstatite} + \text{talc} = \text{anthophyllite} \tag{14}$$

with reactions (11) and (12) gives rise to one or two (mirror-image) invariant points. Chernosky (1976) suggests a pressure ($P_{H_2O} = P_{solids}$) of about 5 kbar for an invariant point placing an *upper* pressure limit on anthophyllite + forsterite, whereas Hemley, Montoya & Shaw (1977) suggest 600°C and 500 bars for an invariant point placing a *lower* pressure limit on anthophyllite + forsterite. These estimates are not mutually exclusive, however, if the topology drawn by Greenwood (1971, Figure 2A) is correct. That this may be so is made more plausible by the probable positive curvature of reaction (14) on the PT diagram (e.g. Evans & Trommsdorff 1974b, Figure 5), as suggested by its small reaction entropy, which declines with increasing temperature (Zen 1971, Table 2). Because of the low ΔS and ΔV, equilibrium (14) will be extremely sensitive to additional components, such as Fe^{2+}, and therefore of limited petrologic value.

The new phase diagram (Figure 3) has reduced the width of the talc + forsterite field to about 140°C, which is in much better agreement with geological estimates than the earlier 200°C +.

Reactions (11) and (12) are strictly valid only for the CaO-absent section of our system, the place of anthophyllite in many metaperidotites being taken by magnesio-cummingtonite (Rice, Evans & Trommsdorff 1974) in the presence of tremolite. These relations await undoubtedly difficult experimental investigation (Cameron 1975).

Fawcett & Yoder (1966) showed that, at H_2O pressures respectively below and above $3\frac{1}{2}$ kbar, chlorite breaks down according to

$$5 \text{ clinochlore} = 10 \text{ forsterite} + \text{cordierite} + 3 \text{ spinel} + 20 \text{ } H_2O \tag{15}$$

and

$$\text{clinochlore} = \text{forsterite} + 2 \text{ enstatite} + \text{spinel} + 4 \text{ } H_2O. \tag{16}$$

Refinements of these reaction curves by Chernosky (1974) and Zen (1972), and of the H_2O-missing reaction

$$\text{cordierite} + \text{forsterite} = \text{enstatite} + \text{spinel} \tag{17}$$

by Seifert (1974), have been adopted in Figure 3. The reaction

$$2 \text{ chlorite} + 6 \text{ enstatite} = 7 \text{ forsterite} + \text{cordierite} + 8 \text{ } H_2O \tag{18}$$

would theoretically represent the upper limit of chlorite in normal metaharzburgites and metalherzolites. In fact, in nature, chlorite ordinarily reacts out according to reaction (16). This is because equilibrium (17) is displaced to lower pressures by the additional components in natural rocks such as Fe^{2+}, Fe^{3+}, and Cr, which fractionate predominantly into spinel (Frost 1975).

A lower temperature limit for two-pyroxene parageneses is given by the equilibrium

$$\text{tremolite} + \text{forsterite} = 5 \text{ enstatite} + 2 \text{ diopside} + H_2O. \tag{2}$$

The P-T curve for reaction (2) (Figure 3) may be located by combining an estimate of the free energy of the reaction

$$\text{forsterite} + \text{quartz} = 2 \text{ enstatite} \tag{19}$$

from Robie & Waldbaum (1968), or any more recent source (e.g. Skippen 1974, Essene, Wall & Shettel 1973), with the coordinates for the reaction determined by Boyd (1959)

$$\text{tremolite} = 3 \text{ enstatite} + 2 \text{ diopside} + \text{quartz} + H_2O. \tag{20}$$

The P-T curve for (2) generated from Skippen's average equilibrium constants (Skippen 1974, Table 2, # 43) unfortunately appears to be too high in temperature. Reaction (2) as written does not, however, provide an upper temperature limit for olivine + Ca-amphibole in the system $CaO–MgO–Al_2O_3–SiO_2–H_2O$, because of the

412   EVANS

tschermakite (and, in nature, pargasite) substitutions into tremolite that begin to take place at high temperatures.

The intersection of the curves for reaction (2) and the H_2O-missing reaction

$$2 \text{ forsterite} + \text{anorthite} = 2 \text{ enstatite} + \text{diopside} + \text{spinel} \qquad (21)$$

(Kushiro & Yoder 1966, Green & Hibberson 1970) generates three additional dehydration reactions (Frost 1976a), two of which are operative in forsterite parageneses:

$$\text{tremolite} + 2 \text{ spinel} = 3 \text{ forsterite} + \text{enstatite} + 2 \text{ anorthite} + H_2O \qquad (22)$$

and

$$2 \text{ tremolite} + 5 \text{ spinel} + \text{diopside} = 8 \text{ forsterite} + 5 \text{ anorthite} + 2 \ H_2O. \qquad (1)$$

Equilibrium (1) supplies a lower temperature limit for plagioclase peridotite in the presence of H_2O.

Figure 4, from Obata (1976), shows possible locations of univariant reactions

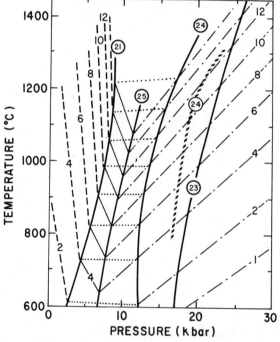

Figure 4 *P-T* diagram for reactions in anhydrous peridotite, from Obata (1976). Shows calculated alumina isopleths (mol % $MgAl_2SiO_6$) for orthopyroxene in the system CaO–MgO–Al_2O_3–SiO_2; dashed lines, equilibrium (37); full lines, equilibrium (39); dotted lines, equilibrium (35); dot-dashed lines, equilibrium (33), see Table 2. The striped curve for reaction (24) is from O'Hara, Richardson & Wilson (1971). Note that (23) is a reaction in the three component system MgO–Al_2O_3–SiO_2.

among anhydrous peridotite and pyroxenite minerals at high pressures and temperatures. The curves are calculated extrapolations of high-temperature experimental data (MacGregor 1964, Kushiro & Yoder 1966), utilizing data on the solubility of Al_2O_3 in orthopyroxene determined in the garnet peridotite field (Boyd & England 1964, MacGregor 1974) and on the diopside-enstatite solvus (Mori & Green 1976), so as to allow for the variation in $(Ca/Ca + Mg)$ ratio and Al contents of both pyroxenes. At 850°C the curve for reaction (21) is within 1 kbar of an experimental determination by K. E. Windom and A. L. Boettcher (personal communication). Curving in the opposite direction are reactions limiting the fields of garnet harzburgite

orthopyroxene + spinel = forsterite + pyrope   (23)

and garnet lherzolite

clinopyroxene + orthopyroxene + spinel = forsterite + garnet,   (24)

apparently separated from each other by 5–7 kbar. Experimental reversals of (24) in aluminous lherzolite compositions at 1200°C and $21\frac{1}{2}$ kbar (Green & Ringwood 1967a), and down to 850°C (Figure 4; O'Hara, Richardson & Wilson 1971), suggest that Obata's curve for (24) might be about 4 kbar too low. There is, in either case, no support for an intersection of (24) with (21), as in earlier extrapolations (MacGregor 1965). The steepness of the phase boundary for garnet peridotite contrasts considerably with slopes for eclogite-related reactions (Green & Ringwood 1967b, Kushiro 1969). A forsterite-absent reaction has also been calculated by Obata

orthopyroxene + anorthite + spinel = garnet + clinopyroxene.   (25)

All the reactions in Figure 4 are strongly influenced by the presence of components outside the system $CaO-Al_2O_3-MgO-SiO_2$ (see below). The widespread occurrence of spinel-plagioclase and spinel-garnet peridotites, occupying divariant fields on the P-T diagram, is a symptom of this problem. Ideally, these equilibria should be used in some such form as $\log K = A/T + B + C(P-1)/T$, with the equilibrium constant suitably corrected to the activity quotient as determined by chemical analyses of the minerals.

System $MgO-SiO_2-H_2O-CO_2$

Increasing amounts of CO_2 in the fluid phase cause the dehydration reactions discussed above to take place at progressively lower temperatures, until the point is reached at which they become metastable because of reactions with CO_2 to produce magnesite (Greenwood 1967, Johannes 1969). For serpentinites, the limiting mole fraction of CO_2 is soon reached: at about 0.05 for metastable chrysotile and 0.20 for antigorite (see below, Figure 5). Fields of stability for the parageneses enstatite + magnesite and anthophyllite + magnesite exist only at high values of X_{CO_2}, over isobarically limited ranges of temperature. With increasing total pressure, the isobaric invariant points magnesite + enstatite + forsterite + talc and magnesite + anthophyllite + forsterite + talc shift to more H_2O-rich fluid compositions (Mel'nik 1972), the former from about $X_{CO_2} = 0.85$ at 2 kbar to about 0.5 at 7 kbar (Evans & Trommsdorff 1974b, Ohnmacht 1974).

System $CaO–MgO–SiO_2–H_2O–CO_2$

Skippen (1974) has shown that a large number of the equilibria in the "siliceous dolomite" system—many of them experimentally determined—are also pertinent to the bulk compositions of ultrabasic rocks. The comparison of equilibria has been facilitated by the use of $CaCO_3$ and $MgCO_3$ as carbonate components rather than $CaCO_3$ and $CaMg(CO_3)_2$. Figure 5 is a $T–X_{CO_2}$ diagram at 2 kbar (Trommsdorff & Evans 1977) for a system containing antigorite, talc, forsterite, tremolite, diopside, calcite, dolomite, and magnesite, calculated from data in Skippen (1974, Table 2) and the new breakdown curve for antigorite from Johannes (1975). It shows, for example, that the paragenesis antigorite + carbonate is represented by partially over-lapping $T–X_{CO_2}$ fields, increasing in X_{CO_2} in the sequence calcite → dolomite → magnesite. This diagram enables thin lenses or layers of metaserpentinite (e.g. talc + magnesite + antigorite rock) occurring within regional metamorphic rocks of low or medium grade to be sensitive monitors of CO_2/H_2O in the fluid in the enveloping schists. It should be stressed that there are many acute intersections among the isobaric univariant curves in Figure 5, and consideration of all the un-certainties in the equilibria (Skippen 1974) could considerably change the diagram, although its topology is in agreement with field data. Obviously, direct experimental determination of key reactions such as

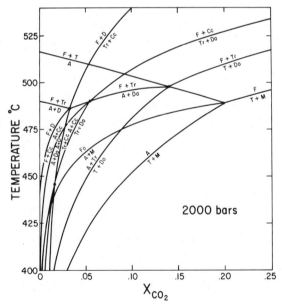

Figure 5 Calculated $T–X_{CO_2}$ phase diagram for antigorite (A), talc (T), tremolite (Tr), diopside (D), calcite (Cc), dolomite (Do), and magnesite (M) in the system $CaO–MgO–SiO_2–H_2O–CO_2$ (Trommsdorff & Evans 1977).

$$3 \text{ tremolite} + 5 \text{ calcite} = 2 \text{ forsterite} + 11 \text{ diopside} + 5 \text{ CO}_2 + 3 \text{ H}_2\text{O} \tag{26}$$

at high X_{H_2O} is needed.

Influence of Additional Components

The application of equilibrium relationships determined in simple systems to multi-component rock systems calls for an understanding of the differences created by the additional components, which, in the case of ultrabasic rocks, are principally Fe^{2+}, Cr, and Al.

In an equilibrium such as (12)

$$\underset{\text{anthophyllite}}{Mg_7Si_8O_{22}(OH)_2} + \underset{\text{olivine}}{Mg_2SiO_4} = 9 \underset{\text{orthopyroxene}}{MgSiO_3} + H_2O,$$

the equilibrium constant can be expressed in the form

$$\ln K = \ln K' + \ln f'_{H_2O},$$

where the standard state for H_2O is 1 bar, and

$$K' = (a^{opx}_{MgSiO_3})^9 (a^{anth}_{Mg_7Si_8O_{22}(OH)_2})^{-1} (a^{ol}_{Mg_2SiO_4})^{-1}.$$

For mole fractions in the region of 0.9, little error is introduced by assuming

$$K' = (X^{M1,opx}_{Mg} X^{M2,opx}_{Mg})^{9/2} (X^{M123,anth}_{Mg})^{-5} (X^{M4,anth}_{Mg})^{-2} (X^{ol}_{Mg})^{-2},$$

where $X_{Mg} = Mg/(Mg + Fe^{2+} + Mn + Ni + Ca)$ and $M1$, $M2$, etc are the non-equivalent octahedral sites in the phases indicated. Then the change in equilibrium fugacity of H_2O at constant P and T caused by the unequal fractionation of impurity components in the minerals is given by $\ln f_{H_2O} - \ln f'_{H_2O} = \ln K'$, where f_{H_2O} is the equilibrium fugacity when the solids are in their standard states ($X_{Mg} = 1$). The solution in terms of a change in P_{H_2O} requires analyses of the minerals in exchange equilibrium and use of the H_2O tables of Burnham, Holloway & Davis (1969).

Alternatively, P and f_{H_2O} may be held constant, and the temperature shift of the equilibrium calculated from the integrated Van't Hoff relation $\Delta \ln K = -(\Delta H^0/R) \cdot \Delta(1/T)$, where, in our case, $\Delta \ln K = -\ln K'$. Typical compositions for equilibrium (12) in natural ultramafic rocks are $X^{ol}_{Mg} = 0.9$, $X^{opx}_{Mg} = 0.907$, and $X^{anth}_{Mg} = 0.89$, giving $\ln K' = 0.18$, using cation-site distribution data for 700°C (Stroh 1976, Figure 6; Seifert & Virgo 1974, Figure 129). If the standard heat of reaction ΔH^0 at 2 kbar and 700°C is taken to be 31 kcal (Evans & Trommsdorff 1974b, Table 1), the temperature shift at 2 kbar compared to the Mg end-member reaction is $+10°C$. This example was purposely chosen because it is an exception to the more usual case in which the addition of Fe^{2+} to dehydration reactions lowers the equilibrium temperature. Some other exceptions in pelitic systems are figured by Thompson (1976).

The effect on temperature of the substitution for Mg by 10 mole % Fe in almost all the other dehydration reactions in Figure 3 is of a comparable or lesser magnitude than the above example. The same is true for devolatilization reactions in the mixed volatile, $CO_2 + H_2O$ system (Figure 5). For *ultrabasic* rocks, therefore, impurity-induced shifts in devolatilization equilibria are comparable to the uncertainties in the experimental determination of equilibrium temperatures. This is also true, contrary to popular belief, for the serpentine dehydration equilibria, provided of course that there are no changes in oxidation state of the iron at the same time (Evans & Trommsdorff 1976). There are unfortunately few *reversed* experiments (e.g., McOnie, Fawcett & James 1975) on compositions intermediate between Fe and Mg end-members for the reactions in Figure 3.

Reaction (11), which defines the low-temperature limit of stability of anthophyllite, is exceptional in that naturally occurring compositions lower the equilibrium temperature 30° to 40°C (Trommsdorff & Evans 1974, p. 344). This result, combined with that for equilibrium (12) discussed above, which defines the upper temperature limit for anthophyllite in metaperidotites, implies a stability field of 70° to 80°C width (at constant P_{H_2O}) for natural ultramafic anthophyllite + olivine assemblages as compared to about 30°C for the synthetic iron-free pair (Figure 3).

Since the impurity-related temperature shifts are inversely proportional to ΔH^0, it follows that reactions that do not involve a change of state, e.g., solid-solid reactions, for which ΔH^0 (and ΔV^0) is generally small, can be strongly affected. For example, it was estimated that the substitution of real compositions for Mg-end-members in equilibrium (14) (4 en + talc = anth) caused a shift of $-180°C$ (Evans & Trommsdorff 1974b), although this figure is highly dependent on the accuracy of the ΔS estimate. The importance of controlling the compositions of phases (not so much the bulk compositions) in experimental studies of solid-solid equilibria cannot be overemphasized. Reactions (21), (23), (24), and (25), delimiting the fields of plagioclase, spinel and garnet peridotite and pyroxenite, will also be highly sensitive to compositional effects. The concentration of Fe^{2+}, Fe^{3+}, and Cr in spinel, and Fe^{2+} in garnet has a large effect on the equilibrium constant for these reactions, particularly if cation solution models are used. MacGregor (1970) has experimentally demonstrated this for equilibrium (24). The effects on equilibrium pressures of given amounts of impurities can be readily estimated using molar volume and partitioning data (e.g. Frost 1976a).

GEOTHERMOMETRY AND GEOBAROMETRY

This subject has always been of central importance to petrologists, but a wider audience has of late taken interest in it because of the direct approach to fossil geothermal gradients made possible through the study of xenolithic and other ultrabasic rocks introduced close to the earth's surface with minimal mineralogic readjustment. In the petrogenetic *P-T* grid of the metamorphic petrologist, univariant curves of gas-liberating reactions unfortunately outnumber those involving only solid phases. Although reactions of the former type, because of their large reaction enthalpies and entropies, tend to proceed in nature with minimum over-stepping of the reaction boundary, the latter possess the advantage of independence from the fugacities of gas species. This advantage is of lesser importance in progressive metamorphic terranes, but is critical in the study of rocks of little-known or complex geological histories. Recent advances in geothermometry and geo-barometry, particularly as applied to ultrabasic rocks, have been considerable. They may be attributed to improved experimental techniques, particularly through the use of the piston-cylinder apparatus, and to a generally improved theoretical understanding by petrologists of equilibria between mineral solid solutions.

Although the equilibrium constant for solid-solid equilibria is a function of both P and T, we may usefully distinguish between equilibria with large values of ΔH^0 and low ΔV^0, which make good potential geothermometers, and those with large ΔV^0, which are more suited to geobarometry. In the former class may be grouped

Table 1   Potential geothermometers and geobarometers for ultrabasic rocks

	Greenschist Blueschist Eclogite	Amphibolite	Pyroxene hornfels (plag. lherzolite)	Granulite (spinel lherzolite)	Garnet lherzolite	Barometer (B) or Thermometer (T)
Intracrystalline Mg–Fe exchange						
Cpx	X		X	X	X	T
Intercrystalline Mg–Fe exchange						
Ol + opx		X	X	X	X	T
Ol + cpx	X		X	X	X	T
Ol + spinel	(X)	X	X	X		T
Ol + garnet					X	T
Opx + cpx			X	X	X	T
Opx + spinel		X	X	X		T
Opx + garnet				(X)	X	T
Cpx + spinel	(X)		X	X		T
Cpx + garnet				(X)	X	T
Spinel + garnet				(X)	(X)	T
Other Ion-exchange						
Al^{VI}/Cr in opx + cpx			X	X	X	T
Solvus; Diopside-enstatite			X	X	X	T
$MgAl_2SiO_6$ in Orthopyroxene						
Opx + cpx + ol + plag			X			B
Opx + ol + spinel		X	X	X		T
Opx + garnet				(X)	X	B
$CaAl_2SiO_6$ in Clinopyroxene						
Cpx + opx + ol + plag			X			B
Cpx + ol + spinel			X	X		T
Cpx + opx + garnet				(X)	X	B
Cpx + opx + plag + spinel			X	X		B

Parentheses indicate usefulness in a restricted compositional range of ultrabasic rocks.

intercrystalline and intracrystalline ion-exchange equilibria and equilibria of the solvus type. In the latter category are sliding equilibria with a sizeable density change, related usually to a change in the oxygen-coordination of a cation such as magnesium or aluminum. Particularly valuable geothermometers are exchange equilibria with pronounced partitioning (large K_D), since large values of ΔG^0 ($= -RT \ln K$) are accompanied in such equilibria by large values of ΔH^0 and consequently good temperature sensitivity [from the relation $d \ln K / d(1/T) = -\Delta H^0/R$]. Unless the temperature or pressure differential of $\ln K$ is large, the accuracy of geothermometry or geobarometry of complex rocks will remain highly dependent on the availability of expressions or models relating measured mineral compositions to activities of the components in the equilibria. These relations themselves may, of course, be functions of $T, P,$ and composition.

A classification of some peridotite geothermometers and geobarometers, their applicability to ultrabasic rocks in the different metamorphic facies, and their P-T roles, is shown in Table 1. It is important to remember that the rate-controlling steps in the listed reactions are in many cases quite different, so that independent results on a sample that disagree may not necessarily indicate an error, but rather different equilibria frozen in under different conditions.

Iron-Magnesium Intercrystalline Exchange Equilibria

Equilibria among the five minerals, olivine, orthopyroxene, clinopyroxene, garnet and spinel, supply ten potential ion-exchange thermometers. Reliable data on $K(P, T)$ are needed only for four of these equilibria, since the remaining six are dependent and can therefore be calculated. Iron-magnesium fractionation is strong when garnet or spinel participate, and so we may expect pairs involving one or the other of these two phases to be the best geothermometers. On the other hand, the temperature dependence of activity coefficients below say 1000°C may be capable of introducing a useful temperature sensitivity to equilibria, even when the equilibrium constant is known to vary little with temperature.

OLIVINE-ORTHOPYROXENE   Experimental studies (Nafziger & Muan 1967, Larimer 1968, Medaris 1969, Williams 1971) have confirmed empirical inferences in showing that K for the exchange equilibrium

$$\tfrac{1}{2} Mg_2SiO_4 + FeSiO_3 = \tfrac{1}{2} Fe_2SiO_4 + MgSiO_3 \tag{27}$$

is close to unity and only slightly temperature-dependent over the range 700°C to 1300°C. At the magnesium-rich end, broad groupings of ultramafic rocks (kimberlites, meteorites, alpine peridotites, ultrabasic xenoliths, etc) can only with difficulty be distinguished on a Roozeboom (X_{Mg}^{ol} vs X_{Mg}^{opx}) diagram. An admittedly over-simplified model of two-site ordering in orthopyroxene and two-site disorder in olivine predicts an increased spread in the distribution curves on the Roozeboom diagram towards lower temperatures (Grover & Orville 1969, Olsen & Bunch 1970, Saxena 1973, Wood 1975), but the difficult task of accurate calibration remains.

OLIVINE-CLINOPYROXENE   A significant temperature and compositional dependence of the distribution coefficient K_D for the equilibrium

$$\tfrac{1}{2} Mg_2SiO_4 + CaFeSi_2O_6 = \tfrac{1}{2} Fe_2SiO_4 + CaMgSi_2O_6 \qquad (28)$$

is clear from empirical data (Mori & Banno 1973, Obata, Banno & Mori 1974), particularly when pairs from antigorite serpentinite (Trommsdorff & Evans 1972) are added. Again, experimental calibration of this potential geothermometer remains to be done. The temperature variation of K_D will be the sum of at least the following factors: the temperature dependency of K, increasing nonideality of olivine at lower temperatures, the intracrystalline exchange-free energy for Fe^{2+} and Mg between the two nonequivalent sites in clinopyroxene (McCallister, Finger & Ohashi 1976), and the fraction of the large $M2$ site in clinopyroxene not blocked by Ca and available for Mg and Fe^{2+} (Blander 1972). It is possible to separate some of these effects graphically using empirical data (Frost 1976b). The uncertainty in Fe^{3+} in microprobe analyses of Mg-rich clinopyroxenes, however, is a problem that should not be overlooked. A calibration by Powell & Powell (1974) unfortunately gives a temperature dependency of K_D for metaperidotites which is opposite to that observed empirically, and also contains unreasonably high values for the regular solution mixing parameters (Wood 1976).

ORTHOPYROXENE-CLINOPYROXENE   Since this equilibrium is given by $(27)-(28)$, the problems underlying calibration are the same as discussed above, with the exception of those arising from nonideality in olivine. Empirical information indicating the pyroxene pairs in exchange equilibrium have temperature significance has been available for some time (Kretz 1963, Bartholomé 1962). Alpine peridotites have values of K_D comparable to those found in igneous rocks. In metaperidotites, of course, the pair is not found in the lower temperature facies (Figure 1).

OLIVINE-SPINEL   This mineral pair can be found in ultrabasic rocks from the green-schist facies to melting temperatures. Complete analysis of the spinel is required, since K for the Fe^{2+}-Mg exchange equilibrium is given by the weighted sum of the standard reaction free energies involving olivine and the Al, Cr, and Fe^{3+} spinel end-members (Irvine 1965, 1967), namely

$$\tfrac{1}{2} Fe_2SiO_4 + Y_{Cr}^{sp}MgCr_2O_4 + Y_{Al}^{sp}MgAl_2O_4 + Y_{Fe^{3+}}^{sp}MgFe_2O_4$$

$$= \tfrac{1}{2} Mg_2SiO_4 + Y_{Cr}^{sp}FeCr_2O_4 + Y_{Al}^{sp}FeAl_2O_4 + Y_{Fe^{3+}}^{sp}Fe_3O_4. \qquad (29)$$

The Y content of spinel is governed by other equilibria, such as Al-spinel with chlorite and two Mg-silicates or the spinel equilibria in Table 2, and the f_{O_2}-dependent equilibrium between Fe^{3+}-spinel component and two MgFe-silicates. The calibration of this geothermometer, based on thermochemical data and ideal spinel and olivine solutions (Jackson 1969), gives reasonable temperatures for layered ultramafic intrusives and alpine spinel peridotites, but excessively high temperatures for volcanic pairs and too low temperatures for amphibolite facies pairs. An empirical correlation among K_D, Y_{Cr}, Y_{Al}, $Y_{Fe^{3+}}$ and T (Evans & Frost 1975) showed that K_D is much more temperature-dependent than found by Jackson. Experimental calibration would be very useful, and in fact probably not too difficult to achieve. The empirical data show clearly that, at the temperatures of the low amphibolite and

Table 2 Equilibria involving solution of Mg-Tschermak's component in orthopyroxene and Ca-Tschermak's component in clinopyroxene

	Equilibrium	ΔV (cm³)
Garnet pyroxenite		
$Mg_2Si_2O_6 + MgAl_2SiO_6 = Mg_3Al_2Si_3O_{12}$ opx　　　　opx　　　　　garnet	(33)	-8.05
$2Mg_2Si_2O_6 + CaAl_2SiO_6 = CaMgSi_2O_6 + Mg_3Al_2Si_3O_{12}$ opx　　　　cpx　　　　cpx　　　　garnet	(34)	-9.13
Spinel peridotite		
$Mg_2SiO_4 + MgAl_2SiO_6 = Mg_2Si_2O_6 + MgAl_2O_4$ olivine　　　opx　　　　opx　　　　spinel	(35)	-0.13
$Mg_2SiO_4 + CaAl_2SiO_6 = CaMgSi_2O_6 + MgAl_2O_4$ olivine　　　cpx　　　　cpx　　　　spinel	(36)	-1.21
Plagioclase lherzolite		
$CaMgSi_2O_6 + MgAl_2SiO_6 = CaAl_2Si_2O_8 + Mg_2SiO_4$ cpx　　　　opx　　　　plagioclase　　olivine	(37)	19.79
$Mg_2Si_2O_6 + CaAl_2SiO_6 = CaAl_2Si_2O_8 + Mg_2SiO_4$ opx　　　　cpx　　　　plagioclase　　olivine	(38)	18.71
Spinel-plagioclase pyroxenite		
$CaMgSi_2O_6 + 2MgAl_2SiO_6 = Mg_2Si_2O_6 + CaAl_2Si_2O_8 + MgAl_2O_4$ cpx　　　　opx　　　　　opx　　　plagioclase　　spinel	(39)	19.66
$(33) - (34) = (35) - (36) = (37) - (38)$		
$CaMgSi_2O_6 + MgAl_2SiO_6 = Mg_2Si_2O_6 + CaAl_2SiO_6$ cpx　　　　opx　　　　opx　　　　cpx	(40)	1.08
$(33) - (35) = (34) - (36)$		
$2Mg_2Si_2O_6 + MgAl_2O_4 = Mg_3Al_2Si_3O_{12} + Mg_2SiO_4$ opx　　　　spinel　　　garnet　　　olivine	(23)	-7.92
$(35) - (37) = (36) - (38)$		
$2Mg_2SiO_4 + CaAl_2Si_2O_8 = CaMgSi_2O_6 + Mg_2Si_2O_6 + MgAl_2O_4$ olivine　　　plagioclase　　cpx　　　　opx　　　spinel	(21)	19.92

Molar volumes refer to 298°K, 1 atm; taken at $X_{Al}^{M1} = 0.05$ for orthopyroxene.

greenschist facies, considerable nonideality in spinel or olivine or both is present (Evans & Frost 1975, Figure 7).

ORTHOPYROXENE-SPINEL, CLINOPYROXENE-SPINEL These equilibria are potential alternatives to the olivine-spinel pair in rocks lacking olivine, such as pyroxenites. The calibration for orthopyroxene-spinel will be almost identical to that for olivine-spinel since $K(29) \gg K(27)$. All Fe-Mg exchange thermometers involving spinel have the advantage that uncertainties arising from nonideality are dwarfed by the actual variation of K_D with temperature.

CLINOPYROXENE-GARNET   The partition coefficient for the equilibrium

$$CaFeSi_2O_6 + \tfrac{1}{3} Mg_3Al_2Si_3O_{12} = CaMgSi_2O_6 + \tfrac{1}{3} Fe_3Al_2Si_3O_{12} \tag{30}$$

varies from about 2 in natural high-temperature rocks (e.g. xenoliths in kimberlite and alkali basalt) to about 40 in the blueschist facies (Banno 1970). This fact, the wide temperature range of applicability, and the wide compositional range for rocks containing this mineral pair, give this geothermometer great potential. Experimental data contributing to its calibration have been supplied by Hensen (1973), Akella & Boyd (1974), and Råheim & Green (1974). The pressure effect on K (30) calculated from 298°K molar volume data ($\Delta V = 0.91$ cm^3) is much less than apparently found experimentally by Råheim & Green, but failure to separate the effect on K_D of the grossular content of the garnet (itself a function of P) from that caused purely by P seems to be a likely explanation. Above 1000°C at least, a large range in Fe/Mg ratio had little effect on experimental K_D values (Råheim & Green 1974); however, empirical data on a low temperature garnet pyroxenite indicate otherwise (Oka & Matsumoto 1974). Recent work on activity-composition relations in complex garnet solutions (Ganguly & Kennedy 1974; Hensen, Schmid & Wood 1975) is beginning to show the magnitude of the nonideal interactions in garnet, notably between the pyrope and grossularite components. Fortunately, in ultrabasic rocks, uncertainties arising from the influence of impurities such as Na, Al, and Ti in the clinopyroxene will not be large, but unrecognized Fe^{3+} in microprobe analyses remains a problem. The sensitivity of K_D to temperature can be expected to swamp a good many of the analytical uncertainties and all but the most serious nonideal effects.

Other Intercrystalline Ion-Exchange Equilibria

Mysen (1976) has experimentally calibrated a two-pyroxene geothermometer based on the exchange equilibrium

$$CaAl_2SiO_6 + MgCrAlSiO_6 = CaCrAlSiO_6 + MgAl_2SiO_6, \tag{31}$$

where $K_D = (Al^{VI}/Cr)^{opx}/(Al^{VI}/Cr)^{cpx}$. The magnitude of the variation of K_D with T may well be enough to counterbalance analytical uncertainties in determining minor amounts of Cr and six-coordinated Al.

The partitioning of Ni between olivine and pyroxene has been suggested as potentially temperature-sensitive (Häkli 1968). Large values for ΔH^0 and ΔG^0 of opposite sign, unlikely in an ion-exchange equilibrium, suggest that his calibration cannot be totally correct, however.

The strong fractionation of Fe^{2+} and Mg between olivine and ilmenite, and the four-orders-of-magnitude change in K from 298° to 1400°K (Robie & Waldbaum 1968) are encouraging indications of a successful ilmenite-olivine geothermometer, although experimental calibration will be needed (Stormer 1972). The pair occurs in rocks ranging from the greenschist to the granulite facies.

There are many other potential ion-exchange geothermometers for ultrabasic rocks. One involving Mn/Mg partitioning between garnet and clinopyroxene, for example, would avoid the Fe^{3+} problem.

Intracrystalline Exchange Equilibria

Preliminary results from crystal structure refinements of untreated and heat-treated xenolithic clinopyroxenes (McCallister, Finger & Ohashi 1976) suggest that intracrystalline fractionation of Mg and Fe^{2+} in clinopyroxenes can supply useful geothermometric data. A five-fold variation in K_D apparently occurs over a 700°C range in temperature. By contrast, Mg-Fe^{2+} site preference in olivine, a function of Mg/Fe ratio and temperature (Ghose, Wan & McCallum 1976), is extremely slight, and orthopyroxenes and Ca-poor amphiboles are similarly lacking in thermometric value because of very low quenching temperatures for intracrystalline fractionation. Anthophyllite, however, appears to have some potential for estimating cooling rates (Seifert & Virgo 1975).

Diopside-Enstatite Solvus

Accelerated progress in the experimental determination of the diopside-enstatite solvus (Davis & Boyd 1966, Nehru & Wyllie 1974, Warner & Luth 1974, Howells & O'Hara 1975, Mysen & Boettcher 1975, Mori & Green 1976, Lindsley & Dixon 1975) promises soon to lead to a measure of agreement as to the effects of T, P, Fe^{2+}, Al, and silica saturation or undersaturation on the location of the two limbs. At equilibrium, the following energy balance holds:

$$\mu_{Mg_2Si_2O_6} \text{ in cpx} = \mu_{Mg_2Si_2O_6} \text{ in opx.} \tag{32}$$

Except at temperatures below 1000°C, experimental data on the multicomponent two-pyroxene system already agree quite well (Banno 1974) with a $\ln K$ (32) vs T^{-1} plot for the simple binary experimental system, if (a) "blocking" cations are calculated out according to their probable site occupancies (Al, Cr, Fe^{3+}, Ti in $M1$, and Na, Mn, Ca in $M2$), (b) ideal temperature-dependent intracrystalline partitioning in both pyroxenes is assumed (Virgo & Hafner 1969, Takeda 1972, McCallister, Finger & Ohashi 1976), and (c) nonreversed or otherwise unacceptable experimental points are excluded. For a practical geothermometer, Banno (1974) suggests using a best-fit straight line, such as

$$\ln \left(\frac{X_{En}^{cpx}}{X_{En}^{opx}} \right) = -\frac{6630}{T} + 3.04,$$

where $X_{En} = X_{Mg}^{M1} X_{Mg}^{M2}$. Such an expression can be refined on the basis of further experimental work, a pressure term with the form $\Delta V(P-1)/RT$ can be introduced, and if necessary a term to take nonideality into account can be added. An empirical fit to correct for Fe-substitution (Wood & Banno 1973) is not necessary for pyroxenes in mafic and ultramafic rocks (Banno 1974). The mathematical formulation has obvious advantages when working with complex natural pyroxenes. Unfortunately, calibration of the solvus below 1000°C, where the geothermometer is needed for a good many alpine peridotites, is still inadequate.

Alumina Solubility in Pyroxenes

The aluminum content of pyroxene (Mg-Tschermak's molecule in orthopyroxene, and Ca-Tschermak's molecule in clinopyroxene) is a function of P and T in equilibria with one or more other magnesium silicates. The minimum necessary phases for each of the equilibria (33) to (39) (Table 2) to be divariant is variable, since different numbers of components are involved. For example, only ortho-pyroxene and garnet are necessary for equilibrium (33) to be used. The relative P sensitivity of the equilibria may be estimated from the 298°K, 1 atm data (Table 2). Consistent with the increase in oxygen coordination of both Mg and Al, the right-hand sides of (33) and (34) are favored by pressure. Conversely, in the plagioclase lherzolite field, the volume change related to a decrease in the coordination of Ca and Al is positive. In the intervening spinel lherzolite field (Figure 4), the coordination of Mg decreases whereas that of Al increases; the net volume change is small, so that equilibria (35) and (36) can only be useful in geothermometry.

The best experimentally calibrated equilibrium to date is (33), where ortho-pyroxene is saturated in pyrope garnet (Boyd & England 1964, Boyd 1970, MacGregor 1974). Wood (1974) has expressed the experimental isopleths of Al in orthopyroxene coexisting with pyrope in the system $MgO-Al_2O_3-SiO_2$ in the useful form

$$\ln\left[\frac{1}{(X_{A1}^{M1})_{opx}(X_{Mg}^{M1})_{opx}}\right] = -\frac{\Delta H^0}{RT} + \frac{\Delta S^0}{R} - \frac{\Delta V}{RT}(P-1), \text{(calories, bars, °K)}$$

where the standard enthalpy of reaction (1 bar) is -7010 cal and the standard entropy of reaction is -3.89 cal deg^{-1}. Reaction volume ΔV can be read as a function of (X_{A1}^{M1}) in orthopyroxene in Table 1 of Wood (1974).

Recognizing the interdependency of the reactions listed in Table 2, for example that (35) = (33) − (23), Wood (1975) has shown that isopleths for Al in orthopyroxene in spinel peridotite (33) have small values of dT/dP, in disagreement with experimental work on that equilibrium (MacGregor 1974). The problems involved in growing homogeneous stable aluminous pyroxene grains sufficiently coarse for microprobe analysis were, however, noted by MacGregor. Obata (1976) extended Wood's calculations and plotted Al isopleths for equilibria (37) and (39) in addition to (33) and (35) (Figure 4). A small dT/dP for equilibrium (36), similar to that for (35), has recently been found experimentally by Herzberg & Chapman (1976). This lends support to the calculated isopleths of Wood and Obata, since close similarity in the reaction entropy and volume terms for equilibria (35) and (36) at high T and P can reasonably be expected.

Successful geobarometry based on the aluminum content of pyroxene can, therefore, only be anticipated in garnet- or plagioclase-bearing ultrabasic rocks (Table 2). As shown in Figure 4, this does provide coverage across the spinel lherzolite field, provided forsterite-absent ultramafics (pyroxenites) can always be found. This is because the isopleths for equilibrium (39) are a continuation of those for (33) across an inflection caused by reaction (25). Obata's (1976) calculated isopleths for plagioclase lherzolites indeed suggest that the possibilities for geo-barometry in high-level ultrabasic rocks, such as ophiolites and layered intrusives of the Stillwater type, are excellent. The flat Al-isopleths for equilibria (35) and (36) (Figure 4) can, on the other hand, be used for temperature estimates that are potentially equally as precise and reliable as those from the diopside-enstatite solvus (e.g. Herzberg & Chapman 1976).

The application of these methods to multicomponent rocks in nature presupposes that the effects of additional components, particularly Fe, Ca, and Cr, can be satisfactorily compensated for. Wood (1974) has shown experimentally that the reaction constants for equilibrium (33) in the $MgO-Al_2O_3-SiO_2$ system can also be used for complex experimental systems and natural ultrabasic rocks by assuming (a) a disordered three-site solution model for garnet, (b) charge-balanced substitution of $Al^{VI}Al^{IV}$ for MgSi in orthopyroxene, (c) all Al^{VI} located on the $M1$ site in orthopyroxene, and (d) Mg and Fe ordered over $M1$ and $M2$ in orthopyroxene according to bulk composition and temperature (Wood & Banno 1973, Figure 2; Stroh 1976, Figure 6). In addition, Al that is not a part of the component $MgAl_2SiO_6$, namely that belonging to molecules such as $NaAlSi_2O_6$ or $CaTiAl_2O_6$, must first be subtracted. For rocks containing roughly equal amounts of Fe and Mg, and hence not ultrabasic, an empirical correction to the simple relation, based on a nonideal interaction between Fe and Al in the $M1$ site of orthopyroxene, may be added if desired (Wood 1974).

Because of the squared and cubic terms in the relation between activities and molecular fractions, and the unequal partitioning of Ca, Fe^{2+}, Fe^{3+}, Cr, etc among spinel, garnet, pyroxene, and olivine, use of plots of "raw" isopleths for Al_2O_3 in pyroxene in the simple experimental systems for natural rocks, including ultrabasic rocks, can give rise to excessively large errors in estimating pressure (Wood 1974, 1975, Stroh 1976). Critical application of mineralogical methods of thermometry and barometry demands attention not only to the petrographic and geologic facts (Wilshire & Jackson 1975), but also to the reality of dealing with multicomponent natural equilibria.

In the case of spinel, calculation of the activity of the $MgAl_2O_4$-component on an ideal cation solution basis seems to overcorrect for the presence of Fe^{2+}, Fe^{3+}, and Cr (Frost 1976a). Use of Obata's (1976) data for equilibrium (35) and Herzberg & Chapman's (1976) data for (36) gives temperature estimates 100° to 300°C higher than those based on the diopside-enstatite solvus in the same rocks. One obvious interpretation is that different frozen equilibria are being measured, but similar excessively high temperatures using Obata's data are found for CrAl spinel-olivine-orthopyroxene assemblages in contact hornfelses that remained in the stability field of chlorite (hence $< 750°C$), which suggests that the problem lies in the model adopted for spinel. Positive excess energy terms due to mixing or ordering in the spinel solid solution would reduce the apparent overcorrection.

MINERALOGY, INCLUDING PROGRESSIVE METAMORPHIC CHANGES

This section reviews the mineralogy of metamorphic ultrabasic rocks, with emphasis on those properties that vary systematically with metamorphic grade.

Of the three rock-forming *serpentine* minerals, it remains correct to regard chrysotile and lizardite as polymorphic, but antigorite, on account of its alternating wave structure, is systematically poorer in $Mg(OH)_2$ than the other two (Wicks & Whittaker 1975). Recent detailed X-ray studies of low-grade serpentinites have generally found lizardite to be more abundant than chrysotile (Page 1967a, Aumento

1970, Dietrich & Peters 1971, Dungan 1974, Wicks & Whittaker 1976). Lizardite tends to replace olivine and pseudomorph pyroxene, whereas chrysotile tends to form slip and cross-fiber veins and also the framework for the typical mesh structure of serpentinites. Both minerals are replaced by antigorite during progressive metamorphism. With the possible exception of lizardite, there is little evidence to suggest that the usual contents of minor elements, such as Fe^{2+}, Fe^{3+}, Al, Cr, and Ni, exert significant influence on serpentinite stabilities. Nor are there any good grounds to suppose that shear stress promotes the stability of antigorite relative to other serpentines. Temperature and favorable kinetics seem to be the key factors. The fractionation of Fe, Mn, Ni, and Cr between antigorite and coexisting metamorphic ultrabasic minerals is given in Trommsdorff & Evans (1974, Table 2; note that the definition of K_D is inverted). Serpentine minerals are typically highly magnesian, e.g. in 40 microprobe analyses of antigorite, $Mg/(Mg + Fe + Ni + Mn)$ ranged from 0.93 to 0.98, averaging 0.956. These values reflect fractionation factors relative to olivine, brucite, talc, and chlorite, and the effects of oxidation during serpentinization; they do not reflect any intrinsic limits to the iron content of serpentine. The distinction between antigorite and chrysotile/lizardite should be made routinely in studies of serpentinite. Not only their stabilities are different, but also many of the important physical properties that they impart to the rock, such as strength, density, and seismic velocities (Coleman 1971b).

The silica-poor environment of metamorphic ultrabasic rocks permits the occurrence of *chlorite* over a wide range of physical conditions, from surface temperatures to those of its breakdown (Figure 1). In the company of two Mg-silicates, its composition is restricted by the phase rule, and it shows a gradual increase in Al^{IV} and Al^{VI} with increasing metamorphic grade: from a formula (minor elements omitted) of approximately $Mg_{11.25}Al_{1.0}Si_{6.5}Al_{1.5}O_{20}(OH)_{16}$ at lowest grades to $Mg_{9.6}Al_{2.4}Si_{5.6}Al_{2.4}O_{20}(OH)_{16}$ at its breakdown temperature (Frost 1975). Its composition is much more variable in monomineralic "blackwall" reaction zones (see below). Particularly in serpentinites, the crystallization of chlorite is obviously favored by microscopically localized regions of high Cr and Al content, as indicated by its spatial association with Cr-magnetite or chromite and local richness in Cr.

Despite a possible range in temperatures of crystallization from $\simeq 400°$ to over $1000°C$, there are virtually no systematic changes with metamorphic grade in the properties of *olivine* in ultramafic rocks. Its composition is obviously a function of the Fe/Mg ratio of the rock, f_{O_2}, and the nature and amounts of coexisting phases. Olivine that is more magnesian than Fo_{95} is not uncommon in rocks that were formerly serpentinized with the production of abundant magnetite. Unusual contents of minor elements (for a given % Fo), for example, very low NiO, or very high MnO (e.g. Frost 1975, p. 289, Vance & Dungan 1977) are a predictable consequence of the low modal abundance of olivine in certain parageneses and the known fractionation factors for these elements (Trommsdorff & Evans 1974, Table 2). However, the vast majority of metamorphic ultramafic olivines do not fall outside the compositional range of olivine in allofacial alpine peridotites, ophiolites, and layered intrusions.

Whereas in volcanic forsteritic olivines there is a slight preference on the part of the larger Fe^{2+} ion for the smaller M1 octahedral site (Finger & Virgo 1971), the octahedral sites are disordered in olivine from ultrabasic rocks in the high amphibolite facies (Brown & Prewitt 1973, Wenk & Raymond 1973) and Fe^{2+} very slightly favors M2 in olivine in antigorite-bearing parageneses (S. Ghose, unpublished). Thus, the petrologic usefulness of site refinements of ultrabasic olivines seems limited.

Talc has the highest Mg/Fe ratio of all the ultrabasic minerals. In well-polished sections, Al_2O_3 is barely detectable by microprobe analysis, even when coexisting with chlorite, as it frequently does.

Both *anthophyllite* and its dimorph *magnesiocummingtonite* (with the primitive cell $P2_1/m$ because of the high Mg-content) occur widely, although their temperature range of stability in the presence of olivine is rather limited (Figure 3). Anthophyllite has a wider distribution in veins and monomineralic contact rocks, where it is not accompanied by olivine. The relationship between anthophyllite and magnesiocummingtonite is probably analogous to that between hypersthene and pigeonite. Magnesiocummingtonite typically contains 0.5% CaO, or 4% of the tremolite molecule, and it invariably occurs with tremolite, often as a homoaxial overgrowth or replacement. Anthophyllite in ultramafic parageneses is generally poorer in CaO and occurs with or without coexisting tremolite. When it has inverted from cummingtonite, thin lamellae of tremolite on $(\bar{1}01)$ may be detected by single crystal X-ray study (Evans et al 1974). Unlike Ca-poor amphiboles in metabasic and metasedimentary rocks (Robinson, Ross & Jaffe 1971), there is minimal solution towards the gedrite composition (Rice, Evans & Trommsdorff 1974), except in contact metasomatic masses.

The recent discovery of Mg-silicates with triple-chain and mixed triple- and double-chain structures (intermediate structurally and chemically between anthophyllite and talc) in association with "blackwall" anthophyllite by Veblen & Burnham (1975) and Hutchison, Irusteta & Whittaker (1975) shows that the total list of structural possibilities in natural rock-forming silicates is still incomplete. It may well be that these structures have already been synthesized by experimental petrologists and have gone unrecognized in powder diffraction patterns. Further and more extensive investigation of natural anthophyllite and anthophyllite-like phases is obviously called for. Structures intermediate between enstatite and anthophyllite may also be discovered.

The *calcic amphibole* in chlorite-olivine-Mg-silicate rocks is a low Al-tremolite, until conditions fairly high in the amphibolite or hornblende hornfels facies are reached (Figure 1). In parageneses with olivine and another Mg-silicate, there is a systematic change across the $Ca_2Mg_5Si_8O_{22}(OH)_2–Mg_7Si_8O_{22}(OH)$ join: from roughly 96% tremolite in antigorite-bearing rocks to less than 80% tremolite in enstatite parageneses (Trommsdorff & Evans 1972, Rice, Evans & Trommsdorff 1974, Frost 1975, Misch & Rice 1975). In the high temperature facies, magnesiohornblendes take the place of tremolite through complex, though linear (Frost 1975, 1976a; Misch & Rice 1975), substitutions of the tschermakite and pargasite endmembers.

Orthopyroxenes in the amphibolite or hornblende hornfels facies are very close to binary $FeSiO_3–MgSiO_3$ solutions, until the higher temperature parts of these facies are reached, where there is a gradual increase in both CaO and Al_2O_3 in suitably buffering assemblages (e.g. tremolite + olivine + enstatite + chlorite). Strongly aluminous orthopyroxenes are, of course, well known in alpine spinel peridotites of the granulite facies. Compositions close to diopside are typical of stable *clinopyroxenes* in serpentinites (Peters 1968, Trommsdorff & Evans 1972, Frost 1975). The structure can obviously accommodate little or no Al or Ti at low temperatures, and Mg–Fe fractionation is such as to make clinopyroxene highly magnesian (see previous section). Clinopyroxenes in the granulite facies contain considerably more Fe, Al, Ti, Na, and less Ca.

Spinel compositions are a function of metamorphic grade in the sense that increased temperatures expand the possible range: from Cr-magnetite alone in low-grade serpentinites, expanding to include ferrit-chromite in antigorite serpentinites, chromite in talc-olivine rocks, and then swiftly through Al-chromite and Cr–Al spinel (picotite) to MgAl-spinel close to the granulite and pyroxene hornfels facies transition (Evans & Frost 1975). There is little evidence in nature to indicate that the sliding equilibria responsible for this trend, principally an H_2O-sensitive equilibrium between MgAl-spinel component, chlorite, and two Mg-silicates, are interrupted by solvi between magnetite and chromite or chromite and MgAl-spinel (cf Cremer 1969). In most serpentinized peridotites, a lack of attainment of equilibrium characterizes chrome-spinel, which is typically zoned from chromite outwards through ferrit-chromite to magnetite (e.g. Bliss & MacLean 1975). Since Fe/Mg fractionation between spinel and the silicates is a strong function of the R^{3+} cation content (29), there is a marked systematic decrease in Fe^{2+}/Mg ratio of spinel accompanying the trend from magnetite through chromite to MgAl-spinel during progressive metamorphism. In peridotites of the granulite facies, the composition of spinel (Medaris 1975), in particular its Cr/Al ratio, will be a function of many factors, including (*a*) bulk composition (Cr/Al-ratio), which will have been determined by an earlier depletion or cumulate process, (*b*) the modal amounts of spinel and pyroxene, and (*c*) *P-T* sensitive equilibria such as (35) and (36).

A fluorine-free, titanium-rich member of the humite group, *titanoclinohumite*, approximately $8M_2SiO_4 \cdot MTi(OH)_2O_2$, is an occasional accessory in metamorphic ultramafics, including antigorite schists (de Quervain 1938, Bearth 1967), metaperidotites in the amphibolite and granulite facies (Heinrich 1963, Möckel 1969) and kimberlite xenoliths (McGetchin, Silver & Chodos 1970). It is clear from the equilibrium

$$\text{titanoclinohumite} = \text{olivine} + \text{geikielite} + H_2O \tag{41}$$

that titanoclinohumite has potential as an indicator of high H_2O pressure as well as a possible host for H_2O in the mantle.

CONTACT METAMORPHISM

Aureoles up to 2 km or more have been generated by intrusive igneous rocks in ultramafic country rock, in most cases formerly strongly serpentinized. Recent studies (Matthes 1971, Trommsdorff & Evans 1972, Springer 1974, Frost 1975, Arai

1975, Irving & Ashley 1976) have established a simple pattern of zonation in the mineral assemblages. In an ideal example, zones defined by the following sequence of critical parageneses are encountered as the intrusive contact is approached:

1. olivine + antigorite + diopside + chlorite,
2. olivine + antigorite + tremolite + chlorite,
3. olivine + talc + tremolite + chlorite,
4. olivine + anthophyllite + tremolite + chlorite,
5. olivine + enstatite + tremolite + chlorite,
6. olivine + enstatite + hornblende + spinel,
7. olivine + enstatite + diopside or anorthite + hornblende + spinel.

Assemblage 5 is usually attained close to contacts with granodioritic and tonalitic intrusives, assemblage 6 close to quartz diorite and gabbro, and assemblage 7 in favorable locations, such as roof pendants and inclusions, in association with gabbro (Frost 1976a). Associated mafic rocks, which in many cases were once dikes, are hornblende hornfels up to and including assemblage 6. Two-pyroxene assemblages develop in mafic rocks (i.e. pyroxene hornfels facies) in association with assemblages 6 and 7, that is, under similar but perhaps slightly cooler conditions than two-pyroxenes in ultramafic rocks.

Reactions tend to be incomplete in the outer zones; for example, chrysotile and lizardite, particularly Al-lizardite (Frost 1975, p. 296), may survive throughout the antigorite zone, and there may be little or no evidence of the outer diopside-bearing zone. The anthophyllite zone tends to be narrower than the others, and locally it is missing. Frequently, anthophyllite overlaps onto the enstatite zone 5. Cumming-tonite, associated with tremolite, seems to be uncommon. Zones 3 and 5 are typically quite broad, the latter as much as $\frac{1}{2}$ to 1 km. Carbonates, if reported at all, tend to occur in late veins or shear zones.

The above observations are broadly consistent with the experimentally determined phase relations discussed earlier. There being a lack of evidence to the contrary (e.g. rarity of carbonates), most authors have made the reasonable assumption that the pore fluid present and escaping at the time of contact metamorphism was rich in H_2O and similar in pressure to the lithostatic load. It follows, then, at a solid and fluid pressure of 1 kbar, temperatures of $\simeq 600°C$, 675°C, and $\simeq 750°C$ (Figure 3 and text) were attained in those aureoles showing assemblages 5, 6 and 7, respectively. At 2 kbar the corresponding temperatures are $\simeq 625°C$, 710°C, and $\simeq 775°C$.

Serpentine minerals carry 12 to 13% H_2O by weight, so that, in a simple episode of contact metamorphism, some visible effects or signs (e.g. channelways; see Rice 1977) of the enormous water loss might be expected. Veining does, in fact, increase tremendously in zone 3 of the Bergell aureole (Trommsdorff & Evans 1972). The veins take the form of irregular, diffuse planes along which the talc-forsterite hornfels has recrystallized to a very coarse, elongate-olivine + talc rock. Obviously, such veins are less likely to have developed in other aureoles where serpentinization of the protolith was less complete in the first place.

Although pressures of less than $3\frac{1}{2}$ kbars should allow the pair forsterite + cordierite to develop in the system $MgO–Al_2O_3–SiO_2–H_2O$ from chlorite breakdown, the higher-pressure assemblage enstatite + forsterite + spinel is invariably found in contact metamorphism. This is undoubtedly due to the strong preference of Fe^{3+}, Fe^{2+}, and Cr for spinel, and the low ΔV for reaction (17). Frost (1975) has made estimates of the decrease in pressure of reaction (17) as a function of Mg and Al in the spinel, based on measured partition coefficients. Significantly, cordierite + forsterite (along with enstatite, spinel, and chlorite, which all together constitute an invariant assemblage in the pure system) has, so far, only been reported from a very high-Mg local assemblage (olivine Fo_{96}) in chromitite (Arai 1975). It is normal for silicates in close association with chromitite seams to be enriched in Mg (Irvine 1965, Jackson 1969).

REGIONAL METAMORPHISM

In contrast to other lithologies, there is a dearth of systematic information on the regional, dynamothermal metamorphism of ultrabasic rocks, a fact that can with little doubt be attributed to their stratigraphically and structurally limited distribution in most crystalline terranes. A further drawback to attempts at petrological interpretation has been the suspicion, in many cases quite justified, that the metamorphic imprint that the rocks now bear was acquired in some different milieu. In addition, at moderate and high grades of metamorphism, where ultrabasic rocks are rich in olivine and pyroxene, some doubt has existed (and still apparently does, e.g. Lensch 1968, cf Cawthorn 1975) as to the igneous or metamorphic origin of the principal minerals, texture, and structure.

A summary of the Alpine metamorphism of ultrabasic rocks in the Pennine units of the Central Alps (Trommsdorff & Evans 1974) provides as complete an account as any currently available. Some of the ultrabasics were originally basal portions of Mesozoic ophiolites, and most were probably considerably serpentinized prior to the regional metamorphism (e.g. Peters 1963). Their present scattered distribution results from the complex tectonic events of the Alpine orogeny, and most of their present mineralogy, textures and structures are due to polyphase metamorphic events synchronous with and post-dating the main phase of deformation. The Lepontine phase (ca 40 to 20 m.y. BP) of Barrovian-style regional metamorphism is responsible for the principal recrystallization in the central part of the region. Isograds mapped in pelitic, mafic, and carbonate rocks postdate the principal folding and nappe-formation events and have served as a framework for the study of assemblages in associated ultrabasic rocks (Figure 6). It is possible to construct isograds for ultrabasic rocks in only a few scattered locations (Trommsdorff & Evans 1974, Figure 9), and these appear to be concordant with the established regional isograds. For most of the ultrabasic isograds, there is textural evidence for the reactions having been prograde (Trommsdorff & Evans 1974, Plate I).

The sequence of mineral assemblages in the Central Alps (Figure 6) is apparently no different from that found in contact metamorphism. There do exist differences, however, in the relative width of zones, drawn on the basis of the first appearance of the higher-temperature paragenesis, and in the degree of overlap of the critical parageneses (Trommsdorff & Evans 1974, Figures 4 to 8). Specifically, the entry of olivine + talc is soon followed by Ca-poor amphibole + olivine, and this in turn very soon by enstatite + olivine. Olivine + talc rocks survive well into both higher-grade zones, and, similarly, rocks carrying Ca-poor amphibole may be found amongst enstatite-olivine-zone rocks. These overlaps are undoubtedly partly due to poly-metamorphism, but some are due to locally, and perhaps regionally, varying fluid phase compositions.

Some of the lower-grade reactions involving serpentine minerals are more readily studied in regional than in contact metamorphism. For example, in the Rhetic Alps of eastern Switzerland it has been possible to correlate the entry of antigorite, in veins, with the disappearance of prehnite in mafic rocks. Further to the south, a sudden increase in rock-forming antigorite, together with brucite, occurs where pumpellyite disappears in the mafic rocks, and where the greenschist facies, *sensu stricto*, begins (Dietrich & Peters 1971). Olivine, often porphyroblastic, first joins antigorite at a somewhat higher grade (further south), a small distance above the biotite isograd (Trommsdorff & Evans 1974, Figure 3). Here, and throughout the Malenco Serpentinite, excluding the contacts with the Bergell intrusive, the calcic phase accompanying olivine and antigorite is a relatively pure diopside.

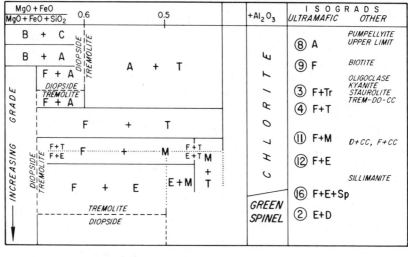

Figure 6 Composite diagram correlating ultrabasic mineral assemblages and isograds in regionally metamorphosed rocks of the Central Alps with isograds for rocks of other bulk compositions (after Trommsdorff & Evans 1974). A, antigorite; B, brucite; C, chrysotile; T, talc; F, olivine; M, magnesiocummingtonite or anthophyllite; E, enstatite; Sp, green spinel; D, diopside; Tr, tremolite. Numbers correspond to reactions in the text.

The frequent occurrence of magnesite in anthophyllite- and enstatite-bearing rocks in the medium and upper zones of the amphibolite facies strongly suggests the regional occurrence of a fluid phase richer in CO_2 than H_2O. The change from H_2O-rich fluids associated with the serpentinites (Figure 5) of the outer metamorphic zones to CO_2-rich fluids in the inner zones may be the explanation for the relatively narrow olivine + talc zone (compare Figures 3 and 6). Calculations along the lines of Greenwood (1975), based on estimates of initial carbonate content in the serpentinites, make it difficult to entertain a concept of buffering from low to high X_{CO_2} by the isobarically univariant assemblage olivine + talc + magnesite, as an explanation for the change in fluid composition with metamorphic grade. A deep-seated source of CO_2 (e.g. Hoefs & Touret 1975) seems more likely than one related to decarbonation reactions in associated carbonate metasediments.

The pair enstatite + talc, in apparent textural and cation exchange equilibrium, was first found in regionally metamorphosed rocks of the Central Alps (Evans & Trommsdorff 1974b). It has since been found also in contact metamorphism (Frost 1975). The rock may be cut by late veins of anthophyllite, indicating a change in P, T, or X conditions in favor of the right-hand side of reaction (14). As discussed in a previous section, the usefulness of this reaction is undermined by its sensitivity to the Mg/Fe ratio. Figure 6 indicates that, in rocks with very high Mg/Fe, the olivine + talc zone passes directly into the olivine + enstatite zone [by reaction (13)], although this needs more detailed study.

A passage into two-pyroxene rocks of the granulite facies is only rarely found in the Lepontine Complex of the Central Alps. The allofacial alpine peridotite is, of course, typically a granulite facies rock. These rocks are characterized by considerable deformation, isoclinal folding, often conspicuous mineral layering (whose exact origin is still debatable), and a range of textures. The diagnostic assemblage is olivine + orthopyroxene + clinopyroxene + spinel. In the gradation from dunite, through harzburgite, to lherzolite and pyroxenite, the spinels range from translucent red, high-Cr to translucent green, high-Al types, and the pyroxenes from low to high-Al (tschermakite substitution). A pargasitic magnesiohornblende is part of the stable paragenesis in some cases, particularly in isofacial peridotites in granulite facies complexes. Olivine-free rocks may contain garnet, ranging from 40 to 70 mol % pyrope, or, alternatively, calcic plagioclase.

Although most of the estimates in the literature of the pressure and temperature of equilibration of alpine spinel peridotites may be questioned, it is probably true that the range 1200–700°C and 5 to 20 kbar, which is broadly consistent with the phase diagram (Figure 4), encompasses most. Those containing augen of pyroxene with high Al_2O_3 or abundant exsolution lamellae of pyroxene, garnet, and spinel (Green 1964, Kornprobst 1969, Dickey 1970, Medaris 1972), or clino-pyroxene with very low CaO-contents (Kornprobst & Conquéré 1972) provide evidence of conditions close to the solidus in the upper mantle. Most spinel peridotites of the alpine-type, isofacial or allofacial, underwent their last major phase of recrystallization and equilibration in the 700°–900°C range. Retrograde reactions producing talc, anthophyllite, antigorite, and chlorite, which are analogous to the prograde reactions in the Central Alps, but indicative of gradients in H_2O

activity rather than temperature, may be found in the marginal parts of allofacial alpine peridotites (e.g. Wolfe 1966, Medaris 1975).

Unlike the considerably more common spinel peridotites of the alpine association, the *alpine garnet peridotites* are invariably enclosed in gneisses of the amphibolite or granulite facies. Examples are mostly from Europe (O'Hara 1967b, Rost & Grigel 1969, Forestier 1971). They are typically accompanied by larger quantities of non-garnetiferous peridotite and eclogite. Chloritic or kelyphitic nodules (fibrous inter-growths of clino- and orthopyroxenes and spinel or amphibole and chlorite) after original garnet porphyroblasts may occur in the nongarnetiferous peridotite. Even where the garnet survives, there is frequently evidence of a retrogressive meta-morphism to the *P-T* conditions of the enclosing gneisses, e.g., kelyphitic rims around garnet, recrystallization of olivine and less calcic and aluminous ortho-pyroxene, crystallization of tremolite, chlorite, and spinel, etc (O'Hara & Mercy 1963, Carswell 1968b, Möckel 1969). Nevertheless, the isofacial metamorphism of some garnet peridotites is a possibility not ruled out by workers in Norway (Brynhi et al 1970, Griffin & Heier 1973) or Central Europe (Rost 1971), particularly where country rocks are kyanite-bearing rocks of the granulite facies.

Textures of alpine garnet peridotites are porphyroclastic, mylonitic, and grano-blastic, and generally not lacking in evidence for a polymetamorphic history. Evidence for solid-state recrystallization includes spherical olivine inclusions inside the garnet porphyroblasts, which in some examples possess concave outward boundaries and show a tendency to form a holly-leaf texture, and idioblastic enstatite partially enclosed by olivine, a texture common in amphibolite facies metaperidotites. In addition, armored relics of green spinel enclosed in garnet (e.g., Forestier 1971, Figure 2; Rost 1971) are clear evidence of a passage from the spinel lherzolite to the garnet lherzolite field across reaction (24), corresponding to either an increase in pressure or a decline in temperature (Figure 4). The latter is clearly responsible for exsolution lamellae of pyroxene, spinel, and garnet observed in the pyroxenes (Carswell 1968a, 1973). Thus, although many are compositionally close to undepleted mantle, the *alpine* garnet peridotites are clearly poor examples of petrographically primitive mantle.

Bulk compositions show a range in FeO/MgO ratio, in Al_2O_3, and in CaO almost as great as in alpine spinel peridotites. The garnet has a distinctive composition: 60 to 75% pyrope, 10 to 15% grossularite, and less than 1% spessartite. Compared to kimberlite garnet peridotites, Cr_2O_3 in pyroxenes and garnet is lower, and the pyroxenes mostly show lower mutual solubility (O'Hara & Mercy 1963, Fiala 1966, O'Hara 1967b, Boyd 1970, Nixon & Boyd 1973, Carswell 1974). Magnesiohorn-blende of edenitic or pargasitic nature (high in Na_2O and Al_2O_3) may belong to the paragenesis in some, but has been excluded in others on textural grounds (O'Hara & Mercy 1963).

A minimum pressure of 15 ± 3 kbar for garnet peridotite (lherzolite) in the system $CaO–MgO–Al_2O_3–SiO_2$ is indicated by Figure 4. Approximately an additional 5 kbar are required to stabilize Ca-free garnet peridotite (MacGregor 1965, Obata 1976), although clinopyroxene-free rocks (harzburgites) are comparatively rare in the alpine garnet peridotite association. Since Cr fractionates strongly into spinel

(Richter 1971) rather than garnet, Cr will dramatically increase the stability field of spinel peridotite (MacGregor 1970) but will not greatly decrease that of garnet peridotite—there will instead be a substantial divariant spinel+garnet peridotite field. The quantitative effect of this and other elements such as Fe^{2+}, Fe^{3+}, Ca, etc, can be evaluated by determining their effect on the equilibrium constant for the equilibrium (23)

$$2\mu_{Mg_2Si_2O_6} + \mu_{MgAl_2O_4} = \mu_{Mg_3Al_2Si_3O_{12}} + \mu_{Mg_2SiO_4},$$

which must hold for all systems. For peridotite minerals at high P and T, we may justifiably use ideal mixing models (Wood 1975), writing

$$K = \frac{(X_{Mg}^{ol})^2(X_{Mg}^{gt})^3(X_{Al}^{gt})^2}{(X_{Mg}^{M1,opx})^2(X_{Mg}^{M2,opx})^2(X_{Mg}^{sp})(X_{Al}^{sp})^2},$$

and obtain the pressure dependence of K from $\Delta \ln K = -(\Delta V/RT)\Delta P$, where $\Delta V = -8$ cm^3. Knowing the various fractionation factors (namely $X_{Mg}^{ol} \simeq X_{Mg}^{opx} > X_{Mg}^{sp} >$ or $< X_{Mg}^{gt}$ [the latter depends on X_{Al}^{sp}], and $X_{Al}^{gt} > X_{Al}^{sp}$), we can conclude that no element normally present besides Ca can make ΔP large and negative, and the effect of Ca is already known (MacGregor 1965, 1970). It is therefore impossible to escape the conclusion that unusual depths, corresponding to 13 kbar or more, were involved in the formation of alpine garnet peridotites. Since crustal thicknesses in zones of plate convergence may perhaps reach as much as 60 km, a mantle origin for the garnet, although likely, is not necessarily required.

Successful geobarometry of garnet peridotite unfortunately requires good geothermometry, and an uncertainty of less than $\pm 100°C$, using the diopside-enstatite solvus or the Fe/Mg partitioning between clinopyroxene and garnet, cannot yet be realistically claimed. The range obtained for alpine garnet peridotites (from analyses of bulk mineral separates) is 700° to 1100°C, values which partially overlap with those for kimberlite garnet peridotites (Carswell 1974). At 827°C, Wood's (1974) method using Al in orthopyroxene gives pressures for alpine garnet peridotites ranging from 25 to 36 kbar; for 727°C they should be reduced by 6 kbar. These figures are possibly high, although they would fit a shield or a subduction geotherm. A serpentinite protolith for garnet peridotite, and hence a crustal origin, is suggested by the discovery of remnants of metarodingite in a small peridotite body in the Central Alps that includes fresh garnet lherzolite (Evans & Trommsdorff 1975). Clearly, we have not yet solved the problem of the origin of alpine garnet peridotites.

OCEANIC METAMORPHISM

Ultrabasic rocks have been dredged from ocean trenches, fracture zones, and median ridges, and comprise a collection of tectonite and layered varieties more or less covering the compositional range found in alpine peridotites, ophiolites, and layered intrusives (Christensen & Salisbury 1975). As yet, there is little to indicate that pervasive ocean crust metamorphism of the ultrabasic rocks involves anything more than cataclastic metamorphism and serpentinization, and related processes such as rodingitization. The metamorphic picture contrasts strongly with that for the mafic rocks, the pillow lavas, diabases, and gabbros. This is consistent, however, with a model of median-ridge hydrothermal metamorphism caused by convective circulation of sea water in the upper few kilometers of crust, perhaps not reaching the base of layer 3, and with the distribution of metamorphism in ophiolites (e.g. Gass & Smewing 1973, Spooner & Fyfe 1973).

Oceanic serpentinization is mostly of the low-grade, lizardite-chrysotile type. Antigorite, confirmed by X-ray diffraction, has been found in a few dredge hauls (Bowin, Nalwalk & Hersey 1966, Miyashiro, Shido & Ewing 1969, Aumento 1970, Aumento & Loubat 1971), and its presence may be used to indicate that a certain threshold temperature, probably between 200° and 300°C, was exceeded. There are only rare indications of an amphibolite facies metamorphism affecting oceanic ultrabasic rocks; for example, Bonatti, Honnorez & Ferrara (1970) report anthophyllite and cummingtonite in dunite. Tremolite and talc have also been observed in serpentinized peridotite, but these minerals alone do not provide useful temperature information.

It must be remembered that most dredged samples have been obtained from regions of high bathymetric relief, and so may be biased in favor of rocks characteristic of zones of dislocation rather than other oceanic tectonic environments. This fact might be responsible for the high degree of cataclasis and serpentinization exhibited by most, but by no means all (Fisher & Engel 1969) samples. The deeper cores which the International Phase of Ocean Drilling (IPOD) plans to obtain will hopefully counteract this bias and provide invaluable new data on the metamorphism of layer 3.

Despite suggestions to the contrary (Aumento & Loubat 1971, Miyashiro, Shido & Ewing 1969), the weight of evidence strongly favors the formation of most oceanic serpentinite at low temperatures, less than 300°C and possibly even less than 200°C. As mentioned earlier, antigorite is quite uncommon, and oxygen-isotope serpentine/magnetite geothermometry of both continental and oceanic lizardite/chrysotile serpentinites consistently gives temperatures of less than 200°C. Independent of this, Wenner & Taylor (1973), having established heated sea water as the dominant aqueous fluid responsible for oceanic serpentinization, used estimated water-serpentine $^{18}O/^{16}O$ fractionation factors to suggest temperatures of less than 250°C for oceanic lizardite/chrysotile samples. Temperatures of serpentinization of 400° to 500°C are not the only possible explanation for the absence of detectable brucite. Other possibilities include unfavorable pyroxene/olivine ratios in the protolith (Hostetler et al 1966), addition of SiO_2 or CO_2 from aqueous solution, oxidation of iron-silicate component to magnetite (which lowers the effective rock $MO:SiO_2$ ratio), and weathering processes.

Serpentinization of oceanic peridotite and ophiolitic peridotite is characterized by distinctly different D/H and $^{18}O/^{16}O$ ratios (Wenner & Taylor 1973). This suggests that caution should be exercised before low temperature events in ophiolites, particularly in the ultramafic portions, are correlated with the oceanic stage in their history. The involvement of sea water in oceanic serpentinization, as indicated by the light isotope data, is supported by other evidence, including a comparison of

B abundances between oceanic serpentinite and fresh peridotite (Thompson & Melson 1970), high (> 0.708) values for $^{87}Sr/^{86}Sr$ in oceanic serpentinite (Subbarao & Hedge 1973, Hart 1972), the association of organic matter with a stage in the serpentinization process (Vdovykin & Dmitriyev 1968), and the occurrence of serpentinite with veins containing calcareous ooze and foraminiferal tests (Aumento & Loubat 1971, Bonatti et al 1974).

The spectrum of ultrabasic samples brought up from the ocean floor, and particularly the rarity of antigorite serpentinite, certainly provides scant support for the hypothesis (Hess 1962) that layer 3 is composed principally of serpentinized peridotite, generated by hydration of mantle peridotite along ridge crests at sites above the 500°C isotherm. Christensen & Salisbury (1975) review in depth other evidence on this question.

Rocks formed by metasomatic processes in association with continental ultrabasics (see next section) also have their counterparts in the ocean basins. Rodingites and "rodingitized" peridotite have been described by Ploshko & Bogdanov (1968), Melson & Thompson (1970), Aumento & Loubat (1971), and Honnorez & Kirst (1975). Possible blackwall alteration rocks have been described by Aumento & Loubat (1971, p. 639) and Bonatti, Honnorez & Ferrara (1970). Shand (1949, p. 91) described a rock which is a mass of tremolite asbestos.

Serpentinite breccias with carbonate veins have been dredged (Quon & Ehlers 1963, Aumento & Loubat 1971, Bonatti et al 1974) and cored (Rabinowitz & Melson 1976). They suggest that the ophicalcite rocks of the Tethyan ophiolites could have a sea floor origin, possibly related to talus formation, transform faulting, or other movements (but cf Folk & McBride 1976).

METASOMATIC ROCKS

Frequently associated with metamorphic ultrabasic rocks is a suite of small bodies of rock of distinctive appearance, mineralogy, and bulk chemistry, whose formation has involved significant mass transfer. They are conveniently subdivided into those formed principally by diffusion metasomatism (stationary fluid) and those formed principally by infiltration metasomatism (moving fluid) (Korzhinskii 1970, Hofmann 1972). The former result from exchange of material down steep gradients in chemical potential caused by the juxtaposition of high Mg, low Si, Fe, K, ultramafic rock, and rock of highly contrasted composition (or a vein-fluid derived from such a rock). The products of infiltration metasomatism associated with ultrabasic rocks are more varied, and include the calc-silicate-rich rodingites, and CO_2, SiO_2, or sulfide-enriched ultramafics.

Diffusion Reaction Zones

The contrast in bulk chemistry, and hence stable mineral assemblage, between ultramafic rock and dissimilar rock (siliceous country rocks, mafic dikes, etc), brought together tectonically or intrusively, generally results at the contact in the development by diffusion of planar zones of nearly monomineralic rocks, from 1 mm to several m in thickness. Such concentrically zoned inclusions have been extensively documented since the first clear descriptions of the phenomenon by Read (1934) and Phillips & Hess (1936). There are minor variations in the typical sequence of minerals, dependent on metamorphic grade. Most commonly found sequences are: in greenschist facies, schist-chlorite-actinolite-talc-serpentine (e.g. Chidester 1962, Jahns 1967); in low amphibolite facies, schist-biotite-chlorite-actinolite-talc-antigorite (e.g. Curtis & Brown 1969); in high amphibolite facies, gneiss-biotite-chlorite-actinolite-anthophyllite-enstatite-olivine (e.g. Carswell, Curtis & Kanaris-Sotiriou 1974). Individual zones may be absent in some cases, for reasons including the chemistry of the country rock (for example, no biotite zone against amphibolite), subsequent deformation, and exhaustion of one or other end of the diffusion couple.

Ultramafic monomineralic reaction zones supply some of the finest examples of the results of diffusion processes in metamorphism. Thompson (1959) used the concept of local equilibrium, which carries with it the implication that reaction rates are much faster than diffusion rates, to explain how a concentration gradient can lead to a number of small, sharp discontinuities in bulk composition, including local concentration maxima and minima. In essence, given independent monotonic gradients in the activities of the diffusing components in a diffusion couple, controlled outside the immediate zone of interest (K-components, Thompson 1970), solution equilibrium requirements between two solid phases made of those components can, at fixed P and T, only be satisfied along an infinitesimal plane perpendicular to the direction of diffusive flux. The number, sequence, and purity (one-phase, two-phase, etc) of reaction zones is predictable in cases where the diffusing components can be identified (Brady 1977). Brady concludes that metasomatic zones at ultramafic contacts are produced by diffusion-imposed chemical potential gradients of several components, notably silica and magnesia, but including CaO, FeO, and K_2O, the number of diffusing and K components varying from zone to zone. Opinion differs as to the location of the original discontinuity (Chidester 1962, Jahns 1967; cf Carswell, Curtis & Kanaris-Sotiriou 1974; Frost 1975), based on petrographic inert markers or a discontinuity in an otherwise constant element ratio such as Cr/Al (Brady 1977). Identification of the original discontinuity is a prerequisite for materials-balance calculations, and, apparently, for determining the possible role of Al as a diffusing component. Qualitatively at least, it is clear that Si, Ca, and Fe are added to the ultramafic, and Mg is lost. Brady (1977) suggests that H_2O is a component with such high "diffusion potential" that its chemical potential ought to be constant across the diffusion zones; if so, then sequences such as serpentine-talc-anthophyllite (Greenwood 1963, Figures 8, 10) and anthophyllite-enstatite-olivine (above) would virtually necessitate gradients in T. Careful study of such simple examples of metasomatic zones in metamorphism, because of the geometrical and compositional constraints it would supply, has considerable potential for a more quantitative understanding of diffusion in metamorphism, including knowledge of the constitution of the intergranular region in metamorphic rocks.

It is a mistake tacitly to assume that the monomineralic contact zones necessarily formed at the most extreme metamorphic conditions reached, or that they formed isothermally and isobarically. Changing T, P, and P_{H_2O} can lead to a change in the number and identity of possible stable minerals, and, by the same token, it can lead to the elimination of some. For example, Frost (1975) and Irving & Ashley (1976) have described the thermal breakdown of chloritic blackwall to olivine + orthopyroxene + spinel + hornblende rock. Granulite facies metamorphism of chlorite + actinolite or similar blackwall should be considered as a possible origin for the high Mg–Al rocks necessary to host sapphirine. Significantly, sapphirine-bearing rocks typically occur in thin zones between ultrabasic rock and rocks of other compositions such as amphibolite, quartzofeldspathic gneiss, or amidst rocks unusually rich in phlogopite, orthopyroxene, etc (e.g. Prider 1945, Segnit 1957, Nalivkina 1961, Hudson & Wilson 1966, Monchoux 1972).

Rodingites

Rodingites are whitish, dense, mostly fine-grained rocks rich in calcium silicates,[1] such as hydrogrossularite, diopside, zoisite, amphibole, vesuvianite, xonotlite, prehnite, etc, that typically occur as dike-like bodies enclosed in serpentinite. They were formerly dikes of diabase, gabbro, diorite, keratophyre, or granite, or tectonic inclusions of metasediment, with, subsequently, Ca added and alkalis removed (e.g. Bilgrami & Howie 1960, Coleman 1966, 1967, Dal Piaz 1967, Vuagnat 1967). Mass transfer is believed to be due to circulation during active serpentinization of an aqueous fluid of high pH, and high Ca^{+2} and OH^{-1} contents, and undersaturation in this fluid of minerals such as olivine, orthopyroxene, plagioclase and alkali feldspar, in immediately adjacent country rock (Barnes & O'Neil 1969). Relict chilled margins, porphyritic or diabase texture, or an unaltered dike center provide valuable clues as to their genesis. A complementary loss of Ca by peridotite as a result of serpentinization is fairly well established (Hess & Otalora 1964, Page 1967b, Coleman & Keith 1971).

The temperatures and pressures at which rodingitization takes place are probably both very low, although there have been few critical studies using all available geologic and phase equilibrium evidence. The significance of rodingite in metamorphism hinges on whether its formation is exclusively related to serpentinization, possibly even only of the lizardite-chrysotile type. If so, the occurrence of metarodingite in medium- and high-grade metamorphic ultrabasic rocks tells us immediately that the terrane was formerly in an environment where serpentinization was possible. In contact metamorphism, there need be little ambiguity as to the origin of "calc-silicate" hornfels associated with metaserpentinite (Frost 1975, Irving & Ashley 1976). In regional metamorphism, geologic evidence such as exclusive association of "metarodingite" with ultramafic bodies, association with amphibolite or eclogite (mafic rocks that escaped rodingitization), distinctive bulk chemistry, and association with oxidized (Fo 95%, En 95%) and low CaO (e.g. <0.05%) ultramafic rock (Trommsdorff & Evans 1969) are useful additional petrogenetic clues. Possibly, many metarodingites in high-grade metamorphic terranes have gone un-

---

[1] The list is still expanding, with vuagnatite for example (Sarp, Bertrand & McNear 1976).

noticed (perhaps Sørensen 1967, p. 207; Mori & Banno 1973; Forbes & Swainbank 1974). A detailed study of their progressive metamorphism would be useful; Frost (1975), for example, recognizes five diagnostic assemblages in metarodingite that can be related to diagnostic assemblages in enclosing metaserpentinite.

CONCLUSIONS

It is hard to single out one aspect of this subject more deserving of immediate study than others. In the area of phase equilibrium, data are needed on equilibria (2) and (22), on anthophyllite and cummingtonite in the system $CaO-MgO-FeO-SiO_2-H_2O$, and on the relative stabilities of lizardite, chrysotile, and antigorite. The possibilities of geothermometry and geobarometry in ultrabasic rocks would be greatly advanced by accurate calibrations of the Mg-Tschermak's equilibria (35) and (37) and of Fe/Mg partitioning between garnet or spinel and olivine, ortho- and clinopyroxene. In order to handle complex natural systems under these conditions, we urgently need data on activity-composition relations, particularly in the FeMg-olivines, $(Mg, Fe)(Al, Cr, Fe^{3+})_2O_4$-spinels, $(Ca, Mg, Fe)_3Al_2Si_3O_{12}$-garnets, and tremolite-pargasite solid solutions. Progress may come soon in the study of mass transfer processes, using direct experimental determination of phenomenological coefficients and fluid phase composition at P and T in buffering assemblages. Finally, the combined use of transmission electron microscopy and more sophisticated experimental techniques promises to lead to an understanding of deformation mechanisms in rocks that constitute the upper mantle.

ACKNOWLEDGMENTS

The author is grateful for critical reviews by P. Misch, B. R. Frost, J. M. Rice, and V. Trommsdorff, and for the financial support of the National Science Foundation (grant no. EAR75-14904).

Literature Cited

Akella, J., Boyd, F. R. 1974. Petrogenetic grid for garnet-peridotites. *Carnegie Inst. Washington Yearb.* 73:269–73

Arai, S. 1975. Contact metamorphosed dunite-harzburgite complex in the Chugoku District, Western Japan. *Contrib. Mineral. Petrol.* 52:1–16

Aumento, F. 1970. Serpentine mineralogy of ultrabasic intrusions in Canada and on the mid-Atlantic ridge. *Geol. Surv. Can., Pap. 69-53* 51 pp.

Aumento, F., Loubat, H. 1971. The Mid-Atlantic Ridge Near 45° N. XVI Serpentinized Ultramafic Intrusions. *Can. J. Earth Sci.* 8:631–63

Ave' Lallement, H. G. 1967. Structural and petrofabric analysis of an "Alpine-type" peridotite: The Lherzolite of the French Pyrenees. *Leidse Geol. Meded.* 42:1–57

Ave' Lallement, H. G. 1975. Mechanism of preferred orientations in tectonite peridotite. *Geology.* 3:653–56

Ave' Lallement, H. G., Carter, N. L. 1970. Syntectonic recrystallization and modes of flow in the upper mantle. *Geol. Soc. Am. Bull.* 81:2203–20

Banno, S. 1970. Classification of eclogites in terms of physical conditions of their origin. *Phys. Earth Planet. Inter.* 3:405–21

Banno, S. 1974. Use of partial solution of multi-component equilibria. Case study on pyroxene-bearing assemblages. *Bull. Soc. Fr. Minéral. Cristallogr.* 97:108–16

Barnes, I., O'Neil, J. R. 1969. The relationship between fluids in some fresh alpine-type ultramafics and possible modern serpentinization, western United States. *Geol. Soc. Am. Bull.* 80:1947–60

Bartholomé, P. 1962. Iron-magnesium ratio in associated pyroxenes and olivines. In Petrologic Studies: A volume in honor of A. F. Buddington, ed. A. E. J. Engel, H. L. James, B. F. Leonard, pp. 1–20. Geol. Soc. Am., Boulder, Colo.

Basu, A. R., MacGregor, I. D. 1975. Chromite spinels from ultramafic xenoliths. Geochim. Cosmochim. Acta. 39:937–45

Battey, M. H. 1960. The relationship between preferred orientation of olivine in dunite and the tectonic environment. Am. J. Sci. 258:716–27

Bearth, P. 1967. Die Ophiolithe der Zone von Zermatt-Saas Fee. Beitr. Geol. Karte Schweiz, N.F. 132:1–130

Benson, W. N. 1926. The tectonic conditions accompanying the intrusion of basic and ultrabasic igneous rocks. Mem. Nat. Acad. Sci. 19:1, 1–90

Bilgrami, S. A., Howie, R. A. 1960. The mineralogy and petrology of a rodingite dike, Hindabagh, Pakistan. Am. Mineral. 45:791–801

Blander, M. 1972. Thermodynamic properties of orthopyroxenes and clinopyroxenes based on the ideal two-site model. Geochim. Cosmochim. Acta. 36:787–99

Bliss, N. W., MacLean, W. H. 1975. The paragenesis of zoned chromite from central Manitoba. Geochim. Cosmochim. Acta. 39:973–90

Bonatti, E., Emiliani, C., Ferrara, G., Honnorez, J., Rydell, H. 1974. Ultramafic-carbonate breccias from the equatorial Mid-Atlantic Ridge. Mar. Geol. 16:83–102

Bonatti, E., Honnorez, J., Ferrara, G. 1970. Equatorial mid-Atlantic ridge: Petrologic and Sr-isotopic evidence for an alpine type rock assemblage. Earth Planet. Sci. Lett. 9:247–56

Boullier, A. M., Nicolas, A. 1975. Classification of textures and fabrics of peridotite xenoliths from South African Kimberlites. Phys. Chem. Earth 9:467–76

Bowen, N. L., Tuttle, O. F. 1949. The System MgO–SiO₂–H₂O. Geol. Soc. Am. Bull. 60:439–60

Bowin, C. O., Nalwalk, A. J., Hersey, J. B. 1966. Serpentinized peridotite from the north wall of the Puerto Rico trench. Geol. Soc. Am. Bull. 77:257–70

Boyd, F. R. 1959. Hydrothermal investigations of amphiboles. In Researches in Geochemistry, ed. P. H. Abelson, 1:377–96. New York: Wiley

Boyd, F. R. 1970. Garnet peridotites and the system $CaSiO_3$–$MgSiO_3$–Al_2O_3. Mineral. Soc. Am. Spec. Pap. 3:63–75

Boyd, F. R., England, J. L. 1964. The system

enstatite-pyrope. Carnegie Inst. Washington Yearb. 63:157–61

Brady, J. B. 1977. Metasomatic Zones in Metamorphic Rocks. Geochim. Cosmochim. Acta 41:113–26

Brown, G. E., Prewitt, C. T. 1973. High-temperature crystal chemistry of hortonolite. Am. Mineral. 58:577–87

Brynhi, I., Green, D. H., Heier, K. S. Fyfe, W. S. 1970. On the occurrence of eclogite in Western Norway. Contrib. Mineral. Petrol. 26:12–19

Bucher, K., Pfeifer, H. R. 1973. Über Metamorphose und Deformation der ostlichen Malenco Ultramafitite und deren Rahmengesteine (Prov. Sondrio, N-Italien). Schweiz. Mineral. Petrogr. Mitt. 53:231–41

Burnham, C. W., Holloway, J. R., Davis, N. F. 1969. Thermodynamic properties of water to 1,000°C and 10,000 bars. Geol. Soc. Am. Spec. Pap. 132:1–96

Cameron, K. L. 1975. An experimental study of actinolite-cummingtonite phase relations with notes on the synthesis of Fe-rich anthophyllite. Am. Mineral. 60:375–90

Carswell, D. A. 1968a. Picritic magma–residual dunite relationships in garnet peridotite at Kalskaret near Tafjord, South Norway. Contrib. Mineral. Petrol. 19:97–124

Carswell, D. A. 1968b. Possible primary upper mantle peridotite in Norwegian basal gneiss. Lithos. 1:322–55

Carswell, D. A. 1973. Garnet pyroxenite lens within Ugelvik layered garnet peridotite. Earth Planet. Sci. Lett. 20:347–52

Carswell, D. A. 1974. Comparative equilibration temperatures and pressures of garnet lherzolite. Lithos. 7:113–21

Carswell, D. A., Curtis, C. D., Kanaris-Sotiriou, R. 1974. Vein Metasomatism in Peridotite at Kalskaret, near Tafjord, South Norway. J. Petrol. 15:383–402

Carter, N. L. 1971. Static deformation of silica and silicates. J. Geophys. Res. 76:5514–40

Carter, N. L. 1975. High-temperature Flow of Rocks. Rev. Geophys. Space Phys. 13:344–49

Carter, N. L., Ave' Lallement, H. G. 1970. High-temperature flow of dunite and peridotite. Geol. Soc. Am. Bull. 81:2181–2201

Cawthorn, R. G. 1975. The amphibole peridotite-metagabbro complex, Finero, N. Italy. J. Geol. 83:437–54

Chernosky, J. V. 1973. The stability of chrysotile, $Mg_3Si_2O_5(OH)_4$ and the free energy of formation of talc $Mg_3Si_4O_{10}(OH)_2$.

440   EVANS

Geol. Soc. Am. Ann. Meet., Dallas, Texas, p. 575

Chernosky, J. V. 1974. The upper stability of clinochlore at low pressure and the free energy of formation of Mg-cordierite. *Am. Mineral.* 59:496–507

Chernosky, J. V. 1975. Aggregate refractive indices and unit cell parameters of synthetic serpentine in the system MgO–Al_2O_3–SiO_2–H_2O. *Am. Mineral.* 60:200–8

Chernosky, J. V. 1976. The stability of anthophyllite—a re-evaluation based on new experimental data. *Am. Mineral.* 61:1145–55

Christensen, N. I., Salisbury, M. H. 1975. Structure and constitution of the lower oceanic crust. *Rev. Geophys. Space Phys.* 13:57–86

Chidester, A. H. 1962. Petrology and geochemistry of selected talc-bearing ultramafic rocks and adjacent country rocks in north-central Vermont. *Prof. Pap. U.S. Geol. Surv.* 345:1–207

Coats, C. J. A. 1968. Serpentine minerals from Manitoba. *Can. Mineral.* 9:322–47

Coe, R. S., Kirby, S. H. 1975. The orthoenstatite to clinoenstatite transformation by shearing and reversion by annealing: mechanism and potential applications. *Contrib. Mineral. Petrol.* 52:29–56

Coleman, R. G. 1966. New Zealand serpentinites and associated metasomatic rocks. *Bull. N.Z. Geol. Surv.* 76:1–102

Coleman, R. G. 1967. Low temperature reaction zones and alpine ultramafic rocks of California, Oregon and Washington. *US Geol. Surv. Bull.* 1247:1–49

Coleman, R. G. 1971a. Plate tectonic emplacement of upper mantle peridotite along continental edges. *J. Geophys. Res.* 76:1212–22

Coleman, R. G. 1971b. Petrologic and geophysical nature of serpentinites. *Geol. Soc. Am. Bull.* 82:897–918

Coleman, R. G., Keith, T. C. E. 1971. A chemical study of serpentinization-Burro Mountain, California. *J. Petrol.* 12:311–28

Cremer, V. 1969. Die Mischkristallbildung im System Chromit-Magnetit-Hercynit zwischen 1,000°C und 500°C. *Neues Jahrb. Petrol.* 24:275–92

Curtis, C. D., Brown, P. E. 1969. The metasomatic development of zoned ultrabasic bodies in Unst, Shetland. *Contrib. Mineral. Petrol.* 24:275–92

Dal Piaz, G. 1967. Le "granatiti" (rodingiti l.s.) nelle serpentine delle Alpi occidentali italiane. *Mem. Soc. Geol. Ital.* 6:267–313

Davis, B. T. C., Boyd, F. R. 1966. The join

$Mg_2Si_2O_6$–$CaMgSi_2O_6$ at 30 kilobars pressure and its application to pyroxenes from kimberlites. *J. Geophys. Res.* 14:3567–76

den Tex, E. 1969. Origin of ultramafic rocks, their tectonic setting and history: a contribution to the discussion of the paper "The origin of ultramafic and ultrabasic rocks" by P. J. Wyllie. *Tectonophysics* 7:457–88

de Quervain, F. 1938. Zur Kenntnis des Titanklinohumites (Titanolivin). *Schweiz. Mineral. Petrogr. Mitt.* 18:591–603

Dewey, J. F., Bird, J. M. 1971. Origin and emplacement of the ophiolite suite: Appalachian ophiolites in Newfoundland. *J. Geophys. Res.* 76:3170–3206

Dickey, J. S. 1970. Partial fusion products in alpine-type peridotites. Serrania de la Ronda and other examples. *Mineral. Soc. Am. Spec. Pap.* 3:33–49

Dietrich, V., Peters, T. 1971. Regionale Verteilung der Mg-Phyllosilikate in den Serpentiniten des Oberhalbsteins. *Schweiz. Mineral. Petrogr. Mitt.* 51:329–48

Dungan, M. A. 1974. *The origin, emplacement and metamorphism of the Sultan Mafic-Ultramafic Complex, Northern Cascades, Snohomish County, Washington.* PhD thesis. Univ. of Washington, Seattle

Essene, E., Wall, V. J., Shettel, D. 1973. Equilibria in CaO–MgO–SiO_2–H_2O. *Trans. Am. Geophys. Union* 54:480

Evans, B. W., Frost, B. R. 1975. Chromespinel in progressive metamorphism—a preliminary analysis. *Geochim. Cosmochim. Acta* 39:959–72

Evans, B. W., Ghose, S., Rice, J. M., Trommsdorff, V. 1974. Cummingtonite-anthophyllite phase transition in metamorphosed ultramafic rocks, Ticino, Switzerland. *Trans. Am. Geophys. Union Abstr.* 55:469

Evans, B. W., Johannes, J., Oterdoom, H., Trommsdorff, V. 1976. Stability of chrysotile and antigorite in the serpentine multisystem. *Schweiz. Mineral. Petrol. Mitt.* 56:79–93

Evans, B. W., Trommsdorff, V. 1969. The stable association enstatite-forsterite-chlorite in amphibolite facies ultramafics of the Lepontine Alps. *Schweiz. Mineral. Petrogr. Mitt.* 49:325–32

Evans, B. W., Trommsdorff, V. 1970. Regional metamorphism of ultramafic rocks in the Central Alps: Parageneses in the system CaO–MgO–SiO_2–H_2O. *Schweiz. Mineral. Petrogr. Mitt.* 50:481–92

Evans, B. W., Trommsdorff, V. 1974a. On elongate olivine of metamorphic origin.

Geology. Vol. 2, No. 3, pp. 131–32

Evans, B. W., Trommsdorff, V. 1974b. Stability of enstatite plus talc, and CO_2-metasomatism of metaperidotite, Val d'Efra, Lepontine Alps. *Am. J. Sci.* 274: 274–96

Evans, B. W., Trommsdorff, V. 1975. Gradation between eclogite and metarodingite, Ticino, Switzerland. *Geol. Soc. Am. Ann. Meet., Salt Lake City, Abstr. with Programs,* Vol. 7, No. 7, pp. 1069–70

Evans, B. W., Trommsdorff, V. 1976. Der Einfluss von Kationenersatz auf die Hydratisierung von Duniten-Korrekturen und Kommentare. *Schweiz. Mineral. Petrol. Mitt.* 55: 457–59

Fawcett, J. J., Yoder, H. S. 1966. Phase relationships of chlorites in the system $MgO-Al_2O_3-SiO_2-H_2O$. *Am. Mineral.* 51: 353–80

Fiala, J. 1966. The distribution of elements in mineral phases of some garnet peridotites from the Bohemian Massif. *Krystallinikum.* 4: 31–53

Finger, L. W., Virgo, D. 1971. Confirmation of Fe/Mg ordering in olivines. *Carnegie Inst. Washington Yearb.* 70: 221–25

Fisher, R. L., Engel, C. G. 1969. Ultramafic and basaltic rocks dredged from the near-shore flank of the Tonga trench. *Geol. Soc. Am. Bull.* 80: 1373–78

Folk, R. L., McBride, E. F. 1976. Possible pedogenic origin of Ligurian ophicalcite: A mesozoic calichified serpentinite. *Geology.* 4: 327–32

Forbes, R. B., Swainbank, R. C. 1974. Garnet clinopyroxenite from the Red Mountain pluton, Alaska. *Geol. Soc. Am. Bull.* 85: 285–92

Forestier, F. H. 1971. Natur und Herkunft der Peridotite in der französischen Kristallinsockeln — Probleme der Granulitfacies. *Fortschr. Mineral.* 48: 75–85

Francis, G. H. 1956. The serpentinite mass in Glen Urquhart, Inverness-shire, Scotland. *Am. J. Sci.* 254: 201–26

Frost, B. R. 1975. Contact Metamorphism of Serpentinite, Chloritic Blackwall and Rodingite at Paddy-Go-Easy Pass, Central Cascades, Washington. *J. Petrol.* 16: 272–313

Frost, B. R. 1976a. Limits to the assemblage forsterite + anorthite as inferred from peridotite hornfelses, Icicle Creek, Washington. *Am. Mineral.* 61: 732–50

Frost, B. R. 1976b. Olivine-clinopyroxene geothermometer: some thermodynamic constraints. *Trans. Am. Geophys. Union Abstr.* 57: 1026

Ganguly, J., Kennedy, G. C. 1974. The energetics of natural garnet solid solution.

I. Mixing of the aluminosilicate end-members. *Contrib. Mineral. Petrol.* 48: 137–47

Gass, I. G., Smewing, J. D. 1973. Instrusion, extrusion and metamorphism at constructive margins: evidence from the Troodos Massif, Cyprus. *Nature.* 242: 26–29

Ghose, S., Wan, C., McCallum, I. S. 1976. $Fe^{2+}-Mg^{2+}$ order in an olivine from the lunar anorthosite 67075 and the significance of cation order in lunar and terrestrial olivines. *Indian J. Earth Sci.* 3: 1–8

Green, D. H. 1964. The petrogenesis of the high-temperature peridotite intrusion in the Lizard area, Cornwall. *J. Petrol.* 5: 134–88

Green, D. H., Hibberson, W. 1970. The instability of plagioclase in peridotite at high pressure. *Lithos.* 3: 209–21

Green, D. H., Ringwood, A. E. 1967a. The stability fields of aluminous pyroxene peridotite and garnet peridotite and their relevance in upper mantle structure. *Earth Planet. Sci. Lett.* 3: 151–60

Green, D. H., Ringwood, A. E. 1967b. An experimental investigation of the gabbro to eclogite transformation and its petrologic applications. *Geochim. Cosmochim. Acta.* 31: 767–833

Greenwood, H. J. 1963. The synthesis and stability of anthophyllite. *J. Petr.* 4: 317–51

Greenwood, H. J. 1967. Mineral equilibria in the system $MgO-SiO_2-H_2O-CO_2$. In *Researches in Geochemistry,* ed. P. H. Abelson, pp. 542–67. New York, London, Sydney: Wiley

Greenwood, H. J. 1971. Anthophyllite, corrections and comments on its stability. *Am. J. Sci.* 270: 151–54

Greenwood, H. J. 1975. Buffering of pore fluids by metamorphic reactions. *Am. J. Sci.* 275: 573–95

Griffin, W. L., Heier, K. S. 1973. Petrological implications of some corona structures. *Lithos.* 6: 315–35

Grover, J. 1972. The stability of low-clinoenstatite in the system $Mg_2Si_2O_6$. *Trans. Am. Geophys. Union Abstr.* 53: 539

Grover, J., Orville, P. M. 1969. The partitioning of cations between single and multi-site phases with application to the assemblage: orthopyroxene-clinopyroxene and orthopyroxene-olivine. *Geochim. Cosmochim. Acta.* 33: 205–66

Häkli, T. A. 1968. An attempt to apply the Makaopuhi nickel fractionation data to the temperature determination of a basic intrusive. *Geochim. Cosmochim. Acta.* 32:

449–60

Hart, S. R. 1972. Geochemistry of an ocean-ridge lherzolite. *EOS Trans. Am. Geophys. Union Abstr.* 53 : 536

Hartman, P., den Tex, E. 1964. Piezo-crystalline fabrics of olivine in theory and nature. *Int. Geol. Congr., 22nd, India, 1964,* Part IV, Proc. Sect 4 : 84–114

Heinrich, E. W. 1963. Paragenesis of clino-humite and associated minerals from Wolf Creek, Montana. *Am. Mineral.* 48 : 597–613

Hemley, J. J., Montoya, J. W., Shaw, D. R. 1977. Mineral equilibria in the MgO–SiO$_2$–H$_2$O system : II Talc-antigorite-forsterite-anthophyllite-enstatite stability relations and some geologic implications in the system. *Am. J. Sci.* In press

Hensen, B. J. 1973. Pyroxenes and garnets as geothermometers and barometers. *Carnegie Inst. Washington Yearb.* 72 : 528–34

Hensen, B. J., Schmid, R., Wood, B. J. 1975. Activity-composition relationships for pyrope-grossular garnet. *Contrib. Mineral. Petrol.* 51 : 161–66

Herzberg, C. T., Chapman, N. A. 1976. Clinopyroxene geothermometry of spinel-lherzolites. *Am. Mineral.* 61 : 626–37

Hess, H. H. 1962. History of ocean basins. In *Petrologic Studies* (Buddington Volume), ed. A. E. J. Engel, H. L. James, B. F. Leonard, pp. 599–620. Geol. Soc. Am., Boulder, Colo.

Hess, H. H., Otalora, G. 1964. Mineralogical and chemical composition of the Maya-guez serpentinized peridotite. In a study of serpentinite, the AMSOC core hole near Mayaguez, Puerto Rico, ed. C. A. Burk, *Nat. Acad. Sci.–Nat. Res. Counc. Publ.* No. 1188 : 152–68

Hoefs, J., Touret, J. 1975. Fluid inclusion and carbon isotope study from Bamle granulites (South Norway). *Contrib. Mineral. Petrol.* 52 : 165–74

Hofmann, A. 1972. Chromatographic theory of infiltration metasomatism and its application to feldspar. *Am. J. Sci.* 272 : 60–90

Honnorez, J., Kirst, P. 1975. Petrology of rodingites from the equatorial Mid-Atlantic fracture zones and their geotec-tonic significance. *Contrib. Mineral. Petrol.* 49 : 233–57

Hostetler, P. B., Coleman, R. G., Mumpton, F. A., Evans, B. W. 1966. Brucite in alpine serpentinites. *Am. Mineral.* 51 : 75–98

Howells, S., O'Hara, M. J. 1975. Palegeo-therms and the diopside-enstatite solvus. *Nature.* 254 : 406–8

Hudson, D. R., Wilson, A. F. 1966. A new

occurrence of sapphirine and related anthophyllite from central Australia. *Geol. Mag.* 293–98

Hutchison, J. L., Irusteta, M. C., Whittaker, E. J. W. 1975. High-resolution electron microscopy and diffraction studies of fibrous amphiboles. *Acta Cryst.* A31, 794–801

Irvine, T. N. 1965. Chromian spinel as a petrogenetic indicator : Part I, Theory. *Can. J. Earth Sci.* 2 : 648–72

Irvine, T. N. 1967. Chromian spinel as a petrogenetic indicator : Part II, Petrologic applications. *Can. J. Earth Sci.* 4 : 71–103

Irving, A. J., Ashley, P. M. 1976. Amphibole-olivine-spinel, cordierite-anthophyllite and related hornfelses associated with metamorphosed serpentinites in the Goobarragandra District, near Tumut, New South Wales. *J. Geol. Soc. Aust.* 23 : 19–43

Jackson, E. D. 1969. Chemical variation in coexisting chromite and olivine in chromi-tite zones of the Stillwater Complex. *Econ. Geol. Mon.* 4 : 41–71

Jackson, E. D., Green, H. W. III, Moores, E. M. 1975. The Vourinos Ophilite, Greece : Cyclic Units of Lineated Cumu-lates Overlying Harburgite Tectonite. *Geol. Soc. Am. Bull.* 86 : 390–98

Jackson, E. D., Thayer, J. P. 1972. Some critera for distinguishing between strati-form, concentric and alpine peridotite-gabbro complexes. *Int. Geol. Congr., 24th, Montreal. Proc. Sect.* 2, 289–96

Jahns, R. H. 1967. Serpentinites of the Roxbury district, Vermont. See Wyllie 1967, pp. 137–59.

Johannes, W. 1968. Experimental investi-gation of the reaction forsterite + H$_2$O = serpentine + brucite. *Contrib. Mineral. Petrol.* 19 : 309–15

Johannes, W. 1969. An experimental investi-gation of the system MgO–SiO$_2$–H$_2$O–CO$_2$. *Am. J. Sci.* 267 : 1083–1104

Johannes, W. 1975. Zur Synthese und thermischen Stabilität von Antigorit. *Fortschr. Mineral.* 53 : 1–36

Kornprobst, J. 1969. Le massif ultrabasique des Beni Bouchera (Rif. Interne, Maroc); Etude des péridotites de haute tempera-ture et de haute pression, et des pyro-xenolites, à grenat ou sans grenat. *Contrib. Mineral. Petrol.* 23 : 283–322

Kornprobst, J., Conquéré, F. 1972. Les Pyroxenolites à Grenat du massif de lherzolite de Moncaup (Haute Garonne-France). *Earth Planet. Sci.* 16 : 1–14

Korzhinskii, D. S. 1970. Theory of Metaso-matic Zoning, Transl. J. Agrell. Oxford : Clarendon. 162 pp.

Kretz, R. 1963. Distribution of magnesium and iron between orthopyroxene and calcic pyroxene in natural mineral assemblages. *J. Geol.* 71:773–85

Kunze, G. 1961. Antigorit. Strukturtheoretische Grundlagen und ihre praktische Bedeutung für die weitere Serpentin-Forschung. *Fortschr. Mineral.* 39:206–324

Kushiro, I. 1969. Clinopyroxene solid solution formed by reactions between diopside and plagioclase at high pressures. *Mineral. Soc. Am. Spec. Pap.* 2:179–91

Kushiro, I., Yoder, H. S. 1966. Anorthite-forsterite and anorthite-enstatite reactions and their bearing on the basalt-eclogite transformation. *J. Petrol.* 7:337–62

Lappin, M. A. 1967. Structural and petrofabric studies of the dunites of Almklovalen, Nordfjord, Norway. See Wyllie 1967, pp. 183–90

Lappin, M. A. 1971. The Petrofabric orientation of olivine and seismic anisotropy of the mantle. *J. Geol.* 79:730–40

Larimer, J. W. 1968. Experimental studies on the system Fe–MgO–SiO$_2$–O$_2$ and their bearing on the petrology of chondritic meteorites. *Geochim. Cosmochim. Acta.* 32:1187–1207

Lensch, G. 1968. Die Ultramafitite der Zone von Ivrea und ihre geologische Interpretation. *Schweiz. Mineral. Petrol. Mitt.* 48:91–102

Lindsley, D. H., Dixon, S. A. 1975. Coexisting diopside and enstatite at 20 kbar and 900°–1200°C. *Geol. Soc. Am. Ann. Meet., Salt Lake City, Abstr. with Programs.* p. 1171

Loney, R. A., Himmelberg, G. R. 1976. Structure of the Vulcan Peak alpine-type peridotite, southwestern Oregon. *Geol. Soc. Am. Bull.* 87:259–74

Loney, R. A., Himmelberg, G. R., Coleman, R. G. 1971. Structure and petrology of the alpine-type peridotite at Burro Mt. California, U.S.A. *J. Petrol.* 12:245–309

McCallister, R. H., Finger, L. W., Ohashi, Y. 1976. Intracrystalline Fe^{2+}–Mg equilibria in three natural Ca-rich clinopyroxenes. *Am. Mineral.* 61:671–76

McGetchin, T. R., Silver, L. T., Chodos, A. A. 1970. Titanoclinohumite: a possible mineralogical site for water in the upper mantle. *J. Geophys. Res.* 75:255–60

MacGregor, I. D. 1964. The reaction 4 enstatite + spinel + forsterite + pyrope. *Carnegie Inst. Washington Yearb.* 63:1–157

MacGregor, I. D. 1965. Stability fields of spinel and garnet peridotites in the synthetic system MgO–CaO–Al$_2$O$_3$–SiO$_2$.

Carnegie Inst. Washington Yearb. 64:126–34

MacGregor, I. D. 1970. The effect of CaO, Cr$_2$O$_3$Fe$_2$O$_3$ and Al$_2$O$_3$ on the stability of spinel and garnet peridotites. *Phys. Earth Planet. Int.* 3:372–77

MacGregor, I. D. 1974. The system MgO–Al$_2$O$_3$–SiO$_2$: Solubility of Al$_2$O$_3$ in enstatite for spinel and garnet peridotite compositions. *Am. Mineral.* 59:110–19

McOnie, A. W., Fawcett, J. J., James, R. S. 1975. The stability of intermediate chlorites of the clinochlore-daphnite series at 2 kbar P$_{H_2O}$. *Am. Mineral.* 60:1047–62

Matthes, S. 1971. Die ultramafischen Hornfelse, insbesondere ihre Phasenpetrologie. *Fortschr. Mineral.* 48:109–27

Medaris, L. G. 1969. Partitioning of Fe^{++} and Mg^{++} between coexisting synthetic olivine and orthopyroxene. *Am. J. Sci.* 267:945–68

Medaris, L. G. 1972. High pressure peridotites in southwestern Oregon. *Geol. Soc. Am. Bull.* 83:41–58

Medaris, L. G. 1975. Coexisting spinel and silicates in alpine peridotites of the granulite facies. *Geochim. Cosmochim. Acta.* 39:947–58

Mel'nik, Y. P. 1972. Thermodynamic parameters of compressed gases and metamorphic reactions involving water and carbon dioxide. *Geochem. Int.* 9:419–26

Melson, W. G., Thompson, G. 1970. Layered basic complex in oceanic crust, Romanche Fracture, equatorial Atlantic Ocean. *Science.* 168:817–20

Mercier, J.-C. C., Nicolas, A. 1975. Textures and fabrics of Upper-Mantle Peridotites as illustrated by Xenoliths from Basalts. *J. Petrol.* 16:454–87

Misch, P., Rice, J. M. 1975. Miscibility of tremolite and hornblende in progressive Skagit Metamorphic Suite, North Cascades, Washington. *J. Petrol.* 16:1–21

Miyashiro, A. 1973. *Metamorphism and Metamorphic Belts.* New York: Wiley. 492 pp.

Miyashiro, A. 1975. Classification, characteristics, and origin of ophiolites. *J. Geol.* 83:249–81

Miyashiro, A., Shido, F., Ewing, M. 1969. Composition and origin of serpentinites from the Mid-Atlantic ridge near 24° and 30° north latitude. *Contrib. Mineral. Petrol.* 23:117–27

Möckel, J. R. 1969. Structural petrology of the garnet peridotite of Alpe Arami (Ticino), Switzerland. *Leidse Geol. Meded.* 42:61–130

Monchoux, P. 1972. Roches à sapphirine au contact des lherzolites pyrenéennes. *Con-*

trib. Mineral. Petrol. 37 : 47–64

Moody, J. B. 1976. Serpentinization: a review. *Lithos.* 9 : 125–38

Moores, E. M. 1973. Geotectonic significance of ultramafic rocks. *Earth Sci. Review.* 9 : 241–58

Moores, E. M., MacGregor, I. D. 1972. Types of alpine ultramafic rocks and their implications for fossil plate interactions. *Geol. Soc. Am. Mem.* 132 : 209–23

Mori, T., Banno, S. 1973. Petrology of peridotite and garnet clinopyroxenite of the Mt. Higashi-Akaisi mass, central Sikoku, Japan—subsolidus relation of anhydrous phases. *Contrib. Mineral. Petrol.* 41 : 301–24

Mori, T., Green, D. H. 1976. Subsolidus equilibria between pyroxenes in the CaO–MgO–SiO$_2$ system at high pressures and temperatures. *Am. Mineral.* 61 : 616–25

Mysen, B. O. 1976. Experimental determination of some geochemical parameters relating to conditions of equilibration of peridotite in the upper mantle. *Am. Mineral.* 61 : 677–83

Mysen, B. O., Boettcher, A. L. 1975. Melting of a hydrous mantle: II Geochemistry of crystals and liquids formed by anatexis of mantle peridotite at high pressures and high temperatures as a function of controlled activities of water, hydrogen and carbon dioxide. *J. Petrol.* 16 : 549–93

Nafziger, R. H., Muan, A. 1967. Equilibrium phase compositions and thermodynamic properties of olivines and pyroxenes in the system MgO–"FeO"–SiO$_2$. *Am. Mineral.* 52 : 1364–85

Nalivkina, E. B. 1961. Metasomatic zonality and genesis of sapphirine-bearing rocks in the Bug region. *Int. Geol. Rev.* 3 : 337–49

Nehru, C. E., Wyllie, P. J. 1974. Electron microprobe measurement of pyroxenes coexisting with H$_2$O undersaturated liquid in the join CaMgSi$_2$O$_6$Mg$_2$Si$_2$O$_6$–H$_2$O at 30 kb with application to geothermometry. *Contrib. Mineral. Petrol.* 48 : 221–28

Nicolas, A., Bouchez, J. L., Boudier, F., Mercier, J. C. 1971. Textures, structures and fabrics due to solid state flow in some European lherzolites. *Tectonophysics.* 12 : 55–86

Nicolas, A., Boudier, E., Boullier, A. M. 1973. Mechanisms of flow in naturally and experimentally deformed peridotites. *Am. J. Sci.* 273 : 853–76

Nixon, P. H., Boyd, F. R. 1973. Petrogenesis of the granular and sheared ultrabasic nodule suite in Kimberlites. In *Lesotho Kimberlites,* ed. P. H. Nixon, pp. 48–56. Lesotho Nat. Dev. Corp. Maseru, Lesotho

Obata, M. 1976. The solubility of Al$_2$O$_3$ in orthopyroxenes in spinel and plagioclase peridotites and spinel pyroxenite. *Am. Mineral.* 61 : 804–16

Obata, M., Banno, S., Mori, T. 1974. The iron-magnesium partitioning between naturally occurring coexisting olivine and Ca-rich clinopyroxene: an application of the simple mixture model to olivine solid solution. *Bull. Soc. Fr. Minéral. Cristallogr.* 97 : 101–07

O'Hara, M. J. 1967a. Mineral parageneses in ultrabasic rocks. See Wyllie 1967, pp. 393–401

O'Hara, M. J. 1967b. Garnetiferous ultrabasic rocks of orogenic regions. See Wyllie 1967, pp. 167–72

O'Hara, M. J. 1968. The bearing of phase equilibria studies in synthetic and natural systems on the origin and evolution of basic and ultrabasic rocks. *Earth Sci. Rev.* 4 : 69–133

O'Hara, M. J., Mercy, E. L. P. 1963. Petrology and petrogenesis of some garnetiferous peridotites. *Trans. R. Soc. Edinburgh.* 65 : 251–310

O'Hara, M. J., Richardson, S. W., Wilson, G. 1971. Garnet peridotite stability and occurrence in crust and mantle. *Contrib. Mineral. Petrol.* 32 : 48–68

Ohnmacht, W. 1974. Petrogenesis of carbonate-orthopyroxenites (Sagvandites) and related rocks from Troms, Northern Norway. *J. Petrol.* 15 : 303–24

Oka, Y., Matsumoto, T. 1974. Study on the compositional dependence of the apparent partition coefficient of iron and magnesium between coexisting garnet and clinopyroxene solid solution. *Contrib. Mineral. Petrol.* 48 : 115–21

Olsen, E., Bunch, T. E. 1970. Empirical derivation of activity coefficients for the magnesium-rich portion of the olivine solid solution. *Am. Mineral.* 55 : 1829–42

Page, N. 1967a. Serpentinization at Burro Mountain, California. *Contrib. Mineral. Petrol.* 14 : 321–42

Page, N. 1967b. Serpentinization considered as a constant volume metasomatic process—a discussion. *Am. Mineral.* 52 : 545–49

Peters, T. 1963. Mineralogie und Petrographie des Totalserpentines bei Davos. *Schweiz. Mineral. Petrol. Mitt.* 43 : 529–685

Peters, T. 1968. Distribution of Mg, Fe, Al, Ca and Na in coexisting olivine, orthopyroxene and clinopyroxene in the Totalp serpentinite. *Contrib. Mineral. Petrol.* 18 : 65–75

Phillips, A. H., Hess, H. H. 1936. Metamorphic differentiation at contacts between serpentinite and siliceous country rock. *Am. Mineral.* 21:333–62

Ploshko, V. V., Bogdanov, Y. A. 1968. Ultrabasites of the deep-water Romanche trench. *Izv. Akad. Nauk SSSR.* 12:82–93

Powell, M., Powell, R. 1974. An olivine-clinopyroxene geothermometer. *Contrib. Mineral. Petrol.* 48:249–63

Prider, R. T. 1945. Sapphirine-bearing rocks from Dangin, W. Australia. *Geol. Mag.* 82:49–54

Quon, S. H., Ehlers, E. G. 1963. Rocks of the northern part of the Mid-Atlantic Ridge. *Geol. Soc. Am. Bull.* 74:1–7

Rabinowitz, P. A., Melson, W. G. 1976. Challenger drills on Leg 45. *Geotimes,* Vol. 21, No. 4, pp. 20–23

Ragan, D. M. 1969. Olivine recrystallization textures. *Mineral. Mag.* 37:238–40

Råheim, A., Green, D. H. 1974. Experimental determination of the temperature and pressure dependence of the Fe–Mg partition coefficient for coexisting garnet and clinopyroxene. *Contrib. Mineral. Petrol.* 48:179–203

Raleigh, C. B. 1965a Structure and petrology of an alpine peridotite on Cypress Island, Washington. *Contrib. Mineral. Petrol.* 11:719–41

Raleigh, C. B. 1965b. Glide mechanisms in experimentally deformed minerals. *Science.* 150:739–41

Raleigh, C. B. 1968. Mechanisms of plastic deformation of olivine. *J. Geophys. Res.* 73:5391–5491

Raleigh, C. B., Kirby, S. H. 1970. Creep in the upper mantle. *Mineral. Soc. Am. Spec. Pap.* 3:113–21

Raleigh, C. B., Kirby, S. H., Carter, N. L., Ave'Lallement, H. G. 1971. Slip and the clinoenstatite transformation as competing rate processes in enstatite. *J. Geophys. Res.* 76:4011–22

Read, H. H. 1934. On zoned associations of talc, actinolite, chlorite and biotite in Unst, Shetland Islands. *Mineral. Mag.* 23:519–40

Rice, J. M. 1977. Progressive metamorphism of impure dolomite limestone in the Marysville Aureole, Montana. *Am. J. Sci.* 277:1–24

Rice, J. M., Evans, B. W., Trommsdorff, V. 1974. Widespread occurrence of magnesio-cummingtonite in ultramafic Schists, Cima di Gagnone, Ticino, Switzerland. *Contrib. Mineral. Petrol.* 43:245–51

Richter, W. 1971. Ariégite, Spinell-peridotite und Phlogopit-Klinopyroxenite aus dem Tuff von Tobaj im südlichen Burgenland. *Tschermaks Mineral. Petrol. Mitt.* 16:227–51

Robie, R. A., Waldbaum, D. R. 1968. Thermodynamic properties of minerals and related substances at 298.15°K (25.0PC) and one atmosphere (1.013 bars) pressure and higher temperatures. *Bull. U.S. Geol. Surv.* 1259. 256 pp.

Robinson, P., Ross, M., Jaffe, H. W. 1971. Composition of the anthophyllite-gedrite series, comparisons of gedrite and hornblende, and the anthophyllite-gedrite solvus. *Am. Mineral.* 56:1005–41

Rost, F. 1968. Über die Fazieseinstufung orogenotyper Peridotite und ihre Beziehungen zur Peridotitschale des Erdmantels. *Int. Geol. Congr. Prague* 1:187–96

Rost, F. 1971. Probleme der Ultramafitite. *Fortschr. Mineral.* 48:54–68

Rost, F., Grigel, W. 1969. Zur Geochemie und Genese granatführender Ultramafitite des mitteleuropäischen Grundgebirges. *Chem. Erde.* 28:91–177

Sarp, H., Bertrand, J., McNear, E. 1976. Vuagnatite, CaAl(OH)SiO$_4$, a new natural calcium aluminum nesosilicate. *Am. Mineral.* 61:825–30

Saxena, S. K. 1973. *Thermodynamics of rock-forming crystalline solutions.* Berlin: Springer. 188 pp.

Scarfe, C. M., Wyllie, P. J. 1967. Serpentine dehydration curves and their bearing on serpentine deformation in orogenesis. *Nature.* 215:945–46

Schreyer, W. 1970. Metamorphose pelitischer Gesteine im Modellsystem MgO–Al$_2$O$_3$–SiO$_2$–H$_2$O. *Fortschr. Mineral.* 47:124–65

Seifert, F. 1974. Stability of sapphirine: a study of the aluminous part of the system MgO–Al$_2$O$_3$–SiO$_2$–H$_2$O. *J. Geol.* 82:173–204

Seifert, F., Virgo, D. 1974. Temperature dependence of intracrystalline Fe^{2+}–Mg distribution in a natural anthophyllite. *Carnegie Inst. Washington Yearb.* 73:405–11

Seifert, F. A., Virgo, D. 1975. Kinetics of Fe^{2+}–Mg order-disorder reaction in anthophyllites: quantitative cooling rates. *Science* 188:1107–09

Segnit, E. R. 1957. Sapphirine-bearing rocks from MacRobertson Land, Antarctica. *Mineral. Mag.* 31:690–97

Shand, S. J. 1949. Rocks of the mid-Atlantic ridge. *J. Geol.* 57:89–92

Skippen, G. B. 1974. An experimental model for low pressure metamorphism of siliceous dolomitic marble. *Am. J. Sci.* 274:487–509

Sørensen, H. 1967. Metamorphic and metasomatic processes in the formation of ultramafic rocks. See Wyllie 1967, pp. 204–11

Spooner, E. T. C., Fyfe, W. S. 1973. Sub sea-floor metamorphism, heat and mass transfer. *Contrib. Mineral. Petrol.* 42:287–304

Springer, R. K. 1974. Contact metamorphosed ultramafic rocks in the Western Sierra Nevada foothills, California. *J. Petrol.* 15:160–95

Spry, A. 1969. *Metamorphic Textures.* Oxford: Pergamon. 350 pp.

Stormer, J. C. 1972. Mineralogy and petrology of the Raton-Clayton Volcanic Field, Northeastern New Mexico. *Geol. Soc. Am. Bull.* 83:3299–3322

Stroh, J. M. 1976. Solubility of alumina in orthopyroxene plus spinel as a geobarometer in complex systems—Applications to spinel-bearing alpine-type peridotites. *Contrib. Mineral. Petrol.* 54:173–88

Subbarao, K. W., Hedge, C. E. 1973. K, Rb, Sr, and $^{87}Sr/^{86}Sr$ in rocks from the Mid-Indian Oceanic Ridge. *Earth Planet. Sci. Lett.* 18:223–28

Takeda, H. 1972. Structural studies of rim augite and core pigeonite from lunar rock 12052. *Earth Planet. Sci. Lett.* 15:65–71

Thompson, A. B. 1976. Mineral reactions in pelitic rocks: I Prediction of P-T-X(Fe-Mg) Phase Relations. *Am. J. Sci.* 276:401–24

Thompson, J. B. 1959. Local equilibrium in metasomatic processes. See Boyd 1959, pp. 427–57

Thompson, J. B. 1970. Geochemical reaction and open systems. *Geochim. Cosmochim. Acta.* 34:529–51

Thompson, G., Melson, W. G. 1970. Boron contents of serpentinites and metabasalts in the oceanic crust: implications for the boron cycle in the oceans. *Earth Planet. Sci. Lett.* 8:61–65

Trommsdorff, V., Evans, B. W. 1969. The stable association enstatite-forsterite-chlorite in amphibolite facies ultramafics of the Lepontine Alps. *Schweiz. Mineral. Petrol. Mitt.* 49:325–32

Trommsdorff, V., Evans, B. W. 1972. Progressive metamorphism of antigorite schist in the Bergell tonalite aureole (Italy). *Am. J. Sci.* 272:423–37

Trommsdorff, V., Evans, B. W. 1974. Alpine metamorphism of peridotitic rocks. *Schweiz. Mineral. Petrol. Mitt.* 54:333–52

Trommsdorff, V., Evans, B. W. 1977. Antigorite-ophicarbonates: phase relations in a portion of the system CaO–MgO–SiO$_2$–H$_2$O–CO$_2$. *Contrib. Mineral. Petrol.* In press

Trommsdorff, V., Wenk, H. R. 1968. Terrestrial metamorphic clinoenstatite in kinks of bronzite crystals. *Contrib. Mineral. Petrol.* 19:158–68

Turner, F. J. 1968. *Metamorphic Petrology—Mineralogical and Field Aspects.* New York: McGraw-Hill. 403 pp.

Turner, F. J., Heard, H., Griggs, D. J. 1960. Experimental deformation of enstatite and accompanying inversion to clinoenstatite. *Int. Geol. Congr. Norden.* 188:399–408

Vance, J. A., Dungan, M. A. 1977. Formation of peridotites by deserpentinization in the Darrington and Salton areas, Cascade Mountains, Washington. *Geol. Soc. Am. Bull.* In press

Vdovykin, G. P., Dmitriyev, L. V. 1968. Organic matter in ultramafic rocks of the Mid-Indian Ridge. *Geochem. Int.* 5:828–31

Veblen, D. R., Burnham, C. W. 1975. Triple chain biopyriboles: newly discovered intermediate products of the retrograde anthophyllite-talc transformation, Chester, Vermont. *Trans. Am. Geophys. Union* 56:1–1107

Vernon, R. H. 1970. Comparative grain boundary studies of some basic and ultrabasic granulites, nodules and cumulates. *Scott. J. Geol.* 6:337–51

Vernon, R. H. 1976. *Metamorphic Processes: Reactions and Microstructure Development.* Halstead Press. New York: Wiley

Virgo, D., Hafner, S. S. 1969. Fe^{2+}–Mg order-disorder in heated orthopyroxenes. *Mineral. Soc. Am. Spec. Paper.* 2:67–81

Vuagnat, M. 1967. Quelques réflexions sur les ophispherites et les rodingites. *Rend. Soc. Mineral. Ital.* 23:471–82

Warner, R. D., Luth, W. C. 1974. The diopside-orthoenstatite two-phase region in the system CaMgSi$_2$O$_6$–Mg$_2$Si$_2$O$_6$. *Am. Mineral.* 59:98–109

Wenk, H. R., Raymond, K. N. 1973. Four new refinements of olivine. *Zeit. Kristallogr.* 137:86–105

Wenner, D. B., Taylor, H. P. 1971. Temperatures of serpentinization of ultramafic rocks based on O^{18}/O^{16} fractionation between coexisting serpentine and magnetite. *Contrib. Mineral. Petrol.* 32:165–85

Wenner, D. B., Taylor, H. P. 1973. Oxygen and hydrogen isotope studies of the serpentinization of ultramafic rocks in oceanic environments and continental ophiolite complexes. *Am. J. Sci.* 272:207–39

Wicks, F. J., Whittaker, E. J. W. 1975. A reappraisal of the structures of the serpentine minerals. *Can. Mineral.* 13:227–43

Wicks, F. J., Whittaker, E. J. W. 1976. Serpentine textures and serpentinization. (Unpublished manuscript)

Williams, R. J. 1971. Reaction constants in the system Fe–MgO–SiO$_2$–O$_2$ at 1 atm. between 900°C and 1300°C: Experimental results. *Amer. J. Sci.* 270:334–60

Wilshire, H. G., Jackson, E. D. 1975. Problems in determining mantle goetherms from pyroxene compositions of ultramafic rocks. *J. Geol.* 83:313–29

Winkler, H. G. F. 1974. *Petrogenesis of Metamorphic Rocks.* New York: Springer. 320 pp.

Wolfe, W. J. 1966. *Petrology, Mineralogy and Geochemistry of the Blue River Ultramafic Intrusion, Cassiar District, British Columbia.* PhD thesis. Yale Univ.

Wood, B. J. 1974. The solubility of alumina in orthopyroxene coexisting with garnet. *Contrib. Mineral. Petrol.* 46:1–15

Wood, B. J. 1975. The application of thermodynamics to some subsolidus equilibria involving solid solution. *Fortschr. Mineral.* 52:21–45

Wood, B. J. 1976. An olivine-clinopyroxene geothermometer. A discussion. *Contrib. Mineral. Petrol.* 56:297–304

Wood, B. J., Banno, S. 1973. Garnet-orthopyroxene and orthopyroxene-clinopyroxene relationships in simple and complex systems. *Contrib. Mineral. Petrol.* 42:109–24

Wyllie, P. J. 1967., Ed. *Ultramafic and Related Rocks.* New York: Wiley. 464 pp.

Wyllie, P. J. 1969. The origin of ultramafic and ultrabasic rocks. *Tectonophysics.* 7:437–55

Wyllie, P. J. 1970. Ultramafic rocks and the upper mantle. *Mineral. Soc. Am. Spec. Pap.* 3:3–32

Zen, E-an 1971. Comments on the thermodynamic constants and hydrothermal stability relations of anthophyllite. *Am. J. Sci.* 270:136–50

Zen, E-an 1972. Gibbs free energy, enthalpy, and entropy of ten rock-forming minerals: calculation, discrepancies, implications. *Am. Min.* 57:524–53

Ann. Rev. Earth Planet. Sci. 1977. 5: 449–89

THE EVOLUTION OF THE LUNAR REGOLITH

×10080

Yves Langevin

Centre de Spectrometrie Nucléaire et de Spectrometrie de Masse, Laboratoire René Bernas, Orsay, France

James R. Arnold

Department of Chemistry, University of California, San Diego, La Jolla, California 92093

1 INTRODUCTION

One of the most familiar things on our planet is the layer of soil that covers it. The more or less finely divided particles that make up this layer are produced from rock by the action of wind and water. The presence of soil in the first surface views of Mars is thus no surprise.

The surface of the Moon, like those of Mercury and >99% of the known bodies in the solar system, is not subject to these erosion mechanisms. The absence of an atmosphere, however, exposes it directly to meteoritic impact and to all the fluxes of radiation in our region of space. The existence of a broken-up surface layer (regolith) was known long before the first soft landings, though its properties were not clearly understood. Nearly all our present knowledge derives from samples returned from the Moon, and secondarily from observations made on the surface or from lunar orbit.

Since the Moon is the first object of its class to be well studied, the properties of its surface are the basis for our present ideas about the others.

The lunar regolith is a blanket of broken-up material whose most striking quality is the completeness of its coverage. Exposures of bedrock are rare almost to the point of nonexistence, and none has yet been sampled. The regolith, then, is a thin deposit containing a record of billions of years of the history of the Moon, and of the inner solar system. It is a rich subject for study.

At the moment of writing, soil samples have been returned from nine lunar sites: the six Apollo landing sites and three Soviet lunar stations. This is a respectably large and diverse list, but it is by no means representative. Six of the sites (Apollo 11, 12, 15, and 17; Luna 16 and 24) are entirely or mainly mare sites. Apollo 16 and Luna 20 landed in highland areas near maria, and Apollo 14 sampled the Fra Mauro formation, a high area of unusual chemistry surrounded by Oceanus Procellarum. No "pure" highland area far from mare or KREEP materials has yet been sampled, though such areas make up most of the lunar surface.

One striking difference between mare and highland regions is the much more heavily bombarded surface of the highlands. The highland regions are older (Papanastassiou & Wasserburg 1975), and were apparently subject to much heavier bombardment by large meteorites before 4 b. y. ago. The maria are generally understood to have been formed from about 4.0 to 3.0 b. y. ago by the accumulation of successive flows of basaltic lava from the interior (Papanastassiou & Wasserburg 1971).

A formal definition of the regolith layer is a matter of some difficulty, and it is not possible to be as precise as one would wish. Broadly it is a layer in which most or nearly all the mass is contained in fine particles (less than 1 mm in diameter), but which generally contains some large coherent fragments (rocks). Its depth is typically a few meters or more. The older and deeper broken-up layer, covering at least the bulk of the highland areas, grading perhaps toward coarser blocky material, and extending perhaps to a depth of kilometers, is sometimes called the *megaregolith*. We shall say little about it, since we know little.

Samples of lunar regolith material exhibit certain characteristic properties. There is an exceedingly heterogeneous distribution of particle sizes, from rocks through chips and coarse sandy textures down to micron and submicron particles (Butler & King 1974). The great bulk of the material is below 1 mm in size. The mode of the mass distribution is in the size range from tens to perhaps one hundred micrometers. Structure and composition of the particles are correlated with particle size.

In the "rock" range—a few centimeters and up—there are two broad classes of material, crystalline rocks and breccias. Most of the crystalline rocks we have are mare basalts, found usually within the mare in which they were formed. Some crystalline rocks seem to be impact melts (for example, rock 14310) (Morgan et al 1972). Large lumps (clasts) or free-standing fragments of almost pure Ca-rich feldspar (anorthosite) occur occasionally in highland material ("Genesis Rock" 15415, and several similar objects at Apollo 16).

The dominant types at the nonmare sites of Apollo 14 and 16, and hence, one assumes, over most of the Moon, are breccias. These are heterogeneous assemblages ranging from loose clods to hard, largely remelted masses, and were presumably formed by meteoritic impact on the regolith. Such rocks are common also, but not dominant, at mare sites. It is not unusual to find evidence in one breccia of successive breakup and reforming. Thus, rock 14321 carried a record of at least three generations of brecciated material (Lindstrom et al 1972).

Glass is a common constituent of breccias, and is also often found as a coating on crystalline rocks. Large, purely glassy objects are rare.

The "rocklets" on a centimeter scale and the "coarse fines" between 0.1–1 cm are similar in composition and structure to the rocks. However, fragments, glass, and single "pure" mineral fragments become commoner at smaller sizes.

Particles smaller than 1 mm appear different. Glass becomes very abundant. Mostly it consists of dark, irregular fragments, but bright-colored glass and shiny glass spheres are easy to find. A large fraction of soil consists of crystalline grains of the common transparent and opaque rock-forming minerals. Mature soils (see below) show increasing quantities of agglutinates, multigrain assemblages loosely or tightly bonded together with glass. Occasionally one sees identifiable meteoritic fragments (Frederiksson et al 1970), but most meteoritic material is intimately mixed into the soil. The finest particles, of micron size and below, are often attached more or less tightly to the surface of coarser grains (Arrhenius & Asunmaa 1973).

The regolith blanket taken as a whole has some other important properties. In virtually all samples examined to date, there is a heterogeneous mixture of materials of different origin including some fragments from great distances. There are a few special exceptions of which the most famous is the Apollo 17 orange glass material (Schmitt 1973) apparently formed by an internal lunar event. The surface topography is remarkably smooth, with typical slopes on a centimeter to meter scale of 2–5°. This can be seen most clearly by contrast with the appearance of fresh craters (Swann et al 1972). It is not obvious how formation and gardening of the regolith by meteoritic impact can produce such a surface.

Cores removed from the regolith show that it is not completely homogenized. Layers can be observed by the presence of color and texture differences, by a higher or lower density of coarser fragments, or by other means. Most of these differences are quite subtle, but there are occasional striking transitions. Impact gardening would be expected to produce ejecta blankets, which could appear as layers in a core.

Over what period has the regolith been formed? By what mechanism(s) has it been produced and modified? What is the rate of mixing and turnover (gardening)? How much distant lunar material has been brought in, and how? What is the content of extralunar (meteoritic) material? Is the process adding or removing matter from the Moon at an appreciable rate? How would these mechanisms be expected to operate elsewhere in the solar system (on Mercury or an asteroidal surface, for example)? These are among the major questions we must ask.

To answer them, the simple descriptive properties of the regolith will not suffice. We need tracers for place of origin, for direct surface exposure, and for exposure at various depths within the regolith. We also need ways of measuring time.

Fortunately, there are a rich variety of properties that have been measured and that bear on these questions. Table 1 lists some of these.

Chemical and mineral composition gives information on place of origin, or at least on distance of the sample from its source. Thus Wood et al (1970) deduced after Apollo 11 that fragments of highland material were present at this mare location. Distance of transport is more difficult to measure over shorter scales, but evidence from mare-highland contact regions such as Apollo 15 and 17 is very helpful (Duncan et al 1975).

Direct surface exposure gives rise to several changes in soil particles. Implantation of solar wind and of excess ^{40}Ar and formation of microcraters and of disordered coatings give us measures of this exposure. There follow a series of exposure indices appropriate to increasing depths: solar cosmic-ray (SCR) produced nuclear tracks, SCR-produced spallation products, galactic cosmic-ray (GCR) produced nuclear

Table 1   Useful properties of lunar soil components

Measurement	Minimum Sample Size	Reference
Impact pits (size & abundance)	Rock or grain	Hörz et al (1975a)
Major chemical & mineral composition	Soil grain (usually $d > 20\,\mu$), tens of mg for average	Meyer, McCallister & Tsai (1975)
Fraction of large agglutinates	Few mg	McKay, Fruland & Heiken (1974)
Trace element abundances	Variable—typically tens of mg	Wänke et al (1975)

Rare gas content
Components:
1. solar wind
2. cosmogenic (spallation)
  (a) GCR produced
  (b) SCR produced
3. radiogenic (nuclear decay process)
  (a) existing activities such as ^{40}K
  (b) extinct activities such as ^{129}I
4. trapped (interior)
5. "excess ^{40}Ar"
6. neutron capture
	Over 100 mg for complete study	Pepin et al (1975)
		Drozd, Hohenberg & Morgan (1975)

Cosmogenic radioactivities 1. GCR produced 2. SCR produced	Tens of mg for ^{53}Mn, grams for others	Nishiizumi et al (1976)
Neutron capture products: Gd and Sm isotopes	Tens of mg	Russ (1973)

Nuclear tracks:
1. fission
  (a) existing nuclides such as ^{238}U
  (b) extinct nuclides such as ^{244}Pu
2. high energy nuclei
  (a) SCR produced
  (b) GCR produced
	Individual grain	Poupeau et al (1975)

Magnetic properties	Variable	Gose, Pearce & Lindsay (1975)
Electrostatic properties	Individual grains	Arrhenius & Asunmaa (1973)
Surface coatings 1. Concentration of volatiles and gases		Epstein & Taylor (1975)
2. Disordered layers	Individual grains	Bibring et al (1975)

tracks, GCR-produced spallation productions (stable and radioactive), and finally, GCR-produced neutron capture effects. These last extend to a depth comparable to the thickness of the regolith. However, there is a major variation of production rates with depth (see below, Section 3).

The total density, for example of nuclear tracks, is a measure of total time of exposure. However, for various reasons such as erosion of grains by impact and sputtering, or fading of tracks over long intervals, these measures tend to be model-dependent. In addition to the experimental problems in each case, one must assume the constancy of the bombarding flux over variable periods of time.

The only direct measures of time are the radioactive clocks. These are of two types. The decay of long-lived natural radioactive nuclides (^{87}Rb for example) gives us a chronology extending over the life of the Moon. It is from radioactive "model ages" of the lunar soil that one can conclude that the soil has been a closed system for exchange of radioactive parent and daughter nuclei over nearly the whole life of the Moon. The heavy bombardment "catastrophe" at about 4.0 b. y. ago is also inferred from such data (see below).

The second type of radioactive clock is produced by nuclear reactions in the soil, caused by bombarding GCR and SCR particles. So far the longest-lived such nuclide measured in the regolith is 3.7×10^6 yr ^{53}Mn. From this the time scale has extended down to weeks. The processes being recorded are of course recent ones; one often makes the assumption that the present era is a representative one.

The minimum sample size for a measurement is important, given the heterogeneous nature of any sample of the regolith. The ideal is to use a single grain; this level can be reached for several of the listed properties. We are not of course interested in the history of every grain in our samples, but we do want to know about means and distributions, and about correlations between nearby grains.

An excellent general description of the lunar regolith is given in Chapter 3 of Taylor's book (1975). The bombardment effects listed in Table 1 are reviewed critically by Lal (1972).

The main body of this paper is in three sections. In the next we consider the flux of impacting bodies and the nature of the cratering process. Section 3 deals with the observed and measured properties of the regolith and its components, and with qualitative conclusions that may be drawn from them. The final section describes the regolith evolution models and their results.

2   MECHANISMS OF TRANSPORT

2.1   *The Meteorite Flux*

The impact of stray bodies in the Earth-Moon region is now widely (but not universally) agreed to be the main cause of the development and turnover of the lunar regolith. Its nature and history must be understood, at least in general terms, if we are to account for the observations.

The nature of the population of these bodies is an important subject in its own right (Wetherill 1974). In this article we confine our attention to those aspects of the subject necessary for our topic. The size distribution of objects and the time variation of the flux require most of our attention.

The angular distribution, mean velocity, and bulk density of the bombarding objects can be dealt with more briefly. Over most, probably all, of the vast range of particle masses—about 10^{-13}–10^{17} g or more—the flux seems at present (see below) to be mainly of cometary origin (Wetherill 1974). As such the angular distribution of the flux is at least roughly isotropic and uniform over the lunar surface (Whipple & Hawkins 1959). The impact velocities range from less than 10 km sec^{-1} to as much as 70 km sec^{-1}. Since crater volume is roughly proportional to energy, this is not a negligible variation. However, it is still very small compared to the range of masses, and except for very small particles the velocity distribution seems to be independent of mass. It is usual, therefore, to adopt an average velocity of 20 km sec^{-1} as applying to the entire flux (Gault, Hörz & Hartung 1972).

The range of particle density is still smaller. A value near or somewhat below 1 g cm^{-3} is reasonable for the weak cometary aggregates in the visible size range. Such small cometary aggregates as well as asteroidal materials consist mainly of silicates (Millman 1972), and hence the individual grains have $\rho \sim 3$ g cm^{-3}. There is a small fraction of Fe-Ni metal grains whose $\rho = 8$ g cm^{-3}. A conventional value for ρ in the range 1–3 will not affect our discussion.

The mass distribution of a population of bodies formed by breaking and crushing, or in other ways, can be represented most simply by the law

$$\frac{dn\,(m_{\mathrm{p}})}{dm_{\mathrm{p}}} = km_{\mathrm{p}}^{-\gamma}.$$

The exponent γ does not change rapidly with m_{p}, and is often not far from 2. In industrial grinding practice, for example, freshly broken rock has a somewhat flatter distribution ($\gamma \sim 1.7$), while long grinding makes more fine particles reaching $\gamma \sim 2.3$. It can be easily seen that the value $\gamma = 2$ has the special property that the total mass flux in each logarithmic mass interval is the same. For $\gamma < 2$ most of the mass flux is in the largest projectiles; for $\gamma > 2$ it is in the smallest ones. As a result small changes in γ near 2 can have a profound qualitative effect on the development of the regolith.

When data are limited, the integral mass flux distribution

$$N(>m_{\mathrm{p}}) = [k/(\gamma-1)]m_{\mathrm{p}}^{1-\gamma}$$

is easier to determine with respectable precision, and is often used.

In the following we divide the discussion of the observed mass distribution into three broad regions, according to the nature of the effects on the regolith.

2.1.1   REGION I: $m_{\mathrm{p}} > 10^8$ g   These objects make craters larger than 100 m, which are primarily responsible for the generation of fresh regolith. Such events on the Moon are rare, and very little is known about the present-day flux. The surfaces of the lunar maria, formed 3–4×10^9 yr ago, are not saturated with such craters. Hence the crater population tells us in principle the total integrated flux (fluence) and spectrum of these bodies over long intervals. Neukum et al (1975) give a critical discussion of the crater diameter distributions [see also Greeley & Gault (1970)]. These distributions, not those of projectile mass, are most pertinent to the rate of formation of regolith (see Section 2.2 below). The value of the exponent δ in the integral diameter distribution

$$N(>D) = k'D^\delta$$

is not constant from $D = 0.1$ to 10 km, ranging from about -2 to -3, but the shape seems the same in different maria.

The work on time variation of the flux has mostly referred to the early time period. From the appearance of the highlands and maria, and the ages of the maria, it is already clear that the bombardment was much more severe before 4×10^9 yr ago. From the evidence of the lunar breccias, Tera et al (1974) concluded that the flux, after decreasing from the beginning, underwent a final flareup (catastrophe) at about 3.95×10^9 yr ago. The flux soon after that time was still higher than the long-time average, and it decreased through the interval to 3×10^9 yr ago. The fluctuations in the last 3×10^9 yr are more difficult to follow. Except for the possibility that the present is an unrepresentative time, these need not concern us here. The subject has been reviewed (Wetherill 1974, Neukum et al 1975).

2.1.2   REGION II: $1 \text{ g} < m_p < 10^8 \text{ g}$ The craters formed by these objects range from tens of centimeters to about one hundred meters. They are formed mainly within the regolith, and are chiefly responsible for its large-scale turnover. Our information on the past flux and mass distribution of projectiles on the Moon comes again from crater counts. There is data on the present flux from lunar seismic data and from the frequency of bright meteors and fireballs in the Earth's atmosphere.

Except at the upper end of this range, the crater population is clearly "saturated." That is, such craters have been formed and erased by later impacts repeatedly over the last 3×10^9 yr. The survival time is of course shortest for the smallest craters.

It can be shown (Neukum et al 1975) that, for all but unreasonably flat projectile mass distributions, the saturation population of impact craters varies as D^{-2} (integral distribution) (Neukum & Dietzel 1971). This is in fact approximately the observed distribution (Shoemaker et al 1970a, Greeley & Gault 1970). Unfortunately, then, the exponent δ in this region must be determined indirectly, for example by comparing the crater diameter (mass) at the transition from a slope corresponding to production to one appropriate to saturation on two surfaces of widely different age. Neukum & Dietzel (1971) deduce from earlier experiments the relation

$$D = k'm_p^{1/\beta}$$

for the crater size as a function of projectile mass. The size index δ is defined by β and γ. For a simple mass-energy scaling one would expect $\beta = 3$; their preferred value is 2.8, making larger projectiles more efficient than smaller ones. Their result from the exponent δ is -2.9, yielding a value for γ of 2.04.

It is very hard to estimate the error of these exponents, but it may reach or exceed 0.05 in γ, even if we assume that γ is constant over this mass range. This would be an error of a factor of 2.5 in the ratio $N(>1 \text{ g})/N(>10^8 \text{ g})$. As noted above, for γ near 2 such changes produce rather striking consequences for the regolith. It is our present view that the historical average value for γ or δ can be fixed more closely by the results of the Monte Carlo models discussed below than by any direct method.

What do we know about the present-day flux? The lunar seismometers record the impacts of bodies as small as 50 g on the lunar surface (Dunnebier et al 1975). Some tens of such impacts have been used to derive a value of γ in the range of 2.13 ± 0.06, for projectiles mostly between 10^2 and 10^5 g mass. The absolute flux so determined is lower than that derived by Neukum & Dietzel (1971). The uncertainties of calibration are considerable, and do not permit definite conclusions about time variations of the flux.

The other potentially useful method is the observation of large meteors in the Earth's atmosphere by photographic methods (McCrosky 1968, Millman 1973). The comparison is again made difficult by calibration problems. However, the much flatter mass distribution ($\gamma = 1.62$) recorded by McCrosky (1968) is disturbing.

2.1.3   REGION III: 10^{-7} g $< m_p < 1$ g   The mass distribution is steeper in this region than for larger masses. The lower limit is chosen for two reasons. First, the distribution flattens progressively for masses below this value, so that the mass flux of smaller particles is not very great. Second, since a typical soil grain has a diameter of some tens of micrometers, smaller projectiles tend to interact with individual grains and not with bulk regolith.

The evidence in this region comes from a variety of sources: visual and radar observations in the Earth's atmosphere, satellite observations (the satellite data apply mainly to smaller masses), and lunar crater and microcrater counts. The careful analysis of Gault, Hörz & Hartung (1972) relies mainly on meteor studies, using the satellite data to fix the low end of the curve.

Their integral distribution for this region

$$N(>m_p) = 9.14 \times 10^{-12} m_p^{-1.213} \text{ cm}^2 \text{ yr}^{-1}$$

($\gamma = 2.213$) has been adopted by ourselves and most other workers in the field. The index is probably uncertain by ± 0.03 or so; the error in the coefficient seems to be less than a factor of 2. The flatter distribution of Neukum & Dietzel (1971) at higher masses may then be normalized to it at $m_p = 1$ g.

Note that in doing this one ties together current and historical fluxes. There are two assumptions implied: 1. the present flux for $m_p < 1$ g is close to the average over the period of interest, which may range for various properties from 3×10^9 down to a few thousand years, and 2. the present values of γ meet the same conditions. It would be very useful to check this in both directions, for example by an analysis of bright fireballs to give the current flux and spectrum above 1 g in terms consistent with the analyses at smaller mass, or by using impact pits to obtain at least some fossil values for γ below 1 g. Neukum et al (1975) are not the only workers who have concluded that present fluxes are atypical.

2.2   The Size and Shape of Craters

The general appearance of impact craters is familiar. There is a central bowl-shaped depression and a more or less continuous concentric ejecta blanket. A small mass of ejecta fall beyond this blanket, producing in very large craters the phenomenon of rays.

Craters in the regolith are larger than those in solid rock for a given projectile, presumably because little energy is spent in breaking up the material.

The relative dimensions change only slowly with scale. For the crater itself, measured to the level of the undisturbed material beyond, the measured depth/ diameter ratio is around 1/6 (Gault et al 1974). The soil below the crater is compressed, at least in model experiments, so that in terms of matter removed it is more accurate to use a ratio of 1/8 (Gault et al 1974). Beyond the edge so defined the ejecta blanket rises rather steeply to a rounded lip, and then falls gradually outward back to the undisturbed level. The material near the lip is unusually weak and of low density; thus astronaut footprints are deeper there. The outer limit of the ejecta blanket is not well defined, but is roughly one to three crater diameters beyond the crater rim.

The size and shape of a crater and blanket are significantly affected by the force of gravity and by atmospheric pressure. Hence terrestrial experiments must be compared cautiously with lunar craters, and still more so with those found on low-g objects like the Martian satellites.

The variation of crater size with m_p is discussed above. The smallest projectiles we consider, around 10^{-7} g, must produce smaller craters because the particle size is comparable to that of lunar grains. The largest projectiles penetrate below the regolith into (presumably) more coherent material, and these craters should be undersized for that reason (Oberbeck et al 1973). Some uncertainty still exists in this area.

At the earliest stages of the impact process some lunar target material is vaporized along with the projectile itself. A larger amount (perhaps 5–10 times m_p) is melted or remelted to form glass. Material slightly farther from the point of impact undergoes a lesser degree of heating; here friction may produce melting at contact points and hence give rise to agglutinates. Material less severely heated may still show some effects, especially if the crater scale is large and cooling comparatively slow.

As the references above indicate, our present knowledge of this process comes largely from the work of Gault and his colleagues.

2.3 Other Agents of Mixing

2.3.1 SECONDARY EFFECTS OF COLLISIONS   There are several effects of collisions that may show themselves at remote points (or at a later time), and are not normally considered part of the primary process of crater formation.

The first effect is slumping. Although trenches excavated in the lunar soil or the boot-prints of astronauts may leave vertical walls for some period, slopes greater than about 30° are unstable. They will collapse at once or at some later time (Mitchell et al 1972). On a very large scale this process produces terraces in craters like Copernicus. The slope near the outer rim of a fresh crater on a flat plain somewhat exceeds this critical angle (Gault et al 1972), and the overlapping of two craters may produce still steeper slopes which then slump. This is a secondary process, resulting in much less transport of material than the primary act, but it may give rise to visible effects such as layers in a core.

Another phenomenon is the formation of secondary craters by impact of the rubble ejected in the primary event. If this takes place in the region of the continuous ejecta blanket, within a few crater diameters of the impact, it is seen as part of the primary event. The effect is most obvious in the rays of major craters. A dramatic example is the region photographed by Ranger-7, which shows clusters of shallow irregular craters or grooves. Most students of the subject regard secondary cratering as a phenomenon of limited and local significance only, and it is not incorporated in any present model. However Gault (1976) believes it may be quite important. At least the ray phenomenon needs to be considered as a major mode of transport of debris from one lunar region to another.

One unusual gardening phenomenon that has been used to good effect by Arvidson et al (1975b) is the rolling of boulders down a slope. This process produces grooves, sometimes more than a kilometer in length. Such a groove, unlike a crater, cannot easily be erased by a single later cratering event, but must be progressively worn down by many events. It is thus a useful indicator of mean gardening rates.

There are still other possible indirect effects of collision. One is the vibration (shaking) of the surface in the region around the impact, which may tend to level the surface or perhaps cause upward movement of coarse fragments and rocks (as observed when soil samples are shaken in a pan). Similar effects might occur during moonquakes of internal origin; the tidal correlation of many moonquakes suggests a progressive irreversible process at least in a few local regions. However, the internal seismic energy release rate on the Moon is on the order of 10^{-11} that of the Earth, so this cannot be a large effect.

2.3.2 ELECTROSTATIC PHENOMENA   There has been considerable interest in the idea of electrostatic levitation and transport as significant in the movement of the lunar fine soil. Two suggested mechanisms have received most attention, those of Gold (1971) and Criswell (1972). It has been demonstrated in Gold's laboratory that electrons whose energy is in the range of 1 keV can cause powders in vacuum to undergo large-scale motion. Gold (1971) concludes that this effect may transport very large masses of material on the Moon. The existence of large dry basins on the backside, in contrast to the filled dark maria on the front, is one argument cited by Gold, since keV electrons in the magnetotail would strike preferentially on the front side. He also emphasizes the lack of clear evidence of bedrock a few meters below mare surfaces, and concludes that maria contain mainly material transported from outside, and chemically fractionated in the process. The required rates of mass transport would be very large, though not necessarily so at the present time. This mechanism would appear to require a concentration of quite pure anorthositic material at the highland rims adjacent to maria, where the fractionation would have to take place. This has not yet been observed (Adler et al 1972).

Criswell (1972) discussed evidence from Surveyor photographs of illuminated dust above the lunar horizon after sunset. His proposed mechanism differs from Gold's in that the charging and transport mechanism operates only very close to the terminator, where illuminated and dark patches of ground are close together. The Surveyor data, and observations in orbit by the astronauts on some but not all missions (McCoy & Criswell 1974), are direct evidence of the presence of some levitated particles and therefore of some amount of transport. However, so far we do not know either (a) the mass rate of transport, and how it depends on lunar surface properties, or (b) the particle size dependence of transport efficiency. On general grounds one would expect small particles to move more easily, since charging is a surface effect and weight is proportional to volume. Hence a likely signature of electrostatic transport would be strong local concentrations of fine particles, which however are not seen in the returned samples. This statement, however, does not rest on a quantitative theory.

3 SURFACE PROPERTIES

We now proceed to a detailed examination of the most significant properties of the regolith. The emphasis is on the qualitative understanding that they can yield. We proceed in Section 4 to a discussion of the theoretical models.

3.1 Macroscopic Characteristics of the Regolith

3.1.1 THICKNESS OF THE LUNAR REGOLITH   The first clues on regolith evolution can be found in the thickness distribution of the contemporary regolith, which yields the rate at which it has accumulated. The first studies of this problem were based on photographic data from Surveyor missions (Shoemaker et al 1967, 1969) and Lunar Orbiter missions (Quaide & Oberbeck 1968, Oberbeck et al 1973). The Shoemaker group evaluated the regolith thickness either from the shallowest crater exhibiting blocks on the rim (Shoemaker et al 1967) or from the transition diameter of craters between the "steady-state" and the "production" curves in the local distribution of crater diameters (Shoemaker et al 1969). Thicknesses in excess of 1 m were deduced for mare regions, in contrast to the very low thickness observed on Tycho's slopes (~ 10 cm). Quaide & Oberbeck (1968) (see also Oberbeck & Quaide 1967) used the transition depth between bowl-shaped and flat-floored craters as a measure of the regolith thickness. The resulting distributions for mare regions showed a spread between 5 m and 15 m with a mean thickness ~ 8 m. More recently, active seismic experiments (Kovach, Watkins & Landers 1971, Kovach, Watkins & Talwani 1972) at the Apollo nonmare sites yielded 8.5 m and 12.2 m for Apollo 14 and 16 respectively. The meaning of this value for these sites is unclear. For the mare sites Apollo 11, 12, and 15, passive seismic methods yield thicknesses near 5 m (Nakamura et al 1975). The active seismic experiment performed on the Apollo 17 site is difficult to interpret because of complex layering. Thicknesses of 8.5 m (Dunnebier, Watkins & Kovach 1974) and 4 m (Cooper, Kovach & Watkins 1974) have been obtained. Lower limits can be derived from the lengths of the deep drill core tubes from sites of Apollo 15, 16, and 17, which range from 2.4 m to 3.2 m. To produce the regolith we can estimate a mean growth rate of the order of 2 m/b.y., keeping in mind that regolith growth was certainly faster just after mare solidification than now, because of the progressive shielding of the bedrock from impacts.

3.1.2   THE GEOMETRIC SURFACE OF THE REGOLITH   The main characteristic of the regolith outer surface is its flatness and smoothness. The mean slope on any scale larger than 1 cm is only a few degrees. Thus, any evolution model based on an all-meteoritic hypothesis has to meet the challenge: can we make a surface smooth by firing hypervelocity bullets at it? As the impact craters are the most significant departures from the overall smoothness, this problem resolves into two others: How deep and numerous are the wounds, and how fast do they heal? The crater diameter distribution has already been discussed in Section 2, where it was used to provide the upper part of the meteoritic mass spectrum. The boundary between the "steady-state" and the "production" regions seems to be at a diameter on the order of 100 m. Thus the lifetime of larger craters exceeds 3 b.y. These do not heal at all, but they represent only a few percent of the mare surface. Two mechanisms can account for the erasure of small craters (< 100 m) and permit the steady-state situation: obliteration by later large craters or erosion by smaller impacts. The first mechanism alone would produce a surface saturated with sharp craters. However, while the surface appears saturated with craters at solar elevations lower than 1°, only about 10% of the surface appears covered by craters when the solar elevation is 20°. This shows that erosion processes reduce internal slopes of craters to 1% on the mean before obliteration occurs (Soderblom 1972). When we consider craters on a smaller scale (~ 1 m), erosion predominates still more, as the ratio of smaller events (erosion) to larger events (obliteration) must be larger than for 100 m craters (because the flux spectrum is steeper in region III). This conclusion is strengthened by a study of boulder track erasure on slopes (Arvidson et al 1975b). It was shown that the filling rate of shallow depressions is $(5 \pm 3) \cdot 10^{-6}$ cm yr^{-1} for features a few meters wide, but hundreds of meters long. Associated lifetimes ($\sim 10^7$ yr) are thus much smaller than the obliteration lifetime (~ 1 b.y.). We will see below how one model (Arnold 1975b) deals with this quantitatively.

3.1.3   STRATIGRAPHY OF THE LUNAR REGOLITH   Information on the structure of the regolith along the vertical axis can be found in seismic experiment data (see Section 3.1.1) and in the deep drill core tubes brought back by the Apollo missions. The seismic experiments can be interpreted in terms of a rather homogeneous layer, both from top to bottom and from site to site (Kovach, Watkins & Talwani 1972, Nakamura et al 1975). Homogeneity is also the main characteristic of the core tubes: with few exceptions all markings are subtle. The same physical and chemical effects can be observed over the whole length of the core, although with different intensities. Thus the regolith must have undergone severe mixing during its growth. On the other hand, a careful examination of the cores reveals a great number of layers, which can be distinguished on the basis of slight variations in albedo, color, or texture. When physical and chemical effects are examined, new stratigraphic units can be defined, which sometimes coincide with "visual" layers. Thus, problems arise in defining layers unambiguously, which seriously hamper the usefulness of such quantities as the number of layers and the layer thickness distribution. From the careful discussion of Duke & Nagle (1975) however, we can conclude that most of the length of the cores is contributed by layers thicker than

0.5 cm. As a conclusion, regolith evolution models have to meet the following constraints: (a) The regolith must be mixed to depths of several meters. (b) Many stratigraphic boundaries must be preserved during this process.

3.2 Physical Characteristics of the Lunar Regolith

Many of the visible physical characteristics of the regolith material have been modified during its evolution. These "physical maturation" effects have been active when the constituents were very close to the surface (a few cm for galactic cosmic ray tracks and rock rupture, 1 mm or less otherwise). This part of the maturation record can thus provide information about the near-surface exposures of constituents during their regolith history. Two kinds of effects are then to be distinguished: 1. The "cumulative" effects integrate the corresponding production rate over the whole regolith history of a constituent. They can yield estimates of the total time spent in the production regions, the "exposure age" of the constituent for a given effect. 2. "Steady-state" effects are due either to saturation of the sample or to a competing destructive effect. These will inform us on the last near surface exposure (or exposures) of a constituent. For each effect we discuss production rates, cumulative or steady-state character, and constraints that can be derived for the regolith evolution.

3.2.1   SIZE DISTRIBUTION OF THE REGOLITH'S CONSTITUENTS   The size distribution of crater ejecta is known from Hörz (1969) for small impacts in basalt and from Polatty et al (1965) for large impacts (nuclear explosions). The coarser fractions dominate, and the "graphic" mean grain size[1] depends upon the crater's size (more than 1 mm for $D \sim 1$ cm; more than 1 cm for $D \sim 10$ m). Since Oberbeck et al (1973) showed that craters larger than ~ 100 m were the main contributors of fresh regolith material, we can conclude that most of the regolith mass began its regolith history in fragments larger than 1 cm.

In the actual regolith, the studies of Duke et al (1970), McKay et al (1972) and McKay, Fruland & Heiken (1974), Lindsay (1971, 1973), and King, Butler & Carman (1971) have shown that although large differences appear between samples the graphic mean size is in the range 30–250 μ. The constituents larger than 1 cm represent only a few percent of the regolith mass (Shoemaker et al 1970b). The size distribution of the regolith has thus changed drastically during evolution. Four processes are presently known to modify the constituent size: 1. micrometeoritic catastrophic rupture, 2. micrometeoritic erosion, 3. solar wind sputtering, 4. impact induced melting-agglutination. Solar wind sputtering is shown to be negligible except for very small (1 μ) grains. The micrometeoritic comminution processes have been extensively studied (Gault 1969, Shoemaker et al 1970b, Hörz, Schneider & Hill 1974, Hörz et al 1975c). Computed rock erosion rates are >0.3 mm m.y.$^{-1}$ for rock surfaces (Hörz et al 1975c), in agreement with track results (see next

---

1 When size fractions (by mass) are plotted on a logarithmic size scale, the mean size and the standard deviation of the distribution are considered as representative of the sample's granulometry and called the "graphic" mean size and standard deviation.

paragraph). An "erosion" lifetime <30 m.y. cm^{-1} of size can thus be derived. Catastrophic rupture lifetimes have been evaluated as ~ 1 m.y. cm^{-1} of size (Hörz, Schneider & Hill 1974), for constituents larger than 1 cm. It appears that catastrophic rupture dominates in this size range. When considering fragments <1 cm, it must be noted that the mass influx of micrometeorites sharply drops for masses lower than 10^{-7} g (<40 μ in size) (Gault, Hörz & Hartung 1972). As micrometeorites induce catastrophic rupture in grains ~ 10 times their size (Hörz et al 1974), erosion processes should become relatively unimportant for grains <400 μ in size. However, as the melted volume is a few times the micrometeorite volume (Gault, Hörz & Hartung 1972), catastrophic rupture should become less important than impact melting (see below) for grains <100 μ in size. Thus the very large fraction of crystalline grains with a size of $\lesssim 50$ μ (up to 50%) (McKay, Fruland & Heiken 1974) does not have an obvious explanation. The melting and welding processes active at the lunar surface oppose comminution processes. The large fraction of "glassy agglutinates," i.e. crystalline grains welded together by glass, when compared to the amount of pure glass (10 to 100 times more) was first noted by Duke et al (1970). It was shown (Duke et al 1970, McKay et al 1972) that the glass in these agglutinates had the same overall composition as the bulk soil. They concluded that this glass was the result of soil melting on impact. The agglutinate fraction ranges between 10% and 60% of the size fractions from 50 μ to 250 μ. Dran et al (1970) showed that the smallest fraction (<1 μ) was depleted in agglutinates: $\gtrsim 80$% of the particles were crystals and the rest were mainly glass spherules. As a conclusion, agglutination processes form large (>50 μ) particles and remove most of the material from the smallest fractions (<5 μ).

These textural and granulometry results, together with albedo and track measurements, are central to the notion of maturity: significant correlations were found between agglutinate content and track densities (McKay et al 1971) and between agglutinate content and mean grain size (McKay et al 1972). Other correlations had already been inferred between albedo and impact glass content (Adams & McCord 1970, 1971). Maturity reflects the amount of meteoritic reworking, and should be related to an "exposure age" of the soil at the surface (McKay et al 1972). Several maturity indexes have been proposed: 1. Mean grain size: The graphic mean grain size decreases from 130 μ (very immature soil) to around 60 μ (very mature soil), 2. Graphic standard deviation: Two groups (McKay, Fruland & Heiken 1974, Lindsay 1974) claim it drops when the mean grain size decreases, whereas another (Butler & King 1974) claims it increases. Thus the question of whether or not the standard deviation is a maturity index, and, if so, whether the size spread should decrease or increase with maturity, is not yet settled. 3. Agglutinate content. It increases from 10% (immature) to 50% (mature) (McKay, Fruland & Heiken 1974). 4. Total specific area ($m^2 g^{-1}$): It increases to 0.6 $m^2 g^{-1}$ (mature) (Holmes, Fuller & Gammage 1973, Cadenhead & Stetter 1974, Gammage & Holmes 1975). 5. Albedo: Dark soils are mature, light soils immature (Gold, Bilson & Yerburg 1972). This series of characteristics points toward a steady evolution of a soil, from light, coarse-grained, and glass-poor to dark, fine-grained, and agglutinate-rich. Maturity has thus been linked with a time of exposure to surface effects or "exposure age." In a discussion at the end of the paragraph, we show that although maturity indexes of soils are meaningful parameters, soil exposure ages could be very difficult to define. The use of high magnetic susceptibility as an operational definition of agglutinate material (Rhodes et al 1975) may add to this problem.

3.2.2 NUCLEAR TRACKS The nuclear track record of lunar samples may provide the single most most useful tool to explore the near surface evolution of the regolith. Such tracks are present in every constituent of the regolith, from the grains less than 1 μ up to meter-sized boulders. Moreover, soil samples only a few milligrams in weight contain thousands of grains, so that good statistics can be obtained and good spatial resolution achieved in lunar cores.

These nuclear tracks can have several origins (Fleischer, Price & Walker 1970). The largest number, and the most useful for our purpose, result from cosmic ray bombardment. Nuclei of the Fe group come from the Sun in solar flare events (SCR or "solar cosmic rays"). These particles, with energy of a few MeV nucleon^{-1} and up, come to rest in distances measured in micrometers up to tens of micrometers. Hence gradients of track density are possible in a single grain. The Fe group nuclei, at energies of GeV nucleon^{-1} or more, are found in the "galactic cosmic rays" (GCR). These are more penetrating, and dominate track production at depths of millimeters to centimeters. Fission tracks, and spallation recoil tracks, are present but not relevant for this discussion.

Tracks are observed by a very wide range of techniques: directly by transmission electron microscopy (TEM) ($\gtrsim 10^{10}$ t cm^{-2}) after chemical etching, by scanning electron microscopy (SEM) ($10^8 \lesssim \rho \lesssim 10^{10}$ t cm^{-2}), by optical microscopy (OM) ($\lesssim 10^8$ t cm^{-2}), or by the "replica" technique ($\sim 10^{10}$ t cm^{-2}). We restrict our discussion to experimental results on three sets of samples: crystals in igneous rocks, 100-μ-sized grains, and 1-μ-sized grains, as these kinds of samples have been extensively studied. Track densities in breccias (MacDougall et al 1973, Dran et al 1972) present numerous problems in interpretation, and are mainly relevant to the study of the period 3–4 b.y. ago. It should also be noted that other fractions could yield interesting clues on the regolith evolution [e.g. coarse fines, (Langevin & Maurette 1976)].

Igneous rocks The first studies of the track record in lunar igneous rocks (Crozaz et al 1970, 1972, Barber et al 1971, Fleischer, Hart & Comstock 1971, Fleischer et al 1970) showed that they present a very steep gradient of track densities from $\gtrsim 10^9$ t cm^{-2} in the first 10 μ down to $\sim 10^6$ t cm^{-2} at 1 cm depth. From the shape of the gradient a micrometeoritic erosion rate was derived, between 0.4 and 1 mm m.y.$^{-1}$ of exposure (Crozaz et al 1970).

The track density measurements in rocks seemed also to provide a straightforward means of determining the track production rate versus depth below a plane surface: one divides the measured track densities by the exposure age of the rock on the surface (first assessed from GCR + SCR flux estimations, this age is now determined from rare gas analysis or microcrater counts). An independent determination from the Surveyor glass camera (Crozaz & Walker 1971) and some of the most recent studies (Walker & Yuhas 1973, Bhandari et al 1973, Hutcheon, MacDougall & Price

1974, Blanford, Fruland & Morrison 1975) agree within a factor of 5 in the $\gtrsim 1$ mm region (GCR tracks). However, they disagree by a factor up to 100 in the region < 1 mm (SCR tracks): at a depth of 50 μ, the derived production rates m.y.$^{-1}$ are 5×10^8 (Bhandari et al 1973), 5×10^9 (Crozaz & Walker 1971, Blanford, Fruland & Morrison 1975), and 5×10^{10} (Hutcheon, MacDougall & Price 1974). These discrepancies cannot be entirely due to differences in experimental techniques: cross checks of OM–SEM and SEM–replicas (Crozaz et al 1970, Hutcheon, MacDougall & Price 1974), as well as a joint study on rock 14310 (Yuhas et al 1972), have shown agreement within a factor of 2. The main problems seem to lie in etching calibration, evaluation of the surface erosion, and differences in thermal histories, leading to differences in track annealing (Dran et al 1972). At the present time, a "standard" track production rate cannot be assessed in the less than 1 mm region. For a complete review up to 1974 see Yuhas (1974).

100-μ grains   The track density distribution in a 100-μ grain can be represented by the density at the "center" of the grain, ρ_c, and the density gradients near the edges $r_\lambda = \rho_\lambda/\rho_c$, where ρ_λ is the density measured at a given depth λ below the grain's surface (Comstock et al 1971).

Distributions of ρ_c were first determined in lunar soils by Crozaz et al (1970), Lal et al (1970), and Fleischer et al (1970). Later studies (Crozaz et al 1972, 1974, Borg & Vassent 1972, Bibring et al 1975) showed that in almost every 100-μ grain ($> 95\%$) ρ_c was higher than 10^7 t cm^{-2} and lower than 10^{10} t cm^{-2}. Results of Berdot et al (1972), Fleischer & Hart (1973a), and Goswami & Lal (1974) indicated a larger proportion ($> 10\%$) of grains with $\rho_c < 10^7$ t cm^{-2}. This could be due to a choice of grains in a wider size range [100 μ–600 μ (Berdot et al 1972)], since Crozaz et al (1972) working on the same samples showed that lower densities ($< 10^7$ t cm^{-2}) were observed in the coarser fractions ($> 150 \mu$). Given the sharp decrease with depth of track production rates, this implies that almost every grain in the first two meters of the regolith (maximum sampling depth) has spent some time in the uppermost centimeter. The high proportion ($\sim 50\%$) (Bibring et al 1975) of grains with $\rho_c > 10^9$ t cm^{-2} shows that many grains have been at or near the very surface (within a few hundred μ) (Crozaz et al 1970, Comstock et al 1971). This seems to set a drastic test for regolith evolution models: how to bring nearly every grain in the regolith near the surface? In our opinion, two remarks can help in assessing the problem correctly: 1. The layers to which the grains belong are usually less than 10 cm thick. As "meteoritic" models (see Section 4) identify these layers with old ejecta blankets, layer boundaries are fossil lunar surfaces, so that grains have actually been at least once a few centimeters from the surface. 2. In Section 3.2.1, we saw that every known comminution process is active only at or very near the top surface of the regolith. As 100-μ grains probably do not result from primary impact in solid rock (see Section 3.2.1), they are the result of comminution processes and began their regolith history close to the surface. We can conclude that for meteoritic models the problem of ubiquitous track densities is directly linked with the problems of grain formation and regolith stratification.

Within the limits we have discussed, the track density distributions vary widely from soil to soil, even from the same site [see for instance Figure 5 in Crozaz et al (1974)]. Until 1975, the best depth resolution achieved was one centimeter in the lunar core tubes (Phakey et al 1972, Fleischer & Hart 1973a, Crozaz et al 1974, Goswami & Lal 1974). A recent study of cores impregnated with plastic (Price et al 1975, Goswami, Braddy & Price 1976) revealed large variations in distributions down to the millimeter scale: minilayers (~ 5 mm) and microlayers (~ 1 mm) were found where track densities were much higher (up to a factor of 10) than in adjacent layers. This method has the best depth resolution to date in defining stratigraphic units (it remains to be seen whether these units are "true" ejecta layers or the heavily mixed, heavily irradiated upper parts of larger ejecta layers $\gtrsim 1$ cm); see Bibring et al (1975). In order to characterize the distributions, two parameters have been mainly used: 1. The "minimum" central track density, ρ_{min}, defined either from a single grain (Fleischer, Hart & Giard 1974) or several grains (Crozaz et al 1970). 2. The "quartile" track density, ρ_q, below which 25% of the central track densities lie (Arrhenius et al 1971, Goswami & Lal 1974). These authors showed that ρ_{min} and ρ_q were correlated with the mean grain size and agglutinate content of the soils. However, Bibring et al (1975) showed that they were not appropriate maturity indexes of mature soils. These authors proposed a new parameter, the density $\hat{\rho}$ of the "peak" in the track density distribution. However, great caution is needed when trying to infer "exposure ages" from the ρ_{min}, ρ_q, and $\hat{\rho}$ parameters, for the following reasons. 1. We will see in Section 3.4 that one cannot easily derive meaningful exposure ages for *soils* from a maturity index. 2. The production rates used to translate central track densities into exposure ages are controversial (see above). 3. In a high proportion of crystals (>20%) most track densities seem to have been acquired before their release as individual regolith grains (see gradients discussion). Track data from such grains should be removed (if possible) from the distributions. 4. Shock erasure processes (Fleischer & Hart 1973b) as well as thermal annealing processes (Maurette, Pellas & Walker 1964, Borg et al 1973, Price et al 1973) may significantly lower track densities over periods of several hundred million years. Track accumulation would then no longer be a "cumulative" effect, and rock derived track production rates would thus not be appropriate.

The first studies of the track distribution inside 100-μ grains (Crozaz et al 1970, Comstock et al 1971) showed that few grains (20%) had large gradients between the 5-μ region and the center ($r_{5\mu} = \rho_{5\mu}/\rho_c$). Further studies (Hart, Comstock & Fleischer 1972, Comstock 1976, Bibring et al 1975) can be summarized as follows: 1. Grains with $\rho_c \lesssim 10^8$ t cm^{-2}: few gradients. 2. $10^8 < \rho_c < 10^9$: a low proportion of the grains ($\sim 20\%$) have gradients, which are usually highly anisotropic. 3. $10^9 < \rho_c$: about 50% of the grains have isotropic gradients in the range $1.5 < r_{5\mu} < 2.5$. The other half exhibits either no significant gradient or anisotropic high gradients ($r_{5\mu} \sim 5$). Separate studies by Arrhenius et al (1971) and Goswami & Lal (1974) exhibit grains with very high $r_{5\mu} (\gtrsim 10)$ [these grains can be possibly interpreted as highly "anomalous" grains in the model of Bibring et al (1975) (see below)]. The interpretation of these results is quite straightforward for grains with $\rho_c < 10^9$: grains with $\rho_c < 10^8$ have never (hardly ever) been less than 1 mm from the surface. Grains with $10^8 < \rho_c < 10^9$ have been only a few times near the surface, so that their gradients are highly anisotropic [for the correlation between the number of exposures and isotropy of 5-μ gradients, see Comstock (1976)].

The results for grains with $\rho_c > 10^9$ are difficult to reconcile with the relatively steep slopes of production rates: using a reasonable lower limit to the gradient in the 5–50-μ region, track densities at 5 μ and 50 μ should be in a ratio ~ 100 if they have been accumulated in a single direct exposure. If a large number of exposures (> 10) are postulated, the center receives relatively 5 times more tracks than any given edge, and the "theoretical" gradient is lowered to $\gtrsim 20$. The remaining discrepancy between this result and experimental data leads to one of the two following hypotheses: 1. A very efficient erosion mechanism removes 10–50 μ from the surface during the history of the grain. Track densities near the edge would then be in the "steady-state" situation and would be divided by a factor $\leqslant 10$ (Bibring et al 1975). 2. The grains spend most of their exposure time "at the surface" below a dust covering ~ 10 μ in thickness (Poupeau et al 1975). A difficulty exists in interpreting the high proportion of grains with $\rho_c > 10^9$ and no isotropic gradient. Bibring et al (1975) interpret these grains as having acquired most of their irradiation record during exposure on the outer surface of a parent rock. If such were the case, these "anomalous" grains should be omitted from ρ_c statistics, because their central track densities would have no relation to the evolution of the grain in the soil.

Other studies beginning with the work of Borg & Vassent (1972) showed that in some of the high-track density grains ($\sim 20\%$), the edges presented a 2-μ-thick track-rich zone ($\rho \gtrsim 10^{10}$ t cm^{-2}). This feature is a result of direct exposure of the surface to space, and has been tentatively interpreted as resulting from the high energy tail of the solar wind (Borg et al 1974). An interesting feature is the very high anisotropy of such gradients (usually only on one edge).

Maturity indices using track gradients have been defined by Arrhenius et al (1971), Berdot et al (1972), and Goswami & Lal (1974). As gradients are produced during near surface exposure, these indices should be related to the mean time spent by the grains in the upper millimeter of the surface.

1-μ grains It has been shown (Borg et al 1970) from TEM observation of 1-μ grains that many of them presented very high latent track densities ($\sim 10^{11}$). Further studies (Borg et al 1971, Phakey et al 1972, Borg & Vassent 1972) showed that the proportion of track-rich grains was variable from one soil to another, and that mature soils (as defined by albedo, grain size, agglutinate content) had more than 90% track-rich grains. These results were difficult to interpret when compared with track production rates: very long exposure times on the surface (~ 1 m.y.) were required. However, Bibring et al (1975) showed that track densities measured in a *section* of a rock or large grain were being compared with track densities counted over the whole *volume* of a 1-μ grain (TEM observations). The actual track densities per cm^2 of section are lower by a factor $\gtrsim 10$, and the $\sim 10^{10}$ t cm^{-2} mean track density derived for mature soils is remarkably close to the highest track densities observed in the first 1 μ of 100-μ grains.

In summary, tracks provide our richest single source of data on the history of regolith grains. However, in order to use track parameters properly, the following steps are still needed: 1. To fix a standard track production rate as a function of depth in the first millimeter below a flat surface. 2. To define the conditions of exposure on the surface and the active erosion processes. 3. To assess correctly the degree of track fading due to shock or thermal annealing.

3.2.3 SURFACE TEXTURE   The surface properties of lunar samples are dominated by the micrometeoritic bombardment (rocks) and the solar wind (small grains). We discuss in the following paragraphs the three sets of samples defined in the preceding section.

Rocks   The best-studied features of rock surfaces are the microcraters (Hartung & Hörz 1972, Neukum et al 1973, Schneider et al 1973, Blanford et al 1974, Hartung et al 1975). The distribution of microcraters has been used to infer surface exposure ages for rocks. However, there are again difficulties of interpretation. Microcraters can help in determining the age of a relatively fresh surface on a rock (Hutcheon, MacDougall & Price 1974). Discrepancies between microcrater ages and rare-gas ages can yield the amount of tumbling of the rock during its exposure history. Accretionary particles can also provide a useful exposure index. As discussed in Morrison et al (1973) and Blanford et al (1974), these particles produce a "patina" that significantly lowers the albedo of rock surfaces.

Exposed surfaces are also covered with cracks, produced by impact, but these have not been much studied. As one of us (JRA) can attest, however, it is easy to distinguish an exposed surface from a pristine one by its behaviour on grinding.

100-μ grains   Blanford et al (1974) have discussed the surface characteristics of middle-sized grains. They showed that the surface of many track-rich grains exhibit microcraters ($\gtrsim 0.3$ μ) and accretionary particles, mainly glassy disks ($\lesssim 2$ μ), and that the distributions of microcraters and of glassy disks ("pancakes") are very anisotropic. Detailed statistical counts (Poupeau et al 1975) showed that only 20% of the track-rich grains ($\rho_c \gtrsim 10^9$ cm^{-2}) have microcrater-rich surfaces ($\sim 10^6$ craters cm^{-2}), and that these surfaces represent only a small fraction of the total surface of the grain. Although still highly anisotropic, the distribution of pancakes (present on $\sim 90\%$ of the track-rich grains) is more regular, pancakes being found on every major surface of the grain. Finally, no clear-cut correlation was found between either microcrater or pancake densities, and track densities (at the center or at 5 μ). This data can be interpreted in terms of either one of the two hypotheses about track gradients (see earlier Section on 100-μ grains): 1. If erosion processes remove several tens of microns from the surfaces of high-density grains, the surface characteristics are in "steady-state," and are not correlated with the total exposure or cumulative surface effects (such as central track density). Moreover, if erosion proceeds by flaking off chunks ($\gtrsim 2$ μ thick), fresh surfaces will appear just after such events. This would be consistent with the high anisotropy of crater and pancake densities, and also the anisotropy of 2-μ gradients (see Section 3.3.2). This hypothesis is, however, difficult to fit with the low correlation between crater and pancake densities. 2. A 10-μ dust covering would shield most of the grain from direct exposure effects, but could leave "windows" to the micrometeorites and would not stop the small glass droplets responsible for pancake formation (Poupeau et al 1975). It is, however, difficult to reconcile with the small fraction of particles below $\sim 10\ \mu$ in the bulk soils.

1-μ grains   The surface characteristics of 1-μ grains are dominated by solar wind-damage features. Dran et al (1970) showed that a large proportion of 1-μ feldspars exhibited a rounded habit and a thin "amorphous coating" of metamict material, which covered the grains isotropically. It was shown (Bibring 1972, Bibring et al 1974b) that the proportion of feldspars with an amorphous coating varied from 10% in an immature soil to 90% in a very mature soil. On the other hand, no ilmenite grain, even in very mature soils, was found with an amorphous coating. This was interpreted using simulation experiments (Bibring 1972, Bibring 1974a), which showed that a flux of 5×10^{16} α particles cm^{-2} was needed to metamictize a feldspar surface, whereas 5×10^{17} α cm^{-2} were needed for ilmenite. Taking into account the isotropy of the amorphous coating, a 2000-yr exposure to the solar wind is sufficient to produce an amorphous coating on feldspar, whereas 20,000 yr are needed for ilmenite crystals. An interesting limit is thus set for the mean exposure time T_0 of a 1-μ grain in a very mature soil: $2000 \lesssim T_0 \lesssim 20,000$ yr. Other evidence for exposure can be found in the presence of tiny accretionary particles (Morrison et al 1973).

Solar wind sputtering erodes the grain surfaces. If high enough rates are postulated, the amorphous coating is in a steady state, which would yield saturation values for implanted species concentrations (see Section 3.3.1). Sputtering rates have been determined either by determinations on small grains (Bibring et al 1974b) that yield 0.3 Å yr^{-1} or by experiments on a flat target (Zinner & Walker 1975, McDonnel & Ashworth 1972) that yield much lower rates (~ 0.03 Å). This discrepancy may possibly be due to effects of the angle of incidence. The isotropy of amorphous coatings suggests frequent changes in orientation for 1-μ grains at the surface. Gas blasts from impacts, the "lunar winds" (Rehfuss 1972), can provide efficient enough turn-over rates. The problem of irradiation effects in small grains has been reviewed by Maurette & Price (1975).

3.2.4   OTHER PHYSICAL EFFECTS   We do not discuss in any great detail thermo-luminescence studies (Dalrymple & Doell 1970, Doell & Dalrymple 1976, Crozaz et al 1972), or studies of the magnetic properties of the lunar soil. The former is a "steady-state" effect due to the competition of cosmic-ray-induced thermo-luminescence center formation and thermal erasure. The time scales involved ($\sim 10^4$ yr) (Dalrymple & Doell 1970) are too short to yield significant limits on regolith evolution. Magnetic properties have been used to separate agglutinates from the bulk soil (McKay et al 1972): impact induced $Fe_{II} \rightarrow Fe_0$ reduction creates iron particles embedded in the glass. Ferromagnetic resonance studies of lunar samples (Gose & Morris 1976, Cirlin et al 1976) can be used to define stratigraphic units but the lack of any well-defined production rate makes the interpretation of this data by regolith evolution models difficult.

3.3   Chemical and Isotopic Effects

3.3.1   SOLAR WIND   The most intense particle flux striking the moon is the solar wind. About 10^8 nuclei cm^{-2} with energy on the order of keV nucleon^{-1} reach the surface each second. If retained this would amount to a few moles cm^{-2} over the lifetime of the maria. The composition is at least approximately solar, with a predominance of H and He. Ions are implanted to a depth of the order of 10^{-5} cm (see Section 3.2.3).

The most thoroughly studied solar wind components in the lunar soil are the rare gases (Pepin et al 1974). The flux of He is so large that exposed layers of crystals would contain as many He atoms as lattice atoms in a few thousand years; it is hardly surprising that leakage occurs and the concentration saturates far below this level. Of special interest to us is the fact that the ^{3}He/^{4}He ratio (about 5×10^{-4}) seems to have varied over the history of the Sun (Eberhardt et al 1972). This raises the possibility of defining the period in which solar wind He was deposited in a given soil sample. However, there are a number of difficulties and little progress has yet been made. Solar wind Ne is also comparatively abundant, and leakage has also occurred here.

Observations of Ar show a very high content of ^{40}Ar, beyond anything possibly coming from the Sun. This "excess" ^{40}Ar is thought to have been evolved from the interior of the Moon (Manka & Michel 1970), to have been ionized by charge exchange with the solar wind, swept up in the stream, and implanted with it in the soil. Here again, since the emission of ^{40}Ar from the interior must have decreased from the mare-forming era to the present, the possibility exists of an absolute time marker (Yaniv & Heymann 1972).

The heavier rare gases, Kr and Xe, are the least likely to be lost by diffusion. Their abundances in the stream are small, so they do not build up to high concentrations, and their large size makes diffusion slow.

Other elements in the soil have also been identified, with reasonable certainty, as mainly or entirely of solar wind origin. The three most studied are H, C, and N.

All or nearly all in the H in lunar soil (before its return to Houston) is of solar origin (Epstein & Taylor 1973). It contains very little deuterium. Its concentration must become saturated even more quickly than that of He. However, its chemical activity in reducing Fe may be cumulative. As indicated above, the factors responsible for the concentration of Fe grains are not yet well defined. The carbon content of the regolith also seems to have a large component from solar wind (Epstein & Taylor 1973).

An interesting current case is that of nitrogen (Goel & Kothari 1972, Becker & Clayton 1975). The ratio ^{15}N/^{14}N is increasing with time in the solar wind (Kerridge 1975, Becker & Clayton 1975). This is a possible marker for the date of the exposure of a given layer of lunar soil at the surface. Since older layers should lie lower, the trend of the ratio should be monotonic.

The mechanisms by which saturation is reached in solar wind bombardment have been studied in detail by Bibring et al (1974b). Because of its combination of high abundance and sputtering effectiveness, the sputtering is mostly due to ^{4}He.

Since all species penetrate into the grain to comparable depths, one mean saturation time for surface exposure corresponds to the time for removing material from the mean depth of penetration. This varies for different minerals, and is especially high for ilmenite. Chemically active elements like C and H may come off as compounds.

3.3.2 METEORITIC COMPONENT   It is well known that the lunar soil contains on the order of 1% of extralunar material. The early differentiation of the lunar crust depleted the surface layers in the siderophile (metal-phase) elements, particularly Ni and the noble metals. These are present in mare basalts, for example, in very low concentration; mare soils contain much more. Unaltered meteoritic fragments are rare. Presumably the bulk of this component has been vaporized and redeposited on surfaces (Boynton et al 1976).

The meteoritic component has been studied especially by Anders and coworkers (1973). The percentage of meteoritic material must rise as the soil matures; however, since the regolith grows thicker at the same time the rise in concentration is much less than linear with time. The models, then, are required only to account for the general order of concentration of siderophile elements. Anders et al (1973) have studied the distribution of siderophile elements with a view to identifying material from individual projectiles such as the Imbrian planetesimal. Such early material is a background from which the contribution of recent smaller impacting masses must be distinguished.

3.3.3 SPALLATION-PRODUCED NUCLIDES   The depth of penetration of nuclear particles in the MeV and GeV range into the lunar surface is of course greater than that of the solar wind. The most important bombarding particles for the present discussion are protons, though ^{4}He nuclei play some role.

High-energy particles in this region of space are of two types. The "galactic" cosmic-rays (GCR) come from outside the solar system. Their energies are of the order of 10^9 eV (GeV) typically; the mean interaction length of protons in lunar soils is ~60 cm. The observed nuclear reactions are due mainly to secondary neutrons in the range of MeV up to tens of MeV; a few of these are found even at a depth of 2–3 meters. The Sun also emits particles in major flare events. The observed effects of these so-called SCR particles are caused mainly by 10–100 MeV protons. The penetration depth of these protons is on the order of 0.05–1 cm.

Because even the SCR protons penetrate through soil grains their spallation products tend to be firmly implanted. Only He and to some degree Ne among the spallation products show losses.

The isotopic pattern of spallation-produced rare gases is clearly distinguishable from other components. Thus the three Ne isotopes are produced in nearly equal abundance.

To calculate a cosmic-ray bombardment age using, for example, ^{38}Ar, one proceeds as follows. Any solar wind ^{38}Ar is corrected for, using ^{36}Ar data. The production rate of ^{38}Ar can be estimated from some radioactive species (see below) like ^{36}Cl. More usually, one relies on some a priori calculation such as Reedy's

(1976). The net concentration divided by the production rate gives the age. Since the production rate varies with depth, and in any case is not yet precisely known, this number is model-dependent.

Radioactive nuclides with half-lives shorter than 10^3 yr may be assumed to have been produced at the present location, and hence are useful in making models of production rates (Reedy & Arnold 1972). The longer lived species, especially ^{53}Mn $(t_{1/2} - 3.7 \times 10^6$ yr) and ^{26}Al $(t_{1/2} = 7 \times 10^5$ yr), also show apparently undisturbed profiles below 30 cm in cores so far studied (Imamura et al 1974, Rancitelli et al 1975). Nearer the surface, disturbances have been seen (Nishiizumi et al 1976, Rancitelli et al 1971). These data provide the most definite information we have on the chronology of gardening processes in the regolith.

3.3.4 THERMAL NEUTRON PRODUCTS   The last useful product of nuclear reactions in the lunar surface is a flux of thermal neutrons, on the order of one neutron cm^{-2} sec^{-1}. Since neutrons near the surface have a good chance to escape, those that reach thermal energies are on the average deeper in the regolith than the neutrons at MeV energies. The distributions have been calculated by Lingenfelter, Canfield & Hampel (1972) and measured in the regolith by Burnett & Woolum (1974).

One radioactive nuclide, ^{60}Co $(t_{1/2} = 5.2$ yr) (Wahlen et al 1973), and several stable ones have been measured. Most important are the rare earth species ^{158}Gd, ^{156}Gd, and ^{150}Sm, formed by neutron capture (Russ 1973). In addition, the isotopes ^{80}Kr and ^{82}Kr have been shown in some samples to include a component due to neutron capture in bromine, while heavy Kr and Xe isotopes are produced by neutron-induced fission of ^{235}U. A case of special interest is ^{131}Xe, produced by epithermal neutrons on Ba (Marti, Lightner & Osborn 1973).

Neutron capture species are produced at a mean depth of roughly 1 m, and production is still important at the maximum (2–3 m) depths so far sampled. Thus it is possible to regard the exposure age for these species as at least crudely equivalent to a time of residence in the top few meters of the regolith. Data for the rare earth isotopes are usually reported in terms of the integrated neutron flux (fluence) required to produce them. These data, together with the spallation exposure ages which refer to somewhat shallower depths, are the key parameters in defining the long-term rates of turnover of the regolith.

3.3.5 SURFACE-CORRELATED CHEMICAL SPECIES   In addition to the solar-wind rare gases and excess ^{40}Ar discussed in Section 3.3.1 above, a number of other elements have been shown to be concentrated in finer-grain fractions, and hence presumably to be at or near grain surfaces. In the case of fluorine (Goldberg, Burnett & Tombrello 1975) this surface concentration has been directly observed. The obvious interpretation is that these species have passed through the gas phase, either as emanations from the interior or as products of volatilization on impact. As Criswell (1975) has made clear, there are difficulties with simple models of this kind. While we must await further direct evidence of surface correlation, what we have is reasonably strong (Boynton et al 1976).

A very interesting case is that of the radiogenic lead isotopes, formed by decay of U and Th. Silver (1975) has pointed out that this element, whose isotopic composition changes with time as the parent species decay, can give us at least a mean absolute date of first deposition on grains. In the soil samples he has measured, this is $\sim 2-3 \times 10^9$ yr.

3.3.6 CHEMICALLY DISTINCT LUNAR COMPONENTS   Since there are differences in composition over broad regions of the lunar surface, we can attempt to use these to study the extent of long-distance transport. Thus the finding by Wood et al (1970) of highland material at Apollo 11 can be considered as evidence of transport over hundreds of kilometers or more, after the formation of Mare Tranquillitatis. The Apollo 15 and 17 sites, located close to boundaries of distinct regions, show trends on a kilometer scale that may be interpreted in terms of lateral mixing rates (Schonfeld 1974). These data, among others, have been used by Arvidson et al (1975a) to make an estimate of the fraction of material in average soil that originates at a distance greater than d. For $d > 100$ km or so, however, the interpretation is made ambiguous by the possibility that foreign material may originate in deeper layers (> 1 km depth) below the mare basalt, rather than from distant points.

3.4   Conclusion

As a general conclusion of this section we would like to discuss the problem of maturity indices and the definition of "exposure ages" for soils. As we have seen, the maturity scale has been considered as a substantially linear time scale, in which maturity increases with the exposure of the soil at or near the surface. This point of view was based on a "closed-system" soil picture: grains in a soil stayed together during the whole soil evolution. This was already inconsistent with some data (McKay, Fruland & Heiken 1974) and evidence of mixing over large distances (Heiken & McKay 1974). Before defining soil exposure ages, one should thus clearly define soil. A sample of soil has not only a spatial extension, but also a temporal dimension: it acquired its present identity only after the deposition of the ejecta blanket (the "layer") to which it belongs, having been undisturbed since then. Before that event, the soil grains belonged to several ancient soils that have been destroyed. For cumulative processes, it is impossible to distinguish between effects accumulated by the grain while belonging to its present soil ["parent stratum" in the terminology of Bibring et al (1975)] and those accumulated before (this last contribution is sometimes called a "pre-irradiation level").

One can consider the same problem from a different angle. The spatial distribution of grains that now form a soil was very different just before the deposition event: grains were then scattered over the volume of the crater, and the correlation between the depth histories of the grains was significantly weaker than when they were tightly packed together.

In the heavily reworked mature soils, high pre-irradiation levels as well as rather low correlations between the exposure histories of grains in the same soil preclude the derivation of a meaningful soil exposure age from maturity indices. Only in immature soils, where reworking has been minimum and pre-irradiation levels low, can such exposure ages be given a simple chronological interpretation (Gault, Hörz & Hartung 1972, Bibring et al 1975).

4   REGOLITH EVOLUTION MODELS

Since 1970, a great variety of models describing the dynamical evolution of the regolith have been built. These models were first devised to fit experimental data on a given maturation effect: for example tracks (Arrhenius et al 1971) or neutron fluences (Russ, Burnett & Wasserburg 1972). In the following years, more ambitious models have appeared, trying to interpret a wider range of effects.

No quantitative model has yet been built on the hypothesis of a major non-meteoritic contribution to regolith mixing processes (see section 2.3). One can thus classify the main evolution models built to date in two categories: 1. Correlation models: arbitrary dynamical evolution parameters are chosen, then fitted with experimental data. No specific mass transport mechanism is assumed. 2. "Meteoritic" models: meteoritic impacts are considered the major triggering agent of regolith mixing and accumulation. Assumptions are made about the meteoritic flux, mass and speed distribution, and the cratering process. Evolution parameters or values of maturation effects are then derived and compared to experimental data. We have further distinguished among these models those not using Monte Carlo methods and those that do.

4.1   Correlation Models

We have seen (Section 3.2.3) that the lunar regolith exhibits a dual character: the presence of layers points towards a build-up by "sedimentation," the discrete deposition of strata. On the other hand, the homogeneity of the regolith to the naked eye, as well as many physical and chemical characteristics, points toward very thorough mixing. Moreover, mass balance must be preserved (at least very nearly). The adjustable parameters chosen by correlation model builders depend on which of the two aspects is considered as dominant. Usually only one measured property is fitted.

4.1.1   MIXING MODELS   There are two such models, both of which have been superseded by much more detailed meteoritic models by the same groups. Comstock et al (1971) made such a model to study central track density distributions. Curtis & Wasserburg (1975) discussed their neutron fluence observations, using mixing depths of 50 g cm^{-2}, 500 g cm^{-2}, and >4 kg cm^{-2}. The later work of these groups is found under Section 4.2 below.

4.1.2   SEDIMENTATION MODELS   The model of Arrhenius et al (1971) deals with central track densities, emphasizing sedimentation processes. Using the "quartile" track densities (see Section 3.3.2) and a low track production rate at a depth of 50 μ ($\lesssim 10^9$ cm^{-2} m.y.$^{-1}$) to determine the duration of the last exposure of a top stratum; they obtain a good fit with a rate of sedimentation ~ 1 to 3 mm m.y.$^{-1}$

(balanced by rarer cratering events). This model relies on somewhat uncertain track production rates, and we have seen that the "quartile" method should be correct only when dealing with a track-poor section. However, the rate obtained is in agreement with the determinations of later models (see below), when proper care is taken in defining terms.

The model of Fireman (1974) uses a sedimentation parameter together with some mixing to fit the observed mean neutron fluences and the rare gas spallation ages in the regolith. A remarkable result is obtained, namely, a high *negative* deposition rate of 0.8 mm m.y.$^{-1}$. The following remarks can be made. 1. The most recent determination of neutron capture rates (Burnett & Woolum 1974) is lower than previous determinations. This could significantly modify the conclusion. 2. No known physical process is within a factor 10 of removing so much material from the Moon. 3. Models described below (Arnold 1975b, Langevin & Maurette 1976) obtain a good fit with less extreme hypotheses.

4.2   *Meteoritic Models*

Since the pioneering work of Gault et al (1974)[2] all meteoritic evolution models so far proposed have these common assumptions: 1. The mass distribution of the meteoritic flux is considered constant over a long period ($\lesssim 3 \times 10^9$ yr). 2. The integral mass distribution can be represented by a power function of the mass (see Section 2.1), presenting few "kinks" (changes in γ) over the whole mass range (from 10^{-12} g to 10^{14} g). 3. The bombarding particles have fixed velocity, so that crater characteristics depend only on the impacting mass. 4. The diameter D and depth d of the crater are simple power functions of the impacting mass ($D = km_p^\lambda$ and $d = k'm_p^{\lambda'}$).

The models differ from one another in the mathematical techniques used. We distinguish in our discussion the models that do not use a Monte Carlo method from those that do. However, the main differences in the results of these models arise from distinct choices of the exponents in the relations expressing the mass distribution and the crater characteristics. From the discussion in Section 2, we derive the most important parameters that can be used to characterize a model: 1. The exponents γ, γ', and γ'' in the mass regions $m_p > 10^8$ g, 1 g $< m_p < 10^8$ g, and 1 g $> m_p$ respectively. These mass regions can be considered responsible for the regolith formation (region I), macroscopic mixing (region II), and surface stirring (region III) respectively. 2. The exponents λ and λ'. 3. Two additional parameters: $\delta = (1-\gamma')/\lambda$, the exponent of the integral production distribution of craters whose diameter is greater than ~ 30 cm (regions I and II); and $\chi = (2\lambda+1-\gamma')/\lambda'$, the exponent of the integral production distribution of depths for region II craters including a given point. These two redundant exponents have been chosen as basic parameters in several models (Oberbeck et al 1973, Gold & Williams 1974, Blake & Wasserburg 1975). They are sensitive indices for the relative mixing efficiency of large and small craters.

---

[2] The first Monte Carlo model, that of Oberbeck et al (1973), deals only with the formation of the regolith.

Table 2   Distribution indices in various models

	Region						
	γ	γ'	γ''	λ	λ'	δ	χ
Gault et al (1974)		2.16	2.27	0.28	0.28	−4.14	−2.14
Blake & Wasserburg (1975)		a		a	a	a	χ
Oberbeck et al (1973)	γ			a	b	−3.4[b]	a
Gold & Williams (1974)		γ'		λ	λ^c	−2.93	−0.93
Comstock (1976)			2.213	0.375	0.333	a	a
Duraud et al (1975)		2.06	2.213	0.375	0.333	−2.83	−0.93
Arnold (1975a, b)		2.06	2.213	0.333	0.333	−3.18	−1.18

[a] This index not defined or not used.

[b] The δ parameter is defined for mass region I ($D \gtrsim 100$ m). The depth/diameter ratio is variable, depending on the local regolith thickness.

[c] In Gold & Williams (1974), the depth/diameter ratio is constant. Exponents λ and λ' are thus equal, and $\chi = \delta + 2$.

We have summarized in Table 2 the parameters chosen in the main meteoritic models. One model (Oberbeck et al 1973) deals with regolith formation (region I). Two models deal only with macroscopic mixing (region II) (Gold & Williams 1974, Blake & Wasserburg 1975). One deals only with surface stirring (region III) (Comstock 1976), and three consider both surface and large-scale mixing (region II, III) (Gault et al 1974, Duraud et al 1975, Arnold 1975b). Several use Monte Carlo techniques, one employs direct statistical methods (Gault et al 1974), and one an analytical treatment (Blake & Wasserburg 1975). The two latter will be dealt with first.

4.2.1   NON-MONTE CARLO METEORITIC MODELS

The model of Gault et al (1974)   This model was the first meteoritic model to be published, and has, together with a former paper (Gault et al 1972), strongly influenced the other evolution model builders. The problem of regolith formation from the bedrock is not considered (mass regions II and III only). The basic parameters are: $\gamma' = 2.16$, $\gamma'' = 2.27$, and $\lambda = \lambda' = 0.28$ (for $m \gtrsim 10^{-5}$ g; $d/D = 1/8$ in this mass range). The derived δ and χ parameters are $\delta = -4.14$ (region II) and $\chi = -2.14$.

Gault's model put more emphasis on mixing than on sedimentation. A given point of the lunar surface is considered "turned-over" to a depth d if it has at any time been inside a crater of local depth greater than d. The main parameters discussed are the number of turnovers to the 50% (or 99%) probability level after a time $T, N(T, d, 50\%)$ and $N(T, d, 99\%)$. N is such that a given point has a 50% (99%) probability of having been within at least N distinct craters locally deeper than d during the time T.

Using a statistical method based on the Poisson law, Gault showed that the upper millimeter (the "mixing zone") has much higher turnover rates than deeper regions. It is stirred ten times after only a few hundred thousand years. Problems of interpretation arise from the slow increase with time of the thickness of the turned over layer: after 3×10^9 yr, 50% of the surface would not be stirred even once down to ~ 10 cm. This is a direct consequence of the choice of basic parameters, as it can be shown that this thickness varies roughly as $T^{-1/\chi}$. For $T \lesssim 500$ m.y., we have $\chi = -2.14$ (region II; see Table 2) and $d \sim T^{0.46}$. To account for the value of observed turned-over thicknesses, Gault et al (1974) were compelled to use a much higher meteoritic bombardment in the very first part of the regolith history (>3 b.y. ago). Such an interpretation raises contradictions with the low spallation rare gas ages measured in surface samples: if the samples had stayed in the upper 10 cm for the last 3 b.y., their spallation ages would have been ~ 3 b.y., not ~ 0.5 b.y. If meteoritic mixing is considered dominant, this model seems to require χ values much flatter than Gault's -2.14 to account for both the mixing of the regolith down to depths greater than 2 m and the available data on spallation ages (see Sections 3.2 and 3.3.3).

Gault's model cannot derive theoretical individual trajectories for regolith constituents. It is thus not designed to study the magnitude of depth-dependent effects in grains and rocks. However, it does interpret the trends observed among exposure effects reaching to different depths, showing that the first mm of the regolith is stirred very efficiently by meteoritic impacts. The parameters adopted by Gault et al for region III have not since been much changed by other model builders. Although Gault's choice of region II parameters does not seem to reproduce the macroscopic evolution of the regolith, this model is very useful to test the input parameters of meteoritic models. Hörz, Hartung, Gault, and coworkers have also built a model of rock micrometeoritic abrasion (Hörz, Schneider & Hill 1974) that is in good agreement with experiment, a model for the catastrophic rupture of rocks (Hörz et al 1975b) and recently, an evolution model for the lunar highlands (Hörz et al 1976).

The model of Blake & Wasserburg (1975)   This model is an attempt to derive analytically the mean value at a depth x of a given cumulative effect after a time t of regolith evolution $\psi(x, t)$. The input data are: 1. The production rate versus depth of the effect, $\phi(x)$. 2. The differential distribution $\alpha(\xi)$ of the depths for craters enclosing a given point; if $\alpha(\xi)$ is a power function of ξ, $\alpha(\xi) = K\xi^{\chi-1}$. After a complex mathematical analysis, differential equations are derived for $\psi(x, t)$. The main parameters entering these equations are μ_0, μ_1, μ_2, the first three moments of $\alpha(\xi)$, defined as

$$\mu_i = \int_0^\infty \frac{\alpha(\xi)\xi^i \, d\xi}{i!}$$

This approach is in principle a good one, as analytical treatments are more general than Monte Carlo methods. However, this model seems not well suited to the lunar regolith situation: one can derive μ_0, μ_1, and μ_2 only if $\chi < -2$. As the authors are mainly interested in isotopic effects, the depth scales involved arise from region II impacts (see Section 3.4.3). We have seen that the lowest χ value in the literature, that of Gault et al, was too low to account for large-scale regolith mixing. Neukum et al (1975) derived $\delta \sim -3$ and thus $\chi \sim -1$ in the upper part of the diameter range considered ($D > 100$ m), and qualitatively this must be correct. Hence the largest events (craters and layers) play a crucial role in the evolution of any given site, although the statistically large number of small events determine other properties. This has two implications: 1. The mean profile $\psi(x,t)$ is not a complete description of the depth variation of a property. Distributions of profiles are required. 2. An analytical treatment is not well suited for large rare events. In their derivations, divergences become critical if $\chi > -2$.

As a conclusion, this model should prove useful it situations where the χ index is lower than -2. The contribution of small events would then dominate the evolution, and each peculiar profile would be close to the mean profile $\psi(x,t)$. We suggest it would be interesting to apply this model to near-surface effects, as the χ index in this region is significantly lower than in region II.

4.2.2 MONTE CARLO METEORITIC MODELS   Monte Carlo methods use the probability distributions of impact events affecting regolith evolution. One theoretical evolution ("pass" or "run") is then derived from a particular succession of such events, which are randomly selected according to the probability distributions of impacting mass and location. Mean values and distributions of evolution parameters can then be derived from the analysis of a large number of such theoretical evolutions. As meteoritic impacts are uncorrelated random events, Monte Carlo methods seem particularly adapted to the study of regolith evolution. There are basically two ways to apply these methods to the lunar regolith. One may consider either the history of a site on the lunar surface or the individual depth history of a single grain.

Study of the evolution of a site   One considers a small part of the lunar surface, which is usually described by a square array of points and the surface height at each point [in (Oberbeck et al 1973) the regolith thicknesses at each point]. The following steps are taken: 1. A probability distribution of crater diameters is chosen, $n(>D) = AD^{\delta}$, yielding the mean number of times a crater larger than D will occur on the surface per unit time. 2. A lower size limit Δ is set, below which craters will not be considered individually. This sets a total rate $n(>\Delta)$ of "individual" craters occurring in the surface. 3. A set of initial heights for each array point is fixed, yielding the "initial situation." 4. A time t_1 after which the first crater occurs is chosen at random, in agreement with the total rate $n(>\Delta)$. The crater diameter, D, is chosen at random, in agreement with the distribution $n(>D)$. A crater center is chosen at random over the surface. 5. The depth changes induced by this event, as inferred from the choice of crater geometry, are then added (with proper sign) to the initial heights, yielding a new situation after a time t_1. The effects of smaller events are estimated statistically, and added in. 6. After returning to step 4, a time t_2, etc is chosen, keeping track of the total time t. After a chosen time interval T has been covered, the routine is stopped. One run is complete. From the study of a large number of such runs, mean values and distributions of observed properties can be derived.

Study of the depth history of a grain   One considers a "test grain" embedded in the regolith. The steps are as follows: 1. From the choice of the cratering parameters, one derives two distributions, $n(>H) = MH^\chi$ and $[n'(>h) = Nh^\chi]$, which yield the mean number of times a crater (an ejecta blanket) of depth (thickness) larger than H (h) will occur at the location of the grain per time interval. 2. A lower limit η is set for crater depths and layer thicknesses, below which they will not be considered as individual events. This sets two total rates: $n(>\eta), n'(>\eta)$ for individual craters (layers) affecting the test grain. 3. An initial depth is chosen for the test grain. 4. The time t_1 after which the first individual event happens is chosen at random, in agreement with the total rates $n(>\eta)$ and $n'(>\eta)$. The nature of the event (crater or layer) and its depth H (thickness h) are chosen at random in agreement with the distributions $n(>H)[n'(>h)]$. 5. The crater depth is subtracted, or the layer depth added, to the initial depth of the test grain. If the depth of the crater is greater than the depth of the grain, the grain is "excavated" and a new depth is chosen at random in the whole ejecta blanket of this crater. Either way, this yields a new depth, $d(t_1, d_0)$. 6. A time t_2 etc is chosen, keeping track of a total time t after returning to step 4, with $d(t_1, d_0)$ as the new initial depth. After a chosen time interval T has been covered, the routine is stopped: A calculated trajectory of the test grain, $d(t, d_0)$ has been derived, up to $d_f = d(T, d_0)$. From the analysis of a great number of such calculated trajectories, one can derive the evolution parameters of the regolith. Coupling such trajectories with the production rate of a depth dependent cumulative effect yields a theoretical distribution of this effect in grains.

The model of Oberbeck et al (1973)   This model deals with the problem of regolith formation from bedrock. It is not concerned with the further evolution of the regolith and considers only craters with diameters $D > 10$ m (mainly region I). A value of -3.4 is chosen for the δ index in this size range (see Table 2). Large craters penetrating the bedrock cannot be accurately described by a simple power function of the mass. Different morphologies are considered, from the bowl-shaped craters dug in the regolith only, to flat bottomed, "concentric," and again bowl-shaped craters as the ratio of the depth of the crater to the regolith thickness increases. A "study of a site" routine is used on a very large scale (2.9×10^6 array points distributed over a surface of 260 km^2), the regolith thicknesses replacing the heights of the points. A new step is introduced between the steps 4 and 5: the crater morphology is deduced from the local current regolith thickness at the center.

The following results have been obtained: 1. The model is able to reproduce not only the mean value, but also the distribution of present-day regolith thicknesses. This firmly supports the hypothesis of a dominant role for meteoritic impacts in regolith formation. 2. The contribution of large impacts (>100 m, region I) to regolith building dominates that of smaller ones (10 to 100 m). 3. Regolith growth is not linear with time because of the progressive protection of the bedrock by the growing regolith. It can be. shown that their choice of δ implies that thickness increases with time more slowly than $T^{0.7}$.

This model has thus been very successful in its objective. It could probably be extended to the highland situation, where the integrated meteoritic flux has been much higher than on the maria.

The model of Gold & Williams (1974)   This model deals with the problem of the ubiquitous track content of regolith grains. Region II impact craters are considered, and a value of -2.93 is adopted for the δ exponent. A "study of a site" type of routine is used, which calculates the number of times a grain has resided in the upper 64 cm of the regolith. The authors derive from these computations that 10% of the grains now at a depth of 2 m have never been in this upper region, in a regolith mixed down to a 4 m depth. They conclude that the "meteoritic mixing" hypothesis is in contradiction with track data (see Section 3.3.2).

It is difficult to assess the validity of these conclusions, as a detailed description of both the method and the results has not yet been published. We raise the following questions: 1. What proportion of grains among the 10% has never been excavated? Such grains would still belong to the bedrock, and should not exhibit any tracks. 2. How many of these grains have been deposited in very thick (>2 m) ejecta blankets? As no deep core has yet been drilled in the ejecta blanket of a fresh large crater, such grains have not been sampled. It is probable that the 2-m deep region of such a layer consists mainly of rocky fragments, and that these exhibit very low track densities. 3. Have the authors taken into account the change in geometry for craters excavating bedrock? The volume of "fresh grains" excavated from the bedrock is much lower than that derived from a "normal" geometry (Oberbeck et al 1973).

The model of Comstock (1976)   This model is essentially devoted to the study of the track record in 100-μ grains. A "depth history" type of routine is used, which analyses the movement of the test grain in the upper 0.5 cm of the regolith (surface stirring). Only region III events are considered. This program generates not only depth histories, but also orientation histories of the test grain: each time it is excavated, a new orientation is randomly chosen. The model is thus able to compute both the central track density and the track gradients acquired during a period of 1 m.y. in the upper 0.5 cm of the regolith, a "surface exposure episode" (SEE) [the track production rate used is twice that of Comstock (1971)]. Comstock's model has an important limitation: lacking any "deep" depth history routine, it cannot provide any information on the mean value or the distribution of the number of surface exposures for a regolith grain. This number thus becomes an arbitrary parameter.

The main results obtained are: 1. A "fine scale" burying rate of 0.4 cm m.y.$^{-1}$ is derived. 2. Most soil grains have been exposed within 1 mm of the surface, but have been shielded by a few micron covering at the very surface. 3. Immature soils track data are consistent with 1 to 10 SEE's (10^6 to 10^7 yr in the upper 0.5 cm), and mature soils track data require 10 to 50 SEE ($10^7 \rightarrow 5.10^7$ yr).

Results 1 and 3 are in agreement with the determinations of more complete models (Arnold 1975b, Bibring et al 1975). However, result 2 is based on a "significant" gradient level of 10%, which may be too low (Bibring et al 1975, Goswami & Lal 1974, Goswami, Braddy & Price 1976).

The model of Duraud et al 1975 (see also Bibring et al 1975)   Work on this model was initiated by G. M. Comstock while staying in Orsay. It was then similar to the model of Comstock (1976). Since 1974, it has been modified and extended to include the "deep" history of regolith constituents ($d > 1$ cm). The model uses a "depth history" type of routine, where type I and II impacts are considered. The choice of parameters (see Table 2) is somewhat extreme in region II: $\lambda = 0.375$, $\lambda' = 0.333$, and $\gamma(\text{II}) = -2.06$ imply a χ index of -0.93 (see Table 2). Thus the model emphasizes the role of large events in macroscopic mixing. Compared to other Monte Carlo models, it has several distinctive features: 1. A variable logarithmic depth scale; the η parameter (minimum thickness of events considered individually) depends on the current depth of the test grain. The routine can thus derive with comparable detail and accuracy theoretical trajectories from depths $\lesssim 100$ μ to >5 m. 2. A radius r is attributed to the test grains. The trajectories $d(t,r,d_0)$ differ only when $d \lesssim r$. This radius can be varied from 1 μm to >10 cm, so that the individual history of rocks as well as very small grains in the regolith can be described. 3. The model uses a "backward" mode of operation; the theoretical trajectories are generated backward in time, starting from a "final" depth d_f. Each random event is subtracted in turn from the regolith; the layers are removed and craters are filled up. This mode has a distinct advantage over the direct mode: the sampling depth of a soil or rock can be considered as the starting depth, d_f, of the routine, thus considerably restricting the possible $d(-t,r,d_f)$ trajectories of a rock (or the grains of a soil). The method yields directly the histories of grains in a single volume of soil at present. 4. Two parts are considered in the trajectory of a constituent: the "plateau" history, defined as the period between the constituent formation and its last deposition in an ejecta blanket, and the "burial" history, after this event, which covers the progressive burying of the constituent down to its present depth. This separates the "individual" part of the trajectory from the collective burial in an undisturbed soil.

Results of the analysis of plateau trajectories; near surface maturation effects   The theoretical distribution of near surface effects in regolith constituents is obtained in this model from an analysis of calculated plateau trajectories $[d(-t,r,d_f)]$, and more specifically from an analysis of the parts of these histories spent in the uppermost stratum ($\lesssim 5$ cm thick). There are 1 to >10 such episodes (an average of ~ 3), each lasting an average of 25 m.y. in a plateau trajectory. The results for such a "parent stratum" episode are the following: 1. The mean number of direct exposures of a test grain on the surface is directly proportional to the radius r, up to ~ 50 μ (100-μ grains), and depends strongly on the final depth d at the end of the episode. For $r = 50$ μ, $N = 30$ if $d < 0.5$ cm (this region is defined as the "lunar skin" of the stratum) and $N \gtrsim 10$ for $d > 1$ cm. The mean duration of an exposure is ~ 5000 yr, and does not depend on the radius in this size range (this is due to the sharp drop in the meteorite mass flux in sizes <40 μ). The 100-μ grains of the lunar skin have thus spent about 150,000 yr at the very surface, while more deeply buried grains have spent much shorter times. The overall mean exposure is $\sim 50,000$ yr. Only 50% of the 1-μ grains from the lunar skin and 10% from deeper regions have been exposed once at the surface ($\lesssim 20\%$ on the mean). The corresponding results for maturity effects are as follows.

Track densities: a production rate $\sim 5.10^9$ t cm^{-2} m.y.$^{-1}$ at a depth of 50 μ

has been chosen (Comstock 1972). From the dominant contribution of direct surface exposure to central track densities in a 100-μ grain, it can be concluded that $\sim 150,000$ yr (~ 30 exposures) are required to get central densities in excess of 10^9 t cm^{-2}. Because of this large number of exposures the calculated track gradients in high track-density grains are isotropic. From the analysis in Section 3.3.2, it appears that in $\sim 50\%$ of such grains the high track density observed cannot be interpreted in terms of regolith exposure only. These "anomalous" grains could have received most of their tracks while on the surface of a parent rock or bigger grain and should not be considered in central track-density statistics. The theoretical distribution of central track densities in 100-μ grains has a shape very similar to that of the corrected experimental distribution (Bibring et al 1975); the same spread of values (~ 3 orders of magnitude), the same peak in the logarithmic distribution near the high-density limit, and a "plateau" below. The absolute values are also in good agreement, but, due to the uncertainty in track-production rates, such a fit is not very significant. Differences between the gradients calculated with likely production rates, and the experimental ones, have been explained by either a rapid "flaking" erosion of 100-μ grains or a dust covering exceeding 10 μ. From 1-μ-grain track data, it appears that a 5000 yr exposure produces $\sim 10^{10}$ t cm^{-2} (see Section 3.3.2). A production rate $\sim 2.10^{12}$ t cm$^{-2}/10^6$ yr is thus derived for a depth of 1 μ.

Surface characteristics are as follows: 1. 100-μ grains. High track-density ($> 10^9$ t cm^{-2}) grains of this size have spent $> 150,000$ yr on the surface. Each main face of the grain should have been exposed $\sim 30,000$ yr, due to the isotropy of the surface exposure. This is consistent with the observation that flattened glassy disks are present on every major face. However, microcraters are very anisotropic and are not present on every track-rich grain, and the exposure ages inferred from them are $\leqslant 10,000$ yr (see Section 3.3.3). The contradiction can be solved by either the "flaking" or the "dust shielding" hypothesis. 2. 1-μ grains. The probability of a 1-μ grain having been exposed at least once on the surface increases from $\leqslant 20\%$ for 1 top stratum episode to $\geqslant 90\%$ for ~ 10 episodes. The typical duration of a surface exposure (~ 5000 yr) is long enough to produce an amorphous coating on feldspar grains. These results are thus consistent with the spread of experimental percentages of amorphous coated feldspars (10% to $\sim 90\%$). As it is highly improbable ($< 1\%$) that a grain has been exposed more than 4 times on the surface even after 10 surface episodes, the total exposure almost never exceeds 20,000 yr, which is consistent with the absence of amorphous coatings on ilmenite (see Section 3.3.3) (Borg et al 1976).

Isotopic effects in surface samples: The plateau history not only accumulates the total amount of near-surface effects, but also contributes to the production of "deeper" effects such as spallation rare gases. The calculated values can be compared with the measures in surface samples, as these have clearly not begun their "burial" history. The most extensively studied cumulative "deep" effects are spallation-induced rare gases and rare earth isotopic anomalies. In 100-μ grains, the computed contributions are highly variable. The distributions obtained for ^{21}Ne concentrations are consistent with measurements in individual grains (T. Kirsten, personal communication). However, most experimental measures have been done on soil samples ~ 50 mg (see Table 1 and Section 3.4). Such samples contain thousands of grains, and the theoretical spreads for soils are thus much narrower than the spread for individual grains (no correlation is assumed between the trajectories of grains now in the same soil). The theoretical mean values: 60×10^{-8} cc g^{-1}, 50×10^{-8} cc g^{-1} (Ba), 270×10^{-8} cc g^{-1} (Ba), 1.8×10^{16} n cm^{-2} for ^{21}Ne, ^{126}Xe, ^{131}Xe concentrations and the neutron fluence respectively are in good agreement with experimental determination (55×10^{-8} cc g^{-1}, 60×10^{-8} cc g^{-1} (Ba), 300×10^{-8} cc g^{-1} (Ba), and 2.1×10^{-16} n cm^{-2} respectively; see Section 3.4). The spread in experimental values for bulk soil samples ($\sim 50\%$) might be interpreted as indicating a weak correlation between the histories of neighboring grains. The fit is especially interesting for the surface ratio of ^{131}Xe/^{126}Xe (~ 5 for both experimental and theoretical values), as the production ratio increases very sharply with depth (~ 2.2 at the surface; $\gtrsim 15$ at 2 m depth). The model is also able to deal with the isotopic effects observed in surface rocks from the analysis of $d(-t, r = 5$ cm, $d_f < 5$ cm) trajectories. The values first determined (about the same for rocks and grains) were in agreement with the wide spread in experimental measures, but were higher than the mean values observed in rocks. However, the limited lifetime λ_t of rocks at the very surface is to be taken into account. The rocks now picked on the surface have not been destroyed, and the theoretical trajectories should be weighted in the statistics by the probability, $\exp[-T(0)/\lambda_t]$, of nondestruction of the test rock [$T(0)$ is the time spent at the surface during a theoretical trajectory]. When this is done, the new mean values (40×10^{-8} cc g^{-1} ^{21}Ne and 1.2×10^{16} neutrons cm^{-2}) are in agreement with experimental ones (see Section 3.4). The distribution of ^{21}Ne and ^{131}Xe/^{126}Xe in rocks is also consistent with experimental data. All the results are from Langevin & Maurette (1976).

Results of the analysis of the "burial" trajectories. These results can be compared with the layer structure and the "deep" effects measured in deep drill lunar core tubes brought by missions Apollo 15, 16, and 17. The mean burial evolution characteristics can be summarized in a theoretical 2–3 m core tube : built up in ~ 2 b.y., it results from the deposition of ~ 90 layers > 1 cm, ~ 50 of which have been destroyed. In order to obtain theoretical profiles of a given effect in a core, the following routine was used. 1. 40 points, 15 g cm^{-2} apart are considered in a core tube of 600 g cm^{-2} depth. 2. A theoretical history of the deepest point is computed in the backward mode. Effects are accumulated at each point according to its current depth. 3. Each time one of the points reaches the surface, the soil at this point is considered to have been just deposited in the core. The soil grains have just ended their plateau history, and the constant plateau contribution is added to the acquired concentrations, the burial contribution. 4. When the bottom point in turn gets to the surface, the routine is stopped. Several effects can be computed along one theoretical burial history, and a set of profiles is obtained. The first application of this routine was to show that the calculated profiles of ^{21}Ne and neutron fluences (Langevin & Maurette 1976) had the same general characteristics as those measured in lunar core tubes. A further application is to use a set of four measurements (^{21}Ne, Gd isotopes, ^{126}Xe, ^{131}Xe/^{126}Xe) to decipher the major events in the burial history of a given site (Langevin & Maurette 1976). It can already be noted that although the constant "plateau" contributions correspond with the "pre-irradiation levels" inferred by others (Pepin et al 1975, Russ, Burnett & Wasserburg 1972) from rare gases or fluence profiles, such a constant contribution is only a first-order approximation, which does not take into account the spread in surface values.

This evolution model has achieved consistency with a wide variety of effects from the maturation record. It can also account within reasonable error limits for direct exposure effects (solar-wind-produced amorphous coating), near surface effects (nuclear tracks), and deep effects (isotopic modifications). However, this one-dimensional model cannot in its present state deal in detail with "three-dimensional" problems like the overall smoothness of the surface, the lateral transport of material, and the degree of correlation between the trajectories of grains sampled in the same soil. Finally, it has shown interesting potential in being adapted to the study of the evolution of other atmosphereless solar system bodies.

The model of Arnold (1975b, also see 1975a)   This model is the most thoroughly worked out of the site history type. It attempts to account for track counts (but not track gradients), spallation and neutron effects and other isotopic effects, as well as layering and the observed slopes of the lunar surface. It permits the calculation of distributions in one or several parameters (correlations). The choice of bombardment parameters (Table 2) was originally made to emphasize the importance of large collisions, though it now appears possible that this did not go quite far enough.

The lower size limit of collisions considered individually in this model is that the change of height Δh (either positive or negative) is >0.1 cm (sometimes 0.5 cm). Events below this limit mostly add material and thus they do not remove layers or cause other gross changes. Individual events ($|\Delta h| > 0.1$ cm) occur at a given site about every 4×10^5 yr. The smaller of these are mostly due to craters a few cm in diameter whose effects are extremely local. As $|\Delta h|$ increases, the emphasis shifts to much larger cratering events, which can affect areas tens of meters across or more. One program, LEVARRAY, follows the heights on a six-by-six square array of test points, of variable spacing, to study surface slopes. Other programs develop means and distributions for observable parameters and correlations between them. The program runs forward, for various periods T after exposure of fresh material. Most frequently chosen are $T = 3 \times 10^9$ yr (age of the maria) and 4×10^8 yr (typical spallation gas-exposure age for soil samples). Unlike the model of Duraud et al (1975) there is no shift of mode of calculation depending on present grain depth or other variables.

We discuss first the results of the model for the history of height $h(t)$. The most notable feature of successive runs of $h(t)$ is their diversity. There is a tendency toward slow rises and abrupt falls, which results from the fact that craters are deep and ejecta blankets comparatively thin. However, this effect is much less prominent than older models (Arrhenius et al 1971, Gault et al 1974) would suggest. Since $|\Delta h| > \sim 1$ m is possible but rare, and the slope of the size distribution is flat ($\chi \sim -1$), some histories over 3×10^9 yr show no changes $|\Delta h| > 20$ cm while others may show multiple craters or blankets with $|\Delta h| > 1$ m. The resulting distributions of height must be interpreted with due regard to the effects of sampling: the extremes of lunar surface relief have so far been avoided in actual explorations.

The irregular "stock market" histories of $h(t)$ affect the way we interpret lunar stratigraphy. This model follows individual layers (usually those thicker than 0.5 cm) over the entire history. Surviving layers arise usually as ejecta blankets, which may or may not have been partially removed by later impact. Typically about one third to one half of a column results (in this model) from such layers. Ignoring the effect of overturning of lunar strata near the edge of craters ("reverse stratigraphy," see Figure 3 in Gault et al 1974) not yet considered in any model, the oldest layers should lie lowest. This model makes it clear that large gaps in the record are normal.

There are other possibilities for creating visually distinct layers. The rapid gardening of the top ~ 0.5 cm may homogenize such a layer before it is preserved by the arrival of a covering ejecta blanket. Slumping of material on a steep face can deposit a layer downslope. An important mechanism for creating layers, and smoothing the surface, arises from the behavior of low-velocity ejecta. These effects are considered in the model.

The fact that even craters of < 10 cm show visible rims of ejecta implies that most material moves out of craters with very low velocity. At an ejection velocity ~ 1 m sec^{-1} [considered typical by Gault et al (1974)] a grain can travel no more than 60 cm in the gravitational field of the Moon. Slowly moving particles show a strong preference for downslope transport. The resulting filling of hollows is the apparent solution to the paradox: how can one make a surface smooth by firing bullets at it? It is also an effective method of creating thin layers that are homogeneous (because the source is an average of the whole nearby region). The results of the LEVARRAY program in this area are given in detail by Arnold (1975b). They are confirmed by the analysis of Arvidson et al (1975b) of the time of filling of two long grooves caused by rolling of Apollo 17 boulders.

The layer structure has been studied across arrays of various sizes. Thick layers persist over long distances, but there can be interruptions due to later, smaller craters. Layers tend to become progressively more distinct with thickness according to the model, as they must on the moon, because thicker layers bring up fresher, less exposed material, and also because the material comes from greater distance. The model assumes that material from a crater is homogenized throughout. A more refined treatment would remove this assumption; persistence of structure is observed in model experiments (Gault et al 1974).

The correlation between initial and final depth exhibits the expected features. That is, the mixing of particles originating at different depths is very extensive, especially at smaller depths (< 1 cm), but it is far from complete.

Distributions of layer thicknesses are also in rough agreement with observations (Duke & Nagle 1975); more cannot be expected until we have better experimental criteria for defining layers in cores. Absolute age distributions are available but not yet testable.

Among the properties observed in the laboratory, track density distributions, based on observation of individual grains, can be compared in detail. The model, like that of Duraud et al (1975), accounts for all key features, well within the uncertainty of production functions.

Spallation rare gases are usually measured on bulk samples, but some observations exist on individual glass fragments (Kirsten et al 1972) and small rock fragments (P. Eberhardt, private communication). The model predicts that the mean exposure age should be $\sim 7 \times 10^8$ yr; some exposure ages $> 10^9$ yr should occur, in accord with experiment. The predicted mean ^{21}Ne ages are too high by a factor of 1.8 or so. All models to date have neglected diffusion loss of spallation ^{21}Ne (Eugster et al 1975); the use of ^{38}Ar data would reduce the discrepancy and provide a better comparison.

The model's results for neutron capture phenomena (Gd and Sm isotopes) also show fairly good agreement with experiment, with a tendency to better agreement as the model is refined. The calculated fluences are still too high compared to experiment by a factor of 1.6 or so.

This represents the largest disagreement between the model of Duraud et al (1975), which fits the data quite well, and that of Arnold (1975b). The explanation is obvious: the former uses a flatter distribution of crater size than the latter. Since both are normalized to absolute rates at $m = 1$ g, it produces more gardening at depth. Appropriate changes in χ do in fact produce good fits in the Arnold model. The authors agree that experimental data on crater counts are not yet precise enough to offer a clear choice of input data. At any rate there is no obvious present need to invoke large scale mass removal from the Moon (Fireman 1974) or major nonimpact processes of gardening (Gold & Williams 1974).

The "disturbed depth" $h_d(T)$ is defined in this model as the present thickness of material that has been disturbed or moved in the interval T before the present. The model gives a dependence of the mean h_d on $T^{0.89}$ in the range from cm to m. The absolute values can be tested against the depth distributions of ^{53}Mn (Nishiizumi et al 1976) or ^{26}Al (Rancitelli et al 1971) in cores. Too few cores have yet been studied to say more than this: the order of magnitude is correct, and the model's prediction of a broad distribution of h_d is verified. In time these data will provide very sensitive tests of our understanding.

4.2.3 SUMMARY In considering the various models presented, we should try to keep clear the distinction between the validity of the assumptions and calculations, and that of the input data. The good agreement between the two most detailed models and observation suggests that both models are more or less valid. However, the great sensitivity of results to exact values of χ or other distribution parameters requires that models be compared using the same parameters. We have done this for the models of Duraud et al (1975) and of Arnold (1975b), with results that we must describe as unexpectedly good, given the many opportunities for hidden assumptions and for numerical mistakes in such complex programs. The two models are of different classes, as discussed above. Their "scale" is not quite the same, with the Duraud et al (1975) model devoting more attention to events on a micron scale. Yet there are now no disagreements of any substance, in the very large area of overlap.

As a consequence, further applications of the evolution models seem justified. In lunar studies work is in progress in the following directions: 1. To establish a chronology of deposition for the deep drill core tubes from the Apollo 15, 16, and

17 (and possibly Luna 24) missions. This would permit us to trace back the history of the solar activity and the galactic cosmic rays over a time scale of several billion years. 2. To make quantitative comparisons with the formation of agglutinates and the injection of the "meteoritic" component in the lunar regolith. 3. To investigate possible correlations between the past depth histories of grains now in the same soil, in order to improve our understanding of maturation indices. 4. To extend the models to the highlands area, mainly built up before 4 b.y.

Another development is to generalize the lunar evolution models to other atmosphereless solar system bodies such as Mercury, the asteroids, and so forth. For this purpose, the input parameters in the models have to be assessed for each one of these bodies. A comparison with experimental data can so far only be performed for "gas-rich" meteorites, which apparently sample the regolith of another parent body.

Literature Cited

Adams, J. B., McCord, T. B. 1970. *Proc. Apollo 11 Lunar Sci. Conf.* 3:1937–45

Adams, J. B., McCord, T. B. 1971. *Proc. Lunar Sci. Conf.* 2:2183–95

Adler, I., Trombka, J., Gerard, J., Lowman, P., Schmadebeck, R., Blodget, H., Eller, E., Yin, L., Lamothe, R., Osswald, G., Gorenstein, P., Bjorkholm, P., Gursky, H., Harris, B. 1972. *Science* 177:256–59

Anders, E., Ganapathy, R., Krähenbuhl, U., Morgan, J. W. 1973. *Moon.* 8:3–24

Arnold, J. R. 1975a. *Moon.* 13:159–72

Arnold, J. R. 1975b. *Proc. 6th Lunar Sci. Conf.* 2:2375–95

Arrhenius, G., Asunmaa, S. K. 1973. *Moon.* 8:368–91

Arrhenius, G., Liang, S., MacDougall, D., Wilkening, L., Bhandari, N., Bhat, S., Lal, D., Rajagopalan, G., Tamhane, A. S., Venkatavardan, V. S. 1971. *Proc. 2nd Lunar Sci. Conf.* 3:2583–98

Arvidson, R. E., Burnett, D. S., Drozd, R. J., Hohenberg, C. M., Morgan, C. J., Podosek, F. A. 1975a. *Lunar Sci. VI, Abstr.*, 1:22–24

Arvidson, R., Drozd, R. J., Hohenberg, C. M., Morgan, C. J., Poupeau, G. 1975b. *Moon.* 13:67–79

Barber, D. J., Cowsik, R., Hutcheon, I. D., Price, P. B., Rajan, R. S. 1971. *Proc. 2nd Lunar Sci. Conf.* 3:2705–14

Becker, R. H., Clayton, R. N. 1975. *Proc. 6th Lunar Sci. Conf.* 2:2131–49

Berdot, J. L., Chetrit, G. C., Lorin, J. C., Pellas, P., Poupeau, G. 1972. *Proc. 3rd Lunar Sci. Conf.* 3:2867–81

Bhandari, N., Goswami, J. N., Lal, D., Tamhane, A. S. 1973. *Proc. 13th Int. Cosmic Ray Conf.* 2:1464–69

Bibring, J. P. 1972. *Thèse de 3ème cycle*, Faculté des Sciences d'Orsay, France

Bibring, J. P., Borg, J., Burlingame, A. L., Langevin, Y., Maurette, M., Vassent, B. 1975. *Proc. 6th Lunar Sci. Conf.* 3:3471–93

Bibring, J. P., Burlingame, A. L., Chaumont, J., Langevin, Y., Maurette, M., Wszolek, P. C. 1974a. *Proc. 5th Lunar Sci. Conf.* 2:1747–62

Bibring, J. P., Langevin, Y., Maurette, M., Meunier, R., Jouffrey, B., Jouret, C. 1974b. *Earth Planet. Sci. Lett.* 22:205–14

Blake, M. L., Wasserburg, G. J. 1975. *Geophys. Res. Lett.* 2:477–79

Blanford, G. E., Fruland, R. M., McKay, D. S., Morrison, D. A. 1974. *Proc. 5th Lunar Sci. Conf.* 3:2501–26

Blanford, G. E., Fruland, R. M., Morrison, D. A. 1975. *Proc. 6th Lunar Sci. Conf.* 3:3557–76

Borg, J., Burlingame, A. L., Maurette, M., Wszolek, P. C. 1974. *Solar Wind.* 3:68–70

Borg, J., Comstock, G. M., Langevin, Y., Maurette, M., Jouffrey, B., Jouret, C. 1976. *Earth Planet. Sci. Lett.* 29:161–74

Borg, J., Dran, J. C., Comstock, G. M., Maurette, M., Vassent, B. 1973. *Lunar Sci. IV, Abstr.* 1:82–84

Borg, J., Dran, J. C., Durrieu, L., Jouret, C., Maurette, M. 1970. *Earth Planet. Sci. Lett.* 8:379–86

Borg, J., Durrieu, L., Jouret, C., Maurette, M. 1971. *Proc. 2nd Lunar Sci. Conf.* 3:2027–40

Borg, J., Vassent, B. 1972. In *The Moon*, ed. H. C. Urey, S. K. Runcorn, pp. 298–308. Dordrecht: Reidel

Boynton, W. V., Chou, C. L., Bild, R. W., Baedecker, P. A., Wasson, J. T. 1976. *Earth Planet. Sci. Lett.* 29:21–33

Burnett, D. S., Woolum, D. S. 1974. *Proc. 5th Lunar Sci. Conf.* 2:2061–74

Butler, J. C., King, E. A. Jr. 1974. *Proc. 5th*

Lunar Sci. Conf. 1:829–41

Cadenhead, D. A., Stetter, J. R. 1974. *Proc. 5th Lunar Sci. Conf.* 2:2243–57

Cirlin, E. H., Housley, R. M., Goldberg, I. B., Crowe, H. 1976. *Lunar Sci. VII, Abstr.* 1:152–53

Cooper, M. R., Kovach, R. L., Watkins, J. S. 1974. *Rev. Geophys. Space Phys.* 12:291–308

Comstock, G. M. 1972. In *The Moon,* ed. H. C. Urey, S. K. Runcorn, pp. 330–52. Dordrecht: Reidel

Comstock, G. M. 1976. *Lunar Sci. VII, Abstr.* 1:169–71

Comstock, G. M., Evwaraye, A. O., Fleischer, R. L., Hart, H. R. 1971. *Proc. 2nd Lunar Sci. Conf.* 3:2569–82

Crozaz, G., Drozd, R., Hohenberg, C. M., Hoyt, H. P., Ragan, D., Walker, R. M., Yuhas, D. 1972. *Proc. 3rd Lunar Sci. Conf.* 3:2917–31

Crozaz, G., Drozd, R., Hohenberg, C., Morgan, C., Ralston, C., Walker, R., Yuhas, D. 1974. *Proc. 5th Lunar Sci. Conf.* 3:2475–99

Crozaz, G., Haack, U., Hair, M., Maurette, M., Walker, R. M., Woolum, D. 1970. *Proc. Apollo 11 Lunar Sci. Conf.* 3:2051–80

Crozaz, G., Walker, R. M. 1971. *Science.* 171:1237–39

Criswell, D. R. 1972. *Proc. 3rd Lunar Sci. Conf.* 3:2671–80

Criswell, D. R. 1975. *Proc. 6th Lunar Sci. Conf.* 2:1967–87

Curtis, D. B., Wasserburg, G. J. 1975. *Lunar Sci. VI, Abstr.* 1:172–74

Dalrymple, G. B., Doell, R. R. 1970. *Proc. Apollo 11 Lunar Sci. Conf.* 3:2081–92

Doell, R. R., Dalrymple, G. B. 1971. *Earth Planet. Sci. Lett.* 10:357–60

Dran, J. C. Duraud, J. P., Maurette, M., Durrieu, L., Jouret, C., Legressus, C. 1972. *Proc. 3rd Lunar Sci. Conf.* 3:2883–2903

Dran, J. C. Durrieu, L., Jouret, C., Maurette, M. 1970. *Earth Planet. Sci. Lett.* 9:391–400

Drozd, R., Hohenberg, C., Morgan, C. 1975. *Proc. 6th Lunar Sci. Conf.* 2:1857–77

Duke, M. B., Nagle, J. S. 1975. *Moon.* 13:153–58

Duke, M. B., Woo, C. C., Sellers, G. A., Bird, M. L., Finkelman, R. B. 1970. *Proc. Apollo 11 Lunar Sci. Conf.* 1:347–61

Duncan, A. R., Sher, M. K., Abraham, Y. C., Erlank, A. J., Willis, J. P., Ahrens, L. H. 1975. *Proc. 6th Lunar Sci. Conf.* 2:2309–20

Dunnebier, F., Droman, J., Lammlein, D., Latham, G., Nakamura, Y. 1975. *Proc. 6th Lunar Sci. Conf.* 2:2417–26

Dunnebier, F. K., Watkins, J. S., Kovach, R. L. 1974. *Lunar Sci. V, Abstr.* 183

Duraud, J. P., Langevin, Y., Maurette, M.,

Comstock, G. M., Burlingame, A. L. 1975. *Proc. 6th Lunar Sci. Conf.* 2:2397–2415

Eberhardt, P., Geiss, J., Graf, H., Grögler, N., Mendia, M. D., Mörgeli, M., Schwaller, H., Stettler, A. 1972. *Proc. 3rd Lunar Sci. Conf.* 2:1821–56

Epstein, S., Taylor, H. P. Jr. 1973. *Proc. 4th Lunar Sci. Conf.* 2:1559–75

Epstein, S., Taylor, H. P. Jr. 1975. *Proc. 6th Lunar Sci. Conf.* 2:1771–98

Eugster, O., Eberhardt, P., Geiss, J., Grögler, N., Jungck, M., Mörgeli, M. 1975. *Proc. 6th Lunar Sci. Conf.* 2:1989–2007

Fireman, E. L. 1974. *Proc. 5th Lunar Sci. Conf.* 2:2075–92

Fleischer, R. L., Haines, E. L., Hart, H. R., Woods, R. T., Comstock, G. M. 1970. *Proc. Apollo 11 Lunar Sci. Conf.* 3:2103–20

Fleischer, R. L., Hart, H. R. 1973a. *Earth Planet. Sci. Lett.* 18:420–26

Fleischer, R. L., Hart, H. R. 1973b. *J. Geophys. Res.* 78:4841–51

Fleischer, R. L., Hart, H. R., Comstock, G. M. 1971. *Science.* 171:1240–42

Fleischer, R. L., Hart, H. R., Giard, W. R. 1974. *Geochim. Cosmochim. Acta.* 38:341–84

Fleischer, R. L., Price, P. B., Walker, R. L. 1970. *Nuclear Tracks in Solids: Principles and Applications.* Berkeley: Univ. Calif. Press

Frederiksson, K., Nelen, J., Melson, W. G., Henderson, E. P., Andersen, C. A. 1970. *Science.* 167:664–66

Gammage, R. B., Holmes, H. F. 1975. *Proc. 6th Lunar Sci. Conf.* 3:3305–16

Gault, D. E. 1969. *Trans. Am. Geophys. Union* 50:219

Gault, D. E. 1976. Private communication

Gault, D. E., Hörz, F., Brownlee, D. E., Hartung, J. B. 1974. *Proc. 5th Lunar Sci. Conf.* 3:2365–86

Gault, D. E., Hörz, F., Hartung, J. 1972. *Proc. 3rd Lunar Sci. Conf.* 3:2713–34

Goel, P. S., Kothari, B. K. 1972. *Proc. 3rd Lunar Sci. Conf.* 2:2041–50

Gold, T. 1971. *Proc. 2nd Lunar Sci. Conf.* 3:2675–80

Gold, T., Bilson, E., Yerburg, M. 1972. *Proc. 3rd Lunar Sci. Conf.* 3:3187–93

Gold, T., Williams, G. J. 1974. *Proc. 5th Lunar Sci. Conf.* 3:2365–86

Goldberg, R. H., Burnett, D. S., Tombrello, T. A. 1975. *Proc. 6th Lunar Sci. Conf.* 2:2189–2200

Gose, W. A., Morris, R. V. 1976. *Lunar Sci. VII, Abstr.* 1:319–21

Gose, W., Pearce, G. W., Lindsay, J. F. 1975. *Proc. 6th Lunar Sci. Conf.* 3:3071–80

Goswami, J. N., Braddy, D., Price, P. B. 1976. *Lunar Sci. VII, Abstr.* 1:328–30

Goswami, J. N., Lal, D. 1974. *Proc. 5th Lunar Sci. Conf.* 3 : 2643–62

Greeley, R., Gault, D. E. 1970. *Moon.* 2 : 10–77

Hart, H. R., Comstock, G. M., Fleischer, R. L. 1972. *Proc. 3rd Lunar Sci. Conf.* 3 : 2831–44

Hartung, J. B., Hodges, F., Hörz, F., Storzer, D. 1975. *Proc. 6th Lunar Sci. Conf.* 3 : 3351–71

Hartung, J., Hörz, F. 1972. *Proc. 24th Int. Geol. Congr., Montreal, Canada,* Sect. 15, 48–56

Heiken, G., McKay, D. S. 1974. *Proc. 5th Sci. Lunar Conf.* 5 : 843–60

Holmes, H. F., Fuller, E. C., Gammage, R. B. 1973. *Earth Planet. Sci. Lett.* 19 : 90–96

Hörz, F. 1969. *Contrib. Mineral. Petrol.* 21 : 365–77

Hörz, F., Brownlee, D. E., Fechtig, H., Hartung, J. B., Morrison, D. A., Neukum, G., Schneider, E., Vedder, J. F. 1975a. *Planet. Space Sci.* 23 : 151–72

Hörz, F., Gault, D. E., Schneider, E., Hill, R. E., Hartung, J. B., Brownlee, D. E. 1975b. *Moon.* 13 : 235–58

Hörz, F., Gibbons, R. V., Gault, D. E., Hartung, J. B., Brownlee, D. E. 1975c. *Proc. 6th Lunar Sci. Conf.* 3 : 3495–3508

Hörz, F., Gibbons, R. V., Hill, R. E., Gault, D. E. 1976. *Lunar Sci. VII, Abstr.* 1 : 381–83

Hörz, F., Schneider, E., Hill, E. R. 1974. *Proc. 5th Lunar Sci. Conf.* 3 : 2397–2412

Hutcheon, I. D., MacDougall, D., Price, P. B. 1974. *Proc. 5th Lunar Sci. Conf.* 3 : 2561–76

Imamura, M., Nishiizumi, K., Honda, M., Finkel, R. C., Arnold, J. R., Kohl, C. P. 1974. *Proc. 5th Lunar Sci. Conf.* 2 : 2093–2103

Kerridge, J. F. 1975. *Science.* 188 : 162–64

King, E. A., Butler, J. C., Carman, M. F. 1971. *Proc. 2nd Lunar Sci. Conf.* 1 : 737–46

Kirsten, T., Deubner, J., Horn, P., Kaneoka, I., Kiko, J., Schaeffer, O., Thio, S. K. 1972. *Proc. 3rd Lunar Sci. Conf.* 2 : 1865–89

Kovach, R. L., Watkins, J. S., Landers, T. 1971. *Apollo 14 Prelim. Sci. Rep.,* pp. 163–174 (NASA SP-272)

Kovach, R. L., Watkins, J. S., Talwani, P. 1972. *Apollo 16 Prelim. Sci. Rep.,* pp. 10-1-10-14 (NASA SP-315)

Lal, D. 1972. *Space Sci. Rev.* 14 : 3

Lal, D., MacDougall, D., Wilkening, L., Arrhenius, G. 1970. *Proc. Apollo 11 Lunar Sci. Conf.* 3 : 2295–2303

Langevin, Y., Maurette, M. 1976. *Proc. 7th Lunar Sci. Conf.* 1 : 75–91

Lindsay, J. F. 1971. *J. Sediment. Petrol.* 41 : 780–97

Lindsay, J. F. 1973. *Proc. 4th Lunar Sci. Conf.* 1 : 215–24

Lindsay, J. F. 1974. *Proc. 5th Lunar Sci. Conf.* 1 : 861–78

Lindstrom, M. M., Duncan, A. R., Fruchter, J. S., McKay, S. M., Stoeser, J. W., Goles, G. G., Lindstrom, D. J. 1972. *Proc. 3rd Lunar Sci.* 2 : 1201–14

Lingenfelter, R. E., Canfield, R. H., Hampel, V. E. 1972. *Earth Planet. Sci. Lett.* 16 : 355–69

MacDougall, D., Rajan, S. J., Hutcheon, I. D., Price, P. B. 1973. *Proc. 4th Lunar Sci. Conf.* 3 : 2319–36

McCoy, J. E., Criswell, D. R. 1974. *Proc. 5th Lunar Sci. Conf.* 3 : 2991–3005

McCrosky, R. E. 1968. *Smithson. Astrophys. Obs. Spec. Rep.* 2800, 1–13

McDonnel, J. A. M., Ashworth, D. G. 1972. *Space Res.* 12 : 333–47 Berlin : Akademie

McKay, D. S., Fruland, R. M., Heiken, G. H. 1974. *Proc. 5th Lunar Sci. Conf.* 1 : 887–906

McKay, D. S., Heiken, G. H., Taylor, R. M., Clanton, U. S., Morrison, D. A., Ladle, G. H. 1972. *Proc. 3rd Lunar Sci. Conf.* 1 : 983–94

McKay, D. S., Morrison, D. A., Clanton, U. S., Ladle, G. H., Lindsay, J. F. 1971. *Proc. 2nd Lunar Sci. Conf.* 1 : 755–73

Manka, R. H., Michel, F. C. 1970. *Science.* 167 : 1325–39

Marti, K., Lightner, B. D., Osborn, T. W. 1973. *Proc. 4th Lunar Sci. Conf.* 2 : 2037–48

Maurette, M., Pellas, P., Walker, R. M. 1964. *5th Int. Conf. Nucl. Photogr., CERN*

Maurette, M., Price, P. B. 1975. *Science.* 187 : 121–29

Meyer, H. O. A., McCallister, R. H., Tsai, H. M. 1975. *Proc. 6th Lunar Sci. Conf.* 1 : 595–614

Millman, P. M. 1972. *J. R. Astron. Soc. Can.* 66 : 201

Millman, P. M. 1973. *Moon.* 8 : 228

Mitchell, J. K., Houston, W. N., Scott, R. F., Costes, N. C., Carrier, W. D. III, Bromwell, L. G. 1972. *Proc. 3rd Lunar Sci. Conf.* 3 : 3235–53

Morgan, J. W., Land, J. C., Krähenbühl, U., Ganapathy, R., Anders, E. 1972. *Proc. 3rd Lunar Sci. Conf.* 2 : 1377–95

Morrison, D. A., McKay, D. S., Fruland, R. M., Moore, H. J. 1973. *Proc. 4th Lunar Sci. Conf.* 3 : 3235–53

Nakamura, Y., Dorman, J., Dunnebier, F., Lammlein, D., Latham, G. 1975. *Moon.* 13 : 57–66

Neukum, G., Dietzel, H. 1971. *Earth Planet. Sci. Lett.* 12 : 59–66

Neukum, G., Hörz, F., Morrison, D. A., Hartung, J. B. 1973. *Proc. 4th Lunar Sci. Conf.* 3 : 3255–76

Neukum, G., König, B., Fechtig, H., Storzer,

D. 1975. *Proc. 6th Lunar Sci. Conf.* 3: 2597–2650

Nishiizumi, K., Imamura, M., Honda, M., Russ, G. P., Kohl, C. P., Arnold, J. R. 1976. *Proc. 7th Lunar Sci. Conf.* 1:41–54

Oberbeck, V. R., Quaide, W. L. 1967. *J. Geophys. Res.* 72:4697–4704

Oberbeck, V. R., Quaide, W. L., Mahan, M., Paulson, J. 1973. *Icarus.* 19:87–107

Papanastassiou, D. A., Wasserburg, G. J. 1971. *Earth Planet. Sci. Lett.* 11:37–62

Papanastassiou, D. A., Wasserburg, G. J. 1975. *Proc. 6th Lunar Sci. Conf.* 2:1467–89

Pepin, R. O., Basford, J. R., Dragon, J. C., Coscio, M. R. Jr., Murthy, V. R. 1974. *Proc. 5th Lunar Sci. Conf.* 2:2149–84

Pepin, R. O., Dragon, J. C., Johnson, N. L., Bates, A., Coscio, M. R. Jr., Murthy, V. R. 1975. *Proc. 6th Lunar Sci. Conf.* 2:2027–55

Phakey, P. P., Hutcheon, I. D., Rajan, R. S., Price, P. B. 1972. *Proc. 3rd Lunar Sci. Conf.* 3:2905–15

Polatty, J. M., Houston, B. J., Stone, R. I., Bancks, C. 1965. *Plowshare Rep. P.N.E.* 5003

Poupeau, G., Walker, R. M., Zinner, E., Morrison, D. A. 1975. *Proc. 6th Lunar Sci. Conf.* 3:3433–48

Price, P. B., Hutcheon, I. D., Braddy, D., MacDougall, D. 1975. *Proc. 6th Lunar Sci. Conf.* 3:3449–69

Price, P. B., Lal, D., Tamhane, A. S., Perelygin, V. P. 1973. *Earth Planet. Sci. Lett.* 19:377–95

Quaide, W. L., Oberbeck, V. R. 1968. *J. Geophys. Res.* 73:5247–70

Rancitelli, L. A., Fruchter, J. S., Felix, W. D., Perkins, R. W., Wogman, N. A. 1975. *Proc. 6th Lunar Sci. Conf.* 2:1891–99

Rancitelli, L. A., Perkins, R. W., Felix, W. D., Wogman, N. A. 1971. *Proc. 2nd Lunar Sci. Conf.* 2:1757–72

Reedy, R. C. 1976. *Lunar Science VII, Abstr.* 2:721–23

Reedy, R. C., Arnold, J. R. 1972. *J. Geophys. Res.* 77:537–55

Rehfuss, D. E. 1972. *J. Geophys. Res.* 77:6303

Rhodes, J. M., Adams, J. B., Blanchard, O. P., Charette, M. P., Rodgers, K. V., Jacobs, J. W. Braunon, J. C., Haskin, L. A. 1975. *Proc. 6th Lunar Sci. Conf.* 2:2291–2307

Russ, G. P. III. 1973. *Earth Planet. Sci. Lett.* 16:275–89

Russ, G. P. III, Burnett, D. S., Wasserburg, G. J. 1972. *Earth Planet Sci. Lett.* 15:172–86

Schmitt, H. H. 1973. *Science* 182:681–90

Schneider, E., Storzer, D., Hartung, J., Fechtig, H., Gentner, W. 1973. *Proc. 4th*

Lunar Sci. Conf. 3:3277–90

Shoemaker, E. M., Batson, R. M., Bean, A. L., Conrad, C., Jr., Dahlem, D. H., Goddard, E. N., Hait, M. H., Schaber, G. G., Schleicher, D. L., Sutton, R. L., Swann, G. A., Waters, A. C. 1970a. *Apollo 12 Prelim. Sci. Rep.,* 113–56 (NASA SP-235)

Shoemaker, E. M., Batson, R. M., Holt, H. E., Morris, E. C., Rennilson, J. J., Whitaker, E. A. 1967. *Surveyor III, A Preliminary Report.* NASA SP-146, pp. 9–60

Shoemaker, E. M., Batson, R. M., Holt, H. E., Morris, E. C., Rennilson, J. J., Whitaker, E. A. 1969. *J. Geophys. Res.* 74:6081–6119

Shoemaker, E. M., Hait, M. H., Swann, G. A., Schleicher, D. C., Schaber, G. G., Sutton, R. L., Dahlem, D. H., Goddard, E. N. and Waters, A. C. 1970b. *Proc. Apollo 11 Lunar Sci. Conf.* 3:2399–2412

Schonfeld, E. 1974. *Proc. 5th Lunar Sci. Conf.* 2:1269–86

Silver, L. T. 1975. *Lunar Science VI, Abstr.* 2:738–40

Soderblom, L. A. 1972. *Apollo 15 Prelim. Sci. Rep.,* 25-87 (NASA SP-289)

Swann, G. A. et al. 1972. *Apollo 15 Prelim. Sci. Rep.,* Figure 5-47 (NASA photo A515082-11082)

Taylor, S. R. 1975. *Lunar Science: A Post Apollo View.* New York: Pergamon

Tera, F., Papanastassiou, D. A., Wasserburg, G. J. 1974. *Earth Planet. Sci. Lett.* 22:1–21

Wahlen, M., Finkel, R. C., Imamura, M., Kohl, C. P., Arnold, J. R. 1973. *Earth Planet. Sci. Lett.* 19:315–20

Walker, R. M., Yuhas, D. 1973. *Proc. 4th Lunar Sci. Conf.* 3:2379–89

Wänke, H., Palme, H., Baddenhausen, H., Dreibus, G., Jagoutz, E., Kruse, H., Palme, C., Spettel, B., Teschke, F., Thacker, R. 1975. *Proc. 6th Lunar Sci. Conf.* 2:1313–40

Wetherill, G. W. 1974. *Ann. Rev. Earth Planet. Sci.* 3:303–31

Whipple, F., Hawkins, G. S. 1959. *Handb. Phys.* 52:519

Wood, J. A., Dickey, J. S. Jr., Marvin, U. B., Powell, B. N. 1970. *Science.* 167:602–04

Yaniv, A., Heymann, D. 1972. *Proc. 3rd Lunar Sci. Conf.* 2:1967–80

Yuhas, D. E. 1974. PhD Thesis. Washington Univ., St. Louis, Missouri

Yuhas, D. E., Walker, R. M., Reeves, H., Poupeau, G., Pellas, P., Lorin, J. C., Chetrit, G. C., Price, P. B., Hutcheon, I. D., Hart, H. R., Fleischer, R. L., Comstock, G. M., Lal, D., Goswami, J. N., Bhandari, N. 1972. *Proc. 3rd Lunar Sci. Conf.* 3:2941–47

Zinner, E., Walker, R. M. 1975. *Proc. 6th Lunar Sci. Conf.* 3:3601–17

Ann. Rev. Earth Planet. Sci. 1977. 5:491–513

THEORETICAL FOUNDATIONS OF EQUATIONS OF STATE FOR THE TERRESTRIAL PLANETS

×10081

Leon Thomsen

Department of Geological Sciences and Environmental Studies, State University of New York, Binghamton, NY 13901

Our understanding of the constitution and structure of the terrestrial planets consists of a lengthy chain of inference. The major links of the chain are near-surface sampling, seismological estimation of elastic parameters of the deeper layers, interpretation with the help of laboratory experiment at high pressure and temperature, and theoretical synthesis. This paper reviews the current status of the last link only of this chain, in interpretive fashion. A noninterpretive, but considerably more exhaustive, listing of recent contributions has been given by Ahrens (1975).

The present work represents considerable evolution of thought since a previous review (Thomsen 1971). As it is intended to inspire a similar evolution of the reader's ideas, the reasons for these changes of opinion merit a philosophical summarization at this point. A set of familiar concepts (velocity-density systematics, Voigt-Reuss-Hill averaging, finite strain extrapolation formulas, etc) have been extensively used to discuss the mantles of the terrestrial planets, and are in fact adequate where the rocks are familiar and the pressure and temperature are modest. However, in the lower mantles of Earth and Venus, the uncertainties in the elastic parameters due to scatter about the velocity-density systematic trends, coupled with the large compression and uncertain temperature, restrict the power of realistic conclusions. The crudity of these concepts, by which we estimate the effects of compression, offer a strange contrast to the elegance of the theory of lattice dynamics, by which we calculate the effects of temperature. The reason for this dissimilarity is that temperature affects density, elasticity, etc only through averages over the eigenfrequencies of vibration (which can be understood in terms of harmonic forces between atoms, plus a small perturbation). By contrast, the effects of compression require not only the energy eigenvalues, but also the electronic charge density, which in turn requires the full apparatus of quantum mechanics.

491

Advances over the last decade in computer hardware and software, and in the physics of many-electron exchange and correlation, are beginning to make such calculations feasible for many solids of geophysical interest. Already the first such calculations are appearing, and promise to make the previous set of theoretical tools obsolete.

1   A BRIEF HISTORY

A simplified history of the development of geophysical equations of state might start with the work of Einstein (1907) and Grüneisen (1912), which utilized the new quantum mechanics to establish the effects of temperature upon the specific heat and volume of solids. For technological reasons, temperature remained for many years the chief thermodynamical variable of theoretical and experimental interest. In the 1930s Bridgman achieved a major capability for high experimental pressures (cf e.g. Bridgman 1938), and this excited the first serious theoretical efforts to understand the effects of high pressure on solids. The full quantum treatment of pressure proved to be far less tractable than the quantum treatment of temperature and was essentially abandoned. Instead, the pressure effects were estimated using heuristic atomistic models like that of Born (1923), or continuum mechanics (finite strain) formalisms like that of Birch (1947).

During this period, Hill (1952) established bounds upon the effective isotropic elasticity of an aggregate of randomly oriented anisotropic crystals, thus forging a theoretical connection between accurate single-crystal data and the mantles of terrestrial planets. A high point of all this effort was the classic demonstration by Birch (1952) of the existence of a transition zone starting near 400 km depth in the earth. It is now well documented as a series of phase changes (cf e.g. D. L. Anderson et al 1971), with analogs inferred to exist also in the terrestrial planets (cf e.g. D. L. Anderson 1972). A major question remaining open is whether or not the transition zone also marks a compositional change; we return to this later.

The existence and importance of these high-pressure phases, together with certain military requirements, generated funds during the late 1950s and 1960s for the development of several new types of high-pressure apparatus, among which the Bridgman anvils of Ringwood, Green, and co-workers (cf Ringwood 1970), the shock-wave gun of Ahrens and co-workers (e.g. Ahrens et al 1969), and the diamond cell of Bassett, Takahashi, and co-workers (e.g. Mao et al 1974) have yielded the most information about geophysical solids at high pressure. The introduction of ultrasonic interferometry into geophysics by O. L. Anderson and co-workers (e.g. 1968) increased the reliability of pressure and temperature derivatives used in many theoretical models to evaluate parameters. And the establishment of the World-Wide Seismic Network by Project Vela Uniform increased enormously the precision of the elasticity and density profiles in the earth.

The theoretical advances did not keep step. Birch (1961) had noticed that, among a suite of rocks at low pressure and temperature, the velocities and densities showed systematic trends. His suggestion, that these trends were preserved under moderate increases of pressure and temperature, and even through some phase changes, was roughly confirmed experimentally (O. L. Anderson 1966) and explained heuristically (D. L. Anderson & O. L. Anderson 1970). Hence it became possible to *model* the composition and structure of the earth by

1. assuming a composition and a corresponding mixture of crystal structures for a given layer in the earth,
2. estimating, using crystal chemistry, the density at low pressure and temperature of this assemblage,
3. estimating the velocities, incompressibility, etc of crystals using the velocity-density systematics of Birch (1961), or similar systematics published subsequently,
4. estimating the velocity of an isotropic aggregate of these crystals using Hill's (1952) mixing theory,
5. extrapolating to high pressure and temperature using, for example, the finite strain formalism of Birch (1947),
6. comparing with seismic estimates for velocity and density at that depth, adjusting assumptions (step 1) if necessary and repeating the steps.

Various authors used conceptual paths equivalent, if not identical, to the above, and the result was commonly the same. [The latest example of the comprehensive use of these to discuss the constitution and structure of the earth is provided by D. L. Anderson (1977) in this volume.] The upper mantle is consistent with Ringwood's pyrolite model (cf Ringwood 1970); the lower mantle is also homogeneous, but composed of high-pressure phases of silicates and oxides, with perhaps twice the iron concentration of the upper mantle. This distinction is important because it indicates a gross chemical disequilibrium between lower and upper mantles. This in turn implies that the two layers are not mixed during the convective overturning accompanying sea-floor spreading, with a consequent constraint on the nature of the driving mechanism.

In the past half-dozen years, a number of different workers have cast doubt upon several of these conceptual steps. Liebermann (1975), summarizing a series of important data papers, has shown that the systematic relations needed in step 3 are modified at pressures high enough to change the coordination number of silicon. Major advances (reviewed by Watt et al 1976) in the theory of the elasticity of an isotropic aggregate (step 4) have shown that although microscopic information (usually unavailable) about the detailed distribution of grains in a rock is necessary to understand its elasticity, still the Hashin-Shtrikman (1962) bounds, properly calculated, constitute geophysically useful constraints. Thomsen (1970, 1972) showed that the particular finite strain formalism (step 5) used by Birch was inconsistent with the theory of lattice dynamics, which describes the response of a solid to high temperature. Ahrens & Thomsen (1972) showed that, disregarding this, the ambiguity inherent in finite strain theory precluded its use as a predictor of high-pressure properties. Spetzler (1970) showed that the current ultrasonic laboratory techniques for measuring elastic velocities may result in systematic errors in the pressure derivatives of as much as 25%, thereby reducing the reliability of the finite strain parameters calculated from such data. And, very recently, Liu et al (1976) have shown that the acoustic dispersion accompanying imperfectly elastic wave propagation produces non-negligible differences between velocities measured in the laboratory at several megacycles, seismic propagation at 20 sec, and eigen-periods of an hour.

The impact of these negative findings has been to slow the development of our understanding of the constitution and structure of the deep interiors of the planets. Distrust in the current theoretical tools has led to doubts concerning the validity of many current results (e.g. the iron enrichment of the lower mantle of the earth). Even taking the current tools at face value, the uncertainties in temperature, phase assemblage, and that due to the low-spin electronic configuration of iron prevent strong conclusions about the state of the lower mantle (Davies 1974).

It is my belief, however, that we stand now at the beginning of a major new advance in understanding. The essence of the advance is that the continuing progress in computing hardware and software is now making it feasible to perform the full quantum-mechanical treatment of the equation of state. Recent work by Bukowinski & Knopoff (1976) and Thomsen (1977) are the first applications of quantum theory to geophysical equations of state, and these approaches will soon make the theoretical tools of the last decade obsolete. A major part of the present review is therefore a discussion of the quantum theory, applied to geophysical equations of state. In preparation for this discussion, a number of ancillary topics must first be addressed.

2 THERMODYNAMIC CONSIDERATIONS

The tensor equation of state relates the stress in a solid to the strain. The principle of conservation of energy implies that, for isotropic pressure P on an isotropic solid, the (scalar) equation of state is

$$P(v, T) = -\left(\frac{\partial U}{\partial v}\right)_S, \tag{1a}$$

where v is specific volume, T is temperature, U is internal energy, and S is specific entropy. The tensor generalization of this is

$$\sigma_{ij}(v, T) = \frac{1}{v}\left(\frac{\partial U}{\partial u_{ij}}\right)_S, \tag{1b}$$

where $\mathbf{u}$ is the infinitesimal symmetric strain tensor familiar in seismology. For solids, U is a well-defined state function (dependent only on v and T, and not, e.g., on history) only if the stresses are of the form $\sigma = -P\mathbf{I} + \sigma'$, where $\mathbf{I}$ is the identity tensor and σ' is small. The adiabatic elastic moduli are defined by

$$C_{ijkl}(v, T) = \left(\frac{\partial \sigma_{ij}}{\partial u_{kl}}\right)_S = \left(\frac{\partial^2 U}{\partial u_{ij}\partial u_{kl}}\right)_S \frac{1}{v}. \tag{1c}$$

An elementary result of the theory of lattice dynamics (cf e.g. section II in Born & Huang 1954) is that the energy dependence on v and T is additively separable,

$$U(v, T) = E(v) + E_{\text{vib}}(v, T), \tag{2}$$

where E is the energy of the static lattice (no vibrations of the nuclei), and E_{vib} is the vibrational contribution. Even in a deep planetary interior, it is possible to show that E_{vib} is always much smaller than E, despite the high ambient temperatures. Hence the main part of the present discussion is concerned with E.

We do note here, however, that if the restoring forces on the vibrating nuclei are approximately harmonic, then

$$E_{vib} = \sum_m \omega_m \hbar \left[\frac{1}{2} + \frac{1}{\exp(\omega_m \hbar/kT) - 1} \right] = \sum_m \varepsilon_m \tag{3}$$

(Einstein 1907), where the ω_m are the eigenfrequencies of vibration (independent of v), $\hbar$ is Planck's constant, and k is Boltzmann's constant. If the restoring energy is approximated instead as a quartic polynomial in nuclear displacements, E_{vib} consists of a term like (3) [except with $\omega_m = \omega_m(v)$, plus an explicit anharmonic term, independent of v (Leibfried & Ludwig 1961)]. This case is, of course, more realistic than the harmonic case; it includes the phenomena of thermal expansivity and T-dependent elastic moduli and leads, using (1a), to the equation of state (Grüneisen 1912)

$$P(v, T) = -\frac{dE}{dv} + \frac{1}{v} \sum_m \gamma_m \varepsilon_m, \tag{4}$$

where $\gamma_m \equiv v \, d\omega_m/\omega_m \, dv$ is called the mth Grüneisen parameter. At high temperature, as in a planetary interior, $\omega_m \rightarrow kT$ and the pressure become, without further approximation,

$$P(v, T) = -\frac{dE}{dv} + \gamma(v) \frac{E_{vib}(T)}{v}, \tag{5}$$

where $\gamma(v)$ is the average of the $\gamma_m(v)$. Despite the high temperature inside a planet, the second term of (5) is always much smaller than the first. One reason is that $\gamma(v)$ decreases markedly with increasing pressure, cf e.g. O. L. Anderson (1967). However, in shock-wave experiments on geophysical materials, the temperature may become so high that the vibrational term is quite significant and must be taken into account in a proper analysis. A complete equation of state must therefore provide accurate expressions for $E(v)$ certainly, and for $\gamma(v)$ if possible.

3   CONTINUUM THEORY OF THE EQUATION OF STATE

The energy $E(v)$ is, of course, fundamentally determined by the forces between the atoms, i.e. the attractive and repulsive electrostatic forces between electrons and nuclei, and the repulsive forces of exchange and correlation of electrons. However, these must be found from the electronic charge distribution, which requires the full apparatus of quantum mechanics. Hence it is worthwhile to attempt easier approaches, such as those of continuum mechanics. The best-known such formalism is that of Birch (1947), who assumed that, for a cubic or isotropic solid,

$$E(v) = B_0 + B_1 \varepsilon + B_2 \varepsilon^2 + \cdots, \tag{6}$$

where the B_1 are constants to be determined empirically, and ε is an Eulerian finite strain scalar given by

$$E(v) = -\tfrac{1}{2}[(v_0/v)^{2/3} - 1]. \tag{7}$$

Differentiation of (6) and evaluation of the constants B_1 leads to the Birch-Murnaghan equation of state,

$$P(v) = -3K_0(v/v_0)^{-5/3}(\varepsilon + \cdots), \tag{8}$$

where K_0 is the adiabatic incompressibility, evaluated at zero pressure. This expression, and variants of it, have been used extensively in the geophysical literature.

An alternative finite strain formalism (Thomsen 1970, 1972) is based on the assumption [in place of (6)] that

$$E(v) = A_0 + A_1\eta + A_2\eta^2 + \cdots, \tag{9}$$

where the A_1 are empirical constants, and η is a Lagrangian finite strain scalar given by

$$\eta(v) = \tfrac{1}{2}[(v/v_0)^{2/3} - 1]. \tag{10}$$

The associated equation of state is

$$P(v) = -3K_0(v/v_0)^{-1/3}(\eta + \cdots). \tag{11}$$

The expressions (8) and (11) are two (of an infinite variety) of equations of state derivable from the formalism of continuum mechanics. All of these are based on expansion of the energy in a power series in some strain function $\zeta(\eta)$ or $\zeta(\varepsilon)$. If the constants are evaluated in forms of low-pressure data, and P derivatives thereof, these equations amount to sophisticated extrapolation schemes that mutually diverge from each other at high P. It is therefore necessary to establish which, if any, of these expressions is useful in geophysics.

It is well known that the Eulerian formalism leading to (8) is not valid for anisotropic materials (cf e.g. Davies 1973b), since it leads to results that depend upon the (arbitrary) choice of orientation of the coordinate frame. [This objection is sufficient to proscribe its usage in a macroscopically isotropic rock, if it is composed of (randomly oriented) anisotropic grains.] The formalism is mathematically valid, however, even for anisotropic materials, if the stress is isotropic (i.e. a pure pressure). But it is clear on physical grounds that the energy of a solid should be described in terms that are independent of the stress to be placed on it. The expression for $E(P)$ should be derivable from that for $E(\sigma)$, as a special case, even if $\sigma = -P\mathbf{I}$ is the only stress of interest. After all, the same atomic forces are involved in all cases!

This, however, is a debater's point that might well be ignored if circumstances warranted it; a better criterion for choosing between (8), (11), and the infinity of other expressions is worth seeking. Thomsen (1970) noted that, since the theory of lattice dynamics is based on a Taylor expansion, one could require that $E(v)$ be based upon the same expansion. If one assumes, in calculating the thermal energy

E_{vib}, that the energy field experienced by a vibrating nucleus is a quartic polynomial in the nuclear displacements, then the static energy E of the entire lattice must also be a quartic polynomial in the nuclear separations, with coefficients composed of lattice sums of the lattice-dynamic coefficients. Alternatively, the coefficients may be considered, as in (8) and (11), to be empirical constants. In either case, the quartic polynomial in nuclear separations can be written as a quartic polynomial in the strain tensor **e** defined by

$$\mathbf{e}\mathbf{r}_0 = \mathbf{r} - \mathbf{r}_0. \tag{12}$$

However, since the components of **r** and $\mathbf{r}_0$ depend on the choice of coordinate frame, this quartic polynomial is also apparently frame dependent. Since this is not physically possible, there must be relations among the coefficients to ensure frame independence.

These may be found by noting that the symmetric tensor defined by

$$\boldsymbol{\eta} = \tfrac{1}{2}(\mathbf{e} + \mathbf{e}^\dagger + \mathbf{e}^\dagger\mathbf{e}), \tag{13}$$

(where the dagger indicates transposition) measures the scalar distance change

$$r^2 - r_0^2 = 2\mathbf{r}_0\boldsymbol{\eta}\mathbf{r}_0 \tag{14}$$

and hence is frame independent. The nonlinear transformation (13) is the simplest tensor relation converting **e** into a frame-independent quantity. One can obtain a frame-independent expression for the energy, then, by equating the fourth-order polynomial in **e** to a fourth-order polynomial in $\boldsymbol{\eta}$. The equality will not be exact, owing to the nonlinearity in (13); evidently the quartic term in $\boldsymbol{\eta}$ contains up to eighth-order terms in **e**. These terms (or others like them; see below) must be present in the description of the energy; however, their small magnitude relative to the four leading terms suggests their neglect in the lattice-dynamic development leading to (3) and (5). In turn, the general success of the theory of lattice dynamics justifies the neglect of the higher-order terms.

In the case of simple pressure on a cubic or isotropic solid, the symmetric finite strain tensor $\boldsymbol{\eta}$ reduces to $\boldsymbol{\eta} = \eta\mathbf{I}$, where the scalar η is given by (10). Hence, Thomsen argued, the assumptions of lattice dynamics, coupled with the requirement of frame independence, constrained the analytical form of the static energy E (and the Grüneisen parameter γ) to expansions in $\boldsymbol{\eta}$ of fixed order (fourth-order for E, second-order for γ). Explicit formulas were developed for the full volume and temperature dependence of pressure, and for elastic moduli for the cubic and isotropic cases. Because of the completeness of this fourth-order anharmonic theory, a large number of empirical parameters occur, making its application cumbersome. For example, the pressure on a cubic solid requires six parameters, which must be found by solving six coupled equations iteratively, using six data, which may be taken as v_0, K_0, $\partial K/\partial P$, $\partial^2 K/\partial P^2$, $\partial K/\partial T$, and the thermal expansivity, all measured at zero pressure.

This effort, however, is rather easily programed for a computer, and Ahrens & Thomsen (1972) have made the calculations for all cubic solids for which high-quality, low-pressure data were available for evaluating the parameters, and shock-wave or static compression data were available for comparison with theory. This study covered 17 solids (with both metallic and ionic/covalent bonding) and included for the first time a valid, high-compression comparison of the Lagrangian and Eulerian (Birch-Murnaghan) formalisms. The results were quite discouraging. Neither formalism was valid in general; roughly half of the solids were described best by each of the two formalisms. The disagreement consisted of a divergence of theory from data at high compression ($\Delta v/v_0 \gtrsim 0.3$), with the failures of the Birch-Murnaghan formalism being excessively high predicted pressures, and the failures of the fourth-order anharmonic theory being insufficiently high predicted pressures. This resulted in two conclusions: 1. either formalism was valid at low pressure, and 2. neither formalism could be trusted for predictions at high compression.[1] It seems clear that the failure of these formalisms to describe the general case is due to the fact that different solids are bound in different ways, which are not necessarily compatible with any one finite strain formalism.

During this same time, profound questions were being raised about the validity of the data used to evaluate the constants in the finite strain equations. Spetzler (1970) developed a variant of ultrasonic interferometry that was designed to permit measurement of elastic waves at simultaneously high P and T, through the use of a buffer rod that isolated the transducer from the high temperature. In the first application of the technique, to the well-studied MgO, he found, for example, that K'_0 [$(\partial K^S/\partial P)_T$ measured near zero pressure] was about 20% lower than the current consensus (3.8 vs 4.5), a discrepancy far outside the precision of all current experiments. In later work with the new buffer rod technique, O'Connell & Graham (1971) found a similar result for spinel: values of K'_0 were 20–25% low. The earlier results for MgO were criticized (Spetzler & Anderson 1971) regarding their internal consistency, and blame was tentatively assigned to the use of a bonding agent between sample and transducer. However, further work (E. K. Graham, private communication) with one standard technique has demonstrated that any bond effects are small, and that the internal consistency is good, reaffirming the previous results. Meanwhile a serious comparison of the two techniques (Davies & O'Connell 1975) failed to establish conclusively the reason for the discrepancies. Thus, we have a continuing systematic error in one or several of the current ultrasonic experiments, one which leads to an uncertainty in K'_0 that makes its use in finite strain extrapolations highly unreliable.

Davies (1973a) later pointed out that the requirement of consistency with lattice dynamics did not in fact constrain the choice of a strain tensor as Thomsen had claimed. Since the transformation (13) is nonlinear, extra terms arise that must be disregarded in the lattice-dynamics part of the theory. [The same situation would

---

[1] It is worth remarking that the divergence of the Birch-Murnaghan expression from shock-wave compression data sometimes goes unnoticed because a Birch-Murnaghan adiabat is compared directly with data on a shock adiabat (Hugoniot). The latter has a higher temperature, at any given compression, than the adiabat, hence a larger thermal pressure [the second term in (5)]. If a proper reduction to adiabatic temperature is made in the data, then the reduced data are lowered, at every compression, and the disagreement with the Birch-Murnaghan expression becomes evident.

arise if one expanded $E(v)$ in any tensor strain function $\zeta(\boldsymbol{\eta})$; extra terms would arise in the transformation to the **e** expansion, which would have to be disregarded.] Thus, Thomsen's procedure did in fact involve an additional approximation, albeit one of the same order, beyond those of lattice dynamics.

A better procedure is to make no transformation at all, but merely define a set of auxiliary relations (Leibfried & Ludwig 1961, equations 2.9) among the coefficients of the **e** expansion so that, to fourth-order, the frame independence of the energy is satisfied. These auxiliary relations imply that the lattice energy is a fourth-order polynomial in the symmetric strain $\mathbf{e}^* \equiv (\mathbf{e} + \mathbf{e}^\dagger)$, rather than **e** itself. In the case $\mathbf{e}^* = e\mathbf{I}$, the energy then becomes

$$E = C_0 + C_1 e + C_2 e^2 + C_3 e^3 + C_r e^4 \tag{15}$$

with a corresponding fourth-order equation of state

$$P(v) = -3K_0(v/v_0)^{-2/3}[e - 3/2(K'_0 - 1)e^2 + 3/2(K_0 K''_0 + K'_0(K'_0 - 1) + 2/9)e^3]. \tag{16}$$

This approach, which properly implements Thomsen's original idea of consistency between assumptions for E and E_{vib}, has some empirical merit, as sample calculations show that (16) generally lies between the corresponding expressions (8) and (11). Thus one would expect that its overall agreement with data, in a study such as that of Ahrens and Thomsen, would be better than that of either earlier theory. However, one would still expect some materials to disagree with the theory at modest pressure, simply because their energy fields are not quartic polynomials, as assumed. Further, until the reliability of the data for K'_0 (and K''_0) is established, there seems to be little point in using any finite strain formalism at high pressure.

Nonetheless, an expansion in the tensor $\mathbf{e}^*$ has recently been used by Davies (1973a) to describe $\gamma(v)$, along with an expansion in ε of $E(v)$ in a study of the behavior of MgO. He found good agreement, despite the inconsistencies of his treatment, using Spetzler's (low) value of K'_0. If instead he had used the older (high) value of K'_0, he would have found better agreement using an η expansion. This illustrates the futility of discussing lower mantle constitution and structure using these theoretical tools and current, grossly uncertain data. It is clear that a better theory is required.

4   SIMPLE MODELS OF ATOMIC INTERACTION

The most fundamental of the problems inherent to the continuum theories discussed above is their failure to address directly the complexity of the interaction between atoms in a solid (Knopoff 1963). Thus, even a simple force model based upon physical, rather than geometrical, assumptions might be expected to yield better results.

The most extensive work of this sort has been done in Russia [Kalinin, Pan'kov & Zharkov (1972), Kalinin & Pan'kov (1973, 1974), Pan'kov & Kalinin (1974)]. They use an r^{-1} attractive potential and an exponential repulsive potential, involving three empirical parameters (plus two more for the thermal pressure). They find reasonable agreement with shock data for moderate pressures on many rock-forming minerals, but encounter a familiar problem when the theory is applied to high-pressure phases; the existing low-pressure data do not constrain the parameters very well. For example, in stishovite the scaling parameter b for the exponential is found to lie between 2 and 22! Models more complicated in design have been applied to the alkali halides and to MgO, for which copious data are available (Demarest 1973, 1974), but these have not been extended to the more complicated crystals of geophysical interest.

A similar approach has led to new insight into the Grüneisen parameter and hence into the thermal pressure. Palciauskas (1975) has shown that the assumption of a simple potential (r^{-n} repulsion), when coupled with a realistic description of the spectrum of lattice vibrations (more realistic than the Debye model), yields a closed expression for $\gamma(v)$. The same basic assumptions transform the Lmdemann criterion for melting into a much more accurate predictor of the melting point, as a function of pressure, than it has previously been considered (Palciauskas 1976).

However, these theoretical models suffer from the same excess of simplicity, arbitrariness of assumptions, and dependence upon low-pressure data as do the continuum mechanics formalisms of the previous section. It is clear that a better theory is required.

5 QUANTUM THEORY OF THE EQUATION OF STATE

The final results of this section are to be found in many physics textbooks (cf e.g. Slater 1968), but they are usually imbedded in physical discussions or assumptions that are not relevant to the equation of state. Therefore, I present them without apology, in the hope that the connecting discussion will shed new light on them. It is assumed that the reader has the equivalent of a junior-level course in quantum mechanics.

We concentrate attention on the energy of the static lattice; the thermal term can later be treated with the anharmonic theory. The basic physical assumption behind equation (1) is the conservation of energy; what we need now is a postulate to define the energy. That is provided by the central dogma of quantum mechanics,

$$E = \int \Psi^* \mathbf{H} \Psi \, dx^{6N} = \langle \mathbf{H} \rangle, \tag{17}$$

where $\mathbf{H}$ is the Hamiltonian operator, Ψ is the state function of the total system, Ψ^* is its complex conjugate, and the integration is over the 3 components $\times 2$ spins of each of the N particles in the system. The bracket notation emphasizes that the integral forms an average (over the state function) of $\mathbf{H}$. Since we are considering the static lattice, the nuclei are not part of the system, but merely define a periodic potential energy field in which the N electrons move. For an equation of state we must solve (17) as a function of strain in this periodic lattice, and differentiate numerically to yield the stress, or the elastic moduli, of equation (1) (however, cf Section 6 below).

In the full relativistic version of (17), the operator $\mathbf{H}$ and the state function Ψ are less physically immediate than in the nonrelativistic version. Since, in the atomic or solid-state contexts, the relativistic effects are important mainly for the heavy elements (starting near iron), we discuss here only the nonrelativistic (Schroedinger) case, and ignore distinctions due to spin. For a crystal, the Hamiltonian is given by

$$\mathbf{H} = \mathbf{H}_0 + \sum_{i=1}^{N} \mathbf{H}_i + \sum_{i,j}' \mathbf{H}_{ij}, \qquad (18)$$

where

$$\mathbf{H}_0 = \frac{e^2}{2} \sum_{g,h}' z_g z_h / r_{gh} \qquad (19)$$

is the "nuclear" contribution, r_{gh} is the distance between nuclei g and h, and z_g is the number of elementary charges e on nucleus g. Commonly, the inner cores of the electron cloud (i.e. those electrons forming spherical closed shells and not being affected by neighboring atoms) are included in this term. This reduces the number of electrons in the dynamic system, and is accomplished by decreasing each z_g by the proper number of core electrons. The prime on the summation sign indicates that the term $g = h$ is omitted, and the factor $\frac{1}{2}$ corrects for double counting.

$\mathbf{H}_i$ is the one-electron term,

$$\mathbf{H}_i = \frac{1}{2m} \mathbf{p}_i^2 - e^2 \sum_g z_g / r_{ig}, \qquad (20)$$

where m is the electron mass, $\mathbf{p}_i$ is the momentum operator for the ith electron, and r_{ig} is the distance to the screened nucleus g. $\mathbf{H}_{ij}$ is the two-electron term:

$$\mathbf{H}_{ij} = e^2 / r_{ij}. \qquad (21)$$

Since, in a solid system with very large N, equation (17) is intractable as it stands, it must be reduced to a set of one-electron equations. This is normally done by assuming that Ψ is multiplicatively separable, i.e. that

$$\Psi(\mathbf{x}_1, \mathbf{x}_2, \ldots, \mathbf{x}_N) = u_1(\mathbf{x}_1) u_2(\mathbf{x}_2) \ldots u(\mathbf{x}_N) \qquad (22)$$

or that Ψ consists of a determinant of such functions (to include the effects of exchange antisymmetry). But an elementary theorem of algebra states that such separability is only possible if the operator $\mathbf{H}$ contains no cross-terms, like $\mathbf{H}_{ij}$. Since $\mathbf{H}_{ij}$ is quite important and non-negligible in this problem, we avoid the use of (22), but arrive at the desired one-electron equation anyway, as follows. Using equations (17) and (18),

$$\langle \mathbf{H} \rangle = \langle \mathbf{H}_0 \rangle + \sum \langle \mathbf{H}_i \rangle + \frac{1}{2} \sum' \langle \mathbf{H}_{ij} \rangle, \qquad (23)$$

and since the electrons are indistinguishable,

$$\langle \mathbf{H} \rangle = \mathbf{H}_0 + \sum_i n_i \left(\langle \mathbf{H}_1 \rangle + \frac{N-1}{2} \langle \mathbf{H}_{12} \rangle \right)$$

$$= \sum n_i \int \left\{ \int \Psi^* \left[\frac{1}{N} \mathbf{H}_0 + \mathbf{H}_1 + \frac{N-1}{2} \mathbf{H}_{12} \right] \Psi \, dx^{3(N-2)} \right\} dx_1 \, dx_2, \qquad (24)$$

where n_i is the occupation index $(0 \leq n_i \leq 1)$ of the ith state, and the integrations over x_1 and x_2 are displayed explicitly. We now carry out the integrations over all the other electrons that are not involved in the operators H_1 and H_{12}, defining the result as

$$\int \Psi^*(x_1, x_2, \ldots, x_N) \left[\frac{1}{N} H_0 + H_1 + \frac{N-1}{2} H_{12} \right] \Psi(x_1, x_2, \ldots, x_N) \, dx^{3(N-2)}$$

$$\equiv \psi^*(x_1, x_2) \left[\frac{1}{N} H_0 + H_1 + \frac{N-1}{2} H_{12} \right] \psi(x_1, x_2). \tag{25}$$

The product $(N/2)\psi^*(x_1, x_2)\psi(x_1, x_2)$ is identical to the "generalized density matrix" $\Gamma_{12}(x_1 x_2)$ of Löwdin (1956, Section 2.1.1). Since the integrations do not involve the coordinates x_1 and x_2, $\psi(x_1, x_2)$ is an unweighted average state function for any two electrons of the system. Inserting this definition into (24), we obtain

$$\langle H \rangle = \sum n_i \int \left[\int \psi^* \left(\frac{1}{N} H_0 + H_1 + \frac{N-1}{2} H_{12} \right) \psi \, dx_2 \right] dx_1. \tag{26}$$

We now perform the integrations over x_2, defining the result as

$$\int \psi^*(x_1, x_2) \left[\frac{1}{N} H_0 + H_1 + \frac{N-1}{2} H_{12} \right] \psi(x_1, x_2) \, dx_2$$

$$\equiv u^*(x_1) \left[\frac{1}{N} H_0 + H_1 + V_{12} \right] u(x_1), \tag{27}$$

where V_{12}, as well as $u(x_1)$ is here defined implicitly. The real product u^*u now has the interpretation of being an average state function for a single electron moving in the average field of all the others. The effect of the others is contained in the term $V_{12}(x_1)$, which may be decomposed as

$$V_{12}(x_1) \equiv \frac{e^2}{2} \int \rho(x_2)/r_{12} \, dx_2 + U_{xc}(x_1), \tag{28}$$

where the first term is one-half the classical interaction of an electron with a charge distribution of density $\rho(x_2)$ and U_{xc} is the (nonclassical) remainder, comprising the effects due to the exchange (x) antisymmetry and mutual correlation (c) of the electrons. Collecting together all the classical potential terms as

$$V_1(x) \equiv \frac{1}{N} H_0 - e^2 \sum_g z_g/r_{1g} + \frac{e^2}{2} \int \rho(x_2)/r_{12} \, dx_2, \tag{29}$$

we may write equation (26) as

$$E = \langle H \rangle = \sum_i n_i \int u^*(x_1) \left[\frac{1}{2m} p_1^2 + V_1 + U_{xc} \right] u(x_1) \, dx_1. \tag{30}$$

This is the standard result for the average energy in terms of the one-electron state function. The Hartree equations may be recovered easily from (30) by use of the assumption (22), the Hartree-Fock equations by using a determinant of terms like

(22). The advantage of the present derivation is that it is exact; the two-body complications are all concentrated in the well-defined term U_{xc}, for which a later approximation must be made in all solid-state methods. Furthermore, the nature of the one-electron function $u(\mathbf{x}_1)$ as an average over the rest of the system is explicit. The end result is the same as if (22) had been used, because the $3N$-dimensional space element $\mathbf{dx}^{3N}$ is multiplicatively separable ($\mathbf{dx}^{3N} = \mathbf{dx}_1\,\mathbf{dx}_2\ldots\mathbf{dx}_N$) even though Ψ is not.

We discuss here only a single approximation to U_{xc}, one due to Slater (1951), Gaspar (1954), and Ќohn & Sham (1965) [for a fuller discussion, see Slater (1974)]. It is expressed by

$$U_{xc}(\mathbf{x}) = -9\alpha \left[\frac{3}{8\pi} \rho(\mathbf{x}) \right]^{1/3}, \tag{31}$$

where the charge density $\rho(\mathbf{x})$ is given by

$$\rho(\mathbf{x}) = \sum_i n_i u_i^*(\mathbf{x})u(\mathbf{x}) \tag{32}$$

and the $u_i(\mathbf{x})$ are the different one-electron solutions of (30). At this point in history, α is an empirically adjustable parameter which is normally in the range $2/3 \leqq \alpha \leqq 1$, and its appearance has led to the designation of quantum methods for calculating electronic structures which use (31) as "$X\alpha$" methods. The derivation of (31) uses concepts of statistical averaging, as in the derivation of (30), so a more descriptive label for this form of U_{xc} is the statistical exchange (and correlation) approximation.

Since the total charge density appears in both V_1 and U_{xc}, the proper solution of (30) requires an iteration procedure. One starts with an appropriate guess for the wave functions u_i, forms the charge density ρ and the potential functions V_1 and U_{xc}, and solves (30) for new wave functions, iterating until self-consistency is obtained. In practice, the literature is filled with non-self-consistent calculations, but as computing power has developed, more and more self-consistent results have appeared [Slater (1974) lists over 5000 recent papers].

We have not yet stated the physical principle needed to specify the wave functions Ψ or u. They are defined to be such as to minimize the average energy $\langle \mathbf{H} \rangle$, under the constraint of constant normalization: $\langle \mathbf{I} \rangle \equiv \int u^*u\,dx = 1$. In practice, one approximates the state function as a sum of analytically known functions with undetermined coefficients:

$$u_i(\mathbf{x}) = \sum_{j=1}^{J} C_j^i v_j(\mathbf{x}). \tag{33}$$

The analytic functions $v_j(\mathbf{x})$ may be plane waves, in which case (33) is simply a three-dimensional Fourier expansion, or they may be atomic-like functions, centered on the various lattice sites, etc. The approximation (33) is a noncritical one, as the number of terms may be increased indefinitely ($J \to \infty$) until the desired accuracy is reached. However, a careful choice of the v_j is required for each special problem, in order that the computation be feasible.

With J fixed, one conceptually varies the C_j^i in (33) so that $\langle \mathbf{H} \rangle$ (30) is minimized, while maintaining constant normalization. This is accomplished [cf e.g. section 1 in Slater (1974)] by requiring that

$$\delta \left\{ \int u^*(\mathbf{x}) \left[\frac{1}{2m} \mathbf{p}^2 + V_1 + U_{xc} - \varepsilon \right] u(\mathbf{x}) \, d\mathbf{x} \right\} = 0, \tag{34}$$

where the Lagrangian multiplier ε is introduced to satisfy the constraint of constant normalization. Since all changes δu are mutually independent, (34) is equivalent to the equation

$$\left[\frac{1}{2m} \mathbf{p}^2 + V_{cl}(\mathbf{x}) + V_{xc}(\mathbf{x}) \right] u(\mathbf{x}) = \varepsilon u(\mathbf{x}), \tag{35}$$

where

$$V_{cl}(\mathbf{x}) \equiv \frac{1}{N} \mathbf{H}_0 - e^2 \sum_g z_g / r_{1g} + e^2 \int \rho(\mathbf{x}) / r_{12} \, d\mathbf{x}_2 \tag{36}$$

is the classical potential energy of the electron and

$$V_{xc}(\mathbf{x}) = \tfrac{4}{3} U_{xc}(\mathbf{x}) = -12\alpha \left[\frac{3}{8\pi} \rho(\mathbf{x}) \right]^{1/3} \tag{37}$$

contains the exchange and correlation effects. The altered numerical factors of the terms involving ρ result from the variation $\delta\rho$ implied by (34). Equation (37) is the one-electron Schroedinger equation for this system, and writing the operator in braces as $\mathbf{H}^1$, it assumes the form of an eigenvalue equation,

$$\mathbf{H}^1 u(\mathbf{x}) = \varepsilon u(\mathbf{x}) \tag{38}$$

with eigenvalue ε. This is solved by standard methods, as it constitutes J linear, homogeneous equations in the J unknowns C_j^i, with nontrivial solutions only for certain values of ε. There will be J different eigenvalues ε_i (although some may be degenerate), each corresponding to a different set of coefficients C_j^i, defining a corresponding eigenfunction u_i. If J is infinite, and the $v_j(\mathbf{x})$ form a complete set, these are the exact u_i and ε_i. If J is finite, the lowest ε_i is an upper bound to the lowest exact ε_i; addition of more terms can only lower the calculated value. Furthermore, if the v_j are orthogonal, the next lowest ε_i will be an upper bound to the next lowest exact ε_i, etc.

The states with the lowest N eigenvalues will be occupied ($n_i = 1$) in the ground state and are required for calculating the total energy E of the static lattice. The higher-lying states are required for calculating transport properties (e.g. conductivity), the effects of high temperature, etc, but have $n_i = 0$ in the ground state. The static energy is, from (30),

$$E = \sum_i^N \varepsilon_i - \frac{1}{2} \int \rho(\mathbf{x}) V_{cl}(\mathbf{x}) \, d\mathbf{x} - \frac{1}{4} \int \rho(\mathbf{x}) V_{xc}(\mathbf{x}) \, d\mathbf{x}. \tag{39}$$

This result shows explicitly that the calculation of E requires not only the eigenvalues ε_i, but also the eigenfunctions $u_i(x)$ (in order to form ρ). Because approximations are required for the exchange term, e.g. (31), and the analytic description of the eigenfunction, e.g. (33), different methods (using different approximations) may yield different eigenvalues ε_i, but similar charge densities ρ, or vice versa.

It is clear that it will not be practical to calculate all N eigenvalues for a solid. Instead, we calculate only a few, and deduce the rest from these. In order to do this, however, we need to understand better the nature of the solutions to (38). This is most easily visualized by conceptually expanding the crystal lattice sufficiently so that state functions on neighboring atoms do not overlap, and the energy levels ε_i reduce to those of the isolated atoms. These are well-defined (sharp) levels, each corresponding to a unique set of principal quantum number n, orbital angular momentum l, total (orbital plus spin) angular momentum j, and magnetic quantum number m_j. In a solid containing only one atomic species, with m valence electrons per atom, this expanded problem requires only m distinct eigenvalues ε_i, each of which is obviously (N/m)-fold degenerate. More complicated solids are degenerate in similar, but more complicated, ways.

As the lattice is conceptually recontracted, so that the state functions overlap slightly, each electron obeys a Schroedinger equation with a small additional term, expressing this interaction, in the potential. This perturbation removes the degeneracy, splitting each sharp level into a band of N/m closely spaced levels; in general, the centroid of the band will be lower in energy than the original sharp level, reflecting the fact that the solid is more tightly bound than the extended configuration. As the contraction continues, the bands get lower and broader, and bands corresponding to different original levels may overlap. Because of this (bandwise) organization of the eigenvalues, the solution of (38) for a solid constitutes the "band structure" of the solid. In practice, one calculates only a few levels in each band, and interpolates to find all the others.

Because each band is broadened from a sharp level by the interactions of neighboring atoms, the fine splittings must depend upon the lattice geometry. In fact, each one-electron charge density $u_i^* u_i$ must possess the full symmetry of the crystal, having the same value at $\mathbf{x}$ as it has at $\mathbf{x}$ plus any lattice vector $\mathbf{R}$. This means that $u_i(\mathbf{x})$ itself may differ from such a periodic function only by a complex phase factor. Hence, if the expansion (33) of u_i is a Fourier series, each term must be of the form $v_j(\mathbf{x}) = \exp[i(\mathbf{k} + \mathbf{K}_j) \cdot \mathbf{x}]$, where the Fourier wave vector $\mathbf{K}_j$ is such that $\mathbf{K}_j \cdot \mathbf{R} = 2\pi$ times an integer for all lattice vectors $\mathbf{R}$. (Such wave vectors are determined by the crystal lattice; the infinite set of all such vectors is the reciprocal lattice familiar in crystallography. The unit cell at the origin of the reciprocal lattice is called the first Brillouin zone.) The Bloch vector $\mathbf{k}$ (Bloch 1928) provides the phase difference mentioned previously. Each $\mathbf{k}$ leads to J separate eigenvalues ε_i, one in each calculated band. Since each band has N/m energy levels, this implies that only N/m nonequivalent values of $\mathbf{k}$ exist. These may be taken, without loss of generality, to uniformly fill the first Brillouin zone, and since N is very large, the coverage is essentially continuous. Hence the band structure may be analytically written as $\varepsilon_j(\mathbf{k})$ for each band with index $j \leqq m \leqq J$.

The tops and bottoms of the bands are usually, but not always, at special $\mathbf{k}$ points of high symmetry (for example, at the center, or a corner of the Brillouin zone). The calculation of $\varepsilon_j(\mathbf{k})$ at a few such points may be supplemented with an interpolation procedure [cf e.g. Slater (1965), Appendix 5] to yield a continuous function that may be integrated to find the total band energy, i.e. the first term in (39). The charge density (32) then yields the classical and exchange-correlation terms of E.

6 THE VIRIAL THEOREM

The virial theorem has an identical form in both classical and quantum physics, and may be written (section 20-4 in Slater 1968, or McLellan 1974) for a solid system as

$$E - V = -\tfrac{1}{2}\Big\langle \sum_i \mathbf{x}_i \cdot \mathbf{F}_i \Big\rangle - \tfrac{1}{2}\Big\langle \sum_j \mathbf{X}_j \cdot \mathbf{F}_j^{\text{ext}} \Big\rangle, \tag{40}$$

where V is the total potential energy of the system, the summation over i covers all particles in the system, and $\mathbf{F}_i$ is the force on the ith particle. The second term on the right is the contribution from the external forces, and the summation over j covers only the nuclei. The brackets indicate a time average in classical physics, a state average in quantum physics. If all the internal forces vary as r^{-1}, then (40) reduces to

$$E - V = -\tfrac{1}{2}V + \tfrac{1}{2}Pv$$

or

$$P = (2E - V)/v, \tag{41}$$

providing a formula for calculating the pressure that appears to be more suitable than the differentiation in equation (1a). However, inspection of the Hamiltonian function in (35) reveals that the exchange term V_{xc} does not correspond to inter-atomic forces varying as r^{-1}. Ross (1969) showed that, if the one-electron wave functions u scale as $r^{-3/2}$, then the special form (41) of the virial theorem still holds, even with the statistical V_{xc}. Sham (1970) showed that if all spatial dimensions in the system scale uniformly, the same result will hold.

Of course, the restrictive assumptions of Ross and of Sham will only be good as approximations, and it is easy to estimate their validity. All of the assumptions required to specialize (40) to (41) are simple dimensional arguments, resulting in the conclusion that the potential energy scales with volume as $v^{-1/3}$, and the kinetic energy scales as $v^{-2/3}$. This conclusion is already sufficient to define an equation of state. If the variation with volume of the energy is thus assumed to be

$$E(v) = av^{-2/3} + bv^{-1/3} \tag{42}$$

then the coefficients a and b may be regarded as empirical constants, rather than as quantities to be calculated a priori. Redefining these constants in terms of the zero-pressure volume v_0 and incompressibility K_0, (42) becomes

$$E(v) = 9K_0 v_0 \big[\tfrac{1}{2}(v/v_0)^{-2/3} - (v/v_0)^{-1/3}\big] \tag{43}$$

with a corresponding equation of state, using (1a),

$$P(v) = 3K_0[(v/v_0)^{-5/3} - (v/v_0)^{-4/3}]. \tag{44}$$

This was first written by Bardeen (1938) and has been discussed in a geophysical context by Knopoff (1963). It is, in general, inadequate to describe the high-pressure behavior of solids. This may be seen immediately by forming two pressure derivatives of (44) and evaluating at zero pressure, yielding $K'_0 = 3$ for all solids. This is, of course, contrary to experiment, which finds that K'_0 varies from 3 to 6 among the common solids (cf e.g. O. L. Anderson et al 1968). It is clear that this equation of state is not sufficiently flexible to describe all solids; this lack of flexibility is traceable to the restrictive assumptions leading to the special form of the virial theorem. This result also provides a qualitative measure of the importance of the exchange potential V_{xc} in the theory of the equation of state, since V_{xc} is responsible for the failure of the special form of the virial theorem. Since this special form leads to errors of 25 to 50% in K'_0, i.e. in the curvature of the relation $P(v)$, one may expect errors in P of about 5%, at compressions of $(\Delta v/v_0) \simeq 0.2$, even if K_0 and v_0 were calculated accurately. The effects of this approximation upon the calculated K_0 and v_0 are probably greater than this, but must be found by a detailed calculation.

As a final point of curiosity, Bardeen's formula (41) may be written as

$$P(v) = -3K_0(v/v_0)^{-5/3}e = -9K_0(d\varepsilon/dv)e, \tag{45}$$

i.e. as a peculiar combination of the finite strain parameters ε and e. The significance of this result is probably negligible.

7 QUANTUM METHODS IN GEOPHYSICS

Calculation of specific electron transition energies, or of conductivity characteristics, requires only the band structure $\varepsilon_j(\mathbf{k})$, usually at special $\mathbf{k}$ points only. But the equation of state requires not only a sum over $\mathbf{k}$ of eigenvalues, but also the charge density ρ, i.e. the eigenfunctions. Hence it is not surprising that most band calculations have not calculated E, but only a few $\varepsilon_j(\mathbf{k})$.

However, several recent papers have shown that E can be calculated with some success [cf e.g. DeCicco 1967 (KCl), Averill 1972 (alkali metals), Surratt et al 1973 (diamond), Snow 1973 (Cu)]. Each of these papers uses different methods, and each is well suited to the crystal at hand, but not all are necessarily adaptable to more geophysical crystals.

Recently, however, Bukowinski & Knopoff (1976) and Bukowinski (1976, 1977) have made the first band structure calculations of direct geophysical interest, on elemental iron and potassium. They used the self-consistent augmented plane wave (SC-APW) method, which deserves some elaboration. The crystal is divided into spheres that are centered on the atomic sites and are mutually tangent. Inside each sphere, the one-electron Schroedinger equation (38) is solved self-consistently, using methods adapted from atomic calculations. That is, each v_j is taken as a product of a spherical harmonic times a tabulated radial function. If the radial functions for the free atom, for example those of Herman & Skillman (1963), are used as a starting point, then the crystal calculation (confined to the sphere, with no overlap from adjacent spheres) can be run to self-consistency in less than a minute of third-generation computer time. Just as for a free atom, the resulting potential is taken as spherically symmetrical.

The interstices between the spheres are assumed to have a constant potential, leading to the description of this configuration as a "muffin-tin potential." The u_i in the interstices are taken as a Fourier series, and continuity with the "atomic" functions is established at the boundaries of the muffin-tin spheres. This requires that the atomic functions contain spherical harmonics of order l higher than the free atom possesses, but it is a reasonable approximation to include only the atomic l's (e.g. $l_{max} = 2$ for iron). Of course, the gradient in u_i is not continuous; however, the discontinuity decreases with large J and l_{max}.

It is clear that the muffin-tin approximation has its best justification for close-packed structures of one element only. When more than one atom is present, the radii of the various spheres are not so easily specified, nor is the constant value of the potential between the spheres. These factors can be adjusted to give good agreement with data at low pressure, but may be subject to unforeseeable changes under high compression.

Hence Bukowinski has concentrated his efforts on iron, with obvious applications to the cores of the terrestrial planets, and on elemental potassium. He calculates the band structure of fcc iron (the stable solid phase at high temperature), and finds that its topology is unchanged at all compressions met in the earth. At much higher compressions ($\rho/\rho_0 \simeq 4$), the band derived from the free-atomic $4s$ level moves above the $3d$ band (implying that the electronic occupancy changes from $3d^6 4s^2$ to $3d^8$), but such compressions do not occur in the terrestrial planets. He concludes that an interpretation of the inner core in terms of such an electronic transition is unrealistic. Bukowinski & Knopoff (1976) discuss the reasons that such a transition would have been interesting.

Bukowinski also calculates (using the virial theorem, cf Section 6) the pressure and the incompressibility of the static lattice of fcc iron at core compressions and finds excellent agreement with current models of those properties of the inner core. By contrast, the outer core is considerably less dense than the calculated iron phase, indicating considerable dilution with a light element such as silicon, sulfur, or potassium. He provides a careful correction for the effects of high temperature, including quasi-harmonic lattice vibrations and electron excitations, but does not discuss the expansion upon melting. He has not investigated the properties of hcp iron, which is the stable phase at high pressure and lower temperatures; if the inner core is solid, it is not obvious (from laboratory data) which structure it should assume. In his present calculations, he uses an exchange-correlation parameter α (cf equation 31) of 0.593, an anomalously low value [in atomic calculations, α is near 0.71 (Slater 1974)]. The lower value is chosen so that the calculation will yield the proper density at zero pressure. He finds that, at high pressure, a 1% uncertainty in α leads to an uncertainty in the pressure of about 1%. The calculation does not contain the effects of electron spin, hence he does not comment on the transition to the low-spin state.

Bukowinski (1977) calculates the band structure and (with the virial theorem) the equation of state of elemental (bcc) potassium. He finds a series of electronic transitions at moderate pressures and a high "ordinary" compressibility that together lead to drastic reductions in the atomic volume. This implies its miscibility in the metallic cores of Earth and Venus, with a consequent implication of deep radioactive heating. These gross features of potassium are probably well modeled by the calculation, even though the potassium deep in a planetary interior will exist as impurities in a lattice determined by the major elements present, rather than in a bcc potassium lattice.

For crystals of more than one element, a different quantum method may be preferable. Dalton (1972) has described a generalized orthogonalized plane waves (OPW) method that the present author is currently implementing (Thomsen 1977). In this method, the wave function expansion consists of two sets of terms. One set is Fourier terms, orthogonalized to the inner core states on each atom. These OPWs describe well those states of the crystal that are continuous throughout the entire crystal, i.e. the conducting states. The restricted OPW method, using only such terms, with $J \simeq 100$, has been generally successful at calculating such states in a wide variety of different materials.

However, many crystals, especially the oxides and silicates of the mantle, contain atoms whose free-atomic functions overlap considerably in the crystalline state. Such free-atomic valence functions are undoubtedly strongly modified in the crystal, yet they remain at once too extended to be considered core states, yet too concentrated to be well described by a small number of plane waves. Good examples are the d states of iron, or the outer p states of oxygen.

It was stated earlier that the lowest calculated eigenvalue is an upper bound for the lowest exact eigenvalue, etc. However, the calculated eigenvalues do not necessarily lie in the same order as the exact eigenvalues. In fact, in the restricted OPW method, the lowest calculated eigenvalue is an excellent approximation to the lowest exact conducting-state eigenvalue, while remaining far above the exact eigenvalues for the valence states. As the number of OPWs is increased, eventually some of the higher-lying calculated eigenvalues will descend, changing order to approximate these exact valence eigenvalues. However, since the valence states oscillate rapidly near the nucleus, it may take many thousands of OPWs to converge properly; the inversion of matrices this size is not yet practicable.

The solution to this dilemma was suggested in the original OPW paper (Herring 1940). The expansion of the u_i should also contain, in addition to OPWs, valence states modified from free-atomic calculations, and also orthogonalized to the core states. Since the valence states overlap both valence and core states on neighboring sites, the expressions for the matrix elements become quite formidable, but in practice the calculation proceeds quite rapidly.

Such terms constitute the generalization of the OPW method mentioned in Dalton's title. The idea is similar to the mixed-basis method of Kunz (1969) and Lipari & Kunz (1971), who applied it to calculate the elastic moduli of several alkali halides. A similar approach was used by Kollar & Solt (1974) in an approximate calculation of the static energy $E(v)$ of copper.

The present author has applied the formalism of Dalton, but including all overlap effects to calculate the band structures of Ge, Cu, and CsCl. The lack of geophysical interest in these solids precludes giving the results here; calculations on MgO are to be done next. The inclusion of valence states in the analytic description of the u_i resolves the valence eigenvalues to within 0.05 eV with only a few (~ 20) OPWs; the conduction states, if any, require ~ 100 OPWs. The muffin-tin approximation is not used, and no artificial boundaries are introduced. The program does not yet calculate the charge density nor the total energy, and is not yet self-consistent, but these shortcomings are being addressed. The results to date encourage optimism that the band structure and the total energy $E(v)$ will be well calculated for virtually all solids, independent of bonding type, crystal symmetry, etc. In addition, we hope to be able to calculate $P(v)$ by differentiation (without use of the simpler form of the virial theorem), and $C_{ijkl}(v)$ by a second differentiation, with sufficient accuracy to realistically discuss the lower mantle. Since the calculation is relativistic, and can be generalized to provide for polarized electronic spin, it should eventually resolve even questions dependent on the spin state of iron in an oxide lattice.

8 RECENT EXPERIMENTAL PROGRESS

To complement this theoretical progress, a number of experimental breakthroughs have appeared in the past several months. Notable among these are the new capability demonstrated by Frankel et al (1976) to measure elastic velocities at very high experimental pressures (270 kbar). This pressure exceeds by a factor of six the previous limit at which elasticity measurements could be made. The initial results published are for NaCl (of course); hopefully the techniques will be applied to more geophysical solids in the near future. Already, however, this work has implications for geophysics: 1. the prediction of the fourth-order anharmonic theory (Thomsen 1972) that the elastic modulus C_{44} vanishes near 290 kbar is experimentally rejected; apparently, for this solid the theory fails at some smaller compression. 2. The ultrasonic data of Spetzler et al (1972), using the new buffer rod technique, predict values of K that are far too low, even at very modest pressures, reemphasizing the need to identify the systematic differences between this and other current ultrasonic experiments.

Another important recent advance is that of Mao & Bell (1976), who made apparently minor changes in the technique of the high-pressure diamond cell and thereby increased its limiting pressure from 300 to 1000 kbar, and perhaps as high as 1500 kbar. This makes static high-pressure (and simultaneous high-temperature) experiments at such pressures (comparable to the core-mantle boundary) possible for the first time. The small size of the sample (10^{-3} mm^3) makes some studies (e.g. acoustic velocity measurements) impossible, but other advances may lie just ahead. Already Weidner et al (1975) have reported the capability of measuring elastic moduli, using the technique of Brillouin scattering, of samples as small as 0.02 mm. The application of this optical technique to the high-pressure diamond cell does not seem impossible. Such developments indicate that we may look forward to a continued period of vigor in high-pressure geophysics.

ACKNOWLEDGMENTS

This work was supported by National Science Foundation Grant Number DES75-15372.

Literature Cited

Ahrens, T. J. 1975. Equations of state of the earth. *Rev. Geophys. Space Phys.* 13:335-39

Ahrens, T. J., Anderson, D. L., Ringwood, A. E. 1969. Equations of state and crystal structures of high-pressure phases of shocked silicates and oxides. *Rev. Geophys. Space Phys.* 7:667-708

Ahrens, T. J., Thomsen, L. 1972. Application of the fourth-order anharmonic theory to prediction of equations of state at high compressions and temperatures. *Phys. Earth Planet. Interiors.* 5:282-94

Anderson, D. L. 1972. Internal constitution of Mars. *J. Geophys. Res.* 77:789

Anderson, D. L. 1977. Composition of the mantle and core. *Ann. Rev. Earth Planet. Sci.* 5:179-202

Anderson, D. L., Anderson, O. L. 1970. The bulk modulus-volume relationship for oxides. *J. Geophys. Res.* 75:3494-795

Anderson, D. L., Sammis, C., Jordan, T. 1971. Composition and evolution of the mantle and core. *Science* 171:1103

Anderson, O. L. 1966. A proposed law of corresponding states for oxide compounds. *J. Geophys. Res.* 71:4963-71

Anderson, O. L. 1967. Equation for thermal expansivity in planetary interiors. *J. Geophys. Res.* 72:3661-68

Anderson, O. L., Schreiber, E. S., Liebermann, R. C., Soga, N. 1968. Some elastic constant data on minerals relevant to geophysics. *Rev. Geophys. Space Phys.* 6:491-524

Averill, F. W. 1972. Calculation of the cohesive energies and bulk properties of the alkali metals. *Phys. Rev. B* 6:3637

Bardeen, J. 1938. Compressibilities of the alkali metals. *J. Chem. Phys.* 6:372

Birch, F. 1947. Finite elastic strain of cubic crystals. *Phys. Rev.* 71:809-24

Birch, F. 1952. Elasticity and constitution of the earth's interior. *J. Geophys. Res.* 57:227-88

Birch, F. 1961. The velocity of compressional waves in rocks to 10 kilobars, Part 2. *J. Geophys. Res.* 66:2199

Bloch, F. 1928. Über die Quantenmechanik der Elektronen in Kristallgittern. *Z. Phys.* 52:555

Born, M. 1923. *Atom-theorie des Festen Zustandes.* Berlin: Teubner. 120 pp.

Born, M., Huang, K. 1954. *Dynamical Theory of Crystal Lattices.* London: Oxford Univ. Press. 420 pp.

Bridgman, P. W. 1938. Rough compressibilities of fourteen substances to 45,000 kg/cm². *Proc. Am. Acad. Arts Sci.* 72:207

Bukowinski, M. S. T. 1976. On the electronic structure of iron at core pressures. *Phys. Earth Planet. Interiors* 13:57-66

Bukowinski, M. S. T. 1977. The effect of pressure on the physics and chemistry of potassium. *J. Geophys. Res.* In press

Bukowinski, M. S. T., Knopoff, L. 1976. Electronic transition in iron and the properties of the core. *Proc. 1974 NATO Conf. Petrophys.*

Dalton, N. W. 1972. Generalizations of the relativistic OPW method including overlapping and non-overlapping atomic orbitals. *Computational Solid State Physics*, ed. F. Herman et al. New York: Plenum. 450 pp.

Davies, G. F. 1973a. Invariant finite strain measures in elasticity and lattice dynamics. *J. Phys. Chem. Solids* 34:841-45

Davies, G. F. 1973b. Quasi-harmonic finite strain equations of state of solids. *J. Phys. Chem. Solids* 34:1417-29

Davies, G. F. 1974. Limits on the constitution of the lower mantle. *Geophys. J. R. Astron. Soc.* 38:479

Davies, G. F., O'Connell, R. J. 1975. An automated ultrasonic interferometer and the effects of transducer and bond phase shifts on sound velocity measurements. *Trans. Am. Geophys. Union (Abstr.)* 56:441

DeCicco, P. D. 1967. Self-consistent energy bands and cohesive energy of KCl. *Phys. Rev.* 153:931

Demarest, H. H. 1973. The breathing shell model and the equation of state of MgO. *EOS Trans. Am. Geophys. Union (Abstr.)* 54:474

Demarest, H. H. 1974. Lattice model calculation of Hugoniot curves and the Grüneisen parameter at high pressure for the alkali halides. *J. Phys. Chem. Solids* 35:1393

Einstein, A. 1907. Die Plancksche Theorie der Strahlung und die Theorie der Specifischen Wärme. *Ann. Phys.* 22:180

Frankel, J., Rich, F. J., Homan, C. G. 1976.

Acoustic velocities in polycrystalline NaCl at 300°K measured at static pressures from 25 to 270 kbar. *J. Geophys. Res.* 81:6357–63

Gaspar, R. 1954. Approximation of the Hartree-Fock potential by a universal potential function. *Acta Phys. Akad. Sci. Hung.* 3:263

Grüneisen, E. 1912. Theorie des festen Zustandes einatomiger Elemente. *Ann. Phys.* 39:257

Hashin, Z., Shtrikman, S. 1962. On some variational principles in anisotropic and nonhomogeneous elasticity. *J. Mech. Phys. Solids* 10:335

Herman, F., Skillman, S. 1963. *Atomic Structure Calculations.* Englewood Cliffs, NJ: Prentice-Hall

Herring, C. 1940. A new method for calculating wave functions in crystals. *Phys. Rev.* 57:1169–77

Hill, R. 1952. The elastic behaviour of a crystalline aggregate. *Proc. Phys. Soc. London* 65A:349–54

Kalinin, V. A., Pan'kov, V. L. 1973. Equations of state of stishovite, coesite, and quartz. *Isv. Akad. Sci. USSR Phys. Solid Earth* 8:495

Kalinin, V. A., Pan'kov, V. L. 1974. Equations of state of rocks. *Isv. Akad. Sci. USSR Phys. Solid Earth* 7:419

Kalinin, V. A., Pan'kov, V. L., Zharkov, V. N. 1972. Equations of state of danites and bronzitites undergoing polymorphic transitions under pressure. *Isv. Akad. Sci. USSR Phys. Solid Earth* 7:472

Knopoff, L. 1963. The theory of finite strain and compressibility of solids. *J. Geophys. Res.* 68:2929–32

Kohn, W., Sham, L. J. 1965. Self-consistent equations including exchange and correlation effects. *Phys. Rev.* 140:A1133–38

Kollar, J., Solt, G. 1974. On the volume dependence of the total energy and the equation of state of copper. *J. Phys. Chem. Solids* 35:1121

Kunz, A. B. 1969. Combined plane-wave tight binding method for energy-band calculations with application to sodium iodide and lithium iodide. *Phys. Rev.* 180:934

Leibfried, G., Ludwig, W. 1961. Theory of anharmonic effects in crystals. *Solid State Phys.* 12:275

Liebermann, R. C. 1975. Elasticity of olivine (α), beta (α), and spinel (β) polymorphs of germanates and silicates. *Geophys. J. R. Astron. Soc.* 42:899

Lipari, N. O., Kunz, A. B. 1971. Energy bands for KCl. *Phys. Rev. B* 4:4639

Liu, H. P., Anderson, D. L., Kanamori, H.

1976. Velocity dispersion due to anelasticity, implications for seismology and mantle composition. *Geophys. J. R. Astron. Soc.* 47:41–58

Löwdin, P. O. 1956. Quantum theory of cohesive properties of solids. *Adv. Phys.* 5:1–171

Mao, H. K., Bell, P. M. 1976. High-pressure physics: The 1-megabar mark on the ruby R, static pressure scale. *Science* 191:851

Mao, H. K., Takahashii, T., Bassett, W. A., Kinsland, G. L., Merrill, L. 1974. Isothermal compression of magnetite to 320 kbar and pressure-induced phase transformation. *J. Geophys. Res.* 79:1165–70

McLellan, A. G. 1974. Vivial theorem generalized. *Am. J. Phys.* 42:239

O'Connell, R. J., Graham, E. K. 1971. Equation of state of stoichiometric spinel to 10 kb and 800°K. *EOS Trans. Am. Geophys. Union (Abstr.)* 52:359

Palciauskas, V. V. 1975. The volume dependence of Grüneisen parameter for monatomic cubic crystals. *J. Phys. Chem. Solids* 36:611

Palciauskas, V. V. 1976. The thermal regime of the lower mantle. *EOS Trans. Am. Geophys. Union (Abstr.)* 57:326

Pan'kov, V. L., Kalinin, V. A. 1974. Equations of state of mineral-forming oxides. *Izv. Akad. Sci. USSR Phys. Solid Earth* 3:143

Ringwood, A. E. 1970. Phase transformations and the constitution of the mantle. *Phys. Earth Planet. Interiors* 3:109

Ross, M. 1969. Pressure calculations and the virial theorem for modified Hartree-Fock solids and atoms. *Phys. Rev.* 179:612

Sham, L. J. 1970. Local exchange approximations and the virial theorem. *Phys. Rev. A* 1:969

Slater, J. C. 1951. A simplification of the Hartree-Fock method. *Phys. Rev.* 81:385

Slater, J. C. 1965. *Quantum Theory of Molecules and Solids, Vol. II.* New York: McGraw-Hill. 563 pp.

Slater, J. C. 1968. *Quantum Theory of Matter.* New York: McGraw-Hill. 763 pp.

Slater, J. C. 1974. *The Self-Consistent Field for Molecules and Solids.* New York: McGraw-Hill. 583 pp.

Snow, E. C. 1973. Total energy as a function of lattice parameter for copper via the self-consistent APW method. *Phys. Rev. B* 8:5391

Spetzler, H. A. 1970. Equation of state of polycrystalline and single-crystal MgO to 8 kilobars and 800°K. *J. Geophys. Res.* 75:2073–87

Spetzler, H. A., Anderson, D. L. 1971. Discrepancies between polycrystalline and single crystal elastic constant data; MgO. *J. Am. Ceram. Soc.* 54:520–25

Spetzler, H. A., Sammis, C. G., O'Connell, R. J. 1972. Equation of state of NaCl: Ultrasonic measurements to 8 kbar and 800°K and static lattice theory. *J. Phys. Chem. Solids* 33:1727–50

Surratt, G. T., Eunema, R. N., Wilhite, D. L. 1973. Hartree-Fock lattice constant and bulk modulus of diamond. *Phys. Rev. B* 8:4019

Thomsen, L. 1970. On the fourth-order anharmonic equation of state of solids. *J. Phys. Chem. Solids* 31:2003–16

Thomsen, L. 1971. Equations of state and the interior of the earth. In *Proc. Int. Sch.*

Phys. Enrico Fermi, Course L, ed. J. Coulomb, M. Caputo. New York: Academic. 277 pp.

Thomsen, L. 1972. The fourth-order anharmonic theory: elasticity and stability. *J. Phys. Chem. Solids* 33:363

Thomsen, L. 1977. A generalized OPW calculation of the band structure of Cu, Ge, CsCl, and MgO. In preparation

Watt, D., Davies, G. F., O'Connell, A. 1976. The elastic properties of composite materials. *Rev. Geophys. Space Phys.* 14:541–64

Weidner, D. J., Swyler, K. S., Carleton, H. R. 1975. Elasticity of microcrystals. *Geophys. Res. Lett.* 2:189

Ann. Rev. Earth Planet. Sci. 1977. 5:515–40

CRATERING AND OBLITERATION HISTORY OF MARS

×10082

Clark R. Chapman

Planetary Science Institute, 2030 East Speedway, Suite 201, Tucson, Arizona 85719

Kenneth L. Jones

Department of Geological Sciences, Brown University, Providence, Rhode Island 02912

INTRODUCTION

The craters on Mars testify to a long history of bombardment by interplanetary objects. Further, the numbers of craters and their morphologies record the effects of erosion and deposition that have been altering the face of Mars throughout the history recorded by observable craters. If this record of cratering and obliteration can be related, through the stratigraphic principles of superposition, to the formation of such structures as volcanoes, stream channels, and lava flows, then we will know the broad outlines of the evolution of the Martian surface. Finally, if we can relate the cratering history of Mars to the histories of the Earth, the Moon, Mercury, and other bodies, then we can compare the evolution of terrestrial planets as a group.

Since the first orbital pictures were returned from Viking I, there has been renewed interest in the complex processes of volcanism, aeolian abrasion and deposition, collapse and sapping of underground frozen volatiles, fluvial processes, and epochs of more clement and rainy weather that have shaped Martian geomorphology in regionally heterogeneous ways. The great increase in resolution of Viking orbital pictures, which far surpasses the resolution of Mariner 9 pictures, and the close-up views from the Viking landers are greatly augmenting our knowledge of Mars (cf Carr et al 1976). Our review is based mainly on the pre-Viking literature. Yet we have been cognizant of preliminary interpretations publicly reported during the first months following the arrival of the first Viking at Mars, and the emphases in this review have been influenced thereby. In particular, because of the great geological complexity of Mars revealed by Viking beyond that appreciated from Mariner 9, we approach our task of a global synthesis of Martian cratering with some trepidation.

Craters have attributes that make them uniquely useful for deciphering the geomorphological history of a planet. With some exceptions, impact craters have the following traits: (*a*) They form a very large sample of similar topographic features that are amenable to statistical analysis. (*b*) They have formed throughout planetary history, although there are very few data constraining the time dependence of the rate of crater formation for Mars. There might be a degree of correlation among the cratering rates on the different terrestrial bodies. (*c*) Impact craters mainly occur at random locations on a planet, despite slight global asymmetries. Formation of doublet or even clustered primary craters has occurred and may be common on Mars (Oberbeck & Aoyagi 1972), but such clustering probably can be taken into account. Formation of secondary craters is also important, especially at small diameters, but they often can be recognized and distinguished from primaries by their morphologies and spatial relationships to primaries. Some endogenic craters (e.g. calderas and cinder cones) are even more easily recognizable. (*d*) The original shapes and traits of newly formed craters of the same size (e.g. diameter/depth ratio, wall terraces, ejecta blankets) are roughly similar. Thus measured departures of a crater's appearance from "fresh" morphology probably represent degrees of modification of the crater. Other adjacent topographic features (such as river channels) of similar scale and stratigraphic age may be expected to have undergone a similar degree of modification. (*e*) Degradation of a crater is generally an irreversible morphological process until the crater finally disappears altogether because of the cumulative erosion or deposition (exhumation is an exception discussed later).

To the degree that these generalizations are true, departures from a uniform

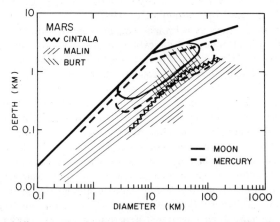

Figure 1   Depth/diameter plots for Martian craters, compared with the Moon and Mercury. Straight lines for the Moon and Mercury are mean fresh crater distributions; regions occupied by degraded craters are encircled below. Mars data, and some of the comparisons, are from Cintala, Head & Mutch (1976), Malin & Dzurisin (1977), and Burt, Veverka & Cook (1976). The Mars data have been acquired with a variety of techniques, most subject to appreciable systematic error. It is uncertain to what degree the apparent difference between the crater depths on Mars and Mercury is established.

distribution of fresh craters around a planet bespeak important processes—whether exogenic or endogenic[1] in origin—which, in the process of modifying the craters, must also have modified other surface topography. For instance, the gross dichotomy of Mars into hemispheres, one rather densely covered by large degraded craters and the other sparsely populated with small fresh craters, reveals that fundamentally different processes have operated on each hemisphere of the planet.

Although Mariner 9 and Viking pictures have revealed a number of cases in which there are exceptions to our generalizations or in which the geological history has been too complex to be accounted for in broad statistical terms, we believe there is a sound basis for attempting to delineate relative and absolute histories of major global regions on Mars. By and large, we believe that the relative chronology is more firmly established than some planetologists believe. On the other hand, at least the more popular accounts seem to us to be too precise in specifying the *absolute* ages of Martian features. The geomorphological evidence for cyclical episodes in Martian geomorphological history, frequently raised in the context of the search for Martian life, is not established; neither is cyclical evolution excluded by present data.

EARLY INTERPRETATIONS

Two cogent observations were made from the Mariner 4 pictures, which were received in 1965. The first is that most Martian craters are highly degraded, indeed much more so than lunar uplands craters. Recently, this obvious fact has been demonstrated quantitatively in terms of crater depth/diameter ratios (Figure 1). The second observation, due to Öpik (1965, 1966) and amplified upon by Hartmann (1966) (see also Binder 1966, and Leighton et al 1967), is that relative to a particular power law fitted to the Martian crater diameter-frequency relation at large diameters, there is a paucity of craters with diameters smaller than ~30 km (estimates varied from 20 to 50 km; Figure 2, left portion). Öpik suggested that older small craters had been entirely obliterated by the same processes responsible for degrading the morphologies of more recent craters less than 30 km in diameter as well as of the older large craters.

It has been assumed that craters were formed by impact of a population of collisionally evolved asteroids, which probably follows a power-law distribution (Dohnanyi 1972) and which, through scaling relationships between asteroid diameter and resultant crater diameter, create a power-law distribution of craters (shown as the right-hand production function, slope $= -3$, in the left portion of Figure 2). Although there are some observational uncertainties in defining the incremental frequency distribution of asteroids, it probably can be approximated by a power law with an exponent between -3 and -3.5, which results in a similar exponent for the crater distribution that increases to about -4 for the largest craters governed by gravity scaling. Öpik showed that a simple obliterative process, such as dust

---

[1] By *exogenic* we mean processes related to impact cratering. All others are *endogenic*, including atmospheric, surficial, and subterranean processes.

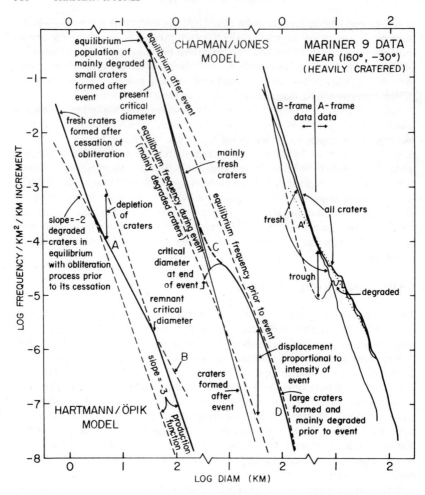

Figure 2 Comparison of two models for Martian cratering and obliteration with data from a heavily cratered unit. Thicker lines represent the present total crater population; other lines represent components. Left panel: Pre-Mariner 9 model due to Öpik (1965), Hartmann (1973), and others. A power-law production function is depleted and reaches equilibrium with an obliteration process at diameters less than a critical diameter; then the obliteration process ends and the subsequent population of mainly small craters retains fresh morphologies. Middle panel: Model due to Jones (1974) and Chapman (1974a), discussed in the text. There was an interval ("event") of greatly augmented obliteration rate that shifted the crater population to a lower equilibrium curve, producing the kink in the total crater curve (near *C*). Right panel: Observed crater frequencies from Jones' (1974) region 32, augmented by some unpublished, small-crater B-frame data due to Chapman (the dashed line indicates inadequate data).

deposition, which continuously degrades and obliterates craters at rates inversely proportional to their depths (for constant depth/diameter ratio), results in an equilibrium frequency distribution of slope one unit shallower than the production function (slope $= -2$ in Figure 2). Öpik's model seemed to fit available Mariner 4 data.

That Öpik's model was incomplete became obvious from Mariners 6 and 7 (Murray et al 1971), although Chapman (1967) recognized disparities even in the Mariner 4 crater data (see Chapman 1974b). At smaller diameters, the *fresh* craters follow a steeper relation than do the more degraded craters; indeed, most craters with diameters less than a few kilometers are fresh (see Figure 2). Therefore, Murray et al (1971), McGill & Wise (1972), and Hartmann (1973) in his Mariner 9 analysis introduced a variant of Öpik's model. They suggested that the processes of erosion have slowed down or ceased so that recent craters, including very small ones, have not been appreciably degraded.

As time progresses, an obliteration process gradually destroys larger and larger members of the first-formed craters on the surface of a planet. As the total crater population moves upward with time on a size-frequency plot, it abuts the equilibrium frequency (B in Figure 2), bending over to follow the equilibrium curve at the "critical diameter." The equilibrium population of craters (A in Figure 2) consists of a spectrum of morphologies; only the most recent craters are fresh. Subsequent to the hypothesized cessation of obliteration, the production function has recratered the surface to the degree shown by the lower production function, which is added to the remaining two-sloped curve; these new fresh craters dominate the population only at small diameters. The break in slope observed by Öpik and others marks the remnant critical diameter from the earlier epoch of active obliterative processes.

Since the northern-hemisphere plains revealed by Mariner 9 contain chiefly fresh, small craters with frequencies similar to or less than the small-crater frequencies in the cratered terrains, early interpretations of Mariner 9 pictures suggested that those units formed contemporaneously with, and subsequent to, the cessation of crater degradation in the cratered terrains (Hartmann 1973). The cessation of degradational processes was first thought to have occurred in relatively recent Martian history but, as we discuss when we deal with absolute chronologies later, this may not be true.

CRATERING/OBLITERATION MODELS

In separate papers based on Mariner 9 crater morphology data (Jones 1974, Chapman 1974a), we derived a new scenario for Martian cratering and obliteration history. We argued that most of the degraded morphologies of moderate-sized Martian craters (e.g. 10–30 km) were caused by a relatively brief episode of obliteration such that, relative to the cratering rate, the degradational processes were augmented by a factor of five to a hundred or more. Subsequently the

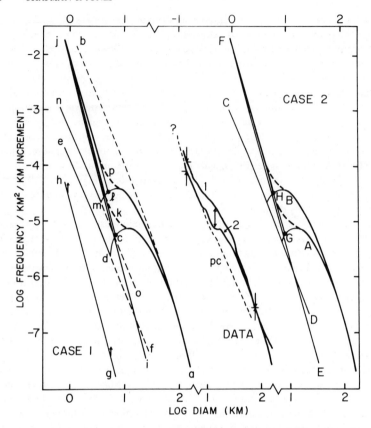

Figure 3   Regional differences in crater frequencies may be interpreted in terms of a crater obliteration event in two different ways. Case 1: Obliteration lasts the same length of time for two regions but is much more intense in one than the other. Before the event, frequencies are in equilibrium along line *ab*. One region undergoes moderate obliteration and a new equilibrium is established along *no,* resulting in frequencies *almn* immediately after the event. The second region undergoes massive obliteration, reaching a lower equilibrium frequency along *ef,* resulting in *acde.* Fresh cratering commences (*gh*) and moves upward to *ij* at present. Observed frequencies for region 1 are the sum of *almn* and *ij* which is *alj* or, for reasons described in the text, *apj.* Observed frequencies for region 2 are, analogously, *acj* or *akj.* Note that *cj* and *lj* are not coincident. Case 2: The obliteration has the same intensity in the two regions but lasts much longer in region 2 than 1. In this case the equilibrium frequency curve during the event is identical for both regions (*CD*), so the addition of the post-event craters (*EF*) results in the same frequency relation for small craters (*GHF*). The region subjected to the shorter event has observable frequencies *BF,* while the other exhibits *AHF.* The middle panel shows total crater data from Mariner 9 A-frame counts for Jones' (1974) regions 1, 24, 29, and 32 (curve 1), regions 19 and 21 (curve 2), and pc units. The data seem more like Case 1 than 2, yet we need better data in the vicinity of the question mark to be sure.

degradation rate declined again to a rate not much greater, and perhaps much less, than the original rate. This final period of low obliteration rate corresponds to the period after cessation of erosion postulated from the Mariner 6, 7, and 9 observations of fresh small-crater populations.

The fundamental difference between our interpretation and earlier ones is our explicit separation of the obliteration rate from the cratering rate prior to what all agree has been a low recent obliteration rate. The earlier models of Öpik (1965) and Chapman, Pollack & Sagan (1969) assumed *constant* obliteration and cratering rates. Several investigators (e.g. Hartmann 1971) argued that the Martian cratering rate has changed drastically with time (it was much higher in the past); Soderblom et al (1974) argued further for a strict association between the obliteration rate and the decreasing cratering rate. If there were such a connection, it would suggest that the obliterative process may have been directly or indirectly *caused* by the cratering process. We, on the other hand, interpreted the data as requiring an obliteration rate that, for a while, varied radically with respect to the cratering rate; we have suggested that the episode of obliteration of moderate-sized craters was largely *subsequent* to the early period of cratering, when the rate of cratering is commonly assumed to have been high. Further, we have suggested that the episode might have been related to the period when rainfall occurred on Mars and carved the widespread system of "furrows," or small channels (see section on aqueous processes, below). The apparent simultaneity of the hypothesized obliterative episode with the formation of the first major plains units in the northern hemisphere is possibly suggestive of an endogenic origin for the obliteration process, uncoupled from the early cratering.

In this section and the one that follows we will, with the aid of Figures 2–4, lead the reader through the observations and cratering theory that led us to our conclusions. Later we describe alternative models. Scientists interested in Martian history should understand our formulation, since any theory for Martian cratering and obliteration can be explicable in equivalent terminology to that which we employ here. Despite the apparent complexity of Figures 2 and 3, the logic is fairly elementary. The basic graph is a log-log plot of incremental crater frequency vs diameter. There are simple mathematical relationships between this plot and a commonly used alternative, a plot of log diameter vs log cumulative frequency (they are compared by Jones 1974); the chief difference is that our incremental relations tend to be one unit steeper in slope than cumulative curves.

The cratering rate c on a planet depends on both time, $c_t(T)$ (e.g. a decreasing flux), and diameter, $c_d(D)$ (e.g. $\propto D^{-3}$). The diameter dependence $c_d(D)$ may itself change with time if the size distribution of the incoming projectiles changes. A general obliteration process, which may be the sum of several separate geological processes, may be described by two simple variables: $o(T)$, the obliteration rate (independent of crater diameter), and $a(D)$, the amount of obliteration required to remove a crater of given diameter (dependent on process but independent of time). Clearly the amount of erosion or deposition necessary to obliterate a large crater is generally more than that which will erase a small crater; thus $a(D)$ is a monotonically increasing function for most geological processes (isostatic compensation is an exception).

Equilibrium

Perhaps the crucial concept in cratering statistics is that of *equilibrium*, in which the observable crater frequencies $N(D)$ do not change with time and equal $F(D)$. Mathematically, the equilibrium frequency $F(D, T)$ at diameter D and time T can be shown (Jones 1974) to equal

$$F(D, T) = [c_t(T)/o(T)] \cdot a(D) \cdot c_d(D). \tag{1}$$

Note that, given some diameter dependence of the cratering and obliterative processes, the equilibrium frequency at any time T depends only on the ratio of rates of cratering to obliteration. (While one can always calculate an equilibrium frequency, the observed crater population $N(D)$ need not yet have attained equilibrium.)

Crater morphologies may be divided into several classes, ranging from fresh to highly degraded (see Figure 5). While craters are classified on the basis of

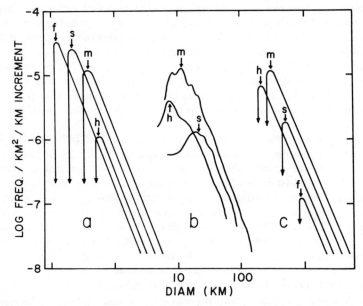

Figure 4   Diameter dependence of maxima in the frequency relations for degraded craters in Martian cu terrains are compared with two models. Curves are shown for fresh (f) craters and successively more degraded craters of classes s, m, and h. (*a*) The effect of observational loss of craters is due to finite resolution. Fresh craters can be observed to smaller diameters than highly degraded craters, yielding the sequence s-m-h with increasing diameter. (*b*) Counts of degraded craters for cratered units 29 and 32 of Jones (1974) show a sequence h-m-s, opposite to that in panel (*a*). (*c*) An equilibrium population of craters is subjected to massive obliteration. Smaller craters are obliterated altogether while the largest ones are relatively unaffected. At intermediate diameters there is the sequence of degraded craters h-m-s, similar to the observations.

Figure 5   Oblique view across the large basin, Argyre Planitia, taken by the Viking 1 Orbiter. Craters of various morphologies, ranging from fresh to highly degraded, surround the basin. South is to the upper right. NASA photo, used courtesy of Viking Orbiter Imaging Team.

morphological criteria (e.g. diameter/depth ratio, sharpness of floor features, terracing, etc), we *define* crater classes here in terms of the "amount" of obliteration $a_i(D)$ necessary to change a crater of given diameter from class i to the next more degraded class. This definition bears a clear intuitive relationship to the observational classes, independent of diameter D, provided plausible morphological criteria are used to classify craters. We employ four classes in this paper: fresh (f), slightly degraded (s), moderately degraded (m), and highly degraded (h). Clearly $a_f(D)+a_s(D)+a_m(D)+a_h(D)$ must equal $a(D)$, the total amount of obliteration that removes the crater entirely. Note that the ratio $a_i(D)/a(D)$ represents the fraction of a crater's lifetime spent in class i, if $c_t(T)/o(T)$ is constant. Given that the classes are properly defined such that $a_i(D)/a(D)$ are constant, then, in equilibrium, there are equilibrium frequencies $F_i(D, T)$ for each class i that are parallel to each other and to the total crater equilibrium curve $F(D, T)$. These are exemplified in Figure 4b, where it is apparent that, by our definitions of morphological classes (Arvidson, Mutch & Jones 1974), craters spend the longest part of their lifetimes in our m class (the shortest times are spent as fresh craters).

A special type of equilibrium may be established if endogenic processes are so ineffective that craters are obliterated primarily by subsequent crater impacts. This is called *saturation equilibrium* and, despite numerous studies (e.g. Gault 1970, Marcus 1970), it remains a poorly understood aspect of cratered surfaces. Saturation equilibrium results from the impossibility of fitting more than a finite number of craters into a given area without the subsequent craters destroying pre-existing ones or covering them with ejecta. Here $o(T)$ is clearly controlled by $c_t(T)$; furthermore, it is known that $a(D)$ depends on $c_d(D)$ in such a way that $F(D, T) \propto D^{-3}$ when the production function $c_d(D)$ has a constant slope steeper than -3. The constant of proportionality, which determines the height of the equilibrium frequency curve on a graph such as Figure 2, is a matter of some dispute (cf Chapman, Mosher & Simmons 1970, Marcus 1970, Woronow 1977a); it depends weakly on $c_d(D)$ and perhaps on the structural characteristics of the ground. Clearly the equilibrium frequency cannot greatly exceed that at which the cumulative area of observable craters exceeds the area in which the craters are formed. It is commonly thought that saturation equilibrium frequencies range from a few percent to perhaps 50% of a given area, for different cases.

The crater production function $c_d(D)$ is commonly represented by a power law or a combination of separate power laws over different diameter ranges. There are both observational and theoretical reasons for using power laws in some cases, but in general the diameter dependence may not be a power law, as recently emphasized by Woronow (1977b). In Figures 2 and 3 we represent production functions by power laws (straight lines on such log-log plots) for purposes of explication only; later we compare with the data.

OBLITERATION EPISODE INTERPRETATION

In order to interpret Martian cratering and obliteration history, it is useful to consider different diameter ranges separately inasmuch as the morphologies of the largest craters tend to have been shaped by the largest-scale processes that occurred in Mars' ancient history while the smallest craters have been shaped by recent processes. It is primarily from the morphologies of intermediate-sized craters that we conclude there was an obliterative episode in intermediate Martian history.

Large Craters

We consider first the population (D in Figure 2) of large, degraded craters which rather densely cover the cratered uplands (cu) units on Mars. The best Mariner 4, 6, and 7 pictures were primarily of cu units, which gave Mars its early, not entirely representative, "moon-like" image. We interpret the largest Martian craters to represent an equilibrium population, as concluded earlier by Chapman, Pollack & Sagan (1969) and Soderblom et al (1974), for two reasons: 1. all the cu units attain nearly identical total crater frequencies at the largest diameters (compare coincidence at large diameters of curves 1 and 2 in Figure 3); 2. there is a spectrum of crater morphologies and the frequency curves for each class roughly parallel the total crater curve (Figure 4).[2] Our interpretation is illustrated in Figure 2 by having $N(D)$ abut an equilibrium frequency curve in the vicinity of D. Since this early period of obliteration (prior to the event discussed below) extended to the largest craters, no remnant critical diameter is visible at large diameters. Our view differs from the Hartmann/Öpik model (note lack of coincidence of observed large craters with hypothesized equilibrium curve B in Figure 2) and from the more recent model of Woronow (1977b) and Strom & Whitaker (1976), in which portion D is interpreted as a production curve $c_d(D)$.

The degradation of the largest Martian craters may have been due to exogenic processes, endogenic processes, or both. Certainly it was assisted by the fact that the largest craters form, or shortly become, shallower in proportion to their diameters than smaller craters. As shown in Figure 1, the depth/diameter ratio even for "fresh" craters decreases towards larger sizes for the Moon, Mercury, and probably for Mars. Figure 1 may be considered for many obliteration processes (e.g. dust filling) as a plot of $a(D)$ since the crater depth is equivalent to the total relief to be eroded or buried by the process. If $a(D)$ levels off to being more nearly constant (depth ~ 2 km) at largest diameters, the shape of the equilibrium frequency curve would approach that of $c_d(D)$, according to Equation (1); this may explain the steep slope near D in Figure 2.

Although Chapman et al (1969) considered that the Martian craters were possibly in saturation equilibrium, most researchers have doubted that possibility because of their low spatial density, which is much lower than many lunar highlands crater populations. Some workers even doubt that the lunar craters are in saturation equilibrium (Woronow 1977a, Strom & Whitaker 1976, Oberbeck et al 1977). If large Martian craters really are formed with much more shallow profiles than their lunar counterparts (Figure 1), perhaps because of high volatile content of the Martian ground plus greater Martian gravity, it is possible that saturation cratering

---

[2] The latter observation need not imply equilibrium if, for some reason, $a(D)$ is a constant for large diameters.

could have been the sole obliterative process on Mars. More likely there was a substantial early endogenic obliterative process in addition to the cratering process itself.

Intermediate Craters

Perhaps the dominant characteristic of frequency relations for Martian cu units is the shallow segment at intermediate diameters. We believe that the shape of this feature, its variation from region to region, and the shapes of the frequency relations for the separate classes of craters are very likely due to the evolution of the crater population to a new, lower equilibrium frequency during an "obliterative event." By this we mean that the ratio $c_t(T)/o(T)$ sharply decreased, lowering $F(D, T)$ by that amount. Small craters formed during the event would abut the new equilibrium line and follow it, just as they follow equilibrium line A in the Hartmann/Öpik model in Figure 2. The "critical diameter," originally very small, evolved to larger diameters and stopped at the point indicated in Figure 2 at the end of the event. Craters somewhat larger than that critical diameter remain highly degraded today, whereas much larger craters (> 50 km diameter) were hardly affected by the event.

To account for the observed population of small fresh craters, we require—as did earlier investigators—that the obliterative process cease or at least diminish to its pre-event level. The addition of the post-event craters to the sparse population of small craters degraded during the tail end of the event yields the observed relation $N(D)$, plotted as a slightly thicker line in the middle panel of Figure 2. (The little kink is smoothed somewhat by the dashed line C to account for statistical variations in the degree to which craters of a given size respond to a given obliteration process.)

As we have explained before, an equilibrium process results in frequency relations for the separate morphological classes that parallel the total crater curve. Such parallel frequencies would be predicted by the Hartmann/Öpik model near A in Figure 2 (also superimposed on the data as dotted line A' to the right). But the degraded craters on all cu units show a prominent trough, as illustrated for one region in Figure 2. This feature, if real, requires, within the constraints of our model, that there was an epoch of *dis*-equilibrium between two equilibrium frequencies, or in other words, an obliterative event.

One might initially suppose that the depletion of degraded craters with diameters between 5 and 15 km is due to observational loss near the lower limit of usable Mariner 9 A-frame (wide-angle) resolution. However, as shown in Figure 4, the frequency relations for the separate classes refute this possibility. Clearly, if there is a limiting resolution, the deep, fresh, bowl-shaped craters will be perceived at smaller sizes than will the most degraded craters; the limiting diameter will increase from f to s to m to h (Figure 4a). But the data show precisely the opposite trend (Figure 4b), which is difficult to explain by resolution effects alone yet is easily explained by obliteration. Figure 4c shows schematically what happens if an equilibrium population of craters is suddenly blanketed, for example, by a layer of dust. Some initially fresh or only slightly degraded craters of small diameters are still visible now as highly degraded craters, whereas the only craters remaining fresh are those so large as to be unaffected by the blanketing. The resulting curves look very similar to the actual data in Figure 4b, emphasizing the plausibility that an obliterative episode would have yielded the observed crater populations.

Frequency relations for the cratered uplands are not identical for all regions. It is noteworthy, however, that they differ in ways that are entirely consistent with an obliteration event. The event must merely have been stronger in some regions than in others. For instance, for total crater populations in two different regions (Figure 3), the kink in curve 1 occurs at a smaller diameter than in curve 2, suggesting that the regions represented by curve 1 suffered less obliteration during the event than those of curve 2. Furthermore, in all cu regions, the diagnostic ordering (h-m-s) of the maxima in Figure 4b is preserved, even though the ensemble is shifted depending on the magnitude of the event. While curves 1 and 2 in Figure 3 are separated substantially near diameters of 15 km, they are nearly coincident at diameters of 3 km with frequencies very similar to frequencies for the cratered plains (pc); apparently the pc plains were formed contemporaneously with the end of the obliteration event.

There is an interesting question that could be resolved if we had good counts of craters with diameters of 1 to 3 km: was the obliteration event more effective in some areas because it lasted longer or was it simply more intense? Case 1 in Figure 3 shows theoretical frequency relations for two different regions suffering different obliteration rates for the same duration, while Case 2 shows two regions suffering the same rates for different durations. The chief difference between the two cases is that the 1 to 3 km frequencies are different for the two regions in Case 1 but the same in Case 2. The data *seem* to resemble Case 1, but a definitive result must await analysis of statistically large samples of 1–3 km craters from Viking photographs.

Small Craters

Virtually all Martian craters smaller than a few kilometers in diameter must have formed relatively recently—just at the end of the obliterative event or later. Older small craters would certainly have been totally obliterated by the processes that so greatly modified the intermediate diameter craters. Of course, the freshness of the small craters means that the degradation rate has been very low since the event. Yet the presence of planet-wide duststorms suggests that *some* erosion and deposition must be continuing. If so, there is an equilibrium frequency that must be attained at *some* small diameter, given the apparent steepness of the production function $c_d(D)$.

The present equilibrium frequency would be evidenced by a bend in slope near a "critical diameter" analogous to that for larger craters in the Hartmann/Öpik model. Crater counts from some narrow-angle Mariner 9 pictures (B frames) do not clearly reveal such a critical diameter, although Chapman (1976a) found a range of degraded crater morphologies at diameters of several hundred meters and smaller, suggesting the onset of obliteration near the lower resolution limit of the frames. This corresponds to a surprisingly slow rate of obliteration, equivalent to a depth of erosion or deposition of the order of 10 m since the end of the obliterative episode. Preliminary studies of B frames by several other investigators suggest a somewhat larger rate of post-event obliteration in some selected cu units. Viking Orbiter pictures have emphasized the widespread distribution of secondary craters (Carr et al 1976); if the small, degraded craters studied by Chapman and others were really secondaries, then the post-event rate of net obliteration has been even less than they inferred.

Summary

Our model for Martian cratering and obliteration is relatively simple and satisfies the data extremely well. That it is a unique solution is more difficult to prove. Our scenario is this: Mars was bombarded by an early rain of projectiles, cratering the uplands units. A large-scale, efficient obliteration process acted contemporaneously with the cratering; it may have been the cratering process itself acting upon abnormally shallow "fresh" Martian craters. More likely there were atmospheric or other endogenic obliterative processes operating as well.

Later, the rate of obliteration increased dramatically with respect to the cratering rate, obliterating craters smaller than 10 km in diameter and modifying those somewhat larger. The high obliteration rate was sufficiently brief that its cumulative effect was insufficient to modify the largest craters. The obliterative episode was more effective in some regions than others, probably because of different obliterative rates, but possibly due to different durations. One might speculate that the episode happened at a time when internally generated heat reached a maximum near the surface of Mars, perhaps generating a thick atmosphere (see Conclusion section below). Whatever the nature of the obliteration, it ceased more or less coincidently with the emplacement of the pc units, which were perhaps due to volcanism associated with the same hypothetical thermal maximum.

During subsequent ages, while portions of the northern plains and the Tharsis Ridge have been resurfaced by volcanic and/or local aeolian or fluvial deposition, the older units (cu and pc) have been subjected to very little erosion.

ALTERNATIVE INTERPRETATIONS OF MARTIAN CRATERING

Soderblom et al (1974) interpreted the Martian cratering record as implying a simultaneity between cratering and obliteration. Our own models (Jones 1974, Chapman 1974a), depicting an obliterative episode occurring *after* the episode of heavy bombardment, might seem incompatible with Soderblom et al, but the models differ only in emphasis. Soderblom et al emphasized the larger Martian craters—a population that we agree evidences a probable equilibrium spectrum of crater morphologies. We have chosen to emphasize the smaller-scale obliterative episode that we believe occurred in intermediate Martian history because of the evidence for its presence and for its possible temporal association with other important geological events on Mars: the aqueous furrowing of the equatorial cu units and the formation of the first extensive plains units remaining today, the pc units. Soderblom et al did not dispute such variations in the ratio of obliteration to cratering rates.

There is a tendency for planetary geologists to avoid "catastrophist" interpretations of planetary histories, so the association of cratering and obliteration rates has seemed more acceptable than some extreme versions of our obliterative event, such as Jones' (1974) suggestion of the possibility that the event might have been an obliterative "spike" in very recent Martian history. We now see little reason to regard the episode in such stark terms; the problem of absolute chronologies is discussed below.

The major alternative models for Martian cratering have been spurred by Mariner 10 and the resulting studies comparing the cratering on Mercury with that on the Moon and Mars. Some fundamental underpinnings of Martian cratering theory are now being questioned. A common thread to the alternatives is that the production function $c_d(D)$ of primary craters differed substantially from the commonly assumed power law, at least during some early epochs.

Studies of the relationships between plains units and cratered units on Mercury and the Moon have led Oberbeck et al (1977) to suggest that $c_d(D)$ was deficient in craters with diameters smaller than 40 km. There is an obvious paucity of such craters on Mercury (Guest & Gault 1976). Oberbeck et al suggest that a similar deficiency exists on the Moon, but is partly masked by the presence of large secondary craters from the basin-forming events. Such basin secondaries are less evident on Mercury because Mercury has fewer basins and because higher Mercurian gravity would constrain such secondaries to the immediate peripheries of the basins (Gault et al 1975). Wilhelms (1976) has independently suggested that many of the larger lunar craters are actually basin secondaries.

If these interpretations are correct, it is reasonable to suppose that $c_d(D)$ for Mars was also deficient in craters less than 40 km in diameter. One current hypothesis (Murray et al 1975) is that all three planets were struck by the same population of projectiles at the same epoch. In fact, Wetherill (1975) has proposed a reasonable scenario for the tidal fragmentation of a large object crossing the orbits of Earth or Venus that would have resulted in a bombardment episode on all of the terrestrial planets. Such a scenario satisfies the interpretations of some researchers (e.g. Tera, Papanastassiou & Wasserburg 1974) that lunar rock-age distributions require a cataclysmic bombardment of the Moon about 4 b.y. ago. Chapman (1976b) offered another version of Wetherill's scenario, involving the collisional fragmentation of a large object in the asteroid belt. If the fragmentation was really tidal, rather than collisional, then the projectile population might well have been dominated by larger bodies and "deficient" in smaller ones.

One need not necessarily accept the cataclysm scenario nor an association between Martian, lunar, and Mercurian cratering in order to hypothesize that $c_d(D)$ for Mars was deficient, relative to a power law, in the sub-40-km range. While collisional fragmentation is usually thought to yield power-law size distributions, the early cratering projectiles need not have been a highly evolved collisional population.

Oberbeck et al have emphasized that tidal stress on very weak incoming projectiles could break them into a cluster, thereby yielding the hypothesized curved $c_d(D)$; such effects would differ for the Moon, Mars, and Mercury.

Strom & Whitaker (1976) have also proposed that the relative absence of smaller craters on the Moon, Mercury, and Mars is characteristic of the production function of an early but now extinct population of impactors rather than due to the effects of obliteration. They emphasize that their suggestion is bolstered by the claims of Woronow (1977a) that the spatial density of craters on even the most heavily cratered regions of the Moon is too low to be accounted for by saturation equilibrium. Woronow's Monte Carlo simulation of saturation cratering reached equilibrium at densities several times higher than observed on the Moon. His model, however, fails to employ a sufficiently large diameter range and is quite elementary. A crater is represented by four points on a virtual surface and the crater ejecta blanket is modelled to remove craters within its limits with 100% effectiveness while not affecting even small craters exterior to it at all.

Moreover, Woronow's simulation fails to match Gault's (1970) physical simulation of saturation cratering; his criticisms of Gault's procedures seem inadequate to account for the discrepancies. Gault's experiments, however, concerned production functions with steep slopes; in the shallower-slope regime, which may be more relevant to the large craters discussed here, the inadequacies of Woronow's simulation may be less serious.

The strongest part of Strom & Whitaker's case is their observation that fresh lunar craters on the Mare Orientale ejecta blanket display a curving $N(D)$ that is deficient in smaller craters and similar in shape to frequency curves displayed by highlands craters. The significance of this observation is that Orientale craters are widely separated and all of reasonably fresh morphology, so they cannot be in equilibrium with saturation cratering nor can they have been affected much by any other lunar obliterative process. Strom & Whitaker suggest that the Orientale craters sample the tail end of cratering by the population responsible for most lunar, Martian, and Mercurian craters. (The *post*-mare lunar craters, as well as plains unit craters on Mars, follow much more nearly a power-law distribution, such as that illustrated for fresh craters in the right-hand side of Figure 2.)

Despite similarities, the models of Oberbeck et al and Strom & Whitaker are partly incompatible. In particular, many of the Orientale ejecta blanket craters counted by Strom & Whitaker are deemed to be Orientale secondaries by Wilhelms (1976), and Oberbeck et al also consider the formation of large secondaries to be important. Strom & Whitaker generally argue for relatively modest influence of basin formation on lunar crater populations, whereas Oberbeck et al regard the effects as pervasive across the lunar frontside, with only a few areas relatively unscathed.

Since these alternative models have been developed chiefly in a lunar and Mercurian context, serious discussions of the implications for Mars have not yet appeared in the literature. We perceive a number of potential difficulties with these hypotheses, however. First, it seems unlikely that all 20- to 30-km craters formed on Mars could remain intact while much larger craters are so highly degraded. It would require a function $a(D)$ very nearly constant from diameters of twenty through several hundred km; that is, one must assume that it is no more difficult to erode or obliterate a 200-km crater than a 20-km crater. While the diameter-depth data do not yet exclude such a possibility for Mars, it would be a most peculiar geological process that would behave in such a fashion if fresh Martian craters bore any resemblance to their lunar or Mercurian counterparts (Figure 1).

A second difficulty is that it is not possible to regard the data shown in Figure 3 as compatible with a single production function; curve 1 is not a simple multiple of curve 2. It is possible to make the ad hoc assumption that the more heavily cratered terrains (curve 1) were struck by a population less deficient in small bodies than moderately cratered terrains (curve 2) in just the proportions that would mimic the behavior of an obliterative episode of variable strength. However, this assumption is not persuasive.

One might argue that curve 1 in Figure 3 is a production function and that curve 2 represents the same population modified slightly by obliteration. But obliteration necessarily modifies crater morphologies, which brings us to a final question, yet to be addressed in the alternative models: Why should a curving production function of craters in the 10- to 30-km diameter range manifest the unusual distribution of morphological classes, shown in Figure 4, that seem so reasonably explained by an obliterative episode? The general question of what processes have been responsible for the degraded appearances of most lunar, Mercurian, and—especially—Martian craters is a matter of great importance for making the proposed alternatives convincing. The question is particularly relevant to Strom & Whitaker; Oberbeck et al at least believe that substantial obliteration accompanied emplacement of basin ejecta [Hartmann & Wood (1971) and Chapman (1974a) interpreted lunar crater frequencies in terms of basin ejecta blanketing].

To summarize, we believe our own model for Martian cratering and obliteration (previous section) well accounts for the data. We cannot assert that the interpretation is unique, however, especially in the face of the fundamental challenge to the long-accepted belief that the lunar upland craters represent a saturation equilibrium population. However, quite apart from debates concerning the plausibility that craters 30 km in diameter can be basin secondaries, the recent models require further development to see if they can prove to be as successful as ours in describing the observed populations of craters of different morphological classes and the variations in those distributions from place to place on Mars.

ABSOLUTE CHRONOLOGY

There is no known way to determine the absolute age (in years) of a geological unit on another planet in the absence of (a) datable rocks or (b) a well-calibrated relationship between cratering flux and crater density. For Mars, of course, we lack datable rocks, so we must rely on the principles of crater-count age-dating first described in detail by Shoemaker, Hackman & Eggleton (1962). In order to find the age of a unit from its crater density one must know the cumulative cratering flux as a function of time.

Figure 6 shows the calibration diagram for Mars. The first widely used Martian chronology was that of Hartmann (1973), which was based on the uncertain hypothesis that the crater-production rate on Mars has been six times that on the Moon (the absolute chronology can then be calculated since the lunar chronology is known, from dated moonrocks). It is now thought that the crater-production rates on the two planets are probably more nearly equal. Soderblom et al (1974) attempted to derive the Martian cratering rate without adopting a particular multiple of the lunar rate. But their calibration curve involved the *assumption* that Martian surface-forming processes occur at a uniform rate—a uniformitarian assumption based on a plausible, but unproven, analogy with the Moon. Murray et al (1975) proposed that the crater-production rates on the Moon, Mars, and Mercury are rather closely equal; that assumption produces a curve on Figure 6 close to Soderblom's, but below it.

The firmest basis for estimating relative cratering rates on the terrestrial planets comes from an inventory of existing small-body populations in the inner solar system and calculations of their subsequent orbital evolution, in particular their collision probabilities with planets. Wetherill (1975, 1976) has emphasized the similarity of the crater production rates on terrestrial bodies, but he does not strongly dispute Chapman's (1976b) assertion that these rates may be uncertain over a range of up to an order of magnitude, especially for Mars. Chapman (1976b) considered the present status of our knowledge of cratering fluxes [based mainly on calculations by Wetherill (1975) and Hartmann (1977)] and proposed two extremes for Mars (shown as alternatives 1 and 2 on Figure 6). None of the cratering rates shown in Figure 6 are sufficient to produce all the observed craters on the cu units in 4.5 b.y. Accordingly there must have been a period of higher flux; most plausibly this occurred ~4 b.y. ago, as on the Moon, but the hypothesis is unproven.

The pc (cratered plains) units are adopted as the standard for comparison of crater densities in Figure 6. These include the eastern part of Solis Planum, Lunae Planum, and Syrtis Major Planitia and appear to be the oldest of the plains units inasmuch as they have small-crater densities comparable to those of the cu units,

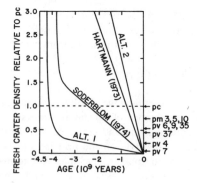

Figure 6   Correlation between small-crater density and age for a variety of cratering fluxes. Crater densities are plotted for a variety of plains units, taken from the data of Jones (1974).

which date the end of the obliterative event. Younger units may be dated from the densities of craters 4 to 10 km in diameter (indicated at the right of Figure 6). The second oldest regions shown in Figure 6 are pm-10, pv-3, and pv-5, which include regions west of Solis Planum. Still younger, with roughly half the pc crater density, are the volcanic plains units pv-6, pv-9, pv-35, and pv-37, which include much of Elysium Planitia and units north of the Tharsis ridge. Plains south-west of Arsia Mons (pv-4) are much younger still and the area surrounding the Tharsis Montes are youngest of all major units for which Jones (1974) tabulated crater frequencies. (The abbreviations cu, pc, pm, etc employed here are for units defined and mapped by Mutch & Head 1975.)

It is clear that the crater densities cannot be translated into ages even as vague as "a couple of billion years" if one agrees that alternatives 1 and 2 provide a reasonable range for the cratering flux on Mars. Elysium Planitia may be only a few hundred million years old, but it might instead date from the tail end of an early bombardment. Of course, presentation of such a broad range of ages for particular features does not adequately convey the rather precise degree to which we can determine the *relative* ages of features. Thus we think that a better way to refer to ages of Martian units is in terms of crater densities relative to the pc units. Attempts to specify ages in years should be avoided until the chronology is better determined.

GEOMORPHOLOGICAL PROCESSES

What we call "the obliterative process" must represent the cumulative effects of innumerable different geological processes. They can be modelled as a single average process for large statistical samples of craters. But verification that our generalization is valid depends upon explicit analyses of separate processes, including detailed photogeological interpretation of specific features.

The literature contains many papers about specific geomorphological processes, mostly based on photogeological interpretation or analogies with familiar terrestrial processes. Some studies, especially of aeolian processes, have included windtunnel experiments and theoretical treatments. It would be premature to treat all of this literature here, since few of the studies have been integrated with the others or applied to a global Martian synthesis. Some recent papers are mentioned below as an introduction to the literature.

Some geomorphological processes are much more general and widespread in their effects than others. Rock sliding and channelcutting may have devastating effects on landforms in certain localities, but such processes of local origins do not substantially affect widespread Martian units. On the other hand, atmospheric processes such as rainfall or aeolian deposition can extend over the entire planet. Perhaps the nonexogenic, nonatmospheric process of greatest potential extent is volcanism; lava flows may spread across vast areas and an epoch of heating causing the permafrost to melt could envelop much of the planet, having similar effects on topography on opposite sides of Mars. Clearly, the processes responsible for most of the crater obliteration discussed in this article have been of the widespread variety.

Exogenic Processes

Destruction of pre-existing topography by the cratering process itself has been extensively studied (see earlier discussion of saturation equilibrium). It had been thought that the planet Mercury provided the most reasonable model for fresh cratering on Mars because of the similarity in gravity on the two planets (Gault et al 1975). However, ejecta blanket morphologies and secondary crater distributions on Mars appear to differ from those on either the Moon or Mercury, suggesting that, at least in certain localities, the atmosphere and/or underground ice deposits have substantially altered the effects of crater formation on nearby topography (Carr et al 1976).

Aeolian Processes

The wind modifies features on Mars in two chief ways: erosion by abrasion or plucking and transport of fine materials by saltation or suspension from one location to another. There is abundant evidence in Mariner 9 and Viking pictures for the local importance of aeolian processes on Mars, including dune fields. McCauley & Grolier (1976) have studied yardangs in the driest terrestrial deserts as analogs of "inverted ship" shaped features on Mars. Scarp recession and formation of pedestal craters also have been ascribed to the wind (King & Riehle 1974, Arvidson et al 1976, and others). There can be no question that dust is moved about on Mars by wind. But inasmuch as we have very little experience on Earth with large-scale terrain that has been affected more by wind than by water, it is uncertain to what degree some of the large-scale Martian features truly are of aeolian origin. Useful studies of Martian aeolian processes and the erosion rate in particular have involved combinations of experimental and theoretical approaches (cf Greeley et al 1974, Sagan & Bagnold 1975, and Iversen et al 1975). Some researchers have calculated surprisingly rapid erosion rates, at least for lateral erosion (e.g. scarp recession) as opposed to deflation.

Many units on Mars have been interpreted as being modified by aeolian deposition. In particular, Soderblom et al (1974) have mapped extensive "dust mantles" poleward of 40° latitude parallels. There can be little doubt that dust deposition has been important in producing layered deposits in polar latitudes. But it remains for Viking Orbiter analysis to determine the extent to which units mapped as dust mantles may in fact reflect poor resolution or lingering atmospheric dust during the Mariner 9 photography.

Aqueous Processes

The flow of liquid water appears to have played a major role in shaping the surface of Mars. First recognized in Mariner 9 photography (Milton 1973), the evidence for fluvial processes has become overwhelming in Viking Orbiter photography (Carr et al 1976). Most striking are the large channels, which are especially prominent near the boundary that separates the cratered and uncratered hemispheres from each other. The first Viking Orbiter pictures showed impressive views of such channels emerging from the chaotic terrains of Mariner 6 and from the canyonlands of Mariner 9, emptying into the Chryse basin where the first Viking landed. While these large channels have been classified into several different types, a common hypothesis is that most large channels were carved by great floods of water derived from ice deposits in the ground. The closest terrestrial analogy may be the channeled scablands of eastern Washington, where catastrophic flooding occurred when the waters of Lake Missoula evidently broke through an ice dam (Baker & Milton 1974).

Still more interesting fluvial evidence is afforded by the "furrows," which are smaller stream and river valleys in the cratered terrains, possibly confined to equatorial and temperate latitudes. While some furrow systems are poorly integrated, there are many well-developed dendritic patterns of great areal extent, implying origin of the waters from widely dispersed areas—most likely by rain (cf Pieri 1976). It is tempting to associate the formation of furrows with the obliterative event we have described. Certainly, the tiny observable furrows could not have been formed prior to the last stages of the event; otherwise they would have been obliterated. But they do not exist on the pc units formed just after the event. Thus they are temporally associated with the episode. Rainfall is certainly known to be a highly effective erosive agent on Earth, so it might well have yielded the high obliteration rate required by the low equilibrium frequencies during the Martian obliterative event.

Other Processes

Volcanism has caused much modification of topography on Mars. It is primarily a constructional process, either laying down vast plains of lava or ash or building giant shields such as Olympus Mons. Carr (1974), however, has discussed some situations in which lava is actually erosive, and he has attributed the origin of some lunar-rille-like channels on Mars to erosive volcanic processes.

There are many tectonic processes that modify the Martian surface. These include the formation of grabens, faults, and fractures due to internal stresses, such as those associated with the comparatively recent formation of the Tharsis uplift. Schultz (1976) has studied explicitly the so-called floor-fractured craters. There are also numerous local processes, including landslides, mudslides, and less spectacular processes of downslope movement that reduce topographic relief on Mars (an example is discussed by Veverka & Liang 1975).

Scarp recession has been the major process of dissecting the fretted terrains and removing many other apparently sedimentary sequences on Mars. Whether the actual erosive agent is mainly windblown particles, as commonly assumed, beach erosion by bodies of water, or slumping due to failure of subsurface ice is not always clear. But the removal of unconsolidated layers often reveals previous topographic horizons. Figure 7 shows a once-buried crater being exhumed as a scarp retreats across it. If processes of exhumation were widespread, our earlier assumption of the one-way direction of crater degradation would be violated.

On Mars, as on the Earth, erosion and deposition cannot be considered independent of physical and chemical weathering processes that render particles subject to fluvial, aeolian, or mass-wasting transport. An early suggestion (Chapman,

Pollack & Sagan 1969) was that the particulates were derived by comminution in cratering impacts. Other processes of weathering have been addressed in other contexts (Malin 1974, Siever 1974, Huguenin 1974, 1976).

CONCLUSION: THE GEOMORPHOLOGICAL EVOLUTION OF MARS

Mars accreted as a planet with substantial volatile content (cf Fanale 1976). The evolution of these volatiles, governed by the internal and external thermal budget of the planet, is evidenced by the cratering record. We have already described the early bombardment of Mars and the probably contemporaneous obliteration process, which was possibly due to saturation cratering alone but was more likely of endogenic origin. Probably Mars had a substantial atmosphere during these early epochs.

Upon losing any accretional heat it may have had, the surface of Mars became very cold. Whatever water vapor was not lost to space must have rapidly condensed onto, and percolated into, the ground and frozen. The evidence for an obliterative episode that affected mid-sized craters (perhaps associated with the furrowing) suggests that the volatiles were released again. Getting water out again onto the surface of Mars in order to form individual large channels requires a subsequent period of heating, perhaps by local volcanic hot spots. But getting water into the

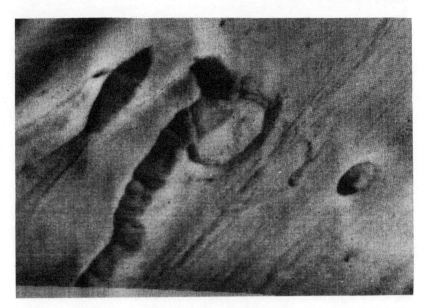

Figure 7   Mariner 9 picture of a crater in the Kasei Vallis region apparently being exhumed from beneath a receding cliff. NASA photograph.

atmosphere, in order to rain and produce dendritic patterns over much of the planet, requires planet-wide reheating of the surface (plus a sufficient amount of available water). That is necessary in order to elevate the temperature of the coldest spot on the planet so that the vapor pressure of water at that spot exceeds the partial pressure of water in the whole atmosphere (Mutch et al 1976). Otherwise any released water would be rapidly (a few years or less) frozen out at the polar cold-traps, which is a problem with some of the heating models based on enhanced solar luminosity (Hartmann 1974a) or cyclical insolation variations associated with obliquity changes (Ward 1973). (The reasoning does not apply to the Earth, because of our planet's superabundance of water, which obviously cannot all be frozen out at the poles.) Planet-wide elevation of surface temperatures on Mars cannot be accomplished by any reasonable level of endogenic heating. An atmospheric green-house effect seems required, perhaps assisted by higher polar temperatures associated with the larger pre-Tharsis obliquities of Mars hypothesized by Burns, Ward & Toon (1977). A period of volcanic activity and crustal heating may have supplied once-frozen underground water to the atmosphere, triggering a rainy spell.

The geological record on Mars provides possible independent evidence for a thermal maximum in crustal temperatures roughly contemporaneous with the obliterative event. The formation of the major volcanic units visible today com-menced with the pc units shortly after the end of the event and has continued, at least near Tharsis, practically to the present day. Thus the onset of major Martian volcanism may have been due to the same planet-wide crustal heating (a radiogenic heat pulse) that led to the generation of a moist Martian atmosphere, producing the obliterative event.

This plausible scenario is only a suggestion. There is evidence for some ancient volcanism on Mars, and Wilhelms (1974) has suspected that it was pervasive; it may even have been a major obliterative process. On the other hand, the northern plains units need not be chiefly volcanic; many may result from aeolian deposition or even sedimentation from fluvial outwash and shallow seas. So the predominance of fresh volcanism on Mars may just reflect the lack of recent obliteration. Thus there may be no temporal association between the thermal evolution of Mars and evolution of its atmosphere. The aqueous period might not even have required new outgassing of water; the rain could have resulted from condensation of water upon cooling of the original atmosphere formed concurrently with accretion.

Nothing in the Martian cratering record requires cyclical variations in climate, such as those that Sagan (1971), Sagan, Toon & Gierasch (1973), and Sagan & Lederberg (1976) have postulated might from time to time provide more clement conditions for organisms "in cryptobiotic repose awaiting the return of wetter and warmer conditions." The polar laminated terrains (Cutts 1973) and evidence that channels flowed more than once (Hartmann 1974b) indicate discontinuous processes, but evidence of anything approaching sinusoidal oscillation is absent. Theoretical calculations (Ward 1973) of cyclical atmospheric evolution yield timescales far too short (e.g. 10^6 years) to be measured by the cratering record, given our present knowledge of the absolute cratering chronology and the slow rates of geological processes evident on Mars today. Instead, the last observable planet-wide crater-obliterative event seems to have occurred prior to half a billion years ago, and perhaps as early as the tail end of the early bombardment period on Mars.

ACKNOWLEDGMENTS

Preparation of this review was supported in part by NASA grant NGR-40-002-088 and NASA contract NASW 2869. We thank A. Woronow, R. Strom, and G. Wetherill for helpful criticisms. This is Planetary Science Institute Contribution 68.

Literature Cited

Arvidson, R. E., Coradini, M., Carusi, A., Coradini, A., Fulchignoni, M., Federico, C., Funiciello, R., Salomone, M. 1976. Latitudinal variation of wind erosion of crater ejecta deposits on Mars. *Icarus* 27:503–16

Arvidson, R. E., Mutch, T. A., Jones, K. L. 1974. Craters and associated aeolian features on Mariner 9 photographs: an automated data gathering and handling system and some preliminary results. *Moon* 9:105–14

Baker, V. R., Milton, D. J. 1974. Erosion by catastrophic floods on Mars and Earth. *Icarus* 23:27–41

Binder, A. B. 1966. Mariner IV: analysis of preliminary photographs. *Science* 152:1053–55

Burns, J. A., Ward, W. R., Toon, O. B. 1977. Mars before Tharsis: much larger obliquity in the past? *Bull. Am. Astron. Soc.* In press

Burt, J., Veverka, J., Cook, K. 1976. Depth-diameter relation for large Martian craters determined from Mariner 9 UVS altimetry. *Icarus* 29:83–90

Carr, M. H. 1974. The role of lava in the formation of lunar rilles and Martian channels. *Icarus* 22:1–23

Carr, M. H., Masursky, H., Baum, W. A., Blasius, K. R., Briggs, G. A., Cutts, J. A., Duxbury, T., Greeley, R., Guest, J. E., Smith, B. A., Soderblom, L. A., Veverka, J., Wellman, J. B. 1976. Preliminary results from the Viking Orbiter imaging experiment. *Science* 193:766–76

Chapman, C. R. 1967. Topics concerning the surfaces of Mercury, Mars, and the moon. Undergraduate thesis (unpublished), Harvard College, Cambridge. 129 pp.

Chapman, C. R. 1974a. Cratering on Mars, I. Cratering and obliteration history. *Icarus* 22:272–91

Chapman, C. R. 1974b. Cratering on Mars, II. Implications for future cratering studies from Mariner 4 reanalysis. *Icarus* 22:292–300

Chapman, C. R. 1976a. Crater morphologies on Mars at Mariner 9 B-frame scales. *NASA Tech. Memo. X-3364*, pp. 175–77

Chapman, C. R. 1976b. Chronology of terrestrial planet evolution: the evidence from Mercury. *Icarus* 28:523–36

Chapman, C. R., Mosher, J. A., Simmons, G. 1970. Lunar cratering and erosion from Orbiter 5 photographs. *J. Geophys. Res.* 75:1445–66

Chapman, C. R., Pollack, J. B., Sagan, C. 1969. An analysis of the Mariner 4 cratering statistics. *Astron. J.* 74:1039–51

Cintala, M. J., Head, J. W., Mutch, T. A. 1976. Craters on the moon, Mars, and Mercury: a comparison of depth/diameter characteristics. *Lunar Science VII*, pp. 149–51

Cutts, J. A. 1973. Nature and origin of layered deposits of the Martian polar regions. *J. Geophys. Res.* 78:4231–49

Dohnanyi, J. S. 1972. Interplanetary objects in review: statistics of their masses and dynamics. *Icarus* 17:1–48

Fanale, F. P. 1976. Martian volatiles: their degassing history and geochemical fate. *Icarus* 28:179–202

Gault, D. E. 1970. Saturation and equilibrium conditions for impact cratering on the lunar surface: criteria and implications. *Radio Sci.* 5:273–91

Gault, D. E., Guest, J. E., Murray, J. B., Dzurisin, D., Malin, M. C. 1975. Some comparisons of impact craters on Mercury and the moon. *J. Geophys. Res.* 80:2444–60

Greeley, R., Iversen, J. D., Pollack, J. B., Udovich, N., White, B. 1974. Wind tunnel studies of Martian aeolian processes. *Proc. R. Soc. Lond. Ser. A* 341:331–60

Guest, J. E., Gault, D. E. 1976. Crater populations in the early history of Mercury. *Geophys. Res. Lett.* 3:121–23

Hartmann, W. K. 1966. Martian cratering. *Icarus* 5:565–76

Hartmann, W. K. 1971. Martian cratering II: asteroid impact history. *Icarus* 15:

396–409

Hartmann, W. K. 1973. Martian cratering, 4, Mariner 9 initial analysis of cratering chronology. *J. Geophys. Res.* 78:4096–4116

Hartmann, W. K. 1974a. Martian and terrestrial paleoclimatology: relevance of solar variability. *Icarus* 22:301–11

Hartmann, W. K. 1974b. Geological observations of Martian arroyos. *J. Geophys. Res.* 79:3951–57

Hartmann, W. K. 1977. Relative crater production rates on planets. *Icarus*. In press

Hartmann, W. K., Wood, C. A. 1971. Moon: origin and evolution of multi-ring basins. *Moon* 3:2–78

Huguenin, R. L. 1974. The formation of goethite and hydrated clay minerals on Mars. *J. Geophys. Res.* 79:3895–3905

Huguenin, R. L. 1976. Mars: chemical weathering as a massive volatile sink. *Icarus* 28:203–12

Iversen, J. D., Greeley, R., White, B. R., Pollack, J. B. 1975. Eolian erosion of the Martian surface, Part 1: erosion rate similitude. *Icarus* 26:321–31

Jones, K. L. 1974. Evidence for an episode of crater obliteration intermediate in Martian history. *J. Geophys. Res.* 79:3917–31

King, J. S., Riehle, J. R. 1974. A proposed origin of the Olympus Mons escarpment. *Icarus* 23:300–17

Leighton, R. B., Murray, B. C., Sharp, R. P., Allen, J. D., Sloan, R. K. 1967. Mariner IV pictures of Mars. *Jet Propul. Lab. Tech. Rep.* 32–884

Malin, M. C. 1974. Salt weathering on Mars. *J. Geophys. Res.* 79:3888–94

Malin, M. C., Dzurisin, D. 1977. Landform degradation on Mercury, the moon, and Mars: evidence from crater depth/diameter relationships. *J. Geophys. Res.* 82:376–88

Marcus, A. H. 1970. Comparison of equilibrium size distributions for lunar craters. *J. Geophys. Res.* 75:4977–84

McCauley, J. F., Grolier, M. J. 1976. Terrestrial yardangs. *NASA Tech. Memo. X-3364*, pp. 110–14

McGill, G. E., Wise, D. U. 1972. Regional variations in degradation and density of Martian craters. *J. Geophys. Res.* 77:2433–41

Milton, D. J. 1973. Water and processes of degradation in the Martian landscape. *J. Geophys. Res.* 78:4037–47

Murray, B. C., Soderblom, L. A., Sharp, R. P., Cutts, J. A. 1971. The surface of Mars, 1. Cratered terrains. *J. Geophys. Res.* 76:313–30

Murray, B. C., Strom, R. G., Trask, N. J., Gault, D. E. 1975. Surface history of Mercury: implications for terrestrial planets. *J. Geophys. Res.* 80:2508–14

Mutch, T. A., Arvidson, R. E., Head, J. W. III, Jones, K. L., Saunders, R. S. 1976. *Geology of Mars.* Princeton: Princeton Univ. Press. 400 pp.

Mutch, T. A., Head, J. W. 1975. The geology of Mars: a brief review of some recent results. *Rev. Geophys.* 13:411–16

Oberbeck, V. R., Aoyagi, M. 1972. Martian doublet craters. *J. Geophys. Res.* 77:2419–32

Oberbeck, V. R., Quaide, W. L., Arvidson, R. E., Aggarwal, H. R. 1977. Comparative studies of lunar, Martian, and Mercurian craters and plains. *J. Geophys. Res.* In press

Öpik, E. J. 1965. Mariner IV and craters on Mars. *Ir. Astron. J.* 7:92–104

Öpik, E. J. 1966. The Martian surface. *Science* 153:255–65

Pieri, D. 1976. Distribution of small channels on the Martian surface. *Icarus* 27:25–50

Sagan, C. 1971. The long winter model of Martian biology: a speculation. *Icarus* 15:511–14

Sagan, C., Bagnold, R. A. 1975. Fluid transport on Earth and aeolian transport on Mars. *Icarus* 26:209–18

Sagan, C., Lederberg, J. 1976. The prospects for life on Mars: a pre-Viking assessment. *Icarus* 28:291–300

Sagan, C., Toon, O. B., Gierasch, P. J. 1973. Climatic change on Mars. *Science* 181:1045–48

Schultz, P. H. 1976. Floor-fractured craters on the moon, Mars, and Mercury. *NASA Tech. Memo. X-3364*, pp. 159–60

Shoemaker, E. M., Hackman, R. I., Eggleton, R. E. 1962. Interplanetary correlation of geologic time. *Adv. Astronaut. Sci.* 8:70–89

Siever, R. 1974. Comparison of Earth and Mars as differentiated planets. *Icarus* 22:312–24

Soderblom, L. A., Condit, C. D., West, R. A., Herman, B. M., Kreidler, T. J. 1974. Martian planetwide crater distributions: implications for geologic history and surface processes. *Icarus* 22:239–63

Strom, R. G., Whitaker, E. A. 1976. Populations of impacting bodies in the inner solar system. *NASA Tech. Memo. X-3364*, pp. 194–96

Tera, F., Papanastassiou, D. A., Wasserburg, G. J. 1974. Isotopic evidence for a terminal lunar cataclysm. *Earth Planet. Sci. Lett.* 22:1–21

Veverka, J., Liang, T. 1975. An unusual

landslide feature on Mars. *Icarus* 24:47–50

Ward, W. R. 1973. Large-scale variations in the obliquity of Mars. *Science* 181:620–62

Wetherill, G. W. 1975. Late heavy bombardment of the moon and terrestrial planets. *Proc. Lunar Sci. Conf., 6th,* pp. 1539–61

Wetherill, G. W. 1976. Comments on the paper by C. R. Chapman: chronology of terrestrial planet evolution—the evidence from Mercury. *Icarus* 28:537–42

Wilhelms, D. E. 1974. Comparison of Martian and lunar geologic provinces. *J. Geophys. Res.* 74:3933–41

Wilhelms, D. E. 1976. Secondary impact craters of lunar basins. *Lunar Sci. VII,* pp. 935–37

Woronow, A. 1977a. Crater saturation and equilibrium: a Monte Carlo simulation. *J. Geophys. Res.* In press

Woronow, A. 1977b. A size-frequency study of large Martian craters. *J. Geophys. Res.* In press

AUTHOR INDEX

CUMULATIVE INDEXES

CONTRIBUTING AUTHORS VOLUMES 1-5

554

CHAPTER TITLES   VOLUMES 1-5

DATE DUE

OCT 1 1 1995			
MAR 1 - 1996			
OCT 6 - 1997			
	261-2500		Printed in USA